PRELIMINARY EDITION

Calculus
VOLUME 1

THE PRINDLE, WEBER & SCHMIDT SERIES IN MATHEMATICS

Althoen and Bumcrot, *Introduction to Discrete Mathematics*
Boye, Kavanaugh, and Williams, *Elementary Algebra*
Boye, Kavanaugh, and Williams, *Intermediate Algebra*
Burden and Faires, *Numerical Analysis, Fourth Edition*
Cass and O'Connor, *Fundamentals with Elements of Algebra*
Cullen, *Linear Algebra and Differential Equations, Second Edition*
Dick and Patton, *Calculus, Volume I*
Dick and Patton, *Calculus, Volume II*
Dick and Patton, *Technology in Calculus: A Sourcebook of Activities*
Eves, *In Mathematical Circles*
Eves, *Mathematical Circles Adieu*
Eves, *Mathematical Circles Squared*
Eves, *Return to Mathematical Circles*
Fletcher, Hoyle, and Patty, *Foundations of Discrete Mathematics*
Fletcher and Patty, *Foundations of Higher Mathematics, Second Edition*
Gantner and Gantner, *Trigonometry*
Geltner and Peterson, *Geometry for College Students, Second Edition*
Gilbert and Gilbert, *Elements of Modern Algebra, Third Edition*
Gobran, *Beginning Algebra, Fifth Edition*
Gobran, *Intermediate Algebra, Fourth Edition*
Gordon, *Calculus and the Computer*
Hall, *Algebra for College Students*
Hall, *Beginning Algebra*
Hall, *College Algebra with Applications, Third Edition*
Hall, *Intermediate Algebra*
Hartfiel and Hobbs, *Elementary Linear Algebra*
Humi and Miller, *Boundary-Value Problems and Partial Differential Equations*
Kaufmann, *Algebra for College Students, Fourth Edition*
Kaufmann, *Algebra with Trigonometry for College Students, Third Edition*
Kaufmann, *College Algebra, Second Edition*
Kaufmann, *College Algebra and Trigonometry, Second Edition*
Kaufmann, *Elementary Algebra for College Students, Fourth Edition*
Kaufmann, *Intermediate Algebra for College Students, Fourth Edition*
Kaufmann, *Precalculus, Second Edition*
Kaufmann, *Trigonometry*
Kennedy and Green, *Prealgebra for College Students*
Laufer, *Discrete Mathematics and Applied Modern Algebra*
Nicholson, *Elementary Linear Algebra with Applications, Second Edition*
Pence, *Calculus Activities for Graphic Calculators*
Pence, *Calculus Activities for the TI-81 Graphic Calculator*
Plybon, *An Introduction to Applied Numerical Analysis*
Powers, *Elementary Differential Equations*
Powers, *Elementary Differential Equations with Boundary-Value Problems*
Proga, *Arithmetic and Algebra, Third Edition*
Proga, *Basic Mathematics, Third Edition*
Rice and Strange, *Plane Trigonometry, Sixth Edition*
Schelin and Bange, *Mathematical Analysis for Business and Economics, Second Edition*
Strnad, *Introductory Algebra*

Swokowski, *Algebra and Trigonometry with Analytic Geometry, Seventh Edition*
Swokowski, *Calculus, Fifth Edition*
Swokowski, *Calculus, Fifth Edition (Late Trigonometry Version)*
Swokowski, *Calculus of a Single Variable*
Swokowski, *Fundamentals of College Algebra, Seventh Edition*
Swokowski, *Fundamentals of College Algebra and Trigonometry, Seventh Edition*
Swokowski, *Fundamentals of Trigonometry, Seventh Edition*
Swokowski, *Precalculus: Functions and Graphs, Sixth Edition*
Tan, *Applied Calculus, Second Edition*
Tan, *Applied Finite Mathematics, Third Edition*
Tan, *Calculus for the Managerial, Life, and Social Sciences, Second Edition*
Tan, *College Mathematics, Second Edition*
Trim, *Applied Partial Differential Equations*
Venit and Bishop, *Elementary Linear Algebra, Third Edition*
Venit and Bishop, *Elementary Linear Algebra, Alternate Second Edition*
Wiggins, *Problem Solver for Finite Mathematics and Calculus*
Willard, *Calculus and Its Applications, Second Edition*
Wood and Capell, *Arithmetic*
Wood and Capell, *Intermediate Algebra*
Wood, Capell, and Hall, *Developmental Mathematics, Fourth Edition*
Zill, *A First Course in Differential Equations with Applications, Fourth Edition*
Zill, *Advanced Engineering Mathematics*
Zill, *Calculus, Third Edition*
Zill, *Differential Equations with Boundary-Value Problems, Second Edition*

THE PRINDLE, WEBER & SCHMIDT SERIES IN ADVANCED MATHEMATICS

Brabenec, *Introduction to Real Analysis*
Ehrlich, *Fundamental Concepts of Abstract Algebra*
Eves, *Foundations and Fundamental Concepts of Mathematics, Third Edition*
Keisler, *Elementary Calculus: An Infinitesimal Approach, Second Edition*
Kirkwood, *An Introduction to Real Analysis*
Ruckle, *Modern Analysis: Measure Theory and Functional Analysis with Applications*
Sieradski, *An Introduction to Topology and Homotopy*

THE OREGON STATE UNIVERSITY
CALCULUS CURRICULUM PROJECT

PRELIMINARY EDITION

Calculus
VOLUME 1

Thomas P. Dick
OREGON STATE UNIVERSITY

Charles M. Patton
HEWLETT-PACKARD COMPANY

PWS-KENT Publishing Company
Boston

PWS-KENT
Publishing Company

20 Park Plaza
Boston, Massachusetts 02116

Copyright © 1992 by PWS-KENT Publishing Company

All rights reserved. No part of this book may be reproduced, stored in a retrieval system, or transcribed, in any form or by any means -- electronic, mechanical, photocopying, recording, or otherwise -- without the prior written permission of the publisher.

PWS-KENT Publishing Company is a division of Wadsworth, Inc.

ISBN 0-534-92391-7
Printed in the United States of America.
91 92 93 94 95 -- 10 9 8 7 6 5 4 3 2 1

Preface

Curriculum revision is generally a process of gradual evolution of scope and sequence, occasionally punctuated by calls for more fundamental changes in content or delivery. The launching of Sputnik precipitated such a call for reform in mathematics education in the 1950's. We find ourselves in the midst of a new period of widespread revitalization efforts in mathematics curriculum and instruction. A forward-looking vision of the entire K-12 mathematics curriculum is outlined in the *Curriculum and Evaluation Standards* of the National Council of Teachers of Mathematics. The Mathematical Sciences Education Board has made an eloquent and urgent case for revitalizing mathematics instruction at all levels in preparation for our country's future workforce needs in *Everybody Counts*. Both of these influential documents recognize the emergence of sophisticated computer and calculator technology as redefining the tools of mathematics education.

Calculus occupies a particularly critical position in mathematics education as the gateway to advanced training in most scientific and technical fields. It is fitting that calculus should receive particular attention as we prepare for the needs of the twenty-first century. The Sloan Conference (Tulane, 1986) and the Calculus for a New Century Conference (Washington, 1987) sounded the call for reform in the calculus curriculum. Now the entire introductory course in calculus is being reexamined under the closest scrutiny that it has received in several years. Through a special funding initiative, the National Science Foundation has made resources available for a variety of calculus curriculum revision efforts to be tried and implemented. The Oregon State University Calculus Curriculum Project is one of these NSF-funded efforts, and this book is one of the major results of the project.

Volume I covers the material for a first year course in calculus of a single variable. (Volume II covers material appropriate for a semester or two quarters of multivariable and vector calculus.) A brief glance at the Table of Contents might suggest that the text does not differ radically from a traditional calculus text in terms of major topics. This is as it should be—calculus reform will not change the importance and vitality of the major ideas of calculus, and any wholesale departure from those ideas should be viewed with great skepticism. What *is possible* is a fresh approach to these important ideas in light of the availability of modern technology. In particular, the technology can invite us to change or adopt new emphases in instruction.

MAJOR THEMES OF THE OREGON STATE PROJECT

Making intelligent use of technology

Computer algebra systems, spreadsheets, and symbolic/graphing calculators are just a few of the readily available technological tools providing students with new windows of understanding and new opportunities for applying calculus. However, technology should not be viewed as a panacea for calculus instruction. This book seeks to take advantage of these new tools, while at the same time alerting the student to their inherent limitations and the care that must be taken to use technology wisely.

While being "technology-aware," the text itself does not assume the availability of any particular machine or software. To do so would invite immediate obsolescence and ignore how quickly technology advances. Rather, the text adopts a language appropriate for the kinds of numerical, graphical, and symbolic capabilities that are found (and will continue to be found) on a wide variety of computer software packages and sophisticated calculators. For example, the language of zooming in on the graph of a function is powerfully suggestive without the need for listing specific keystrokes or syntax.

While technology can provide students new opportunities for understanding calculus, it must be used with care. Numerical computations performed by a machine are subject to magnitude and precision limitations. For example, the calculation of difference quotients is naturally prone to the phenomena of cancellation errors. Machine-generated graphs can also provide misleading information, since graphs consist of a discrete collection of pixels whose locations are computed numerically. Symbolic algebra results need to be interpreted in context. Helping students understand the limitations of technology is a major goal of both the text and the supplement. Students are reminded of the care that must be taken to make intelligent use of technology without becoming a victim of its pitfalls.

In both the text and the student supplements, no specific hardware or software is assumed, so an instructor will need to judge the appropriateness of any particular activity in light of the technology available. However, the exercises activities are designed to be feasible with a very wide variety of available software and hardware. Certainly, a super calculator (symbolic/graphic) has more than enough power to suffice for all the activities, and a simple graphing calculator will be adequate for many of the activities.

PREFACE

Multiple representation approach to functions

The most important concept in all of mathematics is that of *function*, and the function concept is central in calculus. The idea of a function as a process accepting inputs and returning outputs can be captured in a variety of representations—numerically as a table of input-output pairs, graphically as a plot of outputs vs. inputs, and symbolically as a formula describing or modelling the input-output process. The interpretations of the core calculus topics of limits and continuity, differentiation, and integration all have different flavors when approached through different representations. The connections we forge between them enrich our personal concept of function.

All too often, students leave the calculus course with an impoverished mental image of function formed in a context dominated by symbolic forms. This book seeks to take a more balanced three-fold approach to functions. With each new topic or result, an explicit effort is made to interpret the meaning and consequences in a numerical, graphical, and symbolic context. Such an approach does not require the presence of technology, but its availability allows us greater access to numerical and graphical tools, while at the same time reducing the need for heavy emphasis of rote "by hand" symbol manipulation skills.

Visualization and Approximation

Two themes that become increasingly important with the availability of technology are visualization and approximation. The ability to obtain a machine-generated graph as a first step instead of a last one can completely turn around our approach to a variety of calculus topics. Graphical interpretation skills become primary. In particular, graphing can be used as a powerful problem solving aid, both in estimating and monitoring the reasonableness of results obtained numerically or symbolically. Whenever possible, explicit mention is made of the visual interpretation of definitions, theorems, and example solutions, often with direct reference to machine-generated graphs.

Much of calculus grew out of problems of approximation, and many of the key concepts of calculus are best understood as limits of approximations. Numerical tools make once exorbitantly tedious calculations into viable computational estimation strategies. Accordingly, approximation and estimation techniques are given a high priority throughout the text.

Problem solving and mathematical modelling

The application of mathematics to solving "real world" problems may be thought of in terms of the process diagrammed below.

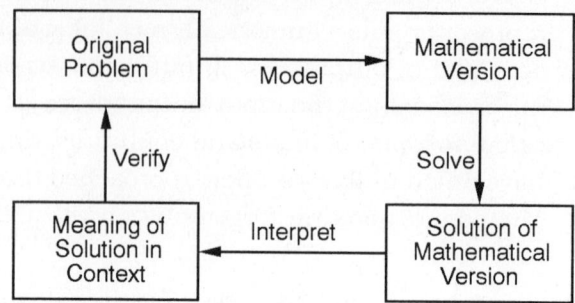

Mathematical training all too often deals only with the part of this process labelled "Solve." With the advent of computation devices capable of the rote work of solving, we can properly include training in the mathematical modelling of problems, interpreting the mathematical solutions, and checking the reasonableness of the results. Much effort has gone into embedding the applications of calculus into natural and plausible real world settings.

ABOUT THE MATERIAL IN VOLUME I

Chapter 0 is a brief introduction to the major ideas of calculus, including a discussion of how the notions of infinite processes and approximation arise naturally in the study of real number measurement.

Chapter 1 introduces functions as input-output processes, with the emphasis on multiple representations that recurs throughout the text. From the outset, transcendental functions are treated, including trigonometric, exponential, logarithmic, and inverse trigonometric functions. With respect to terminology and notation, the distinctions between variables, constants, and parameters are highlighted. The modern arrow notation for function assignment processes ($f : x \longmapsto f(x)$) is used both as a means for distinguishing between *constants* and *constant functions*, and for explaining composition of functions.

Chapter 2 discusses limits and continuity. Numerical and graphical approaches receive equal, if not more emphasis than symbolic techniques. The rigorous epsilon-delta definitions are included, but their meanings are explained with reference to their numerical and graphical consequences, rather than a heavy emphasis on proofs. For example, the definition of continuity of a function has a dynamic interpretation in terms of the scaling of a graphing window. The "δ-hunt" for a particular ϵ becomes a search for a certain horizontal scaling, given a vertical scaling. Numerically, ϵ's and δ's can be interpreted as output and input tolerances.

PREFACE

Chapter 3 starts with a review of linear functions, and then uses the important examples of piece-wise linear functions to discuss the notions of *local slope* and *local linearity*. Differentiable functions can then be considered as *approximately* locally linear functions, an idea visually reinforced by zooming in on the graphs of functions. The physical interpretation of derivative as a rate of change is motivated with the problem of estimating a car's speedometer reading using its odometer and a stopwatch. Computing difference quotients (and symmetric difference quotients) are considered as numerically viable techniques for estimating derivative values, and not just an artifact of the formal definition of derivative. Derivative properties and rules are developed and a dictionary of derivative formulas for all the basic algebraic and transcendental functions are included at the end of the chapter.

Chapter 4 emphasizes the use of the derivative as a measurement tool. The chapter begins by discussing the physical interpretation of derivative as a rate of change. A tangent line is considered as the graph of the best linear approximation of a differentiable function at a point (first-order Taylor forms). The use of the derivative to analyze function behavior, critical points, and extrema, and the consequences of the Mean Value Theorem are interpreted from both physical and graphical perspectives. Higher order derivatives are introduced using the physical example of a car's acceleration, and then used in a discussion of concavity, inflection points, and best quadratic approximations (second-order Taylor forms). The discussion of higher order derivatives concludes with an application to cubic splines.

Chapter 5 starts out with a discussion of problem solving in the spirit of Polya and the role of mathematical modelling in solving real world problems. The use of calculus to solve optimization problems are illustrated with examples and exercises drawn from the context of the day-to-day operations of a manufacturing facility. This chapter also treats implicit differentiation, related rates, and parametric and polar equations.

Chapter 6 motivates the idea of definite integral with the example of approximating π using the method of exhaustion. By using piece-wise linear functions as examples, many of the properties of definite integrals are then explored without the necessity of special summation formulas. The idea of a Riemann sum is then motivated as a reasonable approximation technique for more general functions. Antiderivatives are introduced by reversing the problem of determining a car's speedometer reading from its odometer and clock readings to one of determining distance covered from speed and time readings. *Slope fields* (direction fields) are used as a graphical means of approximating the graph of an antiderivative. Noting that $d = rt$ represents the area under the graph of a car's *constant* speed r over time t leads to the more general conjecture that definite integrals can be used to generate antiderivatives. The two Fundamental Theorems of Calculus tie together differentiation and integration. Chapter 6 closes with a discussion of numerical techniques of integration.

Chapter 7 discusses differential equations from a variety of viewpoints. Slope fields are used to visualize solutions to differential equations. The First Fundamental Theorem of Calculus is reviewed for its use in creating antiderivatives. Exponential and logarithmic functions are both re-examined as the solutions of special differential equations. Applications of exponential functions to problems of growth and decay are included. The method of substitution and integration by parts are discussed as the integral counterparts to the chain rule and the product rule in searching for antiderivatives.

Chapter 8 emphasizes the use of the integral as a measurement tool. The role of Riemann sums in modelling measurements involving continuously varying quantities is highlighted over and over again. Geometric examples include the measurement of area, volume, and arc length. Many of the definite integrals encountered in this chapter require the use of machine numerical integration. The use of polar coordinates in integration closes out the chapter.

Chapter 9 continues the discussion of applications of integration to measuring various averages, including centers of mass, and moving averages. The chapter then turns to physical applications such as velocity, force, and work. Improper integrals are introduced, and the chapter concludes with a section on probability measurement.

Chapter 10 discusses sequences and approximations. Long division is used as a familiar example of an infinite process that can produce an infinite sequence of approximations. Several examples of sequences, including recursive and iterative sequences, are examined. Limits are revisited using derivatives in the study of indeterminate forms (L'Hôpital's Rule). Root-finding methods, including the bisection method and Newton's Method, are also included as examples of techniques yielding sequences of approximations. The chapter concludes with a closer look at iterative methods in general and the Fixed Point Theorem.

Chapter 11 starts out with a discussion of the Archimedean property of real numbers and Zeno's paradox to motivate the idea of a series. A series is then defined as the limit of a sequence of partial sums. Tests of convergence include the Nth term test, comparison and limit comparison tests, the integral test, the alternating series test and the root and ratio test. Absolute and conditional convergence are contrasted. After a discussion of power series, including interval and radius of convergence, this chapter concludes with a study of function approximation techniques, in particular, Taylor polynomials and series.

Accompanying Volume I is a *Student Guide to Using Technology in Calculus*. Like the text, this supplement does not assume the use of any particular hardware or software. The introductory chapter of the supplement outlines some of the important factors one must be aware of in order to make intelligent use of technology. The effects of round-off and cancellation errors in numerical computations, the inherent limitations of machine graphics, and the importance of contextual assumptions are some of the issues addressed. The rest of the student guide consists of activities intended to supplement, extend, or reinforce many of the topics in the text. Of course, as the title suggests, the proper use of technology is a central theme throughout the supplement.

ACKNOWLEDGMENTS

The Oregon State University Calculus Curriculum Project has been made possible with the support of the National Science Foundation, Oregon State University and the Lasells Stewart Foundation, the Hewlett-Packard Corporation, and PWS-KENT Publishers.

This book was typeset on Macintosh computers using Donald Knuth's TeX with Textures 1.2 (Blue Sky Research) and the AmS-TeX - Version 2.0 macro package (American Mathematical Society). The illustrations were produced using MacPaint and MacDraw II (Claris), Illustrator 88 (Adobe), and Grapher 881 Version 1.2 (thanks to Steve Scarborough).

Special thanks are owed to: Steve Quigley, Barbara Lovenvirth, Beverly Jones, and Helen Walden of PWS-KENT Publishers, for their support and expertise in helping bring the work of the project to publication; Bert Waits and Franklin Demana (Ohio State University), Gregory D. Foley (Sam Houston State University), Thomas Tucker (Colgate University), Robert Moore (University of Washington), William Wickes (Hewlett-Packard), John Kenelly and Don LaTorre (Clemson University), for serving at one time or another as advisors and/or consultants to the project; Clain Anderson of Hewlett-Packard, for his efforts in making technology more easily available to teachers and students in both high schools and colleges; Dianne Hart and Howard L. Wilson, of Oregon State University, for their exceptional instructional and in-service efforts with the project; Michelle Jones, Kathy Dukes, and Alison Warr, for preparation of answers to the exercises; all the instructors and students at Oregon State University and the pilot test sites who have shown such enthusiasm and caring (daring?) in trying early drafts of the materials and providing valuable feedback; all the reviewers of early drafts of the materials; Marilyn Wallace, Donna Kent, George and Colleen Dick, for technical typing and preparation of some of the illustrations.

Finally, to Leslie and Colleen, Daniel, Jean, Connor, and Eamon, for tolerating the authors during the temporary insanity that accompanies working on such a project, we dedicate this book.

From the Publisher

It has been said that the only constant in life is change. We see it all around us. From the transformation of seasons ... to the new-found freedoms in Eastern Europe ... to individual growth and development.

The field of mathematics is no exception. Even the notion of *change* itself forms the very foundation of the calculus.

At PWS-KENT Publishing Company we are convinced that the call for reform in calculus instruction is here to stay. It may not alter much in content. It may not usher in a myriad of published product. And it may not appeal to every mathematician. It has received recognition, however, from such venerable institutions as the National Science Foundation, the American Mathematical Society and the Mathematical Association of America. And it has piqued the interest of many dedicated instructors who feel that something new and innovative is needed to revitalize the subject-matter.

Issues and developments in this reform movement have been the focus of several Calculus Reform Workshops which have been jointly sponsored by PWS-KENT, the Oregon State University Calculus Curriculum Project, and the Hewlett-Packard Company (in conjunction with a grant from the NSF). The workshops, whose primary function has been to provide the "grass roots" of the mathematics community with an opportunity to contribute their ideas and concerns on the curriculum, also have contributed greatly to this publication by Thomas Dick and Charles Patton.

As PWS-KENT enters its twenty-sixth year of being "Partners in Education," we hope that the published works of the Oregon State University Calculus Curriculum Project will be recognized as our contribution to an evolving calculus marketplace which has for more than fifteen years been very generous to us.

With this product we invite inquiry, scrutiny, and discovery. For, together, we can seek the security of change.

The Editors

Contents

0. Introduction: What is Calculus? 1

1. **Real Numbers and Functions—The Language of Calculus** **9**
   1.1 ABSOLUTE VALUE, SET AND INTERVAL NOTATION 9
   1.2 FUNCTIONS—
       NOTATION, TERMINOLOGY, AND REPRESENTATIONS 22
   1.3 A DICTIONARY OF FUNCTIONS 33
   1.4 MAKING NEW FUNCTIONS FROM OLD 47

2. **Limits and Continuity** **63**
   2.1 WHAT ARE LIMITS? 63
   2.2 DEFINITION OF LIMIT 71
   2.3 CONTINUITY 82
   2.4 ANALYZING DISCONTINUITIES AND ASYMPTOTIC BEHAVIOR 90

3. **The Derivative** **107**
   3.1 LINEAR FUNCTIONS 108
   3.2 WHAT IS A DERIVATIVE? 125
   3.3 COMPUTING AND ESTIMATING DERIVATIVES 133
   3.4 DERIVATIVE FORMULAS, NOTATIONS, AND PROPERTIES . 147
   3.5 NEW DERIVATIVES FROM OLD 162

4. **Derivative as Measurement Tool** **176**
   4.1 PHYSICAL INTERPRETATION OF DERIVATIVE—
       RATE OF CHANGE 177
   4.2 BEST LINEAR APPROXIMATIONS 184
   4.3 USING THE DERIVATIVE TO ANALYZE FUNCTION BEHAVIOR 192
   4.4 HIGHER ORDER DERIVATIVES 209

5. **Applications of the Derivative** **226**
   5.1 PROBLEM SOLVING AND MATHEMATICAL MODELLING 226
   5.2 SOLVING OPTIMIZATION PROBLEMS—FINDING THE EXTREMA 243
   5.3 IMPLICIT DIFFERENTIATION AND RELATED RATES 263
   5.4 PARAMETRIC AND POLAR EQUATIONS 281

6. **The Integral** **298**
   6.1 DEFINITE INTEGRALS—TERMINOLOGY AND EXAMPLES 304
   6.2 COMPUTING AND ESTIMATING DEFINITE INTEGRALS 313
   6.3 INDEFINITE INTEGRALS—ANTIDERIVATIVES 329
   6.4 THE FUNDAMENTAL THEOREMS OF CALCULUS 347
   6.5 NUMERICAL INTEGRATION TECHNIQUES 360

7. Differential Equations — **374**

7.1 WHAT IS A DIFFERENTIAL EQUATION? — 375
7.2 EXPONENTIAL AND LOGARITHMIC FUNCTIONS — 384
7.3 APPLICATIONS OF THE EXPONENTIAL MODEL — 402
7.4 INVERSE TRIGONOMETRIC FUNCTIONS — 412

8. Integral as Measurement Tool — **428**

8.1 USING DEFINITE INTEGRALS TO MEASURE AREA — 432
8.2 USING DEFINITE INTEGRALS TO MEASURE VOLUME — 440
8.3 USING DEFINITE INTEGRALS TO MEASURE ARC LENGTH — 432
8.4 INTEGRATION USING POLAR COORDINATES — 458

9. Applications of the Integral — **462**

9.1 USING DEFINITE INTEGRALS TO MEASURE AVERAGES — 462
9.2 USING DEFINITE INTEGRALS TO MAKE PHYSICAL MEASUREMENTS — 474
9.3 IMPROPER INTEGRALS — 487
9.4 USING DEFINITE INTEGRALS TO MEASURE PROBABILITY — 498

10. Sequences and Approximations — **514**

10.1 EXAMPLES OF SEQUENCES — 515
10.2 CONVERGENCE AND DIVERGENCE OF SEQUENCES — 527
10.3 FINDING LIMITS—INDETERMINATE FORMS — 537
10.4 ROOT-FINDING APPROXIMATION METHODS — 548

11. Series and Function Approximation — **564**

11.1 SERIES — 565
11.2 CONVERGENCE TESTS FOR SERIES — 576
11.3 RATIO AND ROOT TESTS—POWER SERIES — 595
11.4 TAYLOR POLYNOMIALS AND SERIES — 609

Appendix

ANSWERS TO SELECTED EXERCISES — appendix 1
INDEX — appendix 18

Chapter Titles from Volume II

12. Fundamentals of Vectors and Vector Functions

13. Calculus of Curves

14. Differential Calculus of Multivariate Functions

15. Integral Calculus of Multivariate Functions

16. Vector Fields—Line and Surface Integrals

CHAPTER 0

Introduction: What Is Calculus?

What is calculus? The word itself suggests *calculations*, but to think of calculus as merely a collection of computational techniques does a grave injustice to the subject. Rather, calculus is a network of fundamentally important mathematical *ideas*. These ideas had their origins in attempts to solve particular measurement problems in geometry and physics. Specifically, the problems of measurement of length, area, and volume in geometry and the problems of measuring force, velocity, and acceleration in physics gave birth to the calculus. Today the applications of calculus reach far and wide to quantitative analysis in fields ranging from archaeology to zoology.

There are two major branches of calculus. The *differential calculus* involves measuring the instantaneous rate of change of one quantity relative to the change in another quantity. This *physical* measurement problem corresponds to a *geometric* measurement problem— finding the slope of a graph at a specific point. *Integral calculus* has its origins in the geometric problems of measuring of length, area, and volume. This corresponds to the physical problem of measuring total change from information about rate of change. A realization of the inverse nature of these two problems marked the dawn of the modern age of calculus.

Why is calculus useful? Understanding the dynamics of *change* is a primary goal in the study of any system, whether it be physical, biological, economic, or social. When we analyze processes involving continuous change, calculus can allow us to describe and measure the rate of that change and its total effects. By harnessing the notion of *infinite processes* and their application to approximation, calculus provides both powerful tools and an effective language for describing and measuring change.

Real numbers and measurement problems

Measurement is a fundamental theme in all of mathematics. Two of our earliest mathematical experiences have to do with measurement— counting and using a ruler. These two simple activities capture the essence of two quite different branches of mathematics.

Counting necessarily uses only whole numbers 0, 1, 2, 3, ..., and counting problems occupy much of what is called *discrete* mathematics. We associate ruler measurement with the system of real numbers. The real number line itself can be thought of as an infinite ruler, with each real number associated with exactly one point on the line, and conversely, every point on the line corresponding to a unique real number (Figure 0.1).

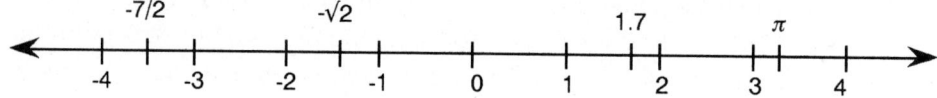

Figure 0.1 The real number line.

While the whole numbers can be thought of as particular real numbers, there is a major distinction between counting problems and real number measurement problems. For example, suppose you were asked to count the number of letters on the page you are now reading. By directly counting, you have every reason to expect your answer to be exactly correct.

On the other hand, suppose you were asked to measure the *height* of the page. If you chose a ruler to make the measurement, your answer could hardly be more accurate than the thickness of the marks on the ruler. Even if you had a sophisticated optical scanner providing a digital readout of the height, there would be some limitation on the precision of the measurement you obtain, as there would be with any physical measuring instrument.

This imprecision is a fact of life when we deal with any physical measuring instrument. Mathematics can provide us with *precise* relationships between quantities. For example, the familiar Pythagorean Theorem tells us that

$$a^2 + b^2 = c^2$$

when a, b, and c represent the leg lengths and hypotenuse length, respectively, of a right triangle. Provided we know the exact values of a and b, we can compute the exact value of c. Or can we?

The square root of 2

One problem of measurement that the ancient Greeks wrestled with had to do with the diagonal of a square with sides of unit length (see Figure 0.2). If we consider the diagonal as being the hypotenuse of a right triangle with unit length legs, then the Pythagorean Theorem tells us that the diagonal has length $\sqrt{2}$. The difficulty the ancient Greeks had was not with the Pythagorean Theorem, but with the number $\sqrt{2}$ itself, which they expected to be a *rational number* (that is, a ratio p/q of two whole numbers p and q). However, $\sqrt{2}$ is an example of an **irrational number**, and it cannot be expressed as the ratio of two integers.

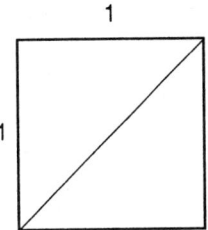

Figure 0.2 Measuring the length of the diagonal of a unit square.

Real numbers can be represented in other ways. Perhaps the most common representation of real numbers is by decimal notation. Any real number has a (possibly infinitely long) decimal representation. The decimal representations of rational numbers either terminate or have a finite block of digits that repeats indefinitely.

For example, the rational number $22/7$ (a common approximation of the *irrational* number π) can be written as

$$\frac{22}{7} = 3.\overline{142857} = 3.142857142857142857142857\ldots$$

where the overline indicates that this block of six digits repeats indefinitely. This notation allows us to communicate the entire decimal expansion, even though it is infinitely long.

As for the decimal representation of $\sqrt{2}$, it will not reveal any such block of repeating digits:

$$\sqrt{2} = 1.41421356\ldots$$

Here, the use of "..." means only that the decimal representation continues, and is not meant to convey that some repeating pattern exists. How can we say that we *know* $\sqrt{2}$ is a perfectly good real number, since we have no way of knowing all of the digits in its decimal representation?

Infinite processes and approximations

The fundamental problem of representing real numbers like $\sqrt{2}$ is at the heart of calculus. While it's physically impossible to write down the entire decimal representation of $\sqrt{2}$, we *can* describe an *infinite process* that would yield a sequence of decimal approximations, each closer to the "exact" value of $\sqrt{2}$. Let's see how.

If x and y are *positive real numbers*, then

$$x^2 < y^2 \quad \text{implies} \quad x < y.$$

In other words, if we can find two positive real numbers x and y such that $x^2 < 2 < y^2$, we can conclude that $x < \sqrt{2} < y$. This allows us to approximate $\sqrt{2}$ to any finite number of decimal places.

First, we must have

$$1 < \sqrt{2} < 2$$

because

$$1^2 = 1 < 2 < 4 = 2^2.$$

This just tells us that $\sqrt{2} = 1.something$. Now, the first decimal place of $\sqrt{2}$ must be 4, since we deduce

$$1.4 < \sqrt{2} < 1.5$$

from checking that

$$(1.4)^2 = 1.96 < 2 < 2.25 = (1.5)^2.$$

Continuing, we have

$$1.41 < \sqrt{2} < 1.42$$

since

$$(1.41)^2 = 1.9881 < 2 < 2.0164 = (1.42)^2,$$

and

$$1.414 < \sqrt{2} < 1.415$$

since

$$(1.414)^2 = 1.999396 < 2 < 2.002225 = (1.415)^2.$$

0. INTRODUCTION: WHAT IS CALCULUS?

If we use the lower of the two approximations at each step of this process, we obtain a sequence of numbers: 1, 1.4, 1.41, 1.414, ..., where each number in the sequence determines another decimal place of $\sqrt{2}$. The upper approximation at each step differs from the lower approximation by one digit in the last decimal place, and these upper approximations also give us a sequence of numbers *converging* on $\sqrt{2}$.

While this process never ends after a finite number of steps, it really can be thought of as *determining* the square root of two. The difference between the upper and lower estimates can be made as small as we like by just continuing the process. Only *one* number can fit between every pair of approximations, and by design, we have guaranteed that $\sqrt{2}$ is between every pair of approximations. Hence, $\sqrt{2}$ is the *unique* number determined by this sequence of inequalities.

Of numbers and symbols

Of course, using a decimal approximation of $\sqrt{2}$ to several thousand digits would be silly in performing computational work with physical measurements, which are precise to only a few decimal places. Is there ever any advantage to using the symbol $\sqrt{2}$ as opposed to simply using a sufficiently accurate decimal approximation? Some might maintain that $\sqrt{2}$ is *exact* or somehow more aesthetically pleasing. But the real advantage to the symbol $\sqrt{2}$ is that it reminds us of the defining characteristic of the number: $\sqrt{2}$ is that *unique* positive real number whose square is 2, and is trapped by the infinite approximation process we just described. Once we use a decimal approximation for $\sqrt{2}$, it becomes anonymous and its defining characteristic is lost in the midst of other calculations. Carrying the true identity of a number through our calculations can prove to be extremely helpful in noticing an important pattern or relationship.

Look at the picture of the right triangles shown in Figure 0.3. The first triangle has legs of unit length and hypotenuse a_1. The second triangle has one leg of unit length and the hypotenuse of the first triangle as its other leg. Similarly, the third triangle has one leg of unit length and the hypotenuse of the second triangle as its other leg. We could imagine building these triangles *ad infinitum*. (Of course, they will eventually "spiral" around and overlap the earlier triangles.) Is there a pattern to the lengths of the hypotenuses, which we have labeled a_1, a_2, a_3, a_4, ..., (where a_n represents the hypotenuse of the *n*th triangle)?

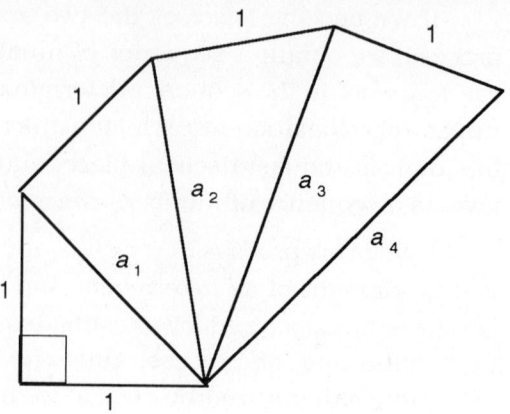

Figure 0.3 A "spiral" of right triangles.

Using a calculator and the Pythagorean Theorem, we obtained the following:

$$a_1 \approx 1.414213562, \ a_2 \approx 1.732050807, \ a_3 \approx 1.999999999, \ a_4 \approx 2.236067977, \ldots$$

but by maintaining the "radical" notation, we obtain

$$a_1 = \sqrt{2}, \ a_2 = \sqrt{3}, \ a_3 = \sqrt{4}, \ a_4 = \sqrt{5}, \ \ldots.$$

From which sequence could you most easily predict the value of a_{99}?

Calculus—past and future

Whether calculus was *invented* or *discovered* is a matter for philosophical debate. Many of the ideas and insights of calculus stem from geometrical and physical observations. In that regard, calculus was discovered. But ideas and insights arise within our minds, and the language and notation with which we communicate them are very much a creative endeavor, so calculus was also invented. In any case, the development of calculus is clearly one of the crowning intellectual achievements of the human race.

The impact calculus has had on technological progress alone makes its development of great historical significance. But calculus continues to be a living, growing body of knowledge whose importance to quantitatively understanding the world around us remains vital. As we move into the next century and address the increasingly complex problems of the future, a knowledge of calculus will continue to be indispensable to analyzing and describing the dynamics of change.

0. INTRODUCTION: WHAT IS CALCULUS?

Some of the technology made possible by calculus is granting us unprecedented access to devices such as calculators and computers. We live in an exciting age where these powerful computational tools enable us to perform complex numerical and symbolic computations and provide tremendous graphics capabilities at our fingertips. In turn, we now have both new ways to understand the ideas of calculus and new opportunities to apply calculus.

However, even the most powerful technology is of little use if we do not know how it can and cannot be applied. We recognize that new technological tools are available and this book was written with the *intelligent* use of those tools in mind. Solving important mathematical problems will always require the inspiration, recognition, and application of the right idea at the right time. The primary goal of this book is to provide you with a working understanding of the important ideas, language, and applications of calculus. Calculus arose in response to the need to solve certain problems, and to understand the *what* and *why* of calculus requires understanding *how, when,* and *where* calculus can be used to solve problems.

EXERCISES

1. Of all positive proper fractions with denominators no larger than 25, exactly which ones have terminating decimal representations?

2. If a rational number p/q is in reduced form (in other words, the p and q have no common integer factors greater than 1), how can one tell whether the number has a terminating decimal representation without actually doing the long division?

3. Give an example of a rational number having two different decimal representations. (Hint: $3 \times \frac{1}{3} = 1$.)

4. In the decimal representation of $22/7$, there is a block of 6 digits that repeats indefinitely. Can the number of digits in the repeating block of the decimal representation of p/q ever be as big or bigger than q, the denominator? If so, give an example. If not, explain why.

5. Is $0.101001000100001000001\ldots$ a rational number? Why or why not?

Exercises 6-13 refer to Figure 0.3.

6. What is the value of a_{99} if we continue the sequence of right triangles?

7. What is the area of the *100th* triangle?

8. What is the value of a_n if we continue the sequence of right triangles?

9. What is the area of the *nth* triangle?

10. Suppose we use each hypotenuse a_n as the diameter of a circle. What would be the area of the *nth* circle?

11. What is the measure of the smallest angle in the *100th* triangle?

12. What is the measure of the smallest angle in the *nth* triangle?

13. For what value of n does the *nth* triangle overlap the first triangle?

---

14. Suppose I want to cut out a cube of wood with a volume of 1000 cm^3. How accurate must my measurements of each side of the cube be (in cm) if I want the volume to be within 1 cm^3 of the desired volume?

15. Suppose I want to cut out a cube of wood with a volume of 1 m^3. Within what accuracy (in cm) must my measurements of each side of the cube be if I want the volume to be within 1 cm^3 of the desired volume?

16. Railroad tracks leave small gaps between sections to allow for expansion and contraction of the metal due to temperature changes. Suppose someone lays a continuous 1 mile stretch of metal track with no gaps between two walls, and on a hot summer day the track expands 1 inch in length for every 100 feet. If one end of the track stays fixed to the ground, how high will the other end of the track have to rise to compensate for the new length?

17. Assume that the circumference of the earth at the equator is exactly 25000 miles. If a rope is wrapped around the earth at the equator so that it was 3 feet off the ground at all times, how much more than 25000 miles of rope is needed?

18. A person charges $1.50 per square foot to refinish wood floors. Suppose she measures a square ballroom floor to be 80 feet on a side within an accuracy of ± 3 inches. What is the most she might inadvertently overcharge (due to her measurement error)?

CHAPTER 1

Real Numbers and Functions— The Language of Calculus

This chapter lays out the the basic terminology and notation we'll use in studying real numbers and functions. Much of the material in this chapter should be familiar to you, but you may find some slight differences in our use of notation, particularly for functions. We'll start the chapter with a discussion of absolute value, set and interval notation, and then turn our attention to the concept of function and the ways we can represent them. Finally, we'll catalog some particularly important examples of functions, and how we can combine functions to produce new ones.

1.1 ABSOLUTE VALUE, SET AND INTERVAL NOTATION

Mathematical notation has a purpose, and is rarely chosen arbitrarily. Its goal is to communicate important information effectively and concisely. Absolute value notation is used to communicate the *distance* between two numbers.

The absolute value of a real number x is usually defined algebraically as follows:

$$|x| = x \text{ if } x \geq 0 \quad \text{and} \quad |x| = -x \text{ if } x < 0.$$

Geometrically, $|x|$ can be thought of as the distance between the point with coordinate x and the origin 0. The absolute value of the difference of two real numbers, $|a-b|$ represents the (nonnegative) distance between the points with coordinates a and b, respectively. This is true regardless of the relative order of a and b on the number line (see Figure 1.1).

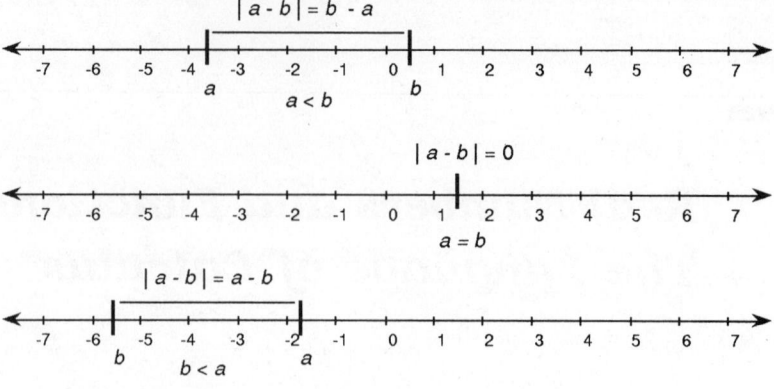

Figure 1.1 Absolute value measures distance between points.

EXAMPLE 1 Suppose that we have measured x in meters to be 23 m with an error tolerance of $\pm.025$ m, so that

$$22.975 \leq x \leq 23.025.$$

One nice advantage of this notation is that we can read off a lower and upper bound for x directly. In terms of the number line (with meters as the unit of measurement), this means that the true value of x must be within a distance of .025 from 23. Using absolute value notation we could write this as

$$|x - 23| \leq .025.$$

The absolute value notation provides a different advantage— we can read off both the measured value 23 and the error tolerance .025 for x directly. ∎

In general, suppose we have measured x (in some given units) to have a measure of a units with an error tolerance *less than* $\pm\delta$ units. The range of possible values for x could be written

$$a - \delta < x < a + \delta$$

to emphasize that the lower bound for x is $a - \delta$ and the upper bound for x is $a + \delta$, or we can write

$$|x - a| < \delta$$

to emphasize the measured value a and the error tolerance δ. (δ is the Greek letter "delta," which corresponds to the letter d, for *distance*.)

1.1 ABSOLUTE VALUE, SET AND INTERVAL NOTATION

We need to take special care with algebraic manipulations that depend on the *sign* of the value of an expression involved. In particular, the absolute value marks cannot be dropped from an expression involving unknowns without a careful analysis of the possible values that expression could have.

EXAMPLE 2 Express $|-2x-3|$ without the use of the absolute value symbols.

Solution It would be *incorrect* to simply drop the "$-$" symbols without any knowledge of the possible replacement values for x. What we *can* say is that

$$|-2x-3| = \begin{cases} -2x-3 & \text{if } -2x-3 \geq 0 \\ 2x+3 & \text{if } -2x-3 < 0. \end{cases}$$

or, after simplifying the inequalities:

$$|-2x-3| = \begin{cases} -2x-3 & \text{if } x \leq -\dfrac{3}{2} \\ 2x+3 & \text{if } x > -\dfrac{3}{2}. \end{cases}$$

This is an example of what we might call a *split formula*, meaning that the rule for evaluating the expression depends on the value of the variable x. In this example, we use $(-2x-3)$ to evaluate $|-2x-3|$ when $x \leq -3/2$, but we use $(2x+3)$ when $x > -3/2$.

Hence, if $x = -7$, then $|-2x-3| = -2(-7) - 3 = 14 - 3 = 11$.

If $x = -1/2$, then $|-2x-3| = 2(-\frac{1}{2}) + 3 = -1 + 3 = 2$. ∎

Set notation

A **set** is simply a collection of objects. The objects in a set are called **elements** or **members** of the set. We use the notations

$$x \in A \qquad x \notin B$$

to denote, respectively, that the object x is a member of the set A, but x is not a member of the set B.

Sets of real numbers can be described in a variety of ways. If a set contains only a few numbers, we could simply enclose the list of these numbers in braces to denote that set.

For example, the set containing the real numbers $-1, \frac{2}{3}, -\frac{17}{8}, \sqrt{2}, 6,$ and π could be denoted as

$$\{-1, \frac{2}{3}, -\frac{17}{8}, \sqrt{2}, 6, \pi\}.$$

This is called the **roster** notation for sets since we are simply listing the elements as on a roster. This method of denoting a set is fine as long as the number of elements in the set is small. We could even use the roster notation for a set with infinitely many elements as long as we could effectively communicate exactly which elements belong to the set.

EXAMPLE 3 The set E of all positive even integers using roster notation can be written $E = \{2, 4, 6, 8, \ldots\}$. ∎

This use of the roster notation depends heavily on the reader discerning the intended pattern. For example, writing $\{2, 4, \ldots\}$ out of any context would make it difficult for the reader to know whether the set intended was the set of positive even integers, the set of positive integer powers of 2, or perhaps some other infinite set.

The **rule** notation for sets gives us much more flexibility and precision for denoting a set of real numbers. The general form for such a set is

$$\{x : \text{a property involving } x\}.$$

We read this notation as "the set of all real numbers x such that x satisfies the given property." (Some books use the vertical bar "|" to denote the "such that" part of this notation.)

EXAMPLE 4 Using rule notation to denote the set in example 3, we have $E = \{x : x = 2n$ for some positive integer $n\}$. ∎

The **set of all real numbers** is denoted by $\mathbb{R}$. If we are using rule notation for a set, and we want to emphasize that the elements are real numbers, then we write

$$\{x \in \mathbb{R} : \text{a property involving } x\}.$$

The set containing no elements whatsoever is called the **empty set** and is denoted by the symbol $\emptyset$ or by a pair of empty braces { }.

1.1 ABSOLUTE VALUE, SET AND INTERVAL NOTATION

 Do not overdo the use of braces. The symbol $\mathbb{R}$ denotes the set of all real numbers and the symbol $\varnothing$ denotes the empty set. However, $\{\mathbb{R}\}$ and $\{\varnothing\}$ are something entirely different. (Each is a set containing exactly one element that also happens to be a set.)

Definition 1

If A and B are two sets of real numbers, then

$$A \cup B = \{x : x \in A \text{ or } x \in B\}$$

and is called the **union** of A and B.

$$A \cap B = \{x : x \in A \text{ and } x \in B\}$$

and is called the **intersection** of A and B. If every element of A is also an element of B, then we write $A \subseteq B$ and say A is a **subset** of B.

Note that if $A = B$, then we could still say $A \subseteq B$. In fact, if we have both $A \subseteq B$ and $B \subseteq A$, then we could conclude that $A = B$. If A is not a subset of B, we write $A \nsubseteq B$.

EXAMPLE 5 Suppose $A = \{2, 4, 6, 8\}$, $B = \{3, 4, 5, 6\}$, and $C = \{2, 4, 6\}$. We note that $A \cup B = \{2, 3, 4, 5, 6, 8\}$, $B \cup C = \{2, 3, 4, 5, 6\}$, and $A \cup C = A$, while $A \cap B = B \cap C = \{4, 6\}$ and $A \cap C = C$.

We can also see that $C \subseteq A$, but $C \nsubseteq B$. ∎

Interval notation

Some special sets of real numbers arise so frequently in our study of calculus that some special shorthand notation has been devised for them. **Intervals** are sets of real numbers containing *no gaps*. Given any two distinct real numbers in an interval, all of the numbers between them also belong to the interval. We can think of intervals geometrically as representing unbroken pieces of the real number line. Here's a glossary of the special notation used for intervals of various types.

Definition 2 | **Glossary of interval notation.** If $a < b$ are real numbers, then

$$[a,b] = \{x : a \le x \le b\} \quad [a,b) = \{x : a \le x < b\} \quad [a,\infty) = \{x : a \le x\}$$
$$(a,b] = \{x : a < x \le b\} \quad (a,b) = \{x : a < x < b\} \quad (a,\infty) = \{x : a < x\}$$
$$(-\infty,b] = \{x : x \le b\} \quad (-\infty,b) = \{x : x < b\} \quad (-\infty,\infty) = \mathbb{R}$$

When they appear as shown here, the real numbers a and b are called **endpoints** of the interval. If an interval has two endpoints, then we call it a **bounded** interval. Otherwise, we say the interval is **unbounded**. The intervals

$$[a,b], \quad (a,b), \quad [a,b), \quad (a,b]$$

are bounded intervals, while

$$[a,\infty), \quad (a,\infty), \quad (-\infty,b], \quad (-\infty,b), \quad (-\infty,\infty)$$

are unbounded intervals.

Note that the endpoints of an interval do not necessarily have to belong to the set of real numbers in that interval. An interval is called **closed** if all of its endpoints are included in the interval. An interval is called **open** if it does *not* include any of its endpoints. The intervals

$$[a,b], \quad [a,\infty), \quad (-\infty,b]$$

are closed, while the intervals

$$(a,b), \quad (a,\infty), \quad (-\infty,b)$$

are open. The intervals $[a,b)$ and $(a,b]$ are called **half-open** (or **half-closed** if you prefer) for obvious reasons. Technically, we'd call the interval $(-\infty,\infty)$ both open and closed, because it has no endpoints at all!

Caution! The symbols ∞ and $-\infty$, which we read as "positive infinity" and "negative infinity," do *not* represent real numbers.

They are simply convenient shorthand symbols used here to indicate when an interval extends infinitely far in one or both directions. If an interval of real numbers is intended, then it is *never* appropriate to use a left square brace "[" adjacent to the symbol $-\infty$ or a right square brace "]" adjacent to the symbol ∞.

We can graph intervals on a number line by marking endpoints with either a small filled-in or empty circle (for included and nonincluded endpoints, respectively) and shading the portion of the number line included in the interval. Figure 1.2 shows a graph for each of the intervals in Definition 2.

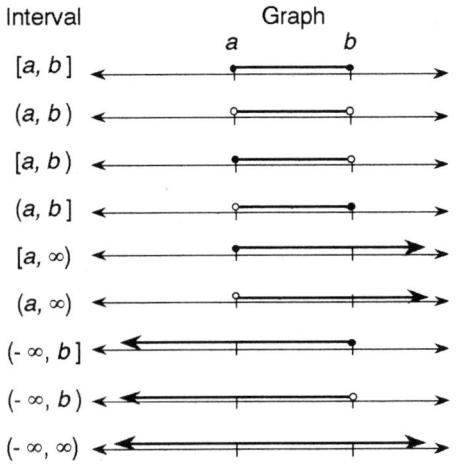

Figure 1.2 Graphs of intervals on the real number line.

Using absolute value notation for intervals

Absolute value notation is sometimes used to express an interval of real numbers. For example, the interval $[-2, 4]$ can be described as the set of all real numbers x at a distance of 3 units or less from 1, the **center** of the interval. We could thus write

$$[-2, 4] = \{x : |x - 1| \leq 3\}.$$

This way of expressing an interval emphasizes the center and the **radius** of the interval.

EXAMPLE 6 Express the open interval $(-7, -2)$ using absolute value notation.

Solution Representing the interval graphically can be very helpful. The graph of the interval $(-7, -2)$ is shown in Figure 1.3.

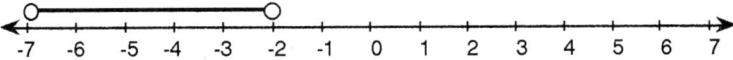

Figure 1.3 Graph of $(-7, 2)$.

The center of the interval $(-7, -2)$ is the average of the endpoints:
$$\frac{(-7)+(-2)}{2} = -\frac{9}{2}.$$

The radius of the interval is half the distance between endpoints:
$$\frac{|(-7)-(-2)|}{2} = \frac{|-5|}{2} = \frac{5}{2}.$$

So the interval can be expressed as the set of all real numbers x at a distance of less than $5/2$ units from $-9/2$, or
$$(-7,-2) = \{x : |x - (-9/2)| < 5/2\} = \{x : |x + 9/2| < 5/2\}.$$

■

EXAMPLE 7 Express $\{x : |x+2| \leq 5\}$ using interval notation.

Solution $|x+2| = |x-(-2)|$, so our set of real numbers all lie 5 units or less from -2. Figure 1.4 below illustrates that -2 is at the center of the interval and the radius of the interval is 5.

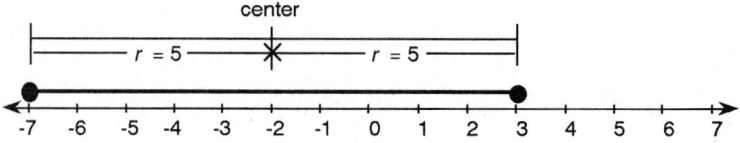

Figure 1.4 Graph of $|x+2| \leq 5$.

The endpoints of the interval are $((-2)+5) = 3$ and $((-2)-5) = -7$, and since these endpoints are included, we have a closed interval
$$\{x : |x+2| \leq 5\} = [-7, 3].$$

■

This use of absolute value notation is particularly useful for communicating *proximity* to a particular real number a. If we want to say that x is within distance δ of a, we could write
$$|x-a| < \delta.$$

The Cartesian plane

An **ordered pair** of objects (a, b) is distinguished from a set containing two objects $\{a, b\}$ by specifying the order of a and b. We call a the **first coordinate** and b the **second coordinate** of the ordered pair (a, b), but there is no specific order intended when we talk of the set $\{a, b\}$.

EXAMPLE 8   The ordered pairs $(0, 1)$ and $(1, 0)$ are considered *different*; the sets $\{0, 1\}$ and $\{1, 0\}$ are *not*. ■

Ordered pairs of real numbers can be represented as points in a plane by making use of a **rectangular coordinate system**. Two perpendicular real number lines provide the **coordinate axes** for the system, and their point of intersection is called the **origin**. Generally, these two axes are placed so that the first coordinate will correspond to the horizontal axis (with numbers increasing from left to right) and the second coordinate will correspond to the vertical axis (with numbers increasing from lower to higher). This arrangement divides the plane into four distinct regions called **quadrants**, and the quadrants are traditionally numbered I-IV starting with the upper right and proceeding counterclockwise.

The first and second coordinates of any point are found by locating the real numbers located perpendicularly from it on the horizontal and vertical axes, respectively. Similarly, we can locate any point given its ordered pair of coordinates (x_0, y_0). Figure 1.5 illustrates a rectangular coordinate system with the quadrants and axes labelled.

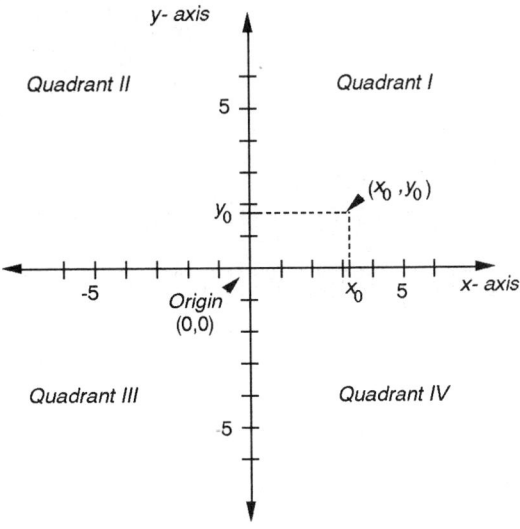

Figure 1.5 A rectangular coordinate system.

A rectangular coordinate system is sometimes called a **Cartesian coordinate system** in honor of the French mathematician and philosopher René Descartes (1590-1650), the founder of analytic geometry.

Cartesian products of intervals

A rectangular region with sides parallel to the coordinate axes can be indicated by the **Cartesian product** of two intervals.

Definition 3

> The **Cartesian product** of two closed intervals is indicated by the notation
>
> $$[a, b] \times [c, d]$$
>
> and is the set
>
> $$\{(x, y) : a \leq x \leq b \text{ and } c \leq y \leq d\}$$
>
> consisting of all ordered pairs (x, y) such that $x \in [a, b]$ and $y \in [c, d]$.

Graphically, $[a, b] \times [c, d]$ corresponds to all the points in the closed box indicated in Figure 1.6.

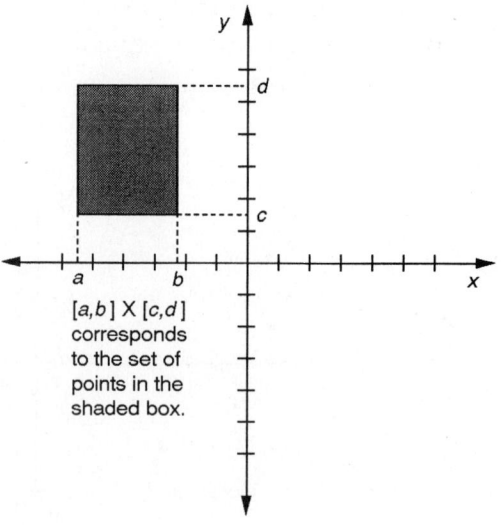

Figure 1.6 Cartesian product of two closed intervals

1.1 ABSOLUTE VALUE, SET AND INTERVAL NOTATION

A Cartesian product of intervals is a convenient way of specifying a *viewing window* for calculator or computer graphics. Another way of specifying a rectangular window of the form $[a,b] \times [c,d]$ would be by specifying two opposite corner points: either (a,c) and (b,d) (lower left and upper right) or (a,d) and (b,c) (upper left and lower right).

EXAMPLE 9 Describe and sketch the set of points represented by $(-\infty, \infty) \times (-1, 1)$.

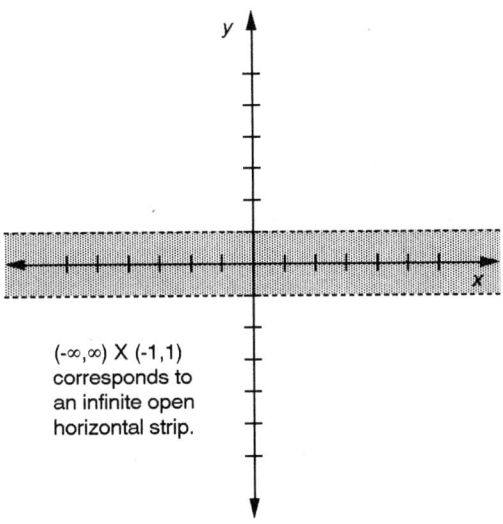

Figure 1.7 Graph of $(-\infty, \infty) \times (-1, 1)$.

Solution   The Cartesian product $(-\infty, \infty) \times (-1, 1)$ would represent an infinite open horizontal strip with the x-axis running down its center as shown below in Figure 1.7. ∎

EXERCISES

In exercises 1-4 express the given inequality using absolute value notation.

1. $-8 \leq x \leq -2$
2. $-1 < x < 4$
3. $x > 3$ or $x < -3$
4. $x \geq 10$ or $x \leq 4$

Find all real values of x satisfying the inequalities in exercises 5-12.

5. $x^2 - 4 \leq 5$
6. $2|x - 9| \geq 1$
7. $\dfrac{x}{1 + 2x} < 3 - x$
8. $|x^2 - 9| \geq 2$
9. $\dfrac{1}{|2x - 3|} < 4$
10. $x^3 - x < 0$
11. $x^2 + 4x + 6 \leq 1$
12. $x^2 - 2x + 3 > 1$

The **triangle inequality** states that for any three real numbers a, b, c (not necessarily distinct), the following inequality is always satisfied:

$$|a - b| + |b - c| \geq |a - c|.$$

In exercises 13-15 you will verify the triangle inequality by way of some particular examples.

13. Take any three different real numbers and label them *six* different ways with the letters a, b, and c. Calculate $|a - b|$, $|b - c|$, and $|a - c|$ for each of the six cases and verify that the triangle inequality is true for each case. For which cases can "$\geq$" be replaced with "$=$" ?

14. Repeat exercise 13, but this time choose only two different real numbers and try all the cases where two of a, b, and c are the same number.

15. Finally, what happens when $a = b = c$?

16. Why do you think the triangle inequality has its name?

In exercises 17-20, express each interval using absolute value notation. Graph each interval on the real number line.

17. $[-5, 0]$
18. $(4, 11)$
19. $[-2, 19]$
20. $(-13, -5)$

In exercises 21-24, express each set of real numbers using interval notation. Graph each interval on the real number line.

21. $\{x : |x - 3| < 7\}$
22. $\{x : |x + 4| \leq 3\}$
23. $\{x : |x| < 1\}$
24. $\{x : |x - a| < \delta\}$ where $\delta > 0$.

1.1 ABSOLUTE VALUE, SET AND INTERVAL NOTATION

In exercises 25-32, graph the following sets of real numbers. If the set is an interval, indicate whether it is open, closed, both, or neither, and indicate whether the interval is bounded or unbounded.

25. $[2, \infty) \cup (-\infty, 2]$
26. $[2, \infty) \cap (-\infty, 2]$
27. $(2, \infty) \cup (-\infty, 2)$
28. $(2, \infty) \cap (-\infty, 2)$
29. $[-2, \infty) \cup (-\infty, 2]$
30. $[-2, \infty) \cap (-\infty, 2]$
31. $(-2, \infty) \cup (-\infty, 2)$
32. $(-2, \infty) \cap (-\infty, 2)$

In exercises 33-40, express the set of real solutions to each of the following inequalities using interval notation, and graph the solution sets on the real number line.

33. $x^2 - 4 \leq 5$
34. $2|x - 9| \geq 1$
35. $x^2 - 2x + 3 > 1$
36. $\dfrac{x}{1 + 2x} < 3 - x$
37. $|x^2 - 9| \geq 2$
38. $\dfrac{1}{|2x - 3|} < 4$
39. $x^3 - x < 0$
40. $x^2 + 4x + 6 \leq 1$

In exercises 41-44, describe and sketch the set of points represented by the given Cartesian product of intervals.

41. The "unit square" $[0, 1] \times [0, 1]$
42. $[-5, 1] \times (2, 4)$
43. $(2, 4) \times [-5, 1]$
44. $(-\infty, -1) \times [2, \infty)$

Quadrant I could be described as the Cartesian product $(0, \infty) \times (0, \infty)$. In exercises 45-52, describe each of the given regions as a Cartesian product of intervals.

45. Quadrant II

46. Quadrant III

47. Quadrant IV

48. the closed upper half-plane (containing all the points in the plane with non-negative y-coordinates)

49. the open infinite strip of width 6 units having the y-axis running down its center

50. a rectangle lying entirely in the third quadrant and has a diagonal of 5 units (there are many)

51. the rectangle with $(-2, 3)$ and $(4, -5)$ at opposite corners, with sides parallel to the coordinate axes

52. the square with the origin at its center and an area of 4 square units

---

53. To achieve acceptable fitting accuracy in a manufactured product, a machine tool setting must be made at 3.72 cm with an allowable error tolerance in either direction of at most 0.003 cm. Express the set of permissible settings x in both interval notation and absolute value notation.

54. Instead of a split formula for the definition of absolute value, someone comes up with the following *algebraic* formula:

$$|x| = \sqrt{(x^2)}$$

where the square root symbol refers, as usual, to the *nonnegative* square root. Does the formula work?

1.2 FUNCTIONS—NOTATION, TERMINOLOGY, AND REPRESENTATIONS

The most important concept in all of mathematics is that of *function*. The idea of function is central not only to mathematics, but to the other disciplines that mathematics serves as a tool and a language. Whether the systems we are studying are physical, biological, social, or economic, we are always on the lookout for significant patterns and connections. The discovery of a functional relationship is a particularly important achievement, for it means we've found a pattern describing a *very strong* connection. Mathematics provides both the means and the terminology to analyze and describe functional relationships.

Like so many other mathematical terms, the word "function" is also used in everyday language. For example, someone might say, "Wisdom is a function of experience" to communicate the idea that one's wisdom is *dependent* on one's experience. Dependence is a key feature of the mathematical notion of *function*. Mathematically, we can think of a function as a special kind of process that accepts certain *inputs* and produces or assigns corresponding *outputs*. Let's make that notion more precise with the following definition:

1.2 FUNCTIONS—NOTATION, TERMINOLOGY, AND REPRESENTATIONS

Definition 4 | A **function** is any correspondence or process that assigns a uniquely determined output value to each member of a given set of inputs. The set of all possible inputs is called the **domain** of the function.

A function could have virtually any kind of inputs as members of its domain, and likewise, the outputs a function assigns to these inputs could be almost anything—numbers, vectors, sets, or even other functions. In calculus we are most interested in those functions having domains consisting of real numbers or vectors of real numbers, and whose values are either real numbers or vectors of real numbers. Accordingly, if we refer to a **real-valued function** or a **vector-valued function**, we have simply indicated whether the function produces real numbers or vectors as outputs. The first part of this book is primarily concerned with real-valued functions, while the second part of this book addresses vector-valued functions. Calculus gives us extremely powerful ways to study these kinds of functions. The differential calculus allows us to measure the rate of change of the output of a function relative to change in its inputs. The integral calculus allows us to measure such things as the *average* output value of a function over a set of inputs. While those may seem to be simple ideas, they have profound consequences for solving many real world problems.

The important idea— function as process

We can use a machine diagram to illustrate a function f (see Figure 1.8).

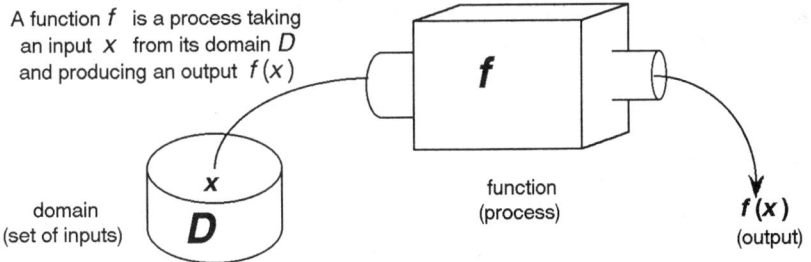

Figure 1.8 Function as process.

A typical input for the function is represented by x, and the corresponding output produced by f is denoted by $f(x)$. The **domain** of the function is pictured as the container of allowable inputs to our function machine f. To *know* a function f requires knowing both the *domain* of the function as well as the output $f(x)$ assigned to each input x. We'll emphasize this in our functional notation.

Definition 5

The notation for a real-valued function f with domain D is

$$f : D \longrightarrow \mathbb{R}$$

The notation for the function assignment process is

$$f : x \longmapsto f(x)$$

Notice that we use two different arrows in this notation. One arrow ($\longrightarrow$) is used to show that the function f takes its inputs from the domain D and assigns real outputs (values in $\mathbb{R}$). The other arrow ($\longmapsto$) is used to denote the actual assignment process, which takes an *individual* input x and assigns it the output $f(x)$. In those instances where we are mainly interested in this assignment process, we may not even give the function a specific name. For example,

$$x \longmapsto x^3$$

might be used all by itself to refer to the function assigning the *cube* of each input as output.

EXAMPLE 10   What is the notation for the function f having the interval $(0,1)$ as its domain and producing the square of each input as its output?

Solution   Using the arrow notation, we write

$$f : (0,1) \longrightarrow \mathbb{R}$$

$$f : x \longmapsto x^2$$

(Figure 1.9 illustrates the function f with a machine diagram.)   ∎

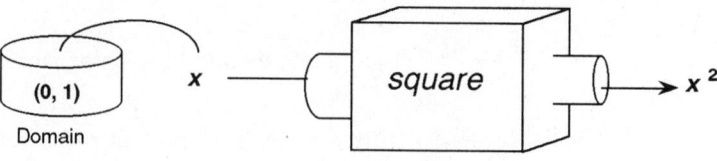

Figure 1.9 Diagram for the squaring function f.

1.2 FUNCTIONS—NOTATION, TERMINOLOGY, AND REPRESENTATIONS

EXAMPLE 11 Find and express with suitable notation the function A that expresses the area of any circle (including a degenerate circle consisting of a single point) as a function of its radius.

Solution The formula for the area of a circle is $A = \pi r^2$. The radius r could be any nonnegative real number. So our notation for the area function is

$$A : [0, \infty) \longrightarrow \mathbb{R}$$
$$A : r \longmapsto \pi r^2.$$

For any particular nonnegative real number r, we have $A(r) = \pi r^2$. ∎

Representations of functions

We have defined a function as a process assigning an output value to each input in the function's domain. We could also define a function in terms of the input-output pairs themselves:

$$f = \{(x, f(x)) : x \in D\}.$$

In other words, f can be thought of as the set of all ordered pairs of the form

$$(x, f(x))$$

where the first coordinate x can be any possible input value in the domain D, and the second coordinate $f(x)$ is the corresponding output value.

This carries essentially the same information as our original definition. Certainly, if we know the domain of a function and the rule or procedure by which we can assign the output to each input, then we could generate the entire set of ordered pairs defining the function. On the other hand, knowing all these ordered pairs certainly tells us both the domain of the function (just read off the set of first coordinates) and the assignment process (just look up the appropriate ordered pair).

If the inputs and outputs of a particular function are always real numbers, then we refer to the set of ordered pairs of inputs and outputs as the **numerical representation** of the function. Whenever you make a table of inputs and outputs for a function, you are utilizing the numerical representation.

The **graph** of a function f is obtained by *plotting* the set of ordered pairs (x, y) where x takes on each value in the domain D and $y = f(x)$ is the corresponding output value of the function f. Whenever we examine the graph or a portion of the graph of a function, we are utilizing its **graphical representation**.

When it is possible to describe a function process by a symbolic formula

$$y = f(x) = some\ expression\ in\ terms\ of\ x,$$

then we say we have a **symbolic representation** of the function. It is common to think of symbolic representations first when someone mentions the word "function." For example,

$$y = x^2$$

might be referred to as a function. This notation is certainly concise, but it has some drawbacks. For one thing, it might lead you to think of a function as simply an equation of this form, or that a function can *only* be described by such an equation. In real life, functions do not always present themselves in such a nice form. For another thing, by itself the notation $y = x^2$ does not give us the needed information about the domain of the function: what exactly are the allowable input values for x? Of course, it's acceptable to square any real number x, but if x represents the numerical value of a physical quantity like length, then we might not want to include negative numbers in the domain. (It would also be essential to know what *units* of measurement are being used. Our formula may look quite different depending on whether x is measured in centimeters, feet, or light-years.)

Interpreting functions in multiple representations

Numeric, graphical, and symbolic representations all have their advantages and disadvantages, and each can give us special insights. Real-valued functions play a prominent role in calculus, and throughout this book we'll try to draw important connections between the numerical and symbolic properties of a function and the visual properties of its graph. Let's illustrate by taking another look at the definition of function and how it can be interpreted in each of the representations.

Numerically, what distinguishes a function's table of input and output values from just any table of paired values?

1.2 FUNCTIONS—NOTATION, TERMINOLOGY, AND REPRESENTATIONS

 Since the output of a function is completely determined by the input, we should never see two lines of a function table having the same input but different outputs.

EXAMPLE 12 In Figure 1.10 we see three selected tables of paired numbers from three different sources. Which tables could have come from a functional relationship?

input	output	input	output	input	output
1	1	1	1	1	1
0	-1	0	0	0	0
-1	-3	1	-1	-1	1
π	$2\pi-1$	0.81	0.9	0.9	0.81
2.3	3.9	4	-2	-2	4
3/2	2	9/4	3/2	3/2	9/4

Figure 1.10 Three selected tables of paired numbers.

Solution The first and third tables could have come from a functional relationship, but the second one could not since we can see at least one input yielding two different outputs. ■

Note that we said that the first and third tables *could* have come from a functional relationship. Unless these tables are complete, meaning that *all possible inputs* are listed, we cannot be sure that for some unlisted input value we might obtain more than one possible output. Since many processes have infinitely many different possible inputs, we can see that it will be impossible to confirm a functional relationship solely on the basis of an incomplete table of values. On the other hand, if we note even a single input that produces more than one possible output, then we can rule out a functional relationship.

Graphically, what distinguishes a function graph from just any plot of ordered pairs? Provided we use the horizontal axis to represent inputs and the vertical axis to represent outputs, then the function definition translates to a requirement that no two points of the graph should lie over the same input value.

A function's graph must pass the vertical line test: Any vertical line must intersect the graph in at most one point.

EXAMPLE 13   Apply the vertical line test to the graphs of the two algebraic relationships $y = x^2$ and $x = y^2$.

Solution   As shown in Figure 1.11, we see that $y = x^2$ describes y as a function of x, but $x = y^2$ does not.   ∎

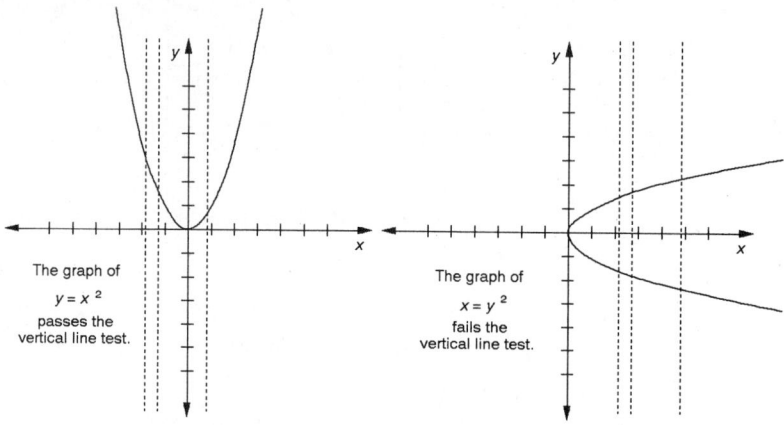

Figure 1.11 The vertical line test for function graphs.

Note that the vertical line test is valid only under the assumption that the horizontal axis represents inputs and the vertical axis represents outputs. The relationship $x = y^2$ *does* describe x as a function of y.

Symbolically, what distinguishes a formula for a function from just any formula?

 The symbolic requirement for an expression to describe a function is that we must be able to solve for a unique output value, given any particular input value from the domain.

EXAMPLE 14   If x represents the input and y the output, then the algebraic relationship $y = x^2$ describes a function since there is exactly one output value determined by each input value. As a non-example, $x = y^2$ yields two possible outputs, $y = \sqrt{x}$ and $y = -\sqrt{x}$ for each positive value of x.

Variables, constants, and parameters

A **variable**, as the name suggests, can vary over some set of values. Suppose we have a function $f : D \longrightarrow \mathbb{R}$ and we let $y = f(x)$ for each $x \in D$. In this case, we call x the **independent variable** and y the **dependent variable**. An independent variable is sometimes called an **argument** of the function.

In conducting an experiment, a scientist may manipulate the value of one quantity and observe its effects on another quantity. For example, in a medical experiment, she might be interested in understanding the effects of a drug dosage on the production of a certain antibody. The scientist is *free* to vary the dosage, and for this reason the dosage is termed the *independent* variable in the experiment. The number of antibodies produced will likely depend on the dosage, and for that reason it is called the *dependent* variable in the experiment. The scientist searches for a functional relationship between the manipulated and observed variables. The motivation for the terminology of independent and dependent variables for functions follows from this practice of the scientific method.

In graphing a function having real numbers as inputs and outputs, it is traditional to use the horizontal axis for the independent variable and the vertical axis for the dependent variable.

It's true that we generally use letters (such as x, y, h, t) to represent variables. But do NOT think of the word "variable" as synonymous with *any* letter used symbolically in mathematics. When used in the context of functions, it refers to either the independent (input) or dependent (output) variable.

We also use letters as symbols in a wide variety of other mathematical contexts. Two other ways letters are used are as *constants* or *parameters*. A **constant** represents a quantity that does not change value. For example, 7, $-\sqrt{2}$, and π are constants, as is the irrational number e (≈ 2.7182, used in the study of exponential and logarithmic functions). The value of a constant may be unknown; what is important is that its value stays fixed.

What is a **parameter**? Perhaps the best explanation is that a parameter is a "constant that varies." That sounds contradictory, but the distinctions between variables, constants, and parameters all depend on context. Here's an example that may make this distinction more clear.

EXAMPLE 15 Consider the function

$$f : \mathbb{R} \longrightarrow \mathbb{R}$$

$$f : x \longmapsto 2x + b$$

where b is an unspecified constant. What effect does varying the value of b have on the graph of f?

Solution If we plot the graph of f for a fixed value b, we obtain a straight line with slope 2 and y-intercept $(0, b)$.

Now, for each value of the constant b, we get an *entirely new line*. In other words, each value of b corresponds to a *different function*. If we plotted each of the functions corresponding to several different values of b, we obtain a family of parallel lines, each with slope 2 (Figure 1.12). ■

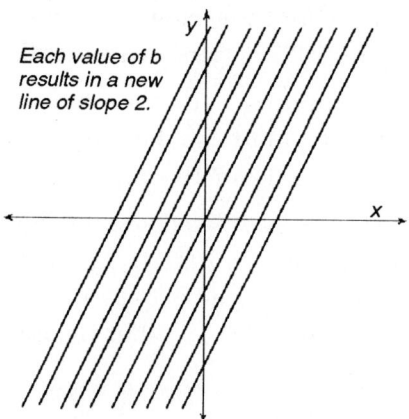

Figure 1.12 The parameter b generates a family of lines $y = 2x + b$.

We call b a **parameter** for this family of lines. This process by which each value of b produces a new line of the form $y = 2x + b$ is itself a function. It takes a real number b as input and produces a *function* as output!

The meaning of mathematical symbolism often depends on the context. For example, $(0, 1)$ could represent either a point or an open interval, depending on the discussion. Whether a particular letter represents a variable, constant, or parameter will also depend on the specific context.

1.2 FUNCTIONS—NOTATION, TERMINOLOGY, AND REPRESENTATIONS

EXERCISES

Write a suitable notation for each function described in exercises 1-3.

1. The function that expresses the perimeter p of an isosceles right triangle as a function of its leg length ℓ, when it is known that the hypotenuse must be at least 5 units long.

2. The function that expresses the volume V of a sphere as a function of its radius r, when it is known that the sphere must have volume of at least one cubic unit.

3. The function that expresses the surface area S of a sphere as a function of its diameter d, when it is known that the radius must be between 2 and 7 units in length (inclusive).

Which of the following sets of ordered pairs in exercises 4-6 could have been generated by a function?

4. $\{(2,3),(4,-1),(5,\sqrt{2}),(\pi,\pi^2)\}$

5. $\{(1,4),(2,4),(3,4),(4,4)\}$

6. $\{(1.5,7),(-4,\sqrt{3}),(\frac{3}{2},1),(0,0)\}$

In each of exercises 7-12, graph the given equation and indicate whether or not the graph could be that of a function of x.

7. $9x^2 + 4y^2 = 36$
8. $9x^2 - 4y^2 = 36$
9. $8y^2 = 2x$
10. $8x^2 = 2y$
11. $xy = 4$
12. $(x+y)^2 = 1$

A family of lines $y = mx - 1$ is parametrized by m, the slope. Exercises 13-18 pertain to this family of lines.

13. Graph several members of this family on the same set of coordinate axes. What point do all the graphs have in common?

14. What value of m guarantees that $(-3, 6)$ lies on the graph of $y = mx - 1$?

15. What value of m guarantees an x-intercept -5?

16. For what value of m is the graph of the line parallel to the line $y = 3x - 7$?

17. For what value of m is the graph of the line perpendicular to the line passing through $(-1, 2)$ and the origin?

18. What value of m guarantees that the point (x_0, y_0) is on the graph? Are there any points for which there is more than one possible answer? Are there any points for which there is no answer at all?

Suppose a family of functions has graphs given by

$$y = ax^2 + bx + c$$

with parameters a, b, and c. Exercises 19-24 refer to this family of functions.

19. When is the graph of f a parabola, and when is the graph a straight line? When the graph of f is a straight line, which parameter gives us the slope and which parameter gives us the y-intercept?

20. What effect on the graph does changing the value of c have, if we keep a and b fixed? In particular, what predictions can we make when c is positive, negative, and zero?

21. What effect on the graph does changing the value of b have, if we keep a and c fixed? In particular, what predictions can we make when b is positive, negative, and zero?

22. What effect on the graph does changing the value of a have, if we keep b and c fixed? In particular, what predictions can we make when a is positive, negative, and zero?

23. The **discriminant** d of the function f is given by $d = b^2 - 4ac$. What does the sign (positive, negative, or zero) of d tell us about the graph of the function? (Examine how many times the graph crosses the x-axis.)

24. What happens when $a = b = 0$? Does this change your answer to exercise 23?

1.3 A DICTIONARY OF FUNCTIONS

Several types of functions occur often enough to deserve special names. In this section we provide a dictionary of some of the important types of functions we study in calculus.

Constant functions

A **constant function** is the simplest type of function, for it assigns exactly the same output value to every input in its domain. A constant function f has the symbolic form

$$f : \mathbb{R} \longrightarrow \mathbb{R}$$
$$f : x \longmapsto c$$

where c is a specific real number.

The identity function

The **identity function** simply returns an output value identical to the input value. The identity function f has the symbolic form

$$f : \mathbb{R} \longrightarrow \mathbb{R}$$
$$f : x \longmapsto x.$$

If the scales on the coordinate axes are the same, then the graph of the identity function is a diagonal line rising at a $45°$ angle to the x-axis.

Figure 1.13 illustrates the graph of the constant function $x \longmapsto 3$ and the graph of the identity function $x \longmapsto x$.

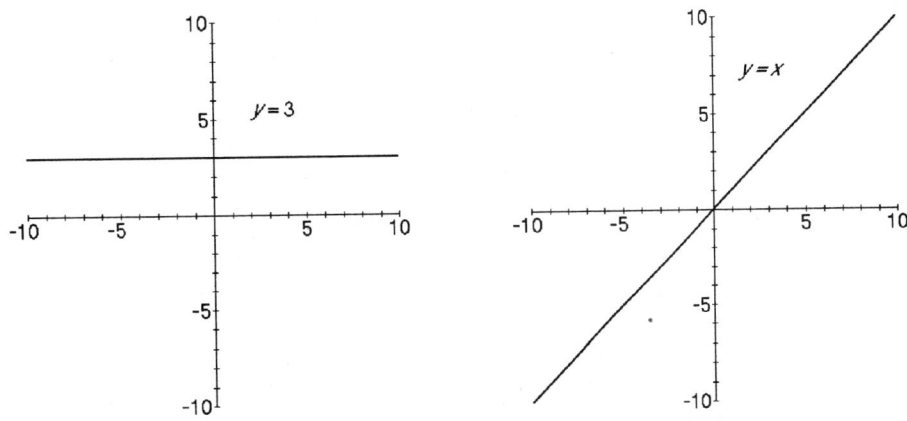

Figure 1.13 Graphs of the functions $x \longmapsto 3$ and $x \longmapsto x$.

What's the difference between a *constant* **and a** *constant function*? **What's the difference between the** *identity function* **and the** *variable x*?

As simple as constant functions and the identity function are, they can be a source of confusion with real number constants and variables. Keep in mind that a function is an *assignment process* taking inputs and producing outputs. A *constant function* has the entire set of real numbers as its domain. Hence, any value x can be input, but the same output value is always produced. The graph of a constant function is a horizontal line. On the other hand, the term *constant* refers to a single fixed real number.

When we feed a value of x to the identity function, the same value is produced as output. Again, it is this *process* of producing an output identical to the input that we call the identity function, and not the variable x by itself.

We suggest using the arrow notation $x \longmapsto c$ and $x \longmapsto x$ whenever a constant function or the identity function is intended, to make clear the distinction between these functions and simple constants and variables.

Linear functions

Constant functions and the identity function are special cases of **linear functions**. A linear function has the symbolic form

$$f : \mathbb{R} \longrightarrow \mathbb{R}$$

$$f : x \longmapsto mx + b$$

where m and b are specific real numbers.

The graph of a linear function is a straight line, and m corresponds to the **slope** ("rise/run") of the graph and b corresponds to the y-intercept. For example, the graph of the constant function $x \longmapsto 3$ has slope $m = 0$ and y-intercept 3. The graph of the identity function has slope $m = 1$ and y-intercept 0. Linear functions have a special importance in the study of calculus, and we will examine them in much greater detail in Chapter 3.

Monomial functions

A **monomial function** has the symbolic form

$$f : \mathbb{R} \longrightarrow \mathbb{R}$$

$$f : x \longmapsto ax^n$$

1.3 A DICTIONARY OF FUNCTIONS

where a is a real number and n is a nonnegative integer. If $a \neq 0$, then n is called the **degree** of the monomial function. If $a = 0$, then we have a *zero* constant function, and its degree is not defined.

EXAMPLE 16   The following are monomial functions:

$$x \longmapsto -2x^7, \quad x \longmapsto -\pi, \quad x \longmapsto x, \quad x \longmapsto \frac{22}{17}x^{100}, \text{ and } x \longmapsto 0$$

In order, the degrees of these five monomial functions are: 7, 0, 1, 100, and *undefined*. ■

The functions $x \longmapsto 5x^{2/3}$ and $x \longmapsto 3x^{-2}$ are *not* monomial functions because the exponent in each case is not a *nonnegative integer*. The linear function $x \longmapsto 2x - 1$ is not a monomial function, because it has more than one term. (We could call it a binomial function.)

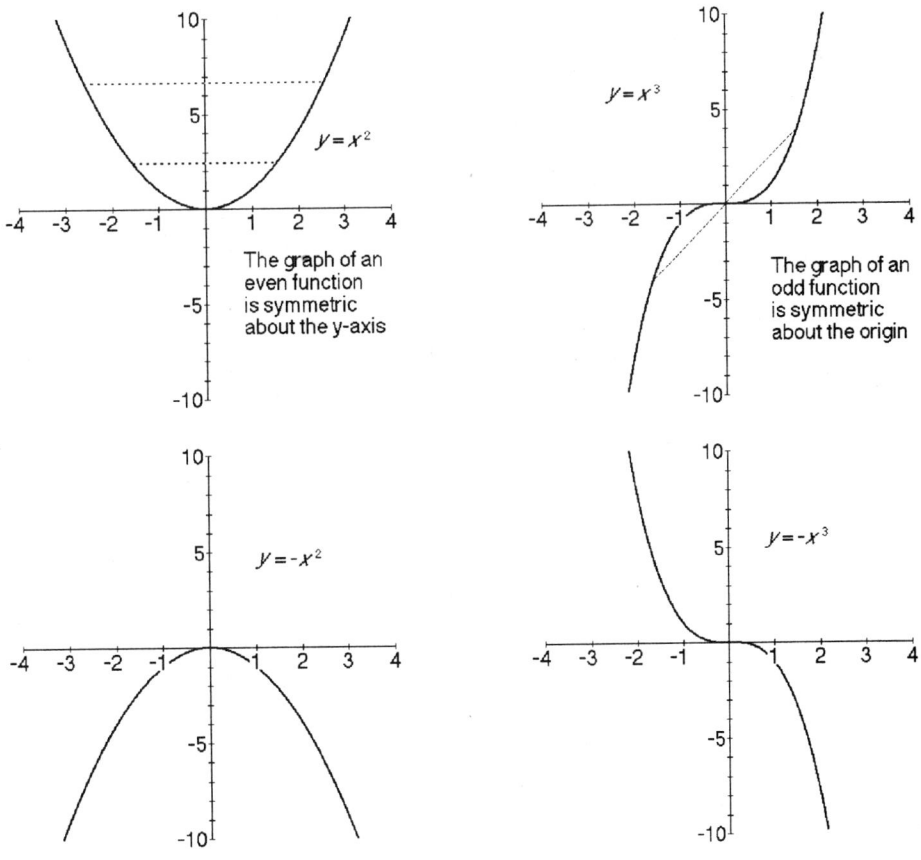

Figure 1.14 Symmetries of even and odd function graphs.

Even and odd functions

The graph of a nonlinear monomial function (in other words, with degree greater than 1) could have one of four characteristic shapes, depending on whether a is positive or negative, and whether n is even or odd. The graphs of the functions $x \longmapsto x^2$, $x \longmapsto -x^2$, $x \longmapsto x^3$, and $x \longmapsto -x^3$ show each of these characteristic shapes (Figure 1.14).

The degree of a monomial function provides the source for the terms *even* and *odd* functions. A function f is called **even** if the outputs satisfy the property

$$f(x) = f(-x)$$

for *every* real number input x in the domain of f. A function f is called **odd** if the outputs satisfy the property

$$-f(x) = f(-x)$$

for *every* real number input x in the domain of f.

Symbolically, replacing x by $-x$ in the formula of an even function should produce the same formula. Note that $x^2 = (-x)^2$ for all real values x. Replacing x by $-x$ in the formula of an odd function should produce the opposite (additive inverse) of the original formula. Note that $-(x^3) = (-x)^3$ for all real values x.

Numerically, an even function always gives the same output value for any input x and its additive inverse $-x$. For example, if $f(x) = x^2$, then

$$(-2)^2 = 4 = 2^2,$$

and

$$5^2 = 25 = (-5)^2.$$

An odd function gives opposite output values for opposite input values. For example, if $f(x) = x^3$, then

$$(-2)^3 = -8 \text{ is the opposite of } 2^3 = 8,$$

and

$$5^3 = 125 \text{ is the opposite of } (-5)^3 = -125.$$

Graphically, an even function will have a graph that is symmetric about the y-axis. That is, the graph of an even function over its positive inputs will appear to be a mirror image of the graph over its negative inputs (see the graph of $y = x^2$ in Figure 1.14). We say that the y-axis is a **line of reflection** for the graph of an even function.

The graph of an odd function also has symmetry, but this time *with respect to the origin*. This means that if we rotated the graph 180° around the origin, the graph would still lie exactly on top of itself. Another way to think of a graph symmetric with respect to the origin is that if we draw a line segment connecting the points $(x, f(x))$ and $(-x, f(-x))$ for any input value x, then the origin will always be the midpoint (see the graph of $y = x^3$ in Figure 1.14).

Polynomial functions

A **polynomial function** has the symbolic form

$$p : \mathbb{R} \longrightarrow \mathbb{R}$$
$$p : x \longmapsto a_n x^n + a_{n-1} x^{n-1} + \cdots + a_1 x + a_0$$

where each of the a_i ($0 \leq i \leq n$) represents a real number and is called a **coefficient** of the polynomial. Any or all of the coefficients could be 0.

All of the functions mentioned so far in this section are special cases of polynomial functions. Additional examples of polynomial functions would be

$$x \longmapsto 3x^4 - 5x^2 + \sqrt{17} \quad \text{and} \quad x \longmapsto \pi x^2 - x^3.$$

Each of the individual nonzero monomials $a_i x^i$ is called a **term** of the polynomial. A monomial itself can be considered a polynomial with a single term. If the leading coefficient a_n (of the highest power term) is *nonzero*, then we say the **degree of the polynomial function** is n.

Any value of x that results in an output value of 0 for a function is called a **root** or **zero** of that function. If r is a root of a polynomial function, then $(x - r)$ is a factor of the polynomial. This means that a polynomial function of degree n could have *at most* n different real roots (if it had more than n different roots, then multiplying all the corresponding factors together would result in a polynomial with a degree larger than n.)

The leading term $a_n x^n$ dominates the relative graphical behavior of a polynomial function—if you zoomed out far enough on the graph of a polynomial function, then it would resemble the graph of the monomial function $x \longmapsto a_n x^n$ and have a shape similar to one shown in Figure 1.14. For example, the graph of

$$y = -2x^3 + 10000x^2 - 5000000x + 987654321$$

will resemble the graph of

$$y = -2x^3$$

if we zoom out sufficiently far. The reason is fairly simple: If the magnitude (absolute value) of the input x is large enough, then the relative size of the leading term will dwarf all the other terms combined. (Compare the size of

$$-2x^3 \quad \text{with} \quad 10000x^2 - 5000000x + 987654321$$

for $x = \pm 10^{10}$.)

Rational functions

A **rational function** has the symbolic form

$$f : D \longrightarrow \mathbb{R}$$

$$f : x \longmapsto \frac{p(x)}{q(x)}$$

where $p(x)$ and $q(x)$ are polynomials. Just as a *rational* number is the quotient of two integers, a *rational* function is the quotient of two polynomial functions. Any polynomial function can be considered a rational function with denominator 1, just as any integer can be considered a rational number.

The domain of any polynomial function is $\mathbb{R}$, the set of all real numbers. Since rational functions involve division, we must take care to avoid division by zero. Accordingly, the domain D of a rational function

$$x \longmapsto \frac{p(x)}{q(x)}$$

is the set of all real numbers x except for those that are roots of $q(x)$. In other words,

$$D = \{x : q(x) \neq 0\}.$$

Some examples of rational functions include

$$f : x \longmapsto \frac{2x+1}{x^2 - 3x + 4} \quad \text{and} \quad g : x \longmapsto \frac{\sqrt{2} - 17x^3}{5x^4 - \pi}.$$

EXAMPLE 17 Find the domain of the functions f and g above.

Solution The domain of f is $D = \{x : x^2 - 4x + 3 \neq 0\} = \{x : x \neq 1, 3\}$.
The domain of g is $D = \{x : 5x^4 - \pi \neq 0\} = \{x : x \neq \pm\sqrt[4]{\frac{\pi}{5}}\}$. ∎

Rational power functions

A **rational power function** has the symbolic form

$$f : D \longrightarrow \mathbb{R}$$
$$f : x \longmapsto x^{p/q}$$

where p/q is a rational number in lowest terms (p and q are integers with no common factors greater than 1 and $q \neq 0$).

To evaluate the output of a rational power function, we say

$$x^{p/q} = \sqrt[q]{x^p} = (\sqrt[q]{x})^p.$$

EXAMPLE 18 Are $x \longmapsto \sqrt[4]{x^2}$ and $x \longmapsto \sqrt[4]{x}^2$ the same function?

Solution NO. Why can't we write both of these functions as

$$x \longmapsto x^{2/4}?$$

The reason is that the two functions have different domains! The order in which we take the fourth root and second power makes a difference in the domain of the function. For $x \geq 0$, both functions have the same outputs as $x \longmapsto x^{1/2} = \sqrt{x}$. However, for $x < 0$, we get different results. Note that $\sqrt[4]{(-4)^2} = \sqrt[4]{16} = 2$, but $(\sqrt[4]{-4})$ is *undefined* (as a real number). The function $x \longmapsto \sqrt[4]{x^2}$ has all real numbers in its domain and the function $x \longmapsto \sqrt[4]{x}^2$ has only nonnegative numbers in its domain. ∎

BEWARE: When both powers and roots are involved in the evaluation of a function's outputs, be careful that the order of evaluation matches the intended meaning for inputs $x < 0$.

This is particularly important if you are using a machine to evaluate a function's output. For example, $x^{2/3}$ is defined for all real numbers under our interpretation of the exponent, but on some machines, evaluating $\sqrt[3]{x^2}$ and $\sqrt[3]{x}^2$ yields different results for negative values x.

The domain D and the general shape of the graph of a rational power function depends on both of the parameters p and q, and not just their ratio p/q.

Trigonometric functions

The **trigonometric functions** can be defined using right triangles, but for our purposes, we will define them as *circular functions*. We start out with the **unit circle**, that is, the set of points in the Cartesian plane

$$\{(x, y) : x^2 + y^2 = 1\}.$$

To avoid confusion with the coordinates used to describe the unit circle, let's use the Greek letter θ (theta) to represent the independent variable for trigonometric functions. (The choice of letters or symbols used to represent the independent and dependent variables in a function process is really arbitrary—later we will use x and y to indicate the inputs and outputs of trigonometric functions as we do for most functions.) The input θ represents the angle in radians measured from the *initial point* $(1, 0)$ on the positive x-axis. Positive angle measure is understood to be in the counterclockwise direction and negative angle measure in the clockwise direction. The angle θ could be any real number— an angle greater than 2π or less than -2π is measured by wrapping around the circle more than one full revolution.

Given a specific input angle θ_0, we find its *terminal point* (x_0, y_0) on the unit circle. The output values of the six trigonometric functions **sine, cosine, tangent, cosecant, secant, cotangent** are determined using one or both of the terminal point coordinates x_0 and y_0 as follows:

$$\cos(\theta_0) = x_0 \qquad \sin(\theta_0) = y_0 \qquad \tan(\theta_0) = \frac{y_0}{x_0}$$

$$\sec(\theta_0) = \frac{1}{x_0} \qquad \csc(\theta_0) = \frac{1}{y_0} \qquad \cot(\theta_0) = \frac{x_0}{y_0}$$

Figure 1.15 illustrates this process.

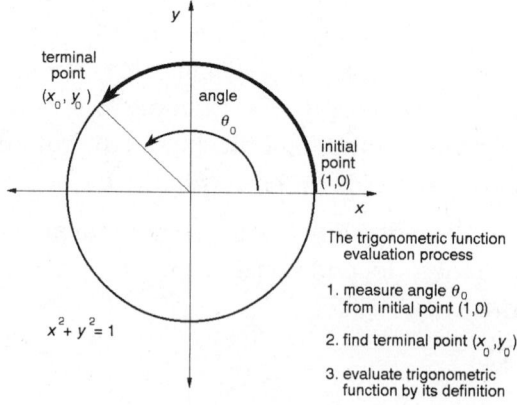

Figure 1.15 The trigonometric function evaluation process.

1.3 A DICTIONARY OF FUNCTIONS

EXAMPLE 19 What are the domains of the trigonometric functions?

Solution We can evaluate the sine and cosine functions for any real input value θ. Because division is involved in the definitions of the other trigonometric functions, we cannot include any value θ in their domains that would result in a division by zero. Hence, we have

domain of sine and cosine functions: $\mathbb{R}$, the set of all real numbers,

domain of tangent and secant functions: $\{\theta : \theta \neq \frac{\pi}{2} + n\pi\}$,

domain of cotangent and cosecant functions: $\{\theta : \theta \neq n\pi\}$. ∎

Degree measure for angles is still commonly used in many applications, and the conversion formulas between degrees and radians are:

$$1° = \frac{\pi}{180} \; radians$$

and

$$1 \; radian = \frac{180°}{\pi}.$$

A table of output values for the trigonometric functions for some common input values (in both degrees and radians) is shown below.

x in radians (in degrees)	sin x	cos x	tan x	csc x	sec x	cot x
0 (0°)	0	1	0	undefined	1	0
$\pi/6$ (30°)	$\frac{1}{2}$	$\frac{\sqrt{3}}{2}$	$\frac{\sqrt{3}}{3}$	2	$\frac{2\sqrt{3}}{3}$	$\sqrt{3}$
$\pi/4$ (45°)	$\frac{\sqrt{2}}{2}$	$\frac{\sqrt{2}}{2}$	1	$\sqrt{2}$	$\sqrt{2}$	1
$\pi/3$ (60°)	$\frac{\sqrt{3}}{2}$	$\frac{1}{2}$	$\sqrt{3}$	$\frac{2\sqrt{3}}{3}$	2	$\frac{\sqrt{3}}{3}$
$\pi/2$ (90°)	1	0	undefined	1	undefined	0
π (180°)	0	-1	0	undefined	-1	undefined
$3\pi/2$ (270°)	-1	0	undefined	0	undefined	0
2π (360°)	0	1	0	undefined	1	undefined

Table 1.1 Common trigonometric values.

The graphs of the trigonometric functions are shown in Figure 1.16.

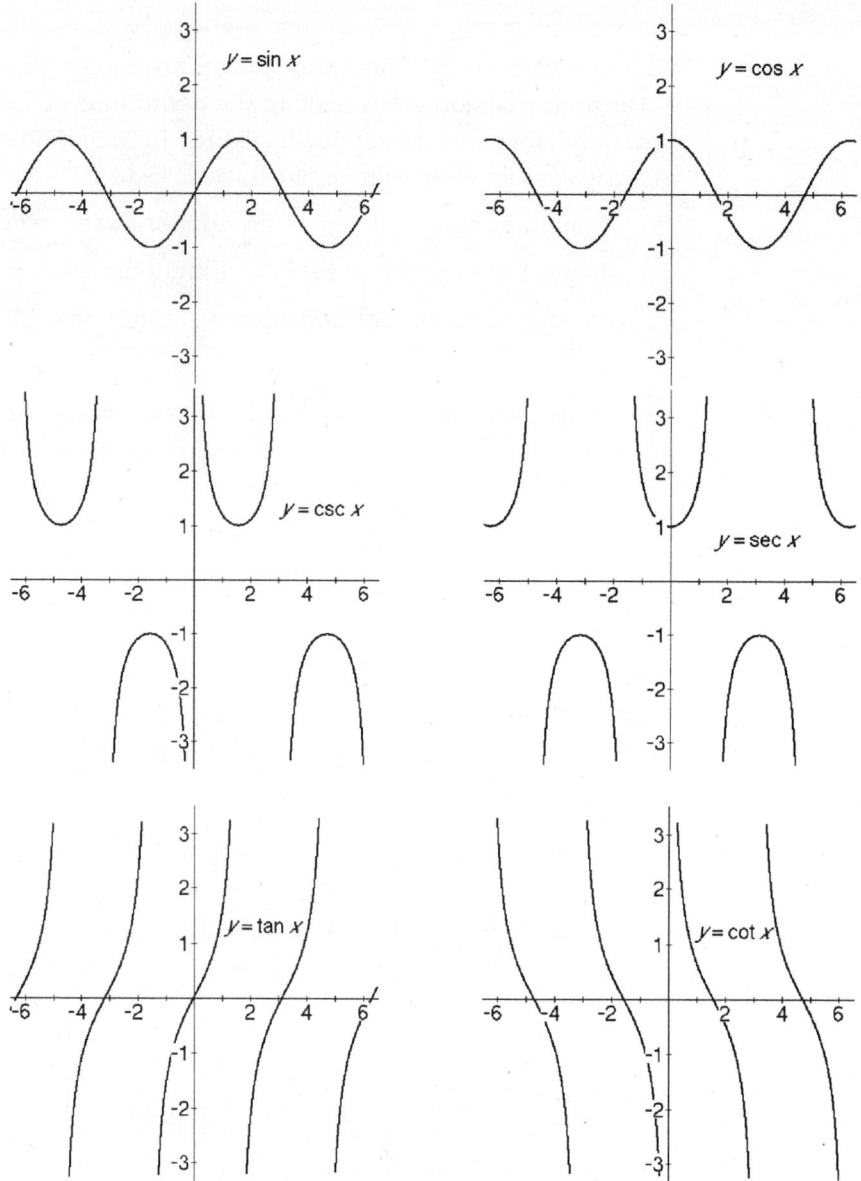

Figure 1.16 Graphs of the six trigonometric functions.

The six trigonometric functions all have one property in common: the outputs *cycle* through the same values over and over again. This shouldn't be too surprising, since the trigonometric outputs are all determined from the coordinates on the unit circle as we wrap around it repeatedly. Functions having this cycling property for their outputs are called *periodic*.

1.3 A DICTIONARY OF FUNCTIONS

In general, a function f is **periodic** if there is a positive real number p such that

$$f(x+p) = f(x)$$

for all x in the domain of f. The smallest *positive* value of p satisfying this property is called the **fundamental period** of the function. The fundamental period of each of sin, cos, csc, and sec is $p = 2\pi$, and the fundamental period of tan and cot is $p = \pi$. Graphically, if we have a plot of points over an interval of one period, then the rest of the graph is determined since we know how to extend it indefinitely by simply pasting copies of one graph cycle end to end.

Because trigonometric functions are used so often in describing cyclic phenomena, their notation is sometimes abbreviated. Take special care to realize that an expression such as

$$\sin^2 \theta$$

means

$$(\sin(\theta))^2.$$

The letters sin by themselves designate a function *name* and not a quantity that can be squared or multipled. On the other hand, $\sin(\theta)$ is the *real number output value* of the sine function for the input θ, and this value can certainly be used in an arithmetic or algebraic expression like any other real number.

A **trigonometric identity** is a relationship between these trigonometric function outputs that holds for all acceptable inputs θ. Some trigonometric identities follow directly from the definitions of the functions. For example, from their definitions, we can see that

$$\tan(\theta) = \frac{\sin(\theta)}{\cos(\theta)}$$

for all values $\theta \neq (\pi/2 + n\pi)$, n an integer. Other identities follow from the geometry of the unit circle used in defining the trigonometric functions. For example,

$$\sin^2(\theta) + \cos^2(\theta) = 1$$

for all real values θ, since $x^2 + y^2 = 1$ for any point (x, y) lying on the unit circle. The most commonly used trigonometric identities are summarized here.

Pythagorean identities:

$$\sin^2(\theta) + \cos^2(\theta) = 1 \qquad \tan^2(\theta) + 1 = \sec^2(\theta) \qquad 1 + \cot^2(\theta) = \csc^2(\theta)$$

Fundamental identities:

$$\tan(\theta) = \frac{\sin(\theta)}{\cos(\theta)} \qquad \cot(\theta) = \frac{\cos(\theta)}{\sin(\theta)} \qquad \sec(\theta) = \frac{1}{\cos(\theta)} \qquad \csc(\theta) = \frac{1}{\sin(\theta)}$$

Even/odd identities:

$$\sin(-\theta) = -\sin(\theta) \qquad \cos(-\theta) = \cos(\theta)$$

Double-angle identities:

$$\sin(2\theta) = 2\sin(\theta)\cos(\theta) \qquad \cos(2\theta) = \cos^2(\theta) - \sin^2(\theta)$$

Half-angle identities:

$$\sin^2(\theta/2) = \frac{1 - \cos(\theta)}{2} \qquad \cos^2(\theta/2) = \frac{1 + \cos(\theta)}{2}$$

Sum and difference formulas:

$$\sin(\alpha+\beta) = \sin(\alpha)\cos(\beta)+\cos(\alpha)\sin(\beta) \qquad \sin(\alpha-\beta) = \sin(\alpha)\cos(\beta)-\cos(\alpha)\sin(\beta)$$

$$\cos(\alpha+\beta) = \cos(\alpha)\cos(\beta)-\sin(\alpha)\sin(\beta) \qquad \cos(\alpha-\beta) = \cos(\alpha)\cos(\beta)+\sin(\alpha)\sin(\beta)$$

Exponential functions

An exponential function has the symbolic form

$$f : \mathbb{R} \longrightarrow \mathbb{R}$$

$$f : x \longmapsto a^x$$

where $a \neq 1$ is a specific *positive* real number. Examples of exponential functions are $x \longmapsto 2^x$ and $x \longmapsto (2/3)^x$, but *not* $x \longmapsto x^3$. Despite a superficial symbolic similarity between them, exponential functions differ greatly from algebraic power functions. In an algebraic function, the input x may be raised to a fixed power, but in an exponential function, x *is* the power.

The graph of $y = a^x$ has one of two general shapes, depending on whether $a < 1$ or $a > 1$. (See Figure 1.17.) Exponential functions arise quite naturally in descriptions of certain growth (like population) or decay (like radioactivity).

1.3 A DICTIONARY OF FUNCTIONS

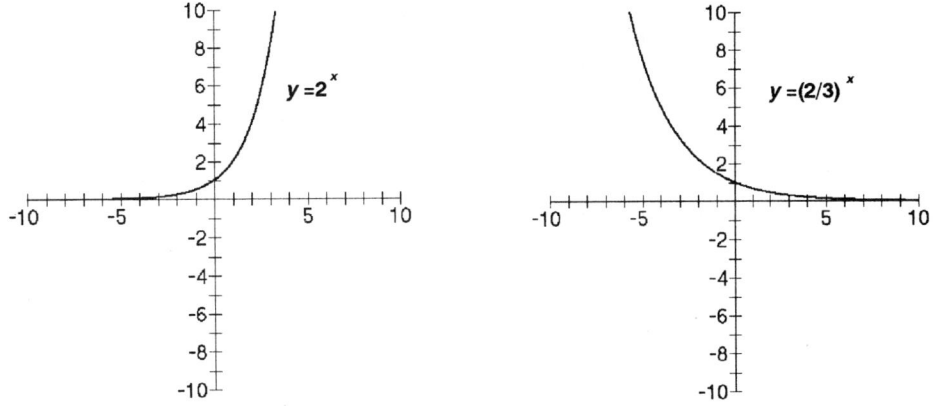

Figure 1.17 Graphs of exponential functions $y = 2^x$ and $y = (2/3)^x$

EXERCISES

1. When we multiply two nonzero polynomials together, how does the degree of the result compare to the degrees of the original two polynomials?

2. Considering exercise 1, why do you think that the constant function $x \longmapsto 0$ has an undefined degree?

3. When we add two nonzero polynomials together, is it possible for the degree of the result to be less than either of the original polynomial degrees?

4. Some books define a linear function to be a function f with the property

$$f(a+b) = f(a) + f(b)$$

for all possible real values of a and b. Which functions mentioned in this section satisfy this property?

5. The graph of *every* rational power function passes through one specific point. What point is it?

6. Which of the trigonometric functions have no zeroes?

Suppose we choose (possibly equal) values for p and q from the set

$$\{-4, -3, -2, -1, 1, 2, 3, 4\}.$$

In exercises 7-18, identify which choices of p and q would result in the given characteristics of the function

$$f : x \longmapsto \sqrt[q]{x^p}.$$

7. domain is $D = \mathbb{R}$.
8. domain is $D = \{x : x \neq 0\}$.
9. domain is $D = \{x : x \geq 0\}$.
10. domain is $D = \{x : x > 0\}$.
11. All outputs are nonnegative.
12. All outputs are nonzero.
13. The graph has a *cusp* (Υ).
14. The graph has a *corner* (V).
15. The graph is a straight line.
16. The graph has a vertical asymptote.
17. The function f is even.
18. The function f is odd.

For exercises 19-22, describe how you might tell whether functions of the indicated type were even, odd, or neither.

19. linear functions
20. monomial functions
21. rational power functions
22. trigonometric functions

---

23. A **bounded** function has all of its outputs contained in a bounded interval (of y-values). Which of the trigonometric functions are bounded and which of the trigonometric functions are unbounded?

24. Is the identity function even, odd, or neither?

25. Can a function be both even and odd? If so, give an example. If not, explain why not.

26. Show that $x \longmapsto 1/x^n$ is an odd function when n is odd, and an even function when n is even.

27. Show that if f is an odd function and $f(0)$ exists, then $f(0) = 0$. (Note that $0 = -0$.)

28. A counterexample is an example showing that a statement is false. Find a *counterexample* to the statements:

a) If the degree of a polynomial function is even, then it is an even function;

b) if the degree of a polynomial function is odd, then it is an odd function.

29. What kind of function is $f : x \longmapsto a^x$ if $a = 1$ or $a = 0$?

30. Consider again the functions from exercises 7-18. Which of these are rational power functions? Is $f : x \longmapsto a^x$ a function when $a < 0$? (Hint: Think about $a = -8$ and try evaluating $f(1/3)$ and $f(2/6)$.)

1.4 MAKING NEW FUNCTIONS FROM OLD

Real-valued functions can be combined algebraically in much the same way that real numbers can be combined arithmetically. Given two functions f and g, we can add, subtract, multiply, and divide them to obtain new functions $f + g$, $f - g$, fg, and f/g, respectively. To evaluate any of these functions numerically at an input x, we simply evaluate $f(x)$ and $g(x)$ and perform the indicated arithmetic on the results.

For an input x to be acceptable for $f + g$, $f - g$, or fg, it must be acceptable for each of the individual functions f and g. In the language of sets, the domain will be the *intersection* of the domains of f and g. For an input x to be acceptable for f/g, we must also be on guard for division by zero. That is, the domain must not include any input value x such that $g(x) = 0$.

EXAMPLE 20 Suppose $f : \{x : x \neq -1\} \longrightarrow \mathbb{R}$ with assignment process

$$f : x \longmapsto \frac{2}{x+1},$$

and suppose $g : \{x : -2 \leq x \leq 2\} \longrightarrow \mathbb{R}$ with assignment process

$$g : x \longmapsto \sqrt{4 - x^2}.$$

Find the domains and formulas for $f + g$, $f - g$, fg, and f/g.

Solution The assignment process formulas are:

$$f + g : x \longmapsto \frac{2}{x+1} + \sqrt{4 - x^2}$$

$$f - g : x \longmapsto \frac{2}{x+1} - \sqrt{4 - x^2}$$

$$fg : x \longmapsto \frac{2\sqrt{4 - x^2}}{x+1}$$

$$\frac{f}{g} : x \longmapsto \frac{2}{(x+1)\sqrt{4 - x^2}}$$

The domains of $f + g$, $f - g$, and fg are all the same:

$$\{x : -2 \leq x \leq 2 \text{ and } x \neq -1\},$$

the intersection of the domains of f and g.

The domain of f/g is

$$\{x : -2 < x < 2 \text{ and } x \neq -1\}.$$

Note that the domain does not include $x = \pm 2$, since these values result in division by zero. ■

We have already seen the algebra of functions at work in the previous section. Using just the simple building blocks of the constant functions and the identity function, we can build the monomials using products of functions (example: $-5x^3 = (-5) \cdot x \cdot x \cdot x$). Then we can build the polynomials from these monomials using sums and differences of functions. Finally, we can build the rational functions from the polynomials using quotients of functions. Similarly, using only the sin and cos functions, we can build the other four trigonometric functions (example: $\tan x = \sin x / \cos x$). A good exercise is to graph these four functions directly from the graphs of $y = \sin x$ and $y = \cos x$.

Composition of functions

Another way of combining functions is through *composition*. When we compose two functions, we do not combine their output values algebraically. Rather, we use the output of one function as the input for the other.

If f and g are two functions, then the **composition** of the function g with f is the function

$$g \circ f$$

whose outputs are determined by the formula

$$(g \circ f)(x) = g(f(x)).$$

Figure 1.18 illustrates the composition of two functions.

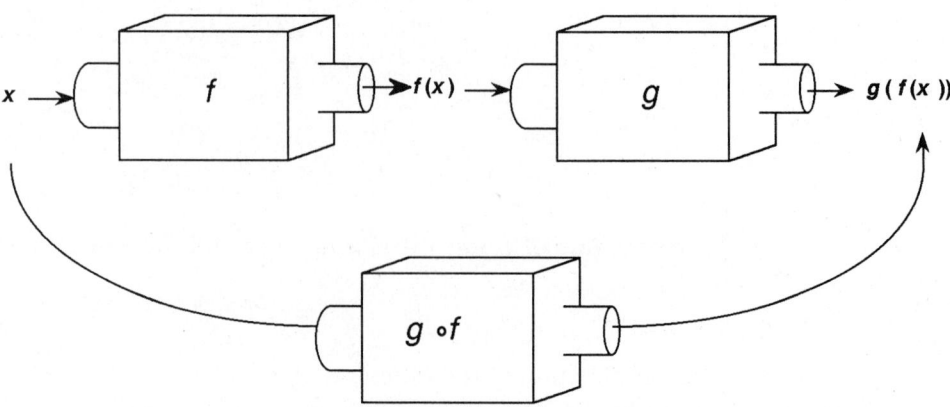

Figure 1.18 Composition of functions.

1.4 MAKING NEW FUNCTIONS FROM OLD

The domain D of $g \circ f$ is the subset of inputs in the domain of f that produce outputs in the domain of g. In set notation,

$$D = \{x : x \in (domain\ of\ f)\ \text{and}\ f(x) \in (domain\ of\ g)\}.$$

EXAMPLE 21 Using f and g as before, find the domains and formulas for $g \circ f$ and $f \circ g$.

Solution To find the formula for $g \circ f$, we simply substitute $2/(x+1)$ for each instance of x in the expression $\sqrt{4-x^2}$ to obtain

$$g(f(x)) = \sqrt{4 - \left(\frac{2}{x+1}\right)^2} = \sqrt{4 - \frac{4}{(x+1)^2}} = 2\sqrt{1 - \frac{1}{(x+1)^2}}.$$

The domain D of $g \circ f$ is the set of all real values x in the domain of f ($x \neq -1$) whose output $f(x)$ is in the domain of g. Because of the square root involved, we must have

$$1 - \frac{1}{(x+1)^2} \geq 0.$$

This means we must have

$$1 \geq \frac{1}{(x+1)^2}$$

or, equivalently,

$$(x+1)^2 \geq 1.$$

Now, $(x+1)^2 \geq 1$ precisely when

$$(x+1) \leq -1 \quad \text{or} \quad (x+1) \geq 1,$$

which in turn means $x \leq -2$ or $x \geq 0$. All of these values are in the domain of f, so we have

$$g \circ f : D \longrightarrow \mathbb{R}$$

$$g \circ f : x \longmapsto 2\sqrt{1 - \frac{1}{(x+1)^2}},$$

where

$$D = \{x : x \leq -2\ \text{or}\ x \geq 0\}.$$

To find the formula for $f \circ g$, we substitute $\sqrt{4-x^2}$ for each instance of x in the expression $2/(x+1)$ to obtain

$$f(g(x)) = \frac{2}{\sqrt{4-x^2}+1}.$$

The domain D of $f \circ g$ is the set of all real values x in the domain of g (in other words, $-2 \leq x \leq 2$) which are also acceptable inputs for this formula. Because of the division involved, we must have

$$\sqrt{4-x^2} + 1 \neq 0.$$

This means we must have

$$\sqrt{4-x^2} \neq -1,$$

but $\sqrt{4-x^2}$ is always positive, so this does not present an added restriction. So,

$$f \circ g : D \longrightarrow \mathbb{R}$$

$$f \circ g : x \longmapsto \frac{2}{\sqrt{4-x^2}+1}$$

where $D = \{x : -2 \leq x \leq 2\}$. ∎

Notice that in this example, the functions $f \circ g$ and $g \circ f$ are different.

The order of composition can definitely make a difference in the final resulting function formula and its domain.

Diagramming compositions and finding domains

To analyze a function with a complicated formula, it can be extremely helpful to think of it as a composition of simpler functions. For example, if we are trying to evaluate a function using the built-in functions on a computer or calculator, we may need to perform the evaluation in steps corresponding to the composition.

Finding the domain of acceptable inputs for a function can also be made easier if we can break down the function into its component parts and follow the path of the evaluation process. This allows us to avoid inputs that would result in a division by zero, a square root (or other even root) of a negative value, or any other unacceptable real number evaluation.

We can diagram a composition of functions by starting with a general input x and connecting the chain of compositions with arrows.

When unraveling a chain of compositions, ask yourself in what order you would perform the computations required by the evaluation of the function.

1.4 MAKING NEW FUNCTIONS FROM OLD

EXAMPLE 22 Diagram the chain of composition for the function $x \longmapsto \sin^2(5x)$ and analyze which functions are being composed.

Solution Suppose we were to evaluate this function at, say $x = \pi$. First we would have to compute 5π, plug this into the sin function to get $\sin(5\pi) = 0$, and finally square this number to get our final output of $0^2 = 0$. This chain of evaluations would take place for any input x, so we diagram the composition as

$$x \longmapsto 5x \longmapsto \sin(5x) \longmapsto (\sin(5x))^2.$$

At the first step we are taking the input x and multiplying it by 5. At the second step we are taking the result from the first step and applying the sine function. At the third and final step, we take the result from the second step and square it. If we had three functions f, g, and h such that

$$f(x) = 5x,$$
$$g(x) = \sin x,$$
$$h(x) = x^2,$$

then

$$(h \circ g \circ f)(x) = h(g(f(x))) = \sin^2(5x),$$

so $h \circ g \circ f$ represents our original function. ∎

EXAMPLE 23 Diagram the chain of composition for the function $x \longmapsto 5\sin(x^2)$ and analyze which functions are being composed.

Solution In this case, the same functions are being composed as in the previous example, but with a different order of composition. The chain is diagrammed as

$$x \longmapsto x^2 \longmapsto \sin(x^2) \longmapsto 5\sin(x^2).$$

With the same f, g, and h as before, we have

$$(f \circ g \circ h)(x) = f(g(h(x))) = 5\sin(x^2),$$

our original function. ∎

EXAMPLE 24 Find the largest possible domain of acceptable real inputs for the function
$$x \longmapsto \frac{\sqrt{7-2x}}{9-x^2}.$$

Solution Evaluation of the numerator takes x through the chain
$$x \longrightarrow (7-2x) \longrightarrow \sqrt{7-2x}.$$

There are no restrictions on evaluating $7-2x$, but at the second step, we must not allow a negative result to be input to the square root function. Hence,
$$7 - 2x \geq 0,$$
which requires
$$x \leq \frac{7}{2}.$$

Evaluation of the denominator $9-x^2$ itself requires no restrictions, but since we will be dividing by this quantity, we must have $9-x^2 \neq 0$, which requires
$$x \neq \pm 3.$$

Combining these restrictions, we see that the domain of our function is
$$\{x : x \leq \frac{7}{2} \text{ and } x \neq \pm 3\}.$$

■

EXAMPLE 25 Find the largest possible domain of acceptable real inputs for the function $x \longmapsto \sqrt{\cot 2x}$.

Solution Evaluation of this function takes us through the chain
$$x \longmapsto 2x \longmapsto \cot 2x \longmapsto \sqrt{\cot 2x}.$$

There are no restrictions in evaluating $2x$, but we cannot feed an input that is any integer multiple of π to the cotangent function. This requires
$$2x \neq n\pi,$$
for all integers n. So we must have
$$x \neq \frac{n\pi}{2}$$

for all integers n. The final step of the chain involves taking the square root of $\cot 2x$, so we also must have

$$\cot 2x \geq 0.$$

Examining the graph of the cotangent function reveals that we need

$$n\pi < 2x \leq n\pi + \frac{\pi}{2}$$

for any integer n. This already includes our previous restriction on x, so the domain can be written as

$$\{x : \frac{n\pi}{2} < x \leq \frac{n\pi}{2} + \frac{\pi}{4} \text{ for any integer } n\}.$$

■

Inverse functions

Sometimes we wish to *reverse* a function process. That is, starting with a function's output value, we would like to recover the original input value.

Definition 6

> A function f has an **inverse** g provided that
>
> $$g(f(x)) = x \text{ and } f(g(y)) = y$$
>
> for each x in the domain of f, and for each y in the domain of g. We usually write
>
> $$g = f^{-1}$$
>
> in the case that f has such an inverse.

Every function must produce a single output for each input. But to reverse the process, we need the *additional* guarantee that no two different inputs produce the same output. In other words, for a function to be invertible, its inputs and outputs must match up *one-to-one*.

Numerically, a function f is one-to-one provided that for any two different inputs $x_1 \neq x_2$, we must have $f(x_1) \neq f(x_2)$.

EXAMPLE 26 The function f defined by the formula $f(x) = x^2$ is not one-to-one, since $2 \neq -2$, but $f(2) = f(-2) = 4$.

The function g given by the formula $g(x) = x^3$ does have an inverse. In fact, we can write a formula for g^{-1}:

$$g^{-1}(x) = \sqrt[3]{x}.$$

■

If a one-to-one function f has an algebraic formula, we can often find a formula for its inverse by solving $x = f(y)$ for y.

EXAMPLE 27 Find f^{-1} if it exists, where

$$f : \mathbb{R} \longrightarrow \mathbb{R}$$

$$f : x \longmapsto x^5 - 3.$$

Solution If f is to have an inverse, we must be able to solve the equation

$$x = y^5 - 3$$

for a single value of y, given any value x. (Note that we switched x and y in the formula for f.) After adding 3 to both sides and taking the fifth root, we obtain

$$y = \sqrt[5]{x + 3}.$$

We check our inverse function by noting that

$$x = (\sqrt[5]{x + 3})^5 - 3$$

and

$$x = \sqrt[5]{(x^5 - 3) + 3}$$

for each real value y. So

$$f^{-1} : \mathbb{R} \longrightarrow \mathbb{R}$$

$$f^{-1} : x \longmapsto \sqrt[5]{x + 3}$$

is the inverse for f.

■

1.4 MAKING NEW FUNCTIONS FROM OLD

EXAMPLE 28 Find g^{-1} if it exists, where

$$g : \mathbb{R} \longrightarrow \mathbb{R}$$
$$g : x \longmapsto |2x - 1|$$

Solution If g is to have an inverse, we must be able to solve the equation

$$x = |2y - 1|.$$

for a single value of y, given any value x. But this is not possible. When $x = 3$, for instance, we could have either $y = 2$ or $y = -1$. So g has no inverse. ∎

Every function has a graph passing the vertical line test. To be one-to-one, the function's graph must also pass the *horizontal line test*: any horizontal line intersects the graph in at most one point. Since f^{-1} reverses the input-output process f, the graph of f^{-1} can be determined by simply switching the coordinates of the ordered pairs of f. If we graph both f and f^{-1}, the graphs will appear to be mirror images through the line $y = x$.

This fact allows us to graph an inverse function f^{-1} directly from the graph of the original function f, even if we are unable or it is impossible to compute a formula for the inverse function. In fact, some inverse functions are simply given new names when there is no algebraic way to describe them. Here is an example.

Logarithmic functions

A **logarithmic function** has the symbolic form

$$f : D \longrightarrow \mathbb{R}$$
$$f : x \longmapsto \log_a x$$

where $a \neq 1$ is a specific *positive* real number, and the domain D is the set of all positive real numbers,

$$D = \{x : x > 0\}.$$

The output value is determined by solving its associated exponential equation for y:

$$y = \log_a x \quad \text{if and only if} \quad a^y = x.$$

The graph of an exponential function $y = a^x$ and the corresponding logarithmic function $y = \log_a x$ look like mirror images of each other through the line $y = x$. Figure 1.19 illustrates this symmetry property between the graphs of logarithmic functions and exponential functions.

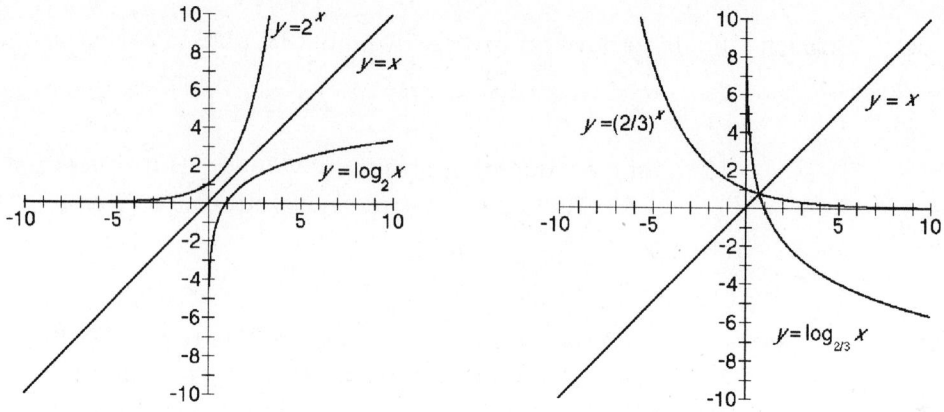

Figure 1.19 Graphs of exponential and logarithmic functions.

Even in the case that a function is not one-to-one, we can obtain a partial inverse by suitably restricting the domain of the function. Remember, if we change the domain of a function, we have effectively created a new function.

EXAMPLE 29 The function $f : x \longrightarrow \mathbb{R}$ such that

$$f : x \longmapsto x^2$$

does not have an inverse because it is not one-to-one (the parabola $y = x^2$ fails the horizontal line test). If we define a new function g by restricting the domain to nonnegative real numbers,

$$g : \{x : x \geq 0\} \longrightarrow \mathbb{R}$$
$$g : x \longmapsto x^2$$

then this function would have an inverse, namely

$$g^{-1} : \{x : x \geq 0\} \longrightarrow \mathbb{R}$$
$$g^{-1} : x \longmapsto \sqrt{x}.$$

∎

This is exactly the technique used to define the inverse trigonometric functions.

1.4 MAKING NEW FUNCTIONS FROM OLD

Inverse trigonometric functions

The **inverse trigonometric functions** are meant to reverse the trigonometric functions in much the same way that exponential and logarithmic functions reverse each other. In other words, if we feed the output of a trigonometric function into its inverse, we should obtain the original input angle back again. However, the since each of the six trigonometric functions is periodic, there are infinitely many different input angles producing the same output. For example, asking for the missing value in

$$\sin(?) = 1/2$$

has infinitely many possible answers (either $\pi/6$, $5\pi/6$, or any other value differing by a multiple of 2π from these). If we want a *unique* answer, we must arbitrarily agree to restrict the choices to some set of possible values. In other words, if we restrict the domain of inputs for each trigonometric function, so that each output value is achieved exactly once, then we can talk about the inverses in a meaningful way. We'll remove this ambiguity by restricting the domain of inputs for each trigonometric function, so that only one choice is possible.

The abbreviations for the inverse trigonometric functions are **arcsin, arccos, arctan, arccsc, arcsec,** and **arccot**. These names are meant to suggest that the *output* of an inverse trigonometric function will be an angle measured along the *arc* of the unit circle. In general, if trig is one of the trigonometric functions, then

$$\text{arctrig}(y) = x$$

will mean the same as

$$y = \text{trig}(x),$$

provided x is in the appropriate range of values. The domains and specific restrictions on outputs for each inverse trigonometric function are given below:

$$0 \leq \arccos(y) \leq \pi \qquad \text{for } -1 \leq y \leq 1$$

$$-\frac{\pi}{2} \leq \arcsin(y) \leq \frac{\pi}{2} \qquad \text{for } -1 \leq y \leq 1$$

$$-\frac{\pi}{2} < \arctan(y) < \frac{\pi}{2} \qquad \text{for all real numbers } y$$

$$0 < \text{arccot}(y) < \pi \qquad \text{for all real numbers } y$$

$$0 \leq \text{arcsec}(y) \leq \pi \text{ (and } \neq \frac{\pi}{2}) \qquad \text{for } |y| \geq 1$$

$$-\frac{\pi}{2} \leq \text{arccsc}(y) \leq \frac{\pi}{2} \text{ (and } \neq 0) \qquad \text{for } |y| \geq 1$$

The graph of any of the inverse trigonometric functions can be obtained by first graphing the corresponding trigonometric function over the restricted domain, and then reflecting the graph over the line $y = x$.

EXAMPLE 30 Graph the inverse sine function $y = \arcsin(x)$.

Solution Figure 1.20 shows the graph of $y = \sin(x)$ over the interval $[-\pi/2, \pi/2]$ along with the graph of $y = \arcsin(x)$ obtained by reflecting over the line $y = x$.

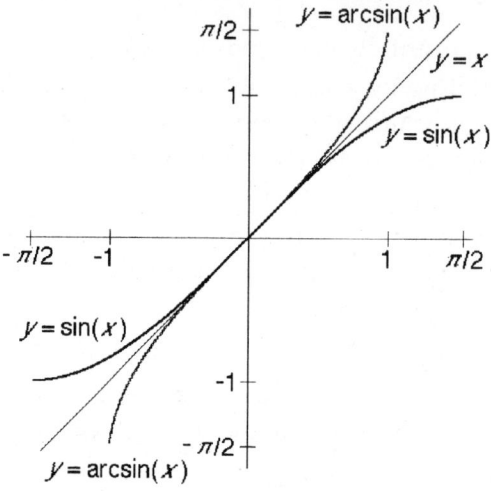

Figure 1.20 Graphs of $y = \sin(x)$ and $y = \arcsin(x)$.

The trigonometric inverses are sometimes labelled: $\sin^{-1}$, $\cos^{-1}$, $\tan^{-1}$, $\csc^{-1}$, $\sec^{-1}$, and $\cot^{-1}$, respectively.

 BEWARE: The notation f^{-1} used for the *inverse* of the function f must not be confused with the similar notation a^{-1} used for the *reciprocal* $1/a$ of the real number a.

We need to take special care with the trigonometric functions because of the common abbreviations used for powers of these functions. For example, $\sin^n x$ stands for $(\sin x)^n$ when n is any power *except* 1. The important exception is

$$\sin^{-1}(x) = \arcsin x,$$

where $\sin^{-1}$ refers to the inverse sine function. Note that when we need it, we already have a special name for the *reciprocal* of $\sin(x)$, namely

$$\frac{1}{\sin(x)} = \csc(x).$$

1.4 MAKING NEW FUNCTIONS FROM OLD

In this book we will usually use the notation arcsin rather than $\sin^{-1}$ in the interest of avoiding this source of confusion.

Figure 1.21 illustrates the graphs of the six inverse trigonometric functions.

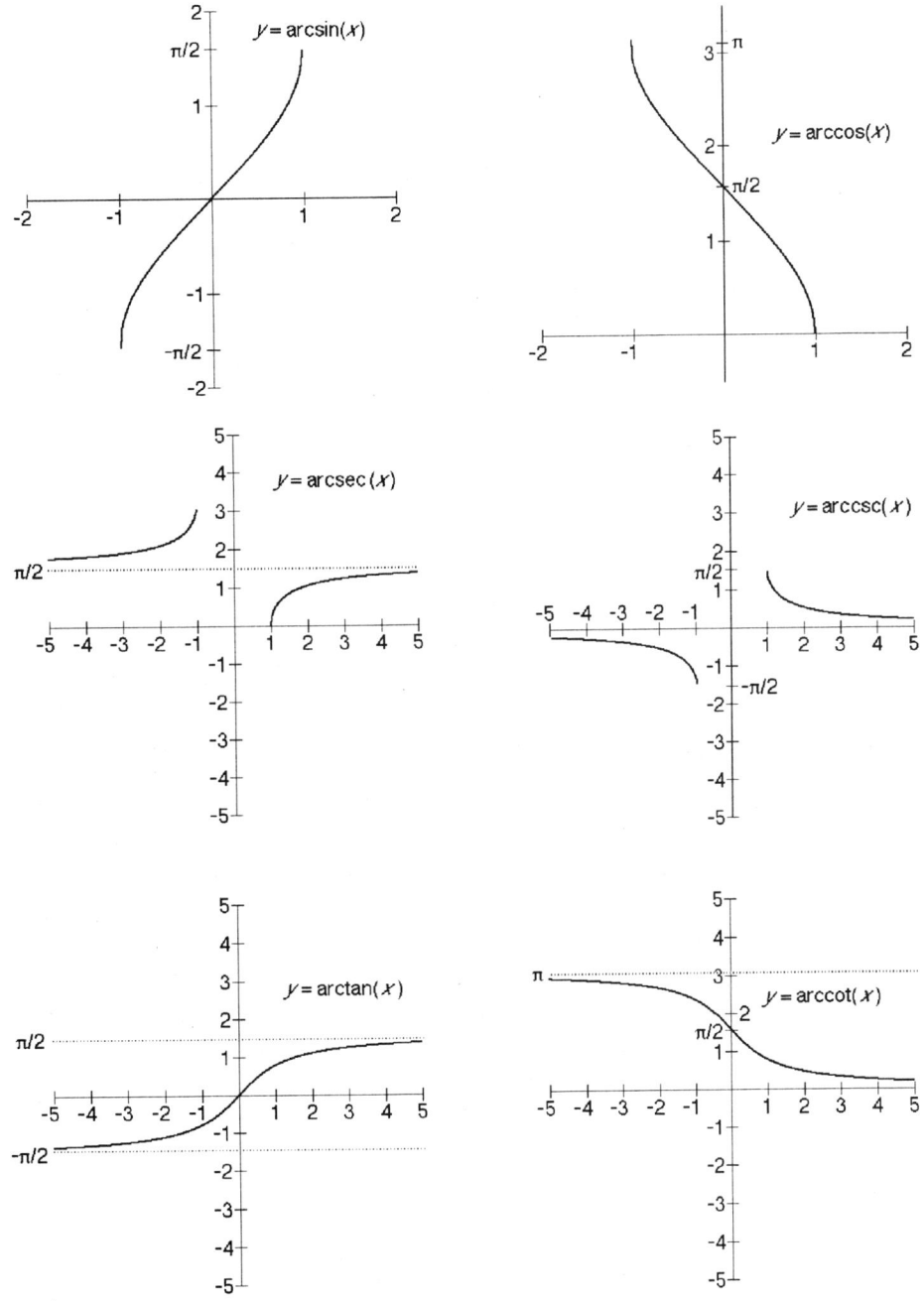

Figure 1.21 Graphs of the inverse trigonometric functions.

What exactly are inverses?

The notion of an *inverse* occurs over and over again in mathematics. Keep in mind that the word inverse is always used in reference to some *operation* and some *identity* for that operation. For example, the *additive* inverse of any real number x is $-x$, since

$$x + (-x) = 0$$

and 0 is the identity for real number addition. The *multiplicative* inverse of a *nonzero* real number x is its reciprocal $1/x$, since

$$x \cdot \left(\frac{1}{x}\right) = 1$$

and 1 is the identity for real number multiplication.

Similarly, the additive and multiplicative inverses for a function f are $-f$ and $1/f$, respectively. The *constant functions* $x \longmapsto 0$ and $x \longmapsto 1$ now play the roles of additive and multiplicative identities for functions.

The inverse function f^{-1} is really an inverse in this same sense. The operation is *function composition*, and the identity is the *identity function* $x \longmapsto x$. Note that $f \circ f^{-1} : x \longmapsto x$ and $f^{-1} \circ f : x \longmapsto x$.

EXERCISES

In exercises 1-6, find the domain D of each function, where D is to include all the real values x for which the output produced would be a real number.

1. $x \longmapsto \dfrac{2x - 1}{x^2 - 3}$
2. $x \longmapsto \sqrt{4 - 3x}$
3. $x \longmapsto \sqrt[3]{4 - 3x}$
4. $x \longmapsto \sec(\pi x)$
5. $x \longmapsto \sqrt{1 + x + x^2}$
6. $x \longmapsto \tan^2(x) + \sec^2(x)$

Express each of the functions in exercises 7-12 as a composition of functions chosen from the five functions

$$f : x \longmapsto x^2, \quad g : x \longmapsto 3x, \quad h : x \longmapsto \sin x, \quad j : x \longmapsto \frac{1}{x - 4}, \quad k : x \longmapsto \sqrt{x}.$$

7. $x \longmapsto 3\sin^2\left(\dfrac{1}{x - 4}\right)$
8. $x \longmapsto \dfrac{1}{\sqrt{x} - 4}$
9. $x \longmapsto \dfrac{3}{\sqrt{x^2 - 4}}$
10. $x \longmapsto 27x^2$
11. $x \longmapsto \sin(9x^2)$
12. $x \longmapsto \dfrac{x - 4}{17 - 4x}$

1.4 MAKING NEW FUNCTIONS FROM OLD

Given the table of values of the following two functions, find the values indicated in exercises 13-26.

Table for f	
input	output
-1	3
0	4
1	-2
2	6
3	2
4	-1

Table for g	
input	output
-1	3
0	1
1	-7
2	0
3	-1
4	2

13. $(f+g)(4)$
14. $(g-f)(-1)$
15. $3f(1)$
16. $(gf)(1)$
17. $(\frac{f}{g})(2)$
18. $(f \circ g)(-1)$
19. $(g \circ f)(4)$
20. $g^2(-1)$
21. $f((-1)^2)$
22. all x such that $f(x) = g(x)$
23. all x such that $(f-g)(x) = 3$
24. all x such that $(f \circ g)(x) = 2$
25. all x such that $(g \circ f)(x) = 2$
26. x such that $(f \circ f \circ f \circ f)(x) = 6$

Given the graphs of the following two functions f and g, graph the indicated functions in exercises 27-40.

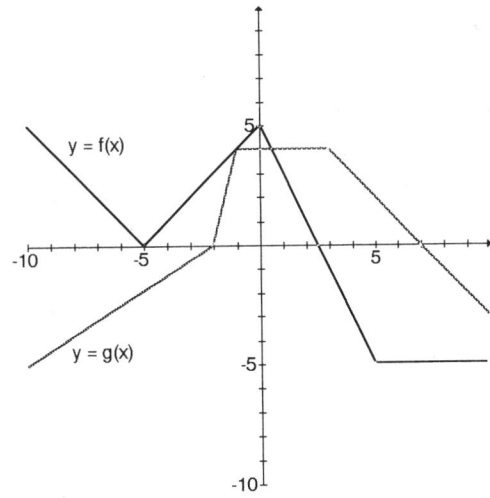

27. $f + g$
28. $g - f$
29. $3f$
30. $y = f(x^2)$
31. $y = f(2/x)$
32. $y = f(x-2)$
33. $y = f(2x)$
34. $y = f(x+2)$
35. $y = f(x) + 3$
36. $\frac{f}{g}$
37. $f \circ g$
38. $g \circ f$
39. g^2
40. gf

Given each of the following functions in exercises 41-47, determine whether it has an inverse, and if so, find and graph its inverse function.

41. the identity function $x \longmapsto x$

42. $f : \mathbb{R} \longrightarrow \mathbb{R},\ x \longmapsto x^3 - x$

43. $g : \{x : x \neq 0\} \longrightarrow \mathbb{R},\ x \longmapsto 1/x$

44. $p : \mathbb{R} \longrightarrow \mathbb{R},\ x \longmapsto x^3 + x + 1$

45. $q : \mathbb{R} \longrightarrow \mathbb{R},\ x \longmapsto x^2 + x + 1$

46. $s : \{x : x \geq 1\} \longrightarrow \mathbb{R},\ x \longmapsto x^2 + x + 1$

47. $h : \mathbb{R} \longrightarrow \mathbb{R},\ x \longmapsto mx + b$ where m and b are real numbers.

CHAPTER 2

Limits and Continuity

Much of calculus concerns describing the behavior of functions, meaning how the outputs act or change relative to the inputs. Limits provide a very effective terminology for describing the behavior of a function, and in this chapter, we introduce the language of limits both visually and through the idea of *error tolerances*. We'll show how limits can be estimated both numerically and graphically. Then we will use the language of limits for discussing the notion of *continuity* and its consequences, and for describing the *asymptotic* behavior of a function. The idea of *limit* also underlies the concept of derivative, as we'll see in the next chapter.

2.1 WHAT ARE LIMITS?

Let's start out with a graphical example by considering the function illustrated in Figure 2.1. Imagine that the graph of f has been put up on a large wall. Now, pretend two people are walking along the x-axis towards a particular value $x = a$, one from the left and one from the right.

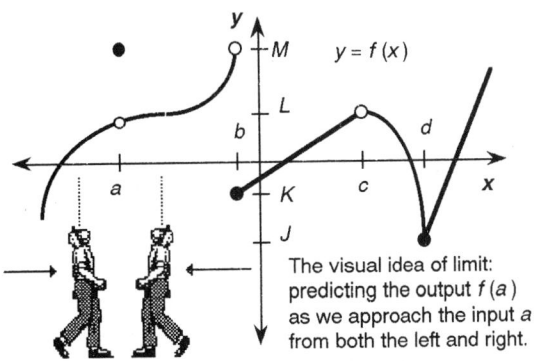

The visual idea of limit: predicting the output $f(a)$ as we approach the input a from both the left and right.

Figure 2.1 Visual picture of the limit process.

Each person watches the graph of f very carefully, taking note of what the output $f(x)$ is for each input x they pass. Now, as they approach very close to a particular value $x = a$, each is asked to make a *prediction* of the real number output value $f(a)$. Neither person is allowed to see the *actual* output value $f(a)$ (if there is one), but they may observe the graph over inputs as close to a as they please. We'll record each person's prediction using the following notation:

$$\lim_{x \to a^-} f(x)$$

represents the prediction from the left side. This is called the **left-hand limit of f as x approaches** a.

$$\lim_{x \to a^+} f(x)$$

represents the prediction from the right side. This is called the **right-hand limit of f as x approaches** a.

☞   **Note that the positive and negative signs used on a^+ and a^- refer only to the** *direction of approach*, **and have nothing to do with the** *value* **of a. If no sign is used as a superscript on a, then this denotes THE limit, meaning a common prediction from both sides.**

For the function in Figure 2.1, the prediction from the left is

$$\lim_{x \to a^-} f(x) = L$$

and the prediction from the right is

$$\lim_{x \to a^+} f(x) = L.$$

Notice that neither prediction matches the *actual* output value $f(a) = M$, but the two predictions match each other. We call the common prediction

$$\lim_{x \to a} f(x) = L$$

and say that the number L is **the limit of f at** a. If the predictions do not match, or if it was impossible for one or both people to even make a prediction, we say the limit **does not exist**.

Another way of denoting this (two-sided) limit is to write

$$f(x) \longrightarrow L \ \text{as} \ x \longrightarrow a.$$

When used in the context of limits, the arrow notation is meant to suggest the idea that the output $f(x)$ gets near the value L as the input x gets near the value a.

2.1 WHAT ARE LIMITS?

EXAMPLE 1 For the function in Figure 2.1, determine visually the limits of f at $x = b$, c, and d, if the limits exist.

Solution For $x = b$, the prediction from the left is

$$\lim_{x \to b^-} f(x) = M$$

and the prediction from the right is

$$\lim_{x \to b^+} f(x) = K.$$

Since the two one-sided limits do *not* match, we conclude that $\lim_{x \to b} f(x)$ *does not exist*.

For $x = c$, the prediction from the left is

$$\lim_{x \to c^-} f(x) = L$$

and the prediction from the right is

$$\lim_{x \to c^+} f(x) = L.$$

Since the two one-sided limits match, we conclude that $\lim_{x \to c} f(x) = L$.

For $x = d$, the prediction from the left is

$$\lim_{x \to d^-} f(x) = J$$

and the prediction from the right is

$$\lim_{x \to d^+} f(x) = J.$$

Since the two one-sided limits match, we conclude that $\lim_{x \to d} f(x) = J$. ∎

As you can see from this example, determining a limit at a point depends only on the left- and right-hand limits, and not on the *actual output* (if any) of the function at that point. It does not matter that the right-hand limit at $x = b$ is the correct output value $f(b) = K$; since the right-hand limit did not match the left-hand limit at b, the limit $\lim_{x \to b} f(x)$ *did not exist*. On the other hand, the output value $f(c)$ is undefined, but since the predictions from the left and right have the same value L, we can say $\lim_{x \to c} f(x) = L$. At $x = d$, we not only have the predictions matching each other, they also match the actual output value $f(d) = J$. In this case we say that f is **continuous at** d (to suggest that there cannot be a break or hole in the graph).

Estimating limits numerically and graphically

The computational power made available by technology gives us a useful numerical tool for investigating limits. We can often get a good idea of whether or not

$$\lim_{x \to a} f(x)$$

exists by computing $f(x)$ for a sequence of values x approaching a from the left side $(x < a)$, and for a sequence of values x approaching a from the right side $(x > a)$. If the outputs seem to close in on a particular value, they might give us a good guess at the limit value.

EXAMPLE 2 Estimate numerically

$$\lim_{x \to 0} \frac{\sin x}{x}$$

where x is understood to be measured in radians.

Solution The function

$$f : \{x : x \neq 0\} \longrightarrow \mathbb{R}$$

$$f : x \longmapsto \frac{\sin x}{x}$$

has no output defined for $x = 0$, but for any other particular input

$$x = x_0 \neq 0,$$

the function has a perfectly good output

$$f(x_0) = \frac{\sin x_0}{x_0}.$$

We'll try a sequence of inputs that approach 0 from both sides, say

$$0.1, 0.01, 0.001, \ldots$$

from the right of 0, and

$$-0.1, -0.01, -0.001, \ldots$$

from the left of 0. The corresponding outputs appear to be getting very close to 1 from either direction, as shown by the first three inputs we try from either side (see Table 2.1).

2.1 WHAT ARE LIMITS?

x	0.1	0.01	0.001
$\sin(x)/x$	0.998334	0.999983	0.999998

x	-0.1	-0.01	-0.001
$\sin(x)/x$	0.998334	0.999983	0.999998

Table 2.1 Values of $(\sin x)/x$ for x near 0.

On the basis of this evidence we could say that *numerically*,

$$\lim_{x \to 0} \frac{\sin x}{x} \text{ appears to be } 1.$$

In fact, if we use machine computation of $\frac{\sin x_0}{x_0}$ for a value x_0 close enough to 0, the round-off precision of the machine may actually result in an output of 1. ∎

With machine graphics, we might be able to estimate the value of a limit by graphing the function f over an interval containing the "target" a, and using a visual analysis as we did for the function in Figure 2.1. In fact, we can approach a in the sense of zooming in by rescaling the horizontal axis.

EXAMPLE 3 Estimate graphically

$$\lim_{x \to 0} \frac{\sin x}{x}.$$

Solution Figure 2.2 shows two machine-generated plots of the graph of $y = \frac{\sin x}{x}$.

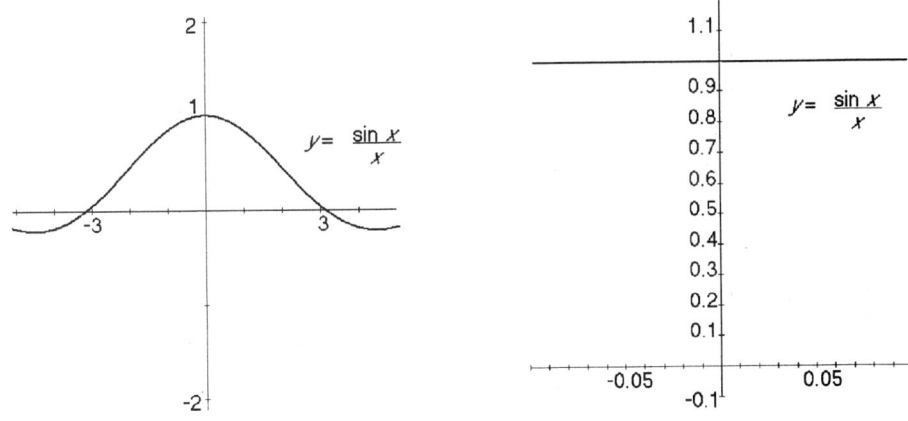

Figure 2.2 Graphs of $y = \frac{\sin x}{x}$.

The first appears to indicate an output value of 1 at $x = 0$ (though we know that 0 is not in the domain of the function). To investigate this further, in the second graph we show a close-up of the graph after we scaled the horizontal axis to plot between $x = -0.1$ and $x = 0.1$. We see that the graph of $y = \dfrac{\sin x}{x}$ resembles the horizontal line $y = 1$. This reinforces our first impression, so we could say that *graphically*,

$$\lim_{x \to 0} \frac{\sin x}{x} \text{ appears to be } 1.$$

■

Since a machine-generated graph is only a finite collection of dots, a hole in the graph may or may not appear. For example, the hole might occur *between* two adjacent plotted values and go undetected. In this example, we do not see the hole in the graph of $y = \dfrac{\sin x}{x}$ because the y-axis itself fills it in. We must emphasize that these examples illustrate some ways we can numerically and graphically *investigate* limits. They do not provide definitive *proof* of the value of the limit. In particular, if we use a machine to help us with either numerical computations or graphical plotting, the precision limitations of that machine will not allow us to look at function outputs $f(x)$ for inputs x arbitrarily close to a given number a. In fact, if we get *too close*, round-off errors may tempt us to some erroneous conclusions.

EXERCISES

In exercises 1-6, sketch the graph of a function satisfying the stated requirements.

1. $\lim\limits_{x \to 1^+} f(x) = 2$ $\quad \lim\limits_{x \to 1^-} f(x) = -1$ $\quad f(1)$ is undefined.

2. $\lim\limits_{x \to -2^-} g(x) = 0$ $\quad \lim\limits_{x \to -2^+} g(x) = 0$ $\quad g(-2) = 1$

3. $\lim\limits_{x \to 2^-} h(x) = -2$ $\quad \lim\limits_{x \to 2^+} h(x) = 2$ $\quad h(2) = 0$

4. $\lim\limits_{x \to 0^-} i(x) = -1$ $\quad \lim\limits_{x \to 0^+} i(x) = -2$ $\quad i(0) = -1$

5. $\lim\limits_{x \to -1^-} j(x) = 3$ $\quad \lim\limits_{x \to -1^+} j(x) = -2$ $\quad j(-1) = -2$

6. $\lim\limits_{x \to -3^-} k(x) = 1$ $\quad \lim\limits_{x \to -3^+} k(x) = 1$ $\quad k(-3)$ is undefined.

2.1 WHAT ARE LIMITS?

The two functions f and g have the graphs shown below.

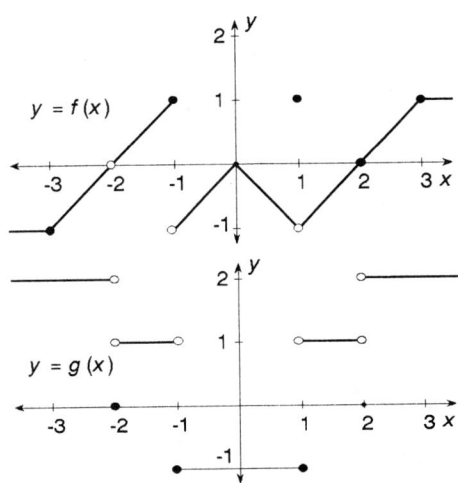

Use these to graph the functions in exercises 7-18 over the interval $[-3, 3]$.

7. $f \circ g$
8. $g \circ f$
9. $x \longmapsto f(x-1)$
10. $x \longmapsto g(x/2)$
11. $f + g$
12. $f - g$
13. fg
14. $2f$
15. $1/g$
16. f/g
17. $1/f$
18. g/f

Referring to your graphs in exercises 7-18, fill out a table similar to the following for each of the functions indicated in exercises 19-30.

input $x = a$	output $f(a)$	left-hand limit $\lim_{x \to a^-} f(x)$	right-hand limit $\lim_{x \to a^+} f(x)$	limit $\lim_{x \to a} f(x)$	Is f continuous at $x = a$? (yes or no)
-3					
-2					
-1					
0					
1					
2					
3					

19. f
20. g
21. $f \circ g$
22. $g \circ f$
23. $x \longmapsto f(x-1)$
24. $x \longmapsto g(x/2)$
25. $f + g$
26. $f - g$
27. fg
28. $2f$
29. $1/f$
30. f/g

For each of the functions defined in exercises 31-46,
a) graph the function over acceptable inputs in the interval $[-5, 5]$
b) graphically estimate the two one-sided limits

$$\lim_{x \to 0^+} f(x) \qquad \lim_{x \to 0^-} f(x)$$

c) numerically estimate $\lim_{x \to 0} f(x)$ if it exists.

31. $x \longmapsto f(x)$ where

$$f(x) = \begin{cases} x^2 & \text{if } x < 0 \\ 2x - 1 & \text{if } x \geq 0 \end{cases}$$

32. $x \longmapsto sign(x)$ where $sign(x)$ denotes the **signum function** and is defined by the split formula

$$sign(x) = \begin{cases} 1 & \text{if } x > 0 \\ 0 & \text{if } x = 0 \\ -1 & \text{if } x < 0 \end{cases}$$

(The signum function essentially tells us whether the input x is positive, negative, or zero.)

33. $x \longmapsto |x|$

34. $x \longmapsto x/x$

35. $x \longmapsto |x|/x$

36. $x \longmapsto \dfrac{\sin(2x)}{5x}$

37. $x \longmapsto 1/x^2$

38. $x \longmapsto 1/x$

39. $x \longmapsto \sin(1/x)$

40. $x \longmapsto x \sin(1/x)$

41. $x \longmapsto \tan(x)$

42. $x \longmapsto \cot(x)$

43. $x \longmapsto \dfrac{1 - \cos x}{x}$

44. $x \longmapsto \sin^2(x) + \cos^2(x)$

45. $x \longmapsto (1 + x)^{1/x}$

46. $x \longmapsto \arctan(1/x)$

2.2 DEFINITION OF LIMIT

If the functions we're analyzing are sufficiently simple in their behavior, numerical and graphical estimation techniques can often give us a good idea of whether the function has a limit at a point. However, while we can make reasonable left- and right-hand predictions of function outputs for some functions, it may be difficult to make the predictions for others. For example, the function outputs might be growing so rapidly or oscillating so wildly that no prediction is possible.

The fact of the matter is that many perfectly good functions can be very difficult or essentially impossible to accurately graph, even with sophisticated graphing calculators or computers. The numerical limitations of a machine effectively prevent us from truly approaching arbitrarily close to a given target a. We really need a tighter definition of limit to handle these situations. The following is the formal definition mathematicians use in judging whether or not a function has a limit.

Definition 1

> The function f **has the limit** L **at** $x = a$, written
> $$\lim_{x \to a} f(x) = L \quad \text{or} \quad f(x) \longrightarrow L \text{ as } x \longrightarrow a$$
> if and only if the following condition holds: Given any $\epsilon > 0$, there is a $\delta > 0$ such that $|f(x) - L| < \epsilon$ whenever $0 < |x - a| < \delta$.

Let's examine this condition more closely. Think of the positive number ϵ (the Greek letter "epsilon") as a desired function *output error tolerance*. The statement $|f(x) - L| < \epsilon$ is just another way of saying that the function output $f(x)$ needs to be within ϵ of the number L. Now, think of the positive number δ (the Greek letter "delta") as the *input error tolerance* required to guarantee our desired output accuracy. The condition $0 < |x - a| < \delta$ means x is within δ of a but $x \neq a$. Hence, if

$$a - \delta < x < a \quad \text{or} \quad a < x < a + \delta,$$

then we must have

$$L - \epsilon < f(x) < L + \epsilon.$$

The formal limit definition says that given *any* positive output error tolerance, we can always find a corresponding positive input error tolerance that guarantees the desired output accuracy. We can think of the definition as providing a universal error tolerance test that L must pass in order to be called the limit of $f(x)$ as $x \longrightarrow a$.

EXAMPLE 4 If we say $\lim_{x \to 2} \dfrac{3x^2 - 3x - 6}{x - 2} = 9$, then for any specific value of ϵ, such as $\epsilon = .001$, we should be able to find some specific positive value for δ satisfying the following test:

$$\left| \frac{3x^2 - 3x - 6}{x - 2} - 9 \right| < .001 \quad \text{whenever} \quad 0 < |x - 2| < \delta.$$

Find such a positive value δ.

Solution Starting with the inequality

$$\left| \frac{3x^2 - 3x - 6}{x - 2} - 9 \right| < .001$$

we can rewrite the left-hand expression as

$$\left| \frac{3x^2 - 3x - 6 - 9(x - 2)}{x - 2} \right| = \left| \frac{3x^2 - 12x + 12}{x - 2} \right| = 3 \left| \frac{(x - 2)(x - 2)}{x - 2} \right|.$$

Now, as long as $x \neq 2$ (we can't divide by 0), we have $\dfrac{x - 2}{x - 2} = 1$. This means that for $x \neq 2$, our desired inequality is equivalent to

$$3|x - 2| < .001.$$

If we use *any* positive value $\delta < \dfrac{.001}{3}$ such as $\delta = .0003$ as our input error tolerance, we will meet our output error tolerance requirement. In other words, as long as

$$1.9997 < x < 2 \quad \text{or} \quad 2 < x < 2.0003,$$

we would certainly have

$$8.999 < \frac{3x^2 - 3x - 6}{x - 2} < 9.001.$$

Note that any smaller value of δ would also guarantee the same output error tolerance. ∎

Graphical interpretation of the formal limit definition

Graphically, the requirements of the formal definition of limit correspond to the graph of $y = f(x)$ being forced to lie between the horizontal lines $y = L - \epsilon$ and $y = L + \epsilon$ provided the inputs x are between the vertical lines $x = a - \delta$ and $x = a + \delta$. The only exception allowed would be at the actual value $x = a$ itself (see Figure 2.3).

2.2 DEFINITION OF LIMIT

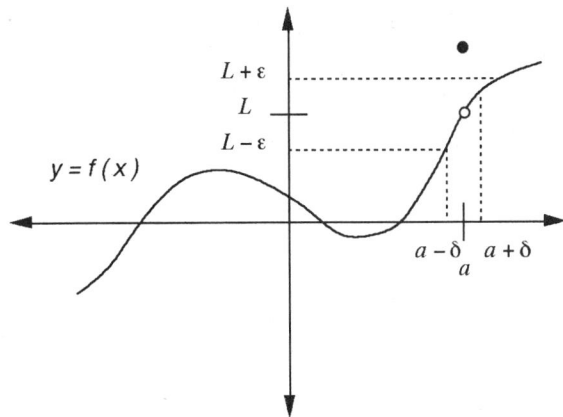

Figure 2.3 Graphical illustration of the limit definition.

We could also restate the formal definition of limit in machine graphical terms. Suppose we are given a value $\epsilon > 0$ as our output tolerance. First, we center the screen at $x = a$, and we scale the *vertical* axis so that the vertical range runs from $L - \epsilon$ to $L + \epsilon$.

Once the vertical axis has been scaled in this way, our challenge is to rescale the *horizontal axis* so that the graph of $y = f(x)$ enters from the left and leaves only from the right (with the possible exception of a hole or jump at a). We are not allowed to tamper with the vertical scaling at all; we must achieve the well behaved graph through horizontal scaling only. If we are successful, then the distance from a to the edge of the graphing window is playing the role of δ in the formal definition. To say that

$$f(x) \longrightarrow L \text{ as } x \longrightarrow a$$

we must be able to achieve this horizontal line graph goal for any given $\epsilon > 0$.

EXAMPLE 5 In the second graph of Figure 2.2, the vertical axis was first scaled so that any output between 0.995 and 1.005 is machine plotted on the same horizontal line $y = 1$. We can think of this line as a very long, thin viewing window with a vertical range corresponding to $\epsilon = .005$. The fact that the graph appears identical to this horizontal line over the interval $(-0.1, 0.1)$ and in particular enters this long, thin viewing window from the left and leaves only from the right suggests that $\delta = 0.1$ is a sufficiently small input tolerance to guarantee that the outputs $\dfrac{\sin x}{x}$ are within $\epsilon = .005$ of 1. ∎

Limit proofs—the epsilon-delta machine

In Example 4, we found an appropriate value $\delta = .0003$ value for a specific value $\epsilon = .001$. To *prove* that

$$\lim_{x \to a} \frac{3x^2 - 3x - 6}{x - 2} = 9$$

would require that we demonstrate that such an appropriate δ-value exists for each possible given ϵ-value. There are infinitely many possible ϵ-values, so we cannot simply list a corresponding δ-value for each one. What we really need is to describe a *process* by which we can always produce the needed δ-value. In other words, we need a ϵ-δ machine taking the given ϵ and producing an appropriate δ (see Figure 2.4).

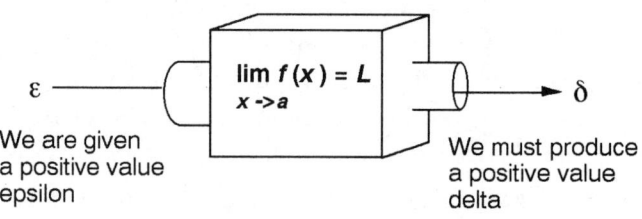

Figure 2.4 The ϵ-δ machine for limits.

Of course, in general, the δ-value also depends on the particular function f, the input target a, and the limit value L.

How does one write a so-called ϵ-δ limit proof? In general, one starts with the assumption that we have already been given some unspecified value $\epsilon > 0$. Then we must either provide some procedure, or somehow demonstrate that there exists a procedure for producing a $\delta > 0$ such that

$$|f(x) - L| < \epsilon \quad \text{whenever} \quad 0 < |x - a| < \delta.$$

EXAMPLE 6 Give an ϵ-δ proof that $\dfrac{3x^2 - 3x - 6}{x - 2} \longrightarrow 9$ as $x \longrightarrow 2$.

Solution PROOF: Let $\epsilon > 0$ be given. Let $\delta = \dfrac{\epsilon}{3}$. Now, whenever

$$0 < |x - 2| < \delta = \frac{\epsilon}{3}$$

2.2 DEFINITION OF LIMIT

we have

$$\left|\frac{3x^2 - 3x - 6}{x - 2} - 9\right| = \left|\frac{3x^2 - 3x - 6 - 9(x - 2)}{x - 2}\right|$$

$$= \left|\frac{3x^2 - 12x + 12}{x - 2}\right|$$

$$= 3\left|\frac{(x - 2)(x - 2)}{x - 2}\right|$$

$$= 3|x - 2| < 3\left(\frac{\epsilon}{3}\right) = \epsilon.$$

∎

Some comments are in order regarding this proof. In obtaining the last line of the proof given, we are able to cancel the factor $(x - 2)$ only because the inequality $0 < |x - 2|$ eliminates the possibility that $x = 2$. The formula $\delta = \epsilon/3$ is our ϵ-δ machine, and it provides a way of picking an appropriate δ for *any* given ϵ. Where did this magical formula come from? If we follow the steps in Example 4 closely, we note that we essentially started out with the desired inequality

$$\left|\frac{3x^2 - 3x - 6}{x - 2} - 9\right| < .001 = \epsilon$$

and then worked backwards algebraically from there to find how small $|x - 2|$ must be in terms of this ϵ. Our limit proof generalizes this, and shows how we can pick δ for any given ϵ.

New limits from old

Once we know some particular limit values, we can often deduce other limit values directly. Suppose two functions f and g each have a limit at a particular point $x = a$, with

$$\lim_{x \to a} f(x) = L_1 \quad \text{and} \quad \lim_{x \to a} g(x) = L_2.$$

Going back to the definition of a limit, we know that provided the input x is sufficiently close to a (but not equal to a), the values of $f(x)$ and $g(x)$ can be guaranteed to be within any predetermined tolerance to the values L_1 and L_2, respectively. It seems natural to expect that we could guarantee that the value of $f(x) + g(x)$ will be close to $L_1 + L_2$, $f(x) - g(x)$ will be close to $L_1 - L_2$, $f(x)g(x)$ will be close to $L_1 L_2$, and $f(x)/g(x)$ will be close to L_1/L_2

(unless $L_2 = 0$). This is indeed the case, and we can formally state that

$$\lim_{x \to a}(f+g)(x) = L_1 + L_2$$

$$\lim_{x \to a}(f-g)(x) = L_1 - L_2$$

$$\lim_{x \to a} fg(x) = L_1 L_2$$

$$\lim_{x \to a} \frac{f}{g}(x) = \frac{L_1}{L_2} \quad \text{provided } L_2 \neq 0.$$

EXAMPLE 7 Suppose $\lim_{x \to 2} f(x) = 3$ and $\lim_{x \to 2} g(x) = -5$. Find the limits of

$$(f+g)(x), \quad (f-g)(x), \quad (fg)(x), \quad \text{and} \quad (f/g)(x)$$

as $x \longrightarrow 2$.

Solution Using the algebra of limits, we have

$$\lim_{x \to 2}(f+g)(x) = -2 \qquad \lim_{x \to 2}(f-g)(x) = 8$$

$$\lim_{x \to 2} fg(x) = -15 \qquad \lim_{x \to 2} \frac{f}{g}(x) = -\frac{3}{5}.$$

■

The key to these results is simply that the ϵ-δ machines for f and g can be used to make ϵ-δ machines for these new functions.

For example, suppose we wanted to guarantee that $f(x)+g(x)$ is within .001 of L_1+L_2. If we choose x close enough to a so that $f(x)$ is within .0005 of L_1 and $g(x)$ is also within .0005 of L_2, then the sum $f(x)+g(x)$ will be within .001 of $L_1 + L_2$. There is nothing special about the output error tolerance .001. Given any positive output error tolerance $\epsilon > 0$, we could make sure $f(x) + g(x)$ is within ϵ of $L_1 + L_2$ by choosing an input tolerance δ small enough so that as long as $0 < |x - a| < \delta$, *simultaneously* $f(x)$ is within $\epsilon/2$ of L_1 and $g(x)$ is within $\epsilon/2$ of L_2. Since $L_1 + L_2$ passes the limit definition test, we can conclude that $(f+g)(x) \longrightarrow L_1 + L_2$ as $x \longrightarrow a$. By a similar analysis we can conclude that $(f-g)(x) \longrightarrow L_1 - L_2$ as $x \longrightarrow a$.

The ϵ-δ machine needed to show that $fg(x) \longrightarrow L_1 L_2$ as $x \longrightarrow a$ is trickier to devise, but it definitely can be done. We can also show that $(f/g)(x) \longrightarrow L_1/L_2$, provided $L_2 \neq 0$. In the case of the product and quotient, the necessary input tolerance (δ) necessary to guarantee a given output error tolerance (ϵ) will depend not only on ϵ, but also on how large L_1 and L_2 are.

The squeezing principle

Another way of deducing a limit value is through comparison with known limit values. The **squeezing principle** is an example of such a comparison technique for determining a limit.

Theorem 2.1

The Squeezing Principle for Limits.
Hypotheses:
1. Two functions f and g have the *same* limit value L at $x = a$:
$$\lim_{x \to a} f(x) = L = \lim_{x \to a} g(x).$$
2. A third function h always has its output $h(x)$ sandwiched in between $f(x)$ and $g(x)$ whenever x is sufficiently close, but not equal to a. That is, for some $c > 0$ we have either
$$f(x) \leq h(x) \leq g(x) \qquad \text{or} \qquad g(x) \leq h(x) \leq f(x)$$
whenever $0 < |x - a| < c$.
Conclusion: We must also have $\lim_{x \to a} h(x) = L$.

Reasoning Assume the hypotheses are true. We want to show that $h(x) \longrightarrow L$ as $x \longrightarrow a$. So, let $\epsilon > 0$ be given. Since both $f(x) \longrightarrow L$ and $g(x) \longrightarrow L$ as $x \longrightarrow a$ (by hypothesis 1), we can choose $\delta > 0$ so that *both* $f(x)$ and $g(x)$ are within ϵ of L whenever $0 < |x - a| < \delta$. (Just choose the smaller of two individual δ-values which work for f and g, respectively.) Furthermore, we can choose this δ small enough so that we also have $\delta < c$. Since $h(x)$ is "squeezed" between $f(x)$ and $g(x)$ for these values of x (by hypothesis 2), we must have $h(x)$ within ϵ of L also. We can conclude that $\lim_{x \to a} h(x) = L$. □

When you analyze a theorem such as this, try to visualize a typical situation where the hypotheses are satisfied. Figure 2.5 illustrates graphically an example of three functions satisfying the hypotheses of Theorem 2.1. Note that the graph of $y = h(x)$ does not always have to lie between the graphs of f and g, but it must be sandwiched between them in some neighborhood of a. (The Squeezing Principle is also known as the *Sandwich Theorem*.)

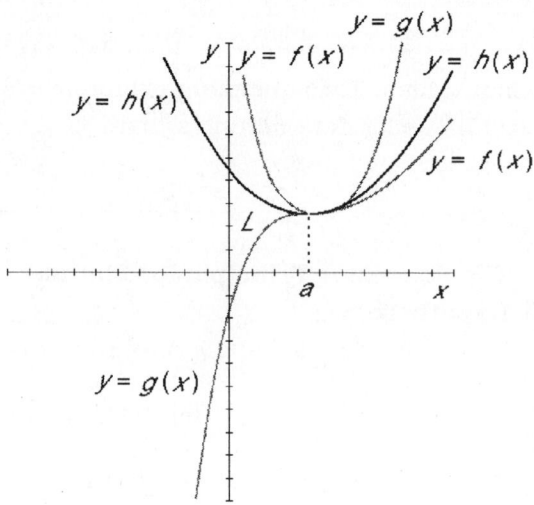

Figure 2.5 Illustration of the Squeezing Principle.

Next, look at the conclusion of the theorem and try to get a feeling for how the hypotheses force this conclusion to be true. In Figure 2.5, we see that near a, the graphs of f and g form a corridor that narrows to a single point at (a, L). Since the graph of h must eventually stay within that corridor, its graph is also forced toward this point.

Finally, ask yourself what happens when any of the hypotheses are relaxed. For example, if we relax the first hypothesis so that f and g do not have the same limit value at $x = a$, then knowing the outputs $h(x)$ are trapped between $f(x)$ and $g(x)$ will not be enough to pin down the limit value. And, if f and g *did* have the same limit value at $x = a$, but we relax hypothesis 2, we certainly could say very little about the limit of h. In this way, we can better see how the hypotheses work together to give us the stated conclusion.

An application of the squeezing principle

If we use the letter θ instead of x, then the numerical and graphical evidence presented earlier would suggest that

$$\lim_{\theta \to 0} \frac{\sin \theta}{\theta} = 1.$$

How can we verify that this limit is indeed correct? We'll find the squeezing principle useful in this discussion.

2.2 DEFINITION OF LIMIT

EXAMPLE 8 Use the squeezing principle to prove that $\lim\limits_{\theta \to 0} \dfrac{\sin \theta}{\theta} = 1$.

Solution In Figure 2.6, we have diagrammed a close-up of the part of the unit circle for positive angles θ close to 0.

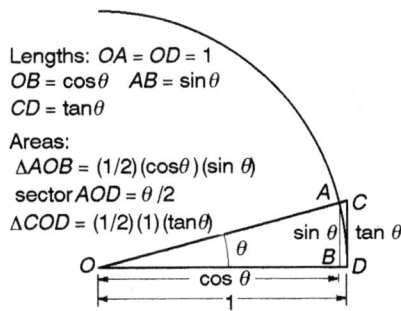

Figure 2.6 Diagram for analyzing $\lim\limits_{\theta \to 0} \dfrac{\sin \theta}{\theta}$.

The relevant lengths and areas are labelled in the picture. With O representing the origin $(0,0)$ and D representing $(1,0)$, we can see that lengths $OB = \cos \theta$ and $AB = \sin \theta$, respectively, from the definitions of the trigonometric functions. Since $\triangle AOB$ and $\triangle COD$ are similar triangles, we have $CD/OD = AB/OB$, from which we obtain $CD = \tan \theta$. The areas of these two triangles in square units are obtained using the usual $A = \dfrac{1}{2}bh$ formula, so that

$$\text{the area of } \triangle AOB = \frac{(\sin \theta)(\cos \theta)}{2}$$

and

$$\text{the area of } \triangle COD = \frac{\tan \theta}{2} = \frac{\sin \theta}{2 \cos \theta}.$$

Trapped between the areas of the two triangles is the area of the circular sector AOD with angle θ. Since the area of the entire unit circle is π square units, and this "piece of pie" represents $\dfrac{\theta}{2\pi}$ of the entire area, we can see that the area of the sector must be $\theta/2$ square units. Thus, we have the inequality

$$\frac{(\sin \theta)(\cos \theta)}{2} < \frac{\theta}{2} < \frac{\sin \theta}{2 \cos \theta}.$$

Multiplying through by $\dfrac{2}{\sin \theta}$ (which is positive for small, nonzero θ) yields

$$(1) \qquad \cos \theta < \frac{\theta}{\sin \theta} < \frac{1}{\cos \theta}$$

and inverting these positive fractions also reverses the inequality to be

(2) $$\cos\theta < \frac{\sin\theta}{\theta} < \frac{1}{\cos\theta}.$$

Now, as $\theta \longrightarrow 0$, the point A geometrically approaches the point $D = (1,0)$. The x-coordinate of A must approach 1. Thus, $\cos\theta \longrightarrow 1$, and likewise $\frac{1}{\cos\theta} \longrightarrow 1$. By the squeezing principle, we conclude that

$$\frac{\sin\theta}{\theta} \longrightarrow 1 \quad \text{as} \quad x \longrightarrow 0^+.$$

If $\theta < 0$ and small, then so is $\sin\theta$, and we would obtain the same inequality of positive quantities in (1) and (2). Thus,

$$\frac{\sin\theta}{\theta} \longrightarrow 1 \quad \text{as} \quad x \longrightarrow 0^-,$$

and we have $\lim_{\theta \to 0} \frac{\sin\theta}{\theta} = 1$. ∎

EXERCISES

Using only the limit tables from exercises 19 and 20 of section 2.1 and the algebra of limits, fill out a similar limit table for each of the functions in exercises 1-6 below. Then check your results against your graphical results from exercises 25-30 of section 2.1.

1. $f + g$ **2.** $f - g$
3. fg **4.** $2f$
5. $1/f$ **6.** f/g

In exercises 7-12 you are given a limit of the form

$$\lim_{x \to a} f(x) = L.$$

a) Find a suitable positive value δ that would guarantee

$$|f(x) - L| < .001 \quad \text{whenever} \quad 0 < |x - a| < \delta$$

b) Verify the stated limit using an $\epsilon - \delta$ proof.

7. $\lim_{x \to 1} x - 7 = -6$ **8.** $\lim_{x \to 3} \frac{x^2 - 9}{x - 3} = 6$
9. $\lim_{x \to -\frac{1}{2}} \frac{4x^3 + 4x + x}{2x^2 + x} = 0$ **10.** $\lim_{x \to -2} |x + 1| = 1$
11. $\lim_{x \to 2} x^2 = 4$ **12.** $\lim_{x \to -93.2} 14.83 = 14.83$

2.2 DEFINITION OF LIMIT

Exercises 13-28 require machine graphics. For each of the functions defined below, determine whether or not $\lim_{x \to 0} f(x)$ exists. If the function has a limit L as $x \to 0$, then set the vertical scaling of a graphing window to $[L-.01, L+.01]$. Then find a horizontal window interval of the form $[-\delta, \delta]$ such that the graph of the function is horizontal (with the possible exception at $x = 0$).

13. $x \longmapsto f(x)$ where

$$f(x) = \begin{cases} x^2 & \text{if } x < 0 \\ 2x - 1 & \text{if } x \geq 0. \end{cases}$$

14. $x \longmapsto sign(x)$ where $sign(x)$ denotes the **signum function** and is defined by the split formula

$$sign(x) = \begin{cases} 1 & \text{if } x > 0 \\ 0 & \text{if } x = 0 \\ -1 & \text{if } x < 0. \end{cases}$$

(The signum function essentially tells us whether the input x is positive, negative, or zero.)

15. $x \longmapsto |x|$
16. $x \longmapsto x/x$
17. $x \longmapsto |x|/x$
18. $x \longmapsto \dfrac{\sin(2x)}{5x}$
19. $x \longmapsto 1/x^2$
20. $x \longmapsto 1/x$
21. $x \longmapsto \sin(1/x)$
22. $x \longmapsto x\sin(1/x)$
23. $x \longmapsto \tan(x)$
24. $x \longmapsto \cot(x)$
25. $x \longmapsto \dfrac{1 - \cos x}{x}$
26. $x \longmapsto \sin^2(x) + \cos^2(x)$
27. $x \longmapsto (1 + x)^{1/x}$
28. $x \longmapsto \arctan(1/x)$

---

29. Use the squeezing principle to show that

$$\lim_{x \to 0} x \sin\left(\frac{1}{x}\right) = 0$$

by comparing $x \longmapsto \sin(1/x)$ to $x \longmapsto -x$ and $x \longmapsto x$.

30. For $-\pi/2 \leq \theta \leq \pi/2$, show that

$$0 \leq 1 - \cos(\theta) \leq \sin^2(\theta).$$

(Hint: $\sin^2(\theta) = 1 - \cos^2(\theta) = (1 + \cos(\theta))(1 - \cos(\theta))$ and $\cos(\theta) \geq 0$ for the values of θ given.)

31. Show that for $0 < \theta < \pi/2$ we have

$$0 < \frac{1 - \cos(\theta)}{\theta} < \frac{\sin^2(\theta)}{\theta}$$

and that for $-\pi/2 < \theta < 0$ we have

$$\frac{\sin^2(\theta)}{\theta} < \frac{1-\cos(\theta)}{\theta} < 0.$$

32. Use the results of the previous exercise along with the squeezing principle to show that

$$\lim_{\theta \to 0} \frac{1-\cos(\theta)}{\theta} = 0.$$

2.3 CONTINUITY

We say a function is **continuous at a point** when the limit of a function matches the output at that point. Using the notation of limits, we have the following definition.

Definition 2

> The function f is **continuous at** $x = a$ if and only if
> $$\lim_{x \to a} f(x) = f(a).$$

More explicitly, there are three requirements for a function f to be continuous at $x = a$:

1) $\lim_{x \to a} f(x)$ **must exist.**
2) $f(a)$ **must be defined.**
3) **These values must match.**

Put simply, if the function f is continuous at $x = a$ then we can predict the *correct* output value $f(a)$ on the basis of the outputs in a neighborhood of a.

These requirements mean that at a point of continuity, we can rule out a multitude of certain types of output behaviors, such as breaks or holes in the graph at that point. This is the motivation for the term continuous.

2.3 CONTINUITY

EXAMPLE 9 Figure 2.7 shows the graph of a function f. At which of the points a, b, c, d is f continuous?

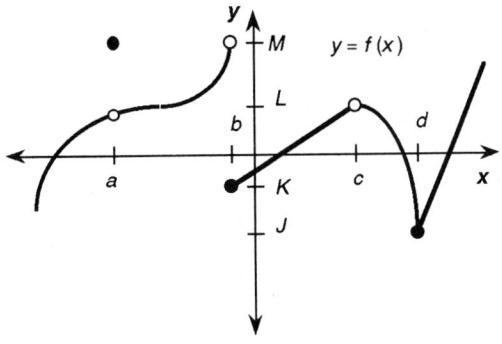

Figure 2.7 Where is f continuous?

Solution At each of the points a, b, c, and d we check each of the three requirements.

At $x = a$, f is not continuous, since $\lim_{x \to a} f(x) = L \neq M = f(a)$. (Requirements 1 and 2 are met, but not Requirement 3.)

At $x = b$, f is not continuous, since $\lim_{x \to b} f(x)$ does not exist. (Requirement 1 is not met.)

At $x = c$, f is not continuous, since $f(c)$ is not defined (Requirement 1 is met, but not Requirement 2.)

At $x = d$, f is continuous, since $\lim_{x \to d} f(x) = J = f(d)$. (All three requirements are met.) ∎

We can define left-hand and right-hand continuity at a point according to whether the left-hand or right-hand limit of the function matches the actual output value there. The function f is **continuous from the left** at $x = a$ if and only if

$$\lim_{x \to a^-} f(x) = f(a).$$

The function f is **continuous from the right** at $x = a$ if and only if

$$\lim_{x \to a^+} f(x) = f(a).$$

EXAMPLE 10 For the function f illustrated in Figure 2.7, determine if f is continuous from the left, right, or neither, at $x = b$.

Solution The function f is continuous from the right at b since $\lim_{x \to b^+} f(x) = K = f(b)$, but f is not continuous from the left at b since $\lim_{x \to b^-} f(x) = M \neq f(b)$. ∎

We can think of a function being continuous at a point if and only if the *predicted* value of the function from both sides matches the *actual* value of the function at that point.

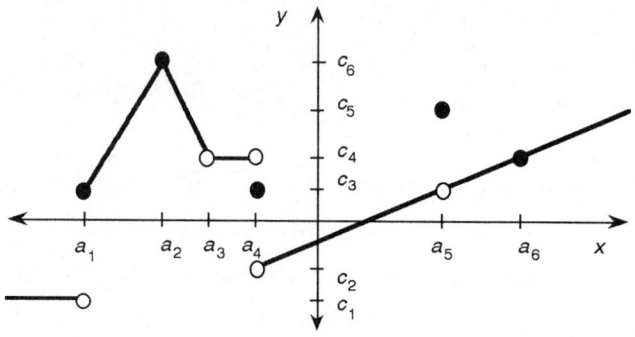

Figure 2.8 Where is the function continuous?

EXAMPLE 11 For the function f illustrated in Figure 2.8, at which of the six points a_1, $a_2, \ldots, a_6$ is f continuous?

Solution Table 2.2 shows the limit and continuity information for each of the inputs $a_1, a_2, \ldots, a_6$. ∎

input $x=a$	output $f(a)$	left-hand limit $\lim_{x \to a^-} f(x)$	right-hand limit $\lim_{x \to a^+} f(x)$	limit $\lim_{x \to a} f(x)$	Is f continuous at $x=a$?
a_1	c_3	c_1	c_3	does not exist	no
a_2	c_6	c_6	c_6	c_6	yes
a_3	undefined	c_4	c_4	c_4	no
a_4	c_3	c_4	c_2	does not exist	no
a_5	c_5	c_3	c_3	c_3	no
a_6	c_4	c_4	c_4	c_4	yes

Table 2.2 Limit table for the function f of Figure 2.8.

The formal definition of continuity is sometimes expressed using ϵ's and δ's directly instead of the limit requirement:

2.3 CONTINUITY

Definition 3
> The function f is **continuous at** $x = a$ if and only if the following condition holds: Given any $\epsilon > 0$, there is a $\delta > 0$ such that $|f(x) - f(a)| < \epsilon$ whenever $|x - a| < \delta$. If f is continuous at all real numbers, then we simply say that f is **continuous**.

To prove that a function is continuous, we must demonstrate that it is continuous at every real number $x = a$.

EXAMPLE 12 Prove that any constant function $h : x \longmapsto c$ (where c is a specific real number) is continuous.

Solution We need to show that h is continuous at every point $x = a$. Using definition 1, this means that for any real number a, we must show

$$\lim_{x \to a} h(x) = h(a) = c.$$

PROOF: Let $\epsilon > 0$ be given. Let $\delta = 1$ (or let δ be *any* positive number, for that matter). Then whenever $|x - a| < \delta$, we certainly have $|h(x) - c| = |c - c| = 0 < \epsilon$. ∎

EXAMPLE 13 Prove that the identity function $i : x \longmapsto x$, is continuous.

Solution We need to show that if a is any real number, then

$$\lim_{x \to a} i(x) = i(a) = a.$$

PROOF: Let $\epsilon > 0$ be given. Let $\delta = \epsilon$. Then whenever

$$|x - a| < \delta = \epsilon,$$

we certainly have

$$|i(x) - a| = |x - a| < \epsilon.$$

∎

Combining continuous functions

If f and g are continuous at a, then they have limit values that match their outputs $f(a)$ and $g(a)$. This would mean that the limits of $f + g$, $f - g$, and fg will also match their output values at $x = a$, so these functions are all continuous at a. The function f/g is also continuous at a (provided $g(a) \neq 0$ so that division by zero does not occur).

As for composition, suppose g is continuous at b and $g(b) = L$. Now suppose
$$\lim_{x \to a} f(x) = b.$$
Since we can guarantee the output $f(x)$ to be as close as we want to b when x is close enough to a, we can also guarantee that $g(f(x))$ is close to L for x close enough to a. In other words,
$$\lim_{x \to a} g(f(x)) = L.$$
If f happens to be also continuous at a so that $b = f(a)$ and $L = g(b) = g(f(a))$, then
$$\lim_{x \to a} g(f(x)) = g(f(a))$$
and $g \circ f$ is continuous at $x = a$.

If f and g are continuous functions, then each is continuous at every real number, and we can conclude that each of the following functions is continuous:

$$f + g \qquad f - g \qquad fg \qquad f \circ g$$

Also, we can say that f/g is continuous at every point a such that $g(a) \neq 0$.

Examples of continuous functions

Using the results
$$\lim_{x \to a} c = c \qquad \text{and} \qquad \lim_{x \to a} x = a$$
and the algebra of limits, we can conclude that every polynomial function p is continuous. In other words, for any real number a, it is true that
$$\lim_{x \to a} p(x) = p(a).$$

2.3 CONTINUITY

EXAMPLE 14 Find $\lim_{x \to -0.5} 8x^3 - 4x^2 + \sqrt{3}x - \pi$.

Solution Since $x \longmapsto 8x^3 - 4x^2 + \sqrt{3}x - \pi$ is a polynomial function, we know that it is continuous. Therefore, we can evaluate this limit by simply evaluating the polynomial at $x = -0.5$:

$$\lim_{x \to -0.5} 8x^3 - 4x^2 + \sqrt{3}x - \pi = 8(-0.5)^3 - 4(-0.5)^2 + \sqrt{3}(-0.5) - \pi = -2 - \frac{\sqrt{3}}{2} - \pi.$$

We have actually used the algebra of continuous functions several times, for we can view the polynomial $x \longmapsto 8x^3 - 4x^2 + \sqrt{3}x - \pi$ as being a sum of products of constant and identity functions:

$$8x^3 - 4x^2 + \sqrt{3}x - \pi = 8 \cdot x \cdot x \cdot x + (-4) \cdot x \cdot x + \sqrt{3} \cdot x + (-\pi).$$

■

Similarly, for a rational function $x \longmapsto p(x)/q(x)$, where p and q are polynomials, we will have

$$\lim_{x \to a} \frac{p(x)}{q(x)} = \frac{p(a)}{q(a)}$$

provided $q(a) \neq 0$.

The trigonometric functions sine and cosine are continuous at every real value because as we geometrically approach any particular angle value $\theta = a$ on the continuous unit circle, the corresponding x-coordinate ($\cos(\theta)$) and y-coordinate ($\sin(\theta)$) must approach $\cos(a)$ and $\sin(a)$ respectively. From the algebra of continuous functions, it follows that the other four trigonometric functions are continuous where they are defined.

EXAMPLE 15 Find the set of values at which the tangent function is continuous.

Solution The set of values at which the tangent function is continuous is

$$\{x : x \neq \frac{\pi}{2} + n\pi, \; n \text{ an integer}\}.$$

This is the domain of the tangent function. ■

A rational power function is also continuous at every point a in its domain, unless a happens to be the *endpoint* of the domain.

EXAMPLE 16 Where are each of the following rational power functions continuous?

$$f: x \longmapsto x^{2/3} \quad g: x \longmapsto x^{1/2} \quad h: x \longmapsto x^{-3} \quad k: x \longmapsto x^{-2}$$

Solution The graphs of these four functions are shown in Figure 2.9.

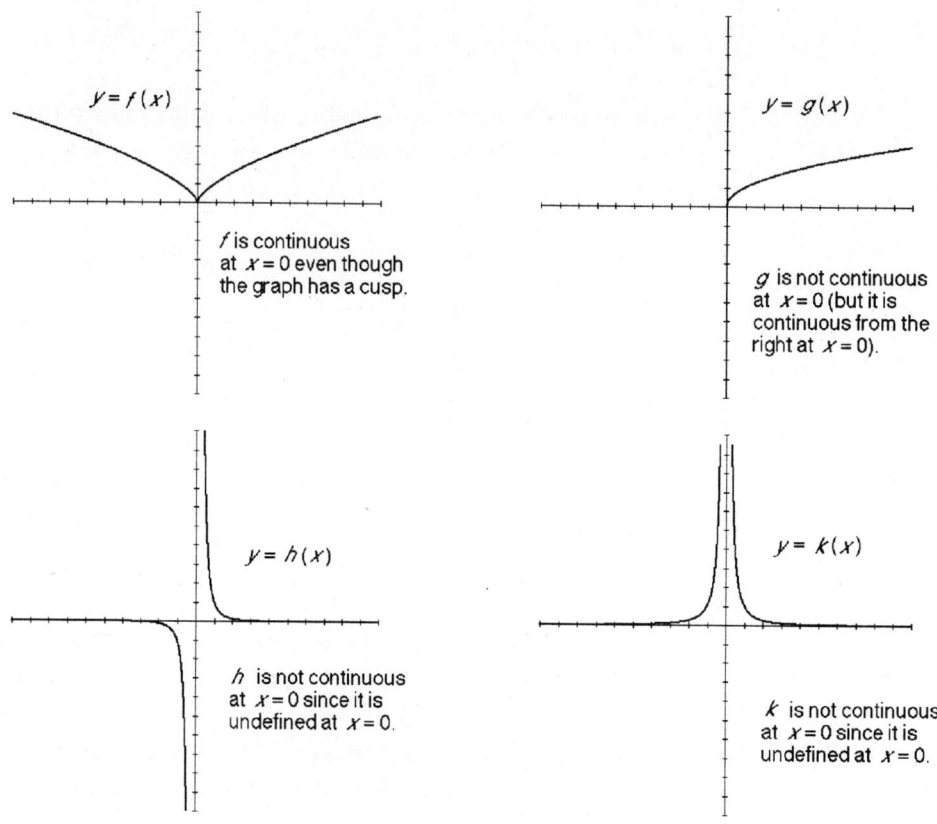

Figure 2.9 Graphs of four rational power functions.

The function f is continuous at every real number, including $x = 0$ where the cusp in the graph occurs.

The function g is continuous at every $x > 0$, but not at $x = 0$ (we cannot approach 0 from the left, so the limit does not exist there). We could only say that g is continuous from the right at 0.

The functions h and k are continuous at every $x \neq 0$. Neither function is defined at 0 (nor does its limit exist at 0.) ∎

Exponential, logarithmic, and *inverse trigonometric* functions are also continuous at each point in their respective domains.

2.3 CONTINUITY

EXERCISES

1. Using Figure 2.8 and Table 2.2, determine whether or not f is continuous from the left at each of the points $a_1, a_2, a_3, a_4, a_5,$ and a_6.

2. Using Figure 2.8 and Table 2.2, determine whether or not f is continuous from the right at each of the points $a_1, a_2, a_3, a_4, a_5,$ and a_6.

The two functions f and g have the graphs shown below.

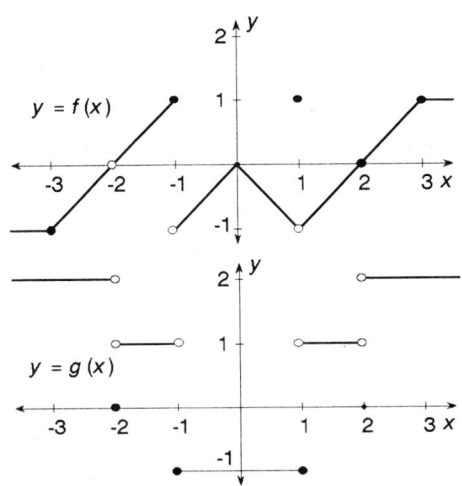

Use these graphs to determine the continuity (including left-hand and right-hand continuity) at the points $-3, -2, -1, 0, 1, 2,$ and 3 of each of the functions in exercises 3-14. You may wish to refer to exercises 19-30 of section 2.1.

3. f
4. g
5. $f \circ g$
6. $g \circ f$
7. $x \longmapsto f(x-1)$
8. $x \longmapsto g(x/2)$
9. $f + g$
10. $f - g$
11. fg
12. $2f$
13. $1/f$
14. f/g

In exercises 15-20, explain what requirement(s) of the continuity definition are violated at the indicated point.

15. $\lim_{x \to 1+} f(x) = 2$ $\quad \lim_{x \to 1-} f(x) = -1$ $\quad f(1)$ is undefined.

16. $\lim_{x \to -2-} g(x) = 0$ $\quad \lim_{x \to -2+} g(x) = 0$ $\quad g(-2) = 1$

17. $\lim_{x \to 2-} h(x) = -2$ $\quad \lim_{x \to 2+} h(x) = 2$ $\quad h(2) = 0$

18. $\lim_{x \to 0-} i(x) = -1$ $\quad \lim_{x \to 0+} i(x) = -2$ $\quad i(0) = -1$

19. $\lim_{x \to -1-} j(x) = 3$ $\quad \lim_{x \to -1+} j(x) = -2$ $\quad j(-1) = -2$

20. $\lim_{x \to -3-} k(x) = 1$ $\quad \lim_{x \to -3+} k(x) = 1$ $\quad k(-3)$ is undefined.

---

21. Prove directly, using an ϵ-δ proof, that the function $f : x \longmapsto 2x + 1$ is continuous.

22. Prove directly, using an ϵ-δ proof, that any linear function of the form $f : x \longmapsto mx + b$ is continuous.

23. Prove directly, using an ϵ-δ proof, that the function $x \longmapsto x^2$ is continuous at 3. (Hint: Split your proof up into two cases, depending on the size of ϵ. First, find a δ that would work for any $\epsilon \geq 1$. Then, for $0 < \epsilon < 1$, find δ in terms of ϵ.)

24. Is it possible for the product of two functions fg to be continuous at a point $x = a$ even if neither function is continuous at $x = a$?

25. At what points x are the trigonometric functions sin, cos, tan and cot not continuous? At what points are the inverse trigonometric functions arcsec and arccsc not continuous?

2.4 ANALYZING DISCONTINUITIES AND ASYMPTOTIC BEHAVIOR

If a function f is continuous at $x = a$, then it behaves predictably at that point. When f is *not* continuous at a then somehow the function's outputs have behaved unpredictably at that point. A point a at which f is not continuous is called a **discontinuity**. Let's look at some of the types of behavior a function could have at a discontinuity.

2.4 ANALYZING DISCONTINUITIES AND ASYMPTOTIC BEHAVIOR

Skips, holes, jumps, and poles

If $\lim_{x \to a} f(x)$ exists, then the only things keeping f from being continuous at a are the possibilities that either

1) $f(a)$ is undefined (leaving a hole in the graph), or

2) $f(a)$ is defined, but does not match the limit value

(the graph skips up or down to an unexpected value).

If we refer back to our original limit illustration function (see Figure 2.10), there is a *hole* at $x = c$ and a *skip* at $x = a$.

We could imagine *fixing* the values of $f(x)$ to match the limit in the case of a hole or a skip, so that the function f would be continuous there. For that reason, whenever f actually has a limit at a discontinuity a, we say that f has a **removable discontinuity** at a. To remove the discontinuity, we would need to *change the function at that point* by either defining or redefining $f(a)$ so that

$$f(a) = \lim_{x \to a} f(x).$$

EXAMPLE 17 For the function f illustrated in Figure 2.10, find the removable discontinuities and determine what new value for f must be defined to remove the discontinuities.

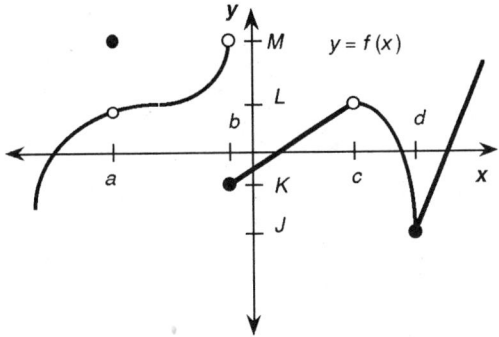

Figure 2.10 Which discontinuities are removable?

Solution   The function f has removable discontinuities at $x = a$ and $x = c$. (f is not continuous at these points, but f does have a limit value at each of these points.)

We need the outputs to match the limit of f at those points, so we should define

$$f(a) = L$$

and redefine

$$f(c) = L$$

in order to make f continuous at a and c.

Since f does not have a limit at the input b, simply defining a new value $f(b)$ cannot remove the discontinuity there. ∎

A **nonremovable discontinuity** is also called an **essential discontinuity**. We should point out that f really is *continuous* at $x = d$, even though the graph makes a sharp turn at that point. One common type of essential discontinuity is the so-called *jump* discontinuity. Jump discontinuities can easily occur with piece-wise defined functions, as in the following example.

EXAMPLE 18   Is the function $x \longmapsto f(x)$ where

$$f(x) = \begin{cases} x^2 & \text{if } x < 0 \\ 2x - 1 & \text{if } x \geq 0 \end{cases}$$

continuous at $x = 0$? If not, is the discontinuity removable or essential?

Solution   As $x \to 0^-$, the first part of the split evaluation formula applies and $f(x) = x^2$ for $x < 0$. From this, we can see that

$$\lim_{x \to 0^-} f(x) = 0.$$

As $x \to 0^+$, the second part of the split evaluation formula applies and $f(x) = 2x - 1$ for $x > 0$. From this, we can see that

$$\lim_{x \to 0^+} f(x) = -1.$$

Since the two one-sided limits do not match, f has no limit at $x = 0$. Hence, the discontinuity there is *essential*. A graph of this function is pictured in Figure 2.11. ∎

2.4 ANALYZING DISCONTINUITIES AND ASYMPTOTIC BEHAVIOR

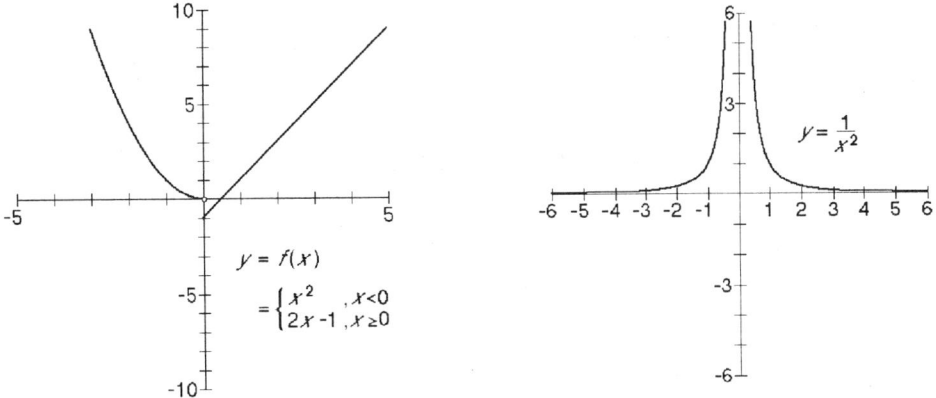

Figure 2.11 Essential discontinuities: jumps and vertical asymptotes.

Another type of nonremovable discontinuity occurs at what are called *vertical asymptotes*. For example, the function $x \longmapsto 1/x$ has a vertical asymptote at $x = 0$. As x approaches 0 through positive values, the values of $1/x$ become larger and larger without bound. As x approaches 0 through negative values, the values of $1/x$ are negative, and their magnitude (size in absolute value) grows larger and larger without bound. Neither the left-hand nor right-hand limit exists, so we have no limit at 0, and $x \longmapsto 1/x$ is definitely not continuous at $x = 0$. The vertical line $x = 0$ is called a **vertical asymptote** for the graph $y = 1/x$.

It is common to use the symbols ∞ and $-\infty$ in cases like this to describe the output behavior near a vertical asymptote. We could write

$$\lim_{x \to 0^-} \frac{1}{x} = -\infty \quad \text{and} \quad \lim_{x \to 0^+} \frac{1}{x} = \infty.$$

Similarly, we could say

$$\frac{1}{x} \longrightarrow -\infty \quad \text{as} \quad x \longrightarrow 0^-$$

and

$$\frac{1}{x} \longrightarrow \infty \quad \text{as} \quad x \longrightarrow 0^+.$$

Again, we emphasize that neither of the two one-sided limits exists. All we are saying with this notation is *why* the limits do not exist. Namely, the function outputs are growing in magnitude without bound in either the positive or negative direction.

BEWARE! The symbols ∞ and $-\infty$ do not represent real numbers. As used here, they are simply a convenient shorthand for describing the behavior of the function.

EXAMPLE 19 Analyze the behavior of the function
$$x \longmapsto \frac{1}{x^2}$$
at $x = 0$.

Solution The graph of $y = 1/x^2$ for $-10 < x < 10$ is shown in Figure 2.11. The value of $1/x^2$ is positive and grows without bound as we approach $x = 0$ from either side. Hence,
$$\lim_{x \to 0^-} \frac{1}{x^2} = \infty \quad \text{and} \quad \lim_{x \to 0^+} \frac{1}{x^2} = \infty.$$

Since $1/x^2 \longrightarrow \infty$ as $x \longrightarrow 0$ from both sides, we could simply write
$$\lim_{x \to 0} \frac{1}{x^2} = \infty.$$

We must emphasize that $1/x^2$ has *no limit* as $x \longrightarrow 0$. ∎

Definition 4

> The graph of a function f has a **vertical asymptote** $x = a$ provided at least one of the following is true:
> $$\lim_{x \to a^-} f(x) = \infty \qquad \lim_{x \to a^-} f(x) = -\infty$$
> $$\lim_{x \to a^+} f(x) = \infty \qquad \lim_{x \to a^+} f(x) = -\infty.$$
> If a function's outputs exhibit unbounded behavior as the inputs approach a from *both* sides, we say f has a **pole** at a.

The functions $x \longmapsto 1/x$ and $x \longmapsto 1/x^2$ each have a pole at 0, and each graph has the vertical asymptote $x = 0$.

A rational function of the form $x \longmapsto p(x)/q(x)$ (where $p(x)$ and $q(x)$ are polynomials) has its points of discontinuity precisely at those values a for which the denominator $q(a) = 0$, or equivalently, those values a for which $(x - a)$ is a factor of $q(x)$. The type of discontinuity (either hole or pole) will depend on the *relative multiplicity* of $(x - a)$ as a factor of the numerator $p(x)$.

2.4 ANALYZING DISCONTINUITIES AND ASYMPTOTIC BEHAVIOR

To put it simply:

If $(x - a)$ appears as a factor in the denominator $q(x)$ *more times* than in the numerator $p(x)$, then the graph of the rational function has a vertical asymptote at $x = a$.

If $(x - a)$ appears as a factor in the denominator $q(x)$ *the same or fewer times* than in the numerator $p(x)$, then the graph of the rational function has a hole at $x = a$.

EXAMPLE 20 Analyze the behavior of the rational function $f : x \longmapsto \dfrac{x^3 - 8x^2 + 16x}{x^2 - 5x + 4}$ near its points of discontinuity.

Solution The denominator factors as $x^2 - 5x + 4 = (x - 4)(x - 1)$ and the numerator factors as $x^3 - 8x^2 + 16x = x(x - 4)(x - 4)$. The points of discontinuity are the zeroes of the denominator:

$$x = 4 \qquad \text{and} \qquad x = 1.$$

Since $(x - 4)$ appears as a factor twice in the numerator and only once in the denominator, there is a removable discontinuity at $x = 4$ (the graph will have a hole there). On the other hand, since $(x - 1)$ appears as a factor in the denominator but not at all in the numerator, there is an essential discontinuity at $x = 1$ (the graph will have a vertical asymptote there). The behavior of f near these two discontinuities is summarized using the language of limits. We have

$$\lim_{x \to 4} f(x) = 0 \qquad \lim_{x \to 1^-} f(x) = +\infty \qquad \lim_{x \to 1^+} f(x) = -\infty.$$

The graph of $y = f(x)$ shown in Figure 2.12 illustrates this behavior. ∎

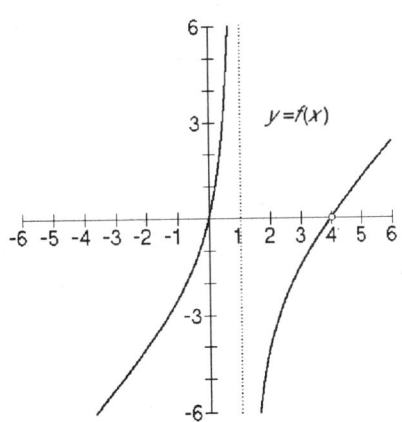

Figure 2.12 Discontinuities of the rational function $f : x \longmapsto \dfrac{x^3 - 8x^2 + 16x}{x^2 - 5x + 4}$.

EXAMPLE 21 Find all the poles of $x \longmapsto \tan x$.

Solution The function $x \longmapsto \tan x$ has a discontinuity at $x = \pi/2$ and, because of its periodic nature, a discontinuity exists at each value $x = \pi/2 + n\pi$ for any integer n. As x approaches $\pi/2$ from the left-hand side with $x < \pi/2$, the values of $\tan x$ grow larger and larger without bound. As x approaches $\pi/2$ from the right-hand side with $x > \pi/2$, the values of $\tan x$ are negative, and their magnitude grows larger and larger without bound. Thus,

$$\lim_{x \to \frac{\pi}{2}^-} \tan x = \infty \quad \text{and} \quad \lim_{x \to \frac{\pi}{2}^+} \tan x = -\infty.$$

So $x \longmapsto \tan x$ has a pole at $x = \pi/2$. Since this behavior is repeated at every other discontinuity, we conclude that the set of poles is

$$\{x : x = \frac{\pi}{2} + n\pi, n \text{ any integer}\},$$

and the graph $y = \tan x$ will have vertical asymptotes at each of these locations. ∎

Not all discontinuities fall neatly into one of the categories of skip, hole, jump, or pole. The formal definition of continuity rules out many other function behaviors. For example, the function $f : x \longmapsto \sin(1/x)$ has an essential discontinuity at $x = 0$ because the values of $f(x)$ wildly oscillate between $y = -1$ and $y = 1$ as $x \longrightarrow 0$. (See Figure 2.13).

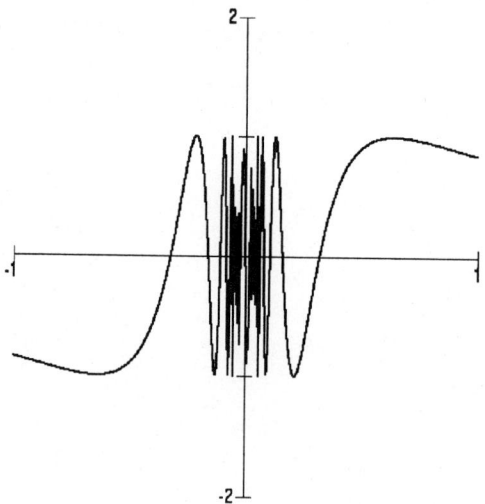

Figure 2.13 The function $x \longmapsto \sin(1/x)$ has an essential discontinuity at $x = 0$.

2.4 ANALYZING DISCONTINUITIES AND ASYMPTOTIC BEHAVIOR

Table 2.3 summarizes the limit and continuity information for several functions, each of which has a discontinuity at the particular point $x = 0$.

function $f(x)$	output $f(0)$	left-hand limit $\lim_{x \to 0^-} f(x)$	right-hand limit $\lim_{x \to 0^+} f(x)$	limit $\lim_{x \to 0} f(x)$	Is the discontinuity removable?
x/x	undefined	1	1	1	yes
$\sin(x)/x$	undefined	1	1	1	yes
$|x|/x$	undefined	-1	1	does not exist	no
$\text{signum}(x)$	0	-1	1	does not exist	no
$1/x$	undefined	$-\infty$ (does not exist)	∞ (does not exist)	does not exist	no
$1/x^2$	undefined	∞ (does not exist)	∞ (does not exist)	∞ (does not exist)	no
$\sin(1/x)$	undefined	does not exist	does not exist	does not exist	no

Table 2.3 Examples of functions with discontinuities at $x = 0$.

Consequences of continuity

If we know that a function f is continuous at $x = a$, then it's easy to compute its limit by simply evaluating the function:

$$\lim_{x \to a} f(x) = f(a).$$

This has some important consequences with both numerical and graphical interpretations.

Numerically, this means that the outputs $f(x)$ cannot be changing drastically near $x = a$. In other words, if x is close to a, and f is continuous at a, then the output $f(x)$ must be close to $f(a)$. What "close to" means here is very relative, but recall that if we are given any output error tolerance $\epsilon > 0$, then we can always specify an input error tolerance $\delta > 0$ so that

$$f(a) - \epsilon < f(x) < f(a) + \epsilon \quad \text{whenever} \quad a - \delta < x < a + \delta.$$

Graphically, this means that for any vertical scaling on a graphing window centered at $(a, f(a))$, we can rescale the horizontal axis so that the graph of $y = f(x)$ enters the window only from the left and leaves only to the right.

The intuitive idea of a continuous function having an unbroken graph seems simple enough, but the property of continuity turns out to be quite crucial in the analysis of functions. Here we want to point out some of the more profound consequences of continuity, and illustrate graphically the reasoning behind them. The mathematical proofs of these statements require using the formal definitions of limit and continuity along with fundamental properties of the real numbers. (These proofs can be found in most books on real analysis.)

Definition 5 | A function f is **continuous on an open interval** (a,b) provided f is continuous at every point of the interval. A function f is **continuous on a closed and bounded interval** $[a,b]$ provided f is continuous at every point in the open interval (a,b) and continuous from the right at a and continuous from the left at b.

Two important properties of a function continuous on a closed and bounded interval are stated below as theorems.

Theorem 2.2 | **The Intermediate Value Theorem.**
Hypothesis: Suppose f is continuous on the closed interval $[a,b]$.
Conclusion: For every real value y_0 between $f(a)$ and $f(b)$ inclusive, there is at least one input $a \leq x_0 \leq b$ such that $f(x_0) = y_0$.

□

Theorem 2.3 | **The Extreme Value Theorem.**
Hypothesis: Suppose f is continuous on the closed interval $[a,b]$.
Conclusion: The function f produces definite minimum and maximum output values over the interval $[a,b]$. That is, there are inputs x_1 and x_2 between a and b inclusive such that for *any* input $a \leq x \leq b$ we have $f(x_1) \leq f(x) \leq f(x_2)$.

□

The hypothesis of a theorem states the requirements needed to guarantee the conclusion. The hypotheses of these two theorems are the same: we have a function f continuous on a closed interval $[a,b]$. If we reason graphically, this suggests that the graph of f must connect the points $(a, f(a))$ and $(b, f(b))$ with no holes, skips, jumps, vertical asymptotes or other such behavior along the way. The conclusion of Theorem 2.2 is simply that such a graph must cross every horizontal line $y = y_0$ that we draw between $y = f(a)$ and $y = f(b)$ at least once. The conclusion of Theorem 2.3 is that we can find definite high and low extreme y-values on our graph between a and b. Figure 2.14 below illustrates the situation in these two theorems.

2.4 ANALYZING DISCONTINUITIES AND ASYMPTOTIC BEHAVIOR

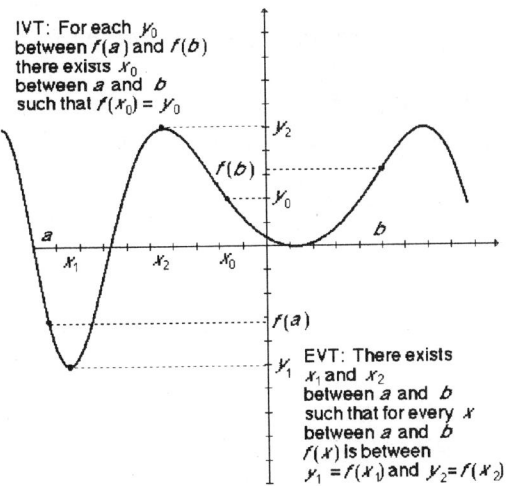

Figure 2.14 Illustration of Intermediate and Extreme Value Theorems.

A *corollary* to a theorem is a proposition which builds on the theorem by adding some additional hypotheses or special cases of the hypotheses to make a special conclusion. The proof of a corollary is usually some easy application of the original theorem. For example, here's a corollary to Theorem 2.2:

Theorem 2.4

Intermediate Zero Theorem.
Hypothesis 1: f is continuous on the closed interval $[a, b]$.
Hypothesis 2: $f(a)$ and $f(b)$ have *opposite signs*.
Conclusion: f has at least one *zero* strictly between a and b.

Reasoning: A *zero* for a function f is any input x_0 such that $f(x_0) = 0$. Since f satisfies the hypothesis of Theorem 2.2 and $y_0 = 0$ is between $f(a)$ and $f(b)$, the conclusion of Theorem 2.2 means that there must be at least one input $a \leq x_0 \leq b$ such that $f(x_0) = 0$. Since neither a nor b is a zero for f, we must have $a < x_0 < b$. □

Theorem 2.4 is also known as the **Location Principle** or **Bolzano's Theorem**, in honor of the mathematician Bernard Bolzano (1781-1848).

Horizontal asymptotes and end behavior

An **asymptotic** property of a function refers to the behavior of its outputs for negative and positive inputs of large magnitude (meaning $|x|$ is large). Asymptotic properties are sometimes referred to as the *end* behavior of a function, even though the x-axis has no ends. The language of limits is also handy for discussing asymptotic behavior. First, we'll talk about the special case of end behavior indicated by *horizontal asymptotes*. We write

$$\lim_{x \to \infty} f(x) = L_1$$

to denote that the values $f(x)$ approach L_1 as the values x take on positive values of larger and larger magnitude, and we write

$$\lim_{x \to -\infty} f(x) = L_2$$

to denote that the values $f(x)$ approach L_2 as the values x take on negative values of larger and larger magnitude. In either case, we would say that the horizontal lines $y = L_1$ and $y = L_2$ are **horizontal asymptotes** of the graph of $y = f(x)$.

EXAMPLE 22 Find the horizontal asymptotes to the graph of $y = \dfrac{1}{x}$.

Solution For large magnitude positive values x, the value of $1/x$ is a small magnitude positive number. For large magnitude negative values x, the value of $1/x$ is a small magnitude negative number. We would write

$$\lim_{x \to \infty} \frac{1}{x} = 0 \quad \text{and} \quad \lim_{x \to -\infty} \frac{1}{x} = 0$$

so that $y = 0$ is a horizontal asymptote for the graph of $y = 1/x$. ■

EXAMPLE 23 Find the horizontal asymptotes, if any, of the rational functions

$$f : x \longmapsto \frac{-2x^2 - 3x + 5}{4x^3 + 6x^2 - 7} \quad g : x \longmapsto \frac{-2x^3 - 3x + 5}{4x^2 + 6x - 7} \quad h : x \longmapsto \frac{-2x^3 - 3x + 5}{4x^3 + 6x^2 - 7}.$$

Solution For each function, we need to analyze the behavior of the outputs y as the inputs $x \longrightarrow \infty$ and $x \longrightarrow -\infty$. We'll make use of the following observation: if n is any positive integer, then no matter what the value of the constant c, we have

$$c/x^n \longrightarrow 0$$

as either $x \longrightarrow \infty$ or $x \longrightarrow -\infty$.

We can use this fact to find the horizontal asymptotes of the rational functions in this example. Since the inputs of interest are definitely *nonzero*, we can first factor out the largest common power of x occurring in both the numerator and denominator:

$$f(x) = \frac{x^2}{x^2} \cdot \frac{-2 - \frac{3}{x} + \frac{5}{x^2}}{4x + 6 - \frac{7}{x^2}} \qquad g(x) = \frac{x^2}{x^2} \cdot \frac{-2x - \frac{3}{x} + \frac{5}{x^2}}{4 + \frac{6}{x} - \frac{7}{x^2}} \qquad h(x) = \frac{x^3}{x^3} \cdot \frac{-2 - \frac{3}{x^2} + \frac{5}{x^3}}{4 + \frac{6}{x} - \frac{7}{x^3}}$$

for nonzero values of x. The leading factor in each case has the value 1 for $x \neq 0$, so we need only characterize the behavior of the trailing factor. Every term in both the numerator and denominator having the form c/x^n will become small in magnitude as x grows large in magnitude. In other words, for $|x|$ large, we have

$$f(x) \approx \frac{-2}{4x} \qquad g(x) \approx \frac{-2x}{4} \qquad h(x) \approx \frac{-2}{4}.$$

Since

$$\lim_{x \to \infty} f(x) = 0 = \lim_{x \to -\infty} f(x)$$

the graph of $y = f(x)$ has a horizontal asymptote $y = 0$. Since

$$\lim_{x \to \infty} g(x) = -\infty \qquad \text{and} \qquad \lim_{x \to -\infty} g(x) = \infty$$

the graph of $y = g(x)$ has no horizontal asymptotes. Finally,

$$\lim_{x \to \infty} h(x) = -\frac{1}{2} = \lim_{x \to -\infty} h(x)$$

so the graph of $y = h(x)$ has a horizontal asymptote $y = -1/2$. ∎

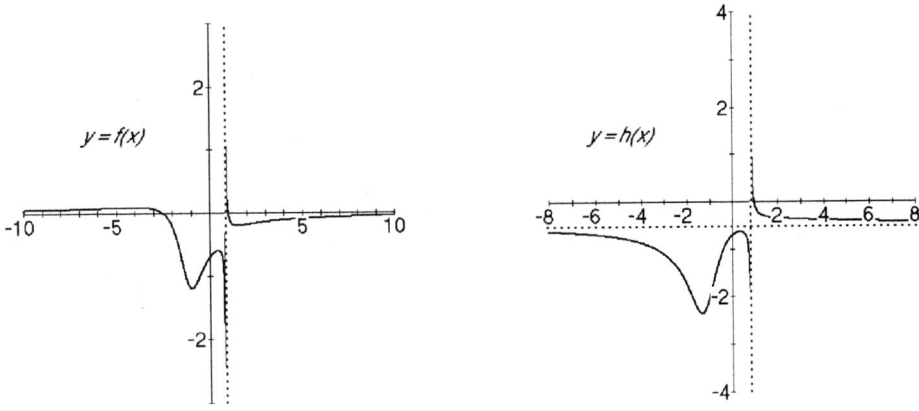

Figure 2.15 Graphs of $y = f(x)$ and $y = h(x)$.

Figure 2.15 shows the graph of $y = f(x)$ and $y = h(x)$ analyzed in Example 23. Beware of a common misconception. It *is possible* for the graph of a function to cross a horizontal asymptote as is the case with $y = f(x)$ shown in Figure 2.15.

We can generalize the results of this example as follows: Suppose the degree (highest power of x) of the numerator $p(x)$ is n, and the degree of the denominator $q(x)$ is m.

If $n < m$, then $y = 0$ is a horizontal asymptote (like f).

If $n > m$, then there's no horizontal asymptote (like g).

If $n = m$, then $y = a/b$ is a horizontal asymptote, where a is the leading coefficient in the numerator and b is the leading coefficient in the denominator (like h).

We can say a little more about the second case. If the degree of the numerator is greater than the degree of the denominator, we can use long division of polynomials to write

$$\frac{p(x)}{q(x)} = s(x) + \frac{r(x)}{q(x)}$$

where $s(x)$ is a polynomial, and $r(x)$ is a polynomial with a degree *less than* the degree of $q(x)$. For $|x|$ large, this remainder will have a very small value. This means for $|x|$ large, the graph of $y = p(x)/q(x)$ will look very much like the graph of $y = s(x)$. We would say $p(x)/q(x) \longrightarrow s(x)$ **asymptotically** in this case.

EXAMPLE 24 Describe the asymptotic behavior of

$$g : x \longmapsto \frac{-2x^3 - 3x + 5}{4x^2 + 6x - 7}.$$

Solution If we long divide $p(x) = -2x^3 - 3x + 5$ by $q(x) = 4x^2 + 6x - 7$, we get

$$\frac{-2x^3 - 3x + 5}{4x^2 + 6x - 7} = -\frac{1}{2}x + \frac{3}{4} + \frac{-11x + 41/4}{4x^2 + 6x - 7}$$

so $g(x) \longrightarrow -\frac{1}{2}x + \frac{3}{4}$ asymptotically.

2.4 ANALYZING DISCONTINUITIES AND ASYMPTOTIC BEHAVIOR

The graph of $y = g(x)$ has vertical asymptotes at both zeroes of the denominator:

$$x = \frac{-3 \pm \sqrt{37}}{4}$$

which can be obtained by applying the quadratic formula to the equation $4x^2 + 6x - 7 = 0$. ∎

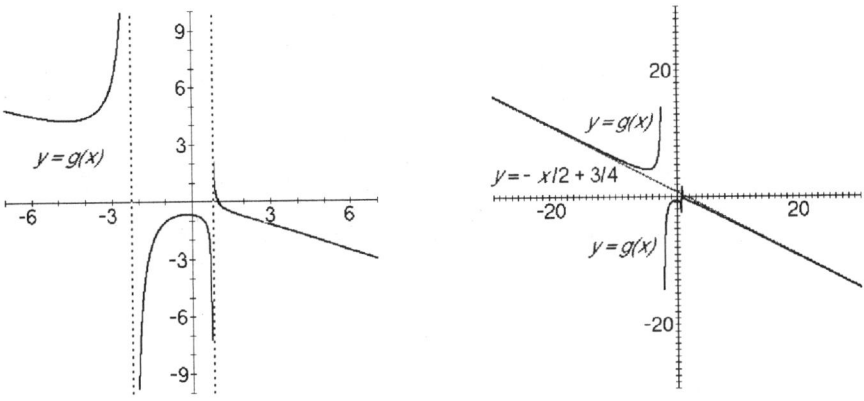

Figure 2.16 Graphs of $y = \dfrac{-2x^3 - 3x + 5}{4x^2 + 6x - 7}$ and $y = -x/2 + 3/4$.

The asymptotic behavior of a rational function can be explored nicely with machine graphics. A horizontal asymptote can be evidenced by the graph of the function resembling a horizontal line when we zoom out in the x-direction. (In other words, if we rescale the x-axis so that the function outputs for only large magnitude values of x are plotted, then a flat graph suggests a horizontal asymptote.) More general asymptotic behavior as in this last example can be checked by plotting both functions at a suitably large scale on both axes.

Figure 2.16 shows two machine plotted graphs of the function g from the last example. In the second plot, we have also graphed the line $y = -x/2 + 3/4$ to illustrate the asymptotic behavior. Notice the *spike* that occurs at one of the *vertical* asymptotes for $y = g(x)$ in the second plot. This is a phenomenon quite common to machine-generated graphs. If a computer graphing utility or calculator is set on *connected* mode, then it assumes that the function being plotted is *continuous*. If the asymptote is not detected (perhaps it lies between two adjacent plot values), then the machine may connect the graph across the asymptote!

EXERCISES

For each of the functions defined in exercises 1-8, determine whether the given function is continuous at $x = 0$ and if not, analyze the type of discontinuity there.

1. $x \longmapsto |x|$
2. $x \longmapsto \dfrac{\sin 2x}{5x}$
3. $x \longmapsto \sin^2 x + \cos^2 x$
4. $x \longmapsto \cot x$
5. $x \longmapsto \dfrac{1 - \cos x}{x}$
6. $x \longmapsto x \sin(1/x)$
7. $x \longmapsto \arctan(1/x)$
8. $x \longmapsto (1 + x)^{1/x}$

Each function in exercises 9-14 has a split or piece-wise defined evaluation formula. For each one, graph the function and determine whether the function is continuous at the split point, or has a jump discontinuity there.

9. $x \longmapsto g(x)$ where $g(x) = \begin{cases} |x| & \text{if } x < -1 \\ x + 2 & \text{if } x \geq -1 \end{cases}$

10. $x \longmapsto h(x)$ where $h(x) = \begin{cases} 4 - x^2 & \text{if } x \leq 2 \\ x^2 - 4 & \text{if } x > 2 \end{cases}$

11. $x \longmapsto i(x)$ where $i(x) = \begin{cases} 0 & \text{if } x \leq -3 \\ (x + 3)^3 & \text{if } x > -3 \end{cases}$

12. $x \longmapsto j(x)$ where $j(x) = \begin{cases} \dfrac{1}{x - 1} & \text{if } x < 1 \\ 2x - 1 & \text{if } x \geq 1 \end{cases}$

13. $x \longmapsto k(x)$ where $k(x) = \begin{cases} \sin x & \text{if } x < \pi \\ x - 2\pi & \text{if } x \geq \pi \end{cases}$

14. $x \longmapsto s(x)$ where $s(x) = \begin{cases} 2 & \text{if } x < -2 \\ x^2 & \text{if } x \geq -2 \end{cases}$

Each function in exercises 15-20 has a split definition containing a parameter, A. In each exercise find a value for A that makes the function continuous across the split point.

15. $x \longmapsto g(x)$ where $g(x) = \begin{cases} |x + A| & \text{if } x \leq 1 \\ 2x^2 & \text{if } x \geq 1 \end{cases}$

16. $x \longmapsto h(x)$ where $h(x) = \begin{cases} 2x^2 + x - A \text{ if } x \leq 0 \\ x^3 + A \text{ if } x > 0 \end{cases}$

17. $x \longmapsto i(x)$ where $i(x) = \begin{cases} A \text{ if } x \leq 0 \\ \arctan(1/x) \text{ if } x > 0 \end{cases}$

18. $x \longmapsto j(x)$ where $j(x) = \begin{cases} (x + A)^2 \text{ if } x \leq 0 \\ (x - 2A)^4 \text{ if } x > 0 \end{cases}$

19. $x \longmapsto k(x)$ where $k(x) = \begin{cases} Ax + 2 \text{ if } x \leq 3 \\ 1 - x \text{ if } x > 3 \end{cases}$

20. $x \longmapsto s(x)$ where $s(x) = \begin{cases} A\dfrac{(x-2)^2(x-1)}{(x-1)(x-3)} \text{ if } x < 1 \\ 2x - 5 \text{ if } x \geq 1 \end{cases}$

Find all the poles of the functions in exercises 21-26.

21. $x \longmapsto \csc x$
22. $x \longmapsto \cot x$
23. $x \longmapsto \sec x$
24. $x \longmapsto \log_{10}(x)$
25. $x \longmapsto \dfrac{1}{(x^2 - 3x + 2)}$
26. $x \longmapsto \dfrac{1}{1 + x + x^2}$

27. Suppose $f : x \longmapsto 1/x$. Then $f(-1) = -1$ and $f(1) = 1$, but there is no value $-1 < x_0 < 1$ such that $f(x_0) = 0$. Why isn't this a contradiction to the Intermediate Value Theorem (and Bolzano's Theorem)?

28. Suppose $f : x \longmapsto 1/x$. Then f is continuous on the closed interval $[1, \infty)$, but f has no definite minimum value on this interval. Why isn't this a contradiction to the Extreme Value Theorem?

29. Suppose $f : x \longmapsto 1/x$. Then f is continuous on the interval $(0, 1]$, but f has no definite maximum value on this interval. Why isn't this a contradiction to the Extreme Value Theorem?

30. A natural question to ask regarding a theorem is to what extent can the hypothesis and conclusion be interchanged. This is the so-called converse of the original theorem. Draw the graph of a function f such that $f(-2) = -1$, $f(2) = 1$, and for each value $-1 \leq y_0 \leq 1$ there is some value $-2 \leq x_0 \leq 2$ such that $f(x_0) = y_0$, but f is not continuous on the interval $[-2, 2]$. This shows that the converse of the Intermediate Value Theorem does not hold.

31. Draw the graph of a function g such that g takes on definite extreme high and low values over the interval $[-2, 2]$, but g is not continuous over this interval. This shows that the converse of the Extreme Value Theorem does not hold.

32. Suppose you start on a hike up a mountain path at 8 a.m., and reach the summit at 5 p.m. that afternoon. After spending the night camping, you set out the next morning at 8 a.m. and head back down the mountain along the same path. Suppose you reach your original starting point at 5 p.m. Is there any place on the mountain path that you pass at exactly the same time of day going down as you did going up?

Find

$$\lim_{x \to -\infty} f(x) \quad \text{and} \quad \lim_{x \to \infty} f(x)$$

for each of the functions f in exercises 33-38.

33. $f : x \longmapsto \arctan(x)$ **34.** $f : x \longmapsto \text{arccot}(x)$
35. $f : x \longmapsto \text{arccsc}(x)$ **36.** $f : x \longmapsto \text{arcsec}(x)$
37. $f : x \longmapsto 2^x$ **38.** $f : x \longmapsto \log_{10}(x)$

For each of the rational functions f in exercises 39-46: a) find all zeroes, b) find all points of discontinuity, c) analyze the behavior of the function at each point of discontinuity, and d) analyze the asymptotic behavior of the function as $x \longrightarrow \infty$ and $x \longrightarrow -\infty$. In particular, if the function has a limit at a point of discontinuity, give the coordinates of the hole in the graph. If the function has a pole, describe the behavior on each side of the vertical asymptote. If the graph has a horizontal asymptote, give its equation $y = c$. Otherwise, find the polynomial p such that $f(x) \longrightarrow p(x)$ asymptotically. Graph the function, indicating clearly its zeroes, holes, and asymptotes.

39. $f : x \longmapsto \dfrac{x^2 - 6x + 5}{x^2 - 5x + 6}$ **40.** $f : x \longmapsto \dfrac{x^2 + 2x - 3}{x^3 + 27}$

41. $f : x \longmapsto \dfrac{x^2 - 3x + 4}{x^2 + 2x - 3}$ **42.** $f : x \longmapsto \dfrac{x^3 - 4x + 7}{2x^2 - 5x + 1}$

43. $f : x \longmapsto \dfrac{x^2 + 3x + 2}{2x^2 + 3x - 2}$ **44.** $f : x \longmapsto \dfrac{7 + 4x - 2x^2}{3x^2 + 2x + 1}$

45. $f : x \longmapsto \dfrac{x^2 - 9}{x^2 - 6x + 9}$ **46.** $f : x \longmapsto \dfrac{4x^3 + 2x - 5}{8 - 9x^3}$

CHAPTER 3

The Derivative

The **derivative** of a function is one of the most useful and powerful concepts in all of calculus. To **differentiate** a function means to find its derivative, and **differential calculus** is one of the two major branches of calculus (the other being **integral calculus**). This chapter motivates the definition of a derivative through the graphical illustration of *slope* and the physical illustration of *speed*. Keeping these two interpretations of derivative in mind can be helpful in understanding and utilizing the information the derivative provides us. In the next two chapters, we will use the derivative to analyze function behavior and apply that analysis to solving problems.

Linear functions—old friends with new importance

Linear functions occupy much of the study of algebra, but they gain a new importance in the study of calculus. A linear function has a straight line for a graph, and the slope m of that line tells us how fast (and in what direction) the y-values (outputs) change as the x-values (inputs) increase. Derivatives allow us to measure the slope of more general function graphs. While a linear function has a single slope value associated with its graph, the slope of a general function graph will vary from point to point. We will start the chapter by reviewing some of the key properties of linear functions. Then we'll extend these ideas to include *locally linear* functions, whose graphs may be made up of pieces of straight lines of different slopes. Remarkably, many functions are approximately locally linear, and it is that fact that allows the derivative to be used so widely as a tool.

The slope of a line can be thought of as the rate of change of output relative to change in input (rise/run), and derivatives can be interpreted in a similar way. After examining the physical idea of speed using the example of a car's motion, we'll develop a formal definition of derivative, and then turn to ways of computing and estimating derivative values. Finally, we'll establish some general formulas and rules that can make finding derivatives a mechanical process for many functions.

3.1 LINEAR FUNCTIONS

Let's start by reviewing some of the important aspects of linear functions.

The **general form** of the equation for a line is

$$Ax + By + C = 0$$

where A, B, and C are real numbers.

If $B \neq 0$, then we can solve for y as a function of x to get the very familiar **slope-intercept form**

$$y = mx + b$$

where the slope $m = -A/B$. Any function f whose evaluation formula can be written as

$$f(x) = mx + b$$

for all x is called a **linear function**.

If $B = 0$ in the standard form of the equation of the line, then the equation can be rewritten in the form $x = a$, and the graph is a vertical line. A vertical line does not represent the graph of y as a function of x, and has *undefined slope*. On the other hand, a horizontal line with an equation of the form

$$y = b$$

represents a constant function, and the line has slope $m = 0$.

If we have *any two distinct* ordered pairs, (x_1, y_1) and (x_2, y_2) from the graph of a linear function, we can compute the slope by using the **two-point formula**

$$m = \frac{y_2 - y_1}{x_2 - x_1}.$$

If we call the linear function f, then we can write this as

$$m = \frac{f(x_2) - f(x_1)}{x_2 - x_1}.$$

The slope of a linear function measures the *rise/run* of the function graph—as we move one unit to the right in the x-direction, the graph will rise m units in the y-direction (a negative *rise* represents a *fall*). Hence, the value of the slope of a linear function tells us instantly whether the function outputs $f(x)$ are strictly increasing (positive slope), strictly decreasing (negative slope), or constant (zero slope) as we increase the inputs x (Figure 3.1).

3.1 LINEAR FUNCTIONS

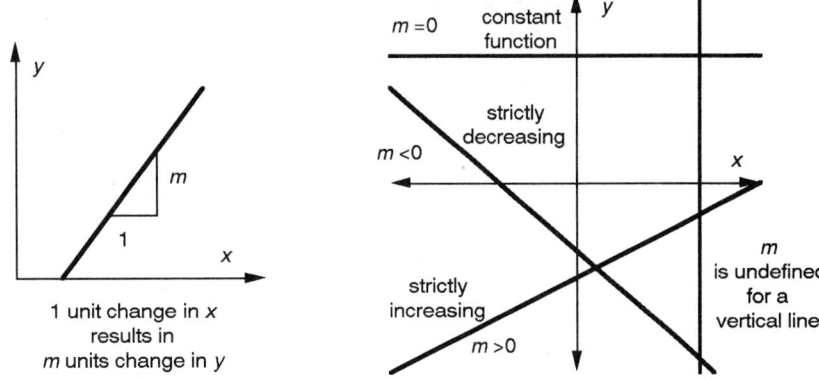

Figure 3.1 Interpreting slope values.

Point-slope and Taylor forms

When we know the slope m and a particular point (x_0, y_0) on the graph of a linear function, a very convenient formula for recovering an equation for the line is the **point-slope form**:

$$y - y_0 = m(x - x_0).$$

Using function notation, we can write $f(x)$ for y and $f(x_0)$ for y_0:

$$f(x) - f(x_0) = m(x - x_0).$$

If we add $f(x_0)$ to both sides of this formula, we obtain what is called the **Taylor form** of a linear function at $x = x_0$:

$$f(x) = m(x - x_0) + f(x_0).$$

The Taylor form is named after the English mathematician Brook Taylor (1685-1731).

EXAMPLE 1 A line has an equation $2x - 3y + 6 = 0$ in standard form. Find its equation in slope-intercept form, its point-slope form at the point $(3, 4)$, and the Taylor form at $x_0 = -4$.

Solution Solving the equation for y gives us the slope-intercept form

$$y = \frac{2}{3}x + 2.$$

Once we know the slope $m = 2/3$, we can write down the point-slope form at $(3, 4)$.

$$y - 4 = \frac{2}{3}(x - 3).$$

When $x_0 = -4$, $y_0 = f(x_0) = \frac{2}{3}(-4) + 2 = -\frac{2}{3}$. Hence,

$$f(x) = \frac{2}{3}(x - (-4)) - \frac{2}{3}$$

is the Taylor form at $x_0 = -4$. ∎

Given a point on a line, the slope m allows us to obtain other points on the line by stepping off m units in the y-direction for every 1 unit we step right in the x-direction. The Taylor form can be thought of as emphasizing a specific starting point $(x_0, f(x_0))$ from which we measure. The slope-intercept form

$$y = mx + b$$

is just the Taylor form at the point $x_0 = 0$.

EXAMPLE 2 The graph of a certain linear function f has slope 2 and passes through the point $(-1, 7)$. Find its Taylor form at $x_0 = -1$.

Solution Using $x_0 = -1$, $f(x_0) = 7$, and $m = 2$, we have $f(x) = 2(x - (-1)) + 7$. ∎

EXAMPLE 3 The function g is such that $g(0) = 3$ and $g(2) = 2$. If we assume that the function is linear, what is its Taylor form at $x_0 = 5$?

Solution If we let $x_1 = 0$ and $x_2 = 2$ as given, then

$$m = \frac{g(2) - g(0)}{2 - 0} = \frac{2 - 3}{2} = -\frac{1}{2}.$$

We have $x_0 = 5$, so the only missing information is the value of $g(5)$. Since g is assumed to be linear, we must have

$$g(5) - g(0) = -\frac{1}{2}(5 - 0) = -\frac{5}{2}$$

and substituting $g(0) = 3$ shows that $g(5) = 1/2$. Now we can write the Taylor form of g at $x_0 = 5$ as

$$g(x) = -\frac{1}{2}(x - 5) + \frac{1}{2}.$$

∎

3.1 LINEAR FUNCTIONS

We can check the Taylor form by substituting 0 and 2 for x and verifying that the outputs are 3 and 2 respectively.

Not just linear functions have Taylor forms. Any polynomial function can be put into Taylor form at $x = x_0$ by rewriting it as a sum of terms involving powers of $(x - x_0)$ instead of x. The following example illustrates how to rewrite a polynomial in Taylor form.

EXAMPLE 4 Find the Taylor form of the quadratic polynomial $5x^2 + 3x - 7$ at $x = 2$.

Solution We need to find a polynomial of the form

$$a(x-2)^2 + b(x-2) + c$$

that is equal to the given quadratic polynomial. Perhaps the simplest algebraic method for accomplishing this is to substitute $((x-2)+2)$ for x, and expand the resulting expression in terms of $(x-2)$:

$$5((x-2)+2)^2 + 3((x-2)+2) - 7$$
$$= (5(x-2)^2 + 20(x-2) + 20) + (3(x-2) + 6) - 7.$$
$$= 5(x-2)^2 + 23(x-2) + 19.$$

As a check on this, substitute the values $x = 0$, $x = 1$, and $x = 2$ into both polynomials and observe that we get the same result in each case. ∎

The important idea—change in output is proportional to change in input

For a linear function, the change in output is always *proportional* to the change in input. In other words, there is a constant m such that

$$f(x_2) - f(x_1) = m(x_2 - x_1)$$

for *any* two inputs x_1 and x_2.

On the surface, it appears that all we have done here is rewrite the slope formula in a way that holds true even when $x_1 = x_2$. But the real importance lies in the notion of *proportional change*. Here *proportional* means that if we multiply the change in input by any factor, then the change in output will be multiplied by the same factor. We could call m the *constant of proportionality* relating the change in output to the change in input. For example, for any linear function with any slope m, *twice as big a change in input produces twice as big a change in output*.

EXAMPLE 5   Suppose the function g is linear, and $g(3) - g(-2) = 7$. Find $g(5) - g(3)$.

Solution   We are told that a change in input $5(= 3 - (-2))$ produces a change in output 7. For a linear function, this means that for every 1 unit of change in input, we'll have a 7/5 units change in output. So

$$g(5) - g(3) = \frac{7}{5}(5 - 3) = \frac{14}{5}.$$

(Note that two-fifths the change in input produces two-fifths the change in output.)   ∎

Let's be quite careful here. Note that we're talking about the *change* in inputs and outputs rather than the *values* of inputs and outputs themselves. This frees the notion of linearity from the units of measurement we use for the dependent and independent variables.

If one says today's temperature is twice as high as yesterday's, this will represent different physical situations depending on whether the Fahrenheit or Celsius temperature scales is used. For example, if yesterday's temperature was at the freezing point (for water), and we are using the Fahrenheit scale ($32°F$), then it is pleasantly warm today ($64°F$). If we are using the Celsius scale ($0°C$), then water will still freeze today! But, if one says that the temperature *rise* from Tuesday to Wednesday was twice the *rise* from Monday to Tuesday, then the physical situation is the same, regardless of the temperature scale.

EXAMPLE 6   Show that $x \longmapsto x^2$ is not linear.

Solution   For $x \longmapsto x^2$ to be linear would require the same change in output for every 1 unit change in input. For a change in input from $x = 0$ to $x = 1$, we have a change in output of $1^2 - 0^2 = 1$. However, for a change in input from $x = 2$ to $x = 3$, we have a change in output of $3^2 - 2^2 = 5$, so $x \longmapsto x^2$ is definitely not linear. The nonlinearity of $x \longmapsto x^2$ is also evident from the graph of $y = x^2$.   ∎

The difference quotient

The slope of a line gives us the ratio of change in output to change in input for a linear function. We can talk about this ratio for any function.

3.1 LINEAR FUNCTIONS

Definition 1 The **difference quotient** for a function f between two distinct inputs x_1 and x_2 is defined to be
$$\frac{\Delta f}{\Delta x} = \frac{f(x_2) - f(x_1)}{x_2 - x_1}.$$

If we use y to denote $f(x)$, with $y_1 = f(x_1)$ and $y_2 = f(x_2)$ then we may also write this difference quotient as
$$\frac{\Delta y}{\Delta x} = \frac{y_2 - y_1}{x_2 - x_1}.$$

The symbol Δ is the capital Greek letter "delta". We read Δy as the *change in y* and Δx as the *change in x*. These symbols are sometimes called the **increments** in y and x, respectively.

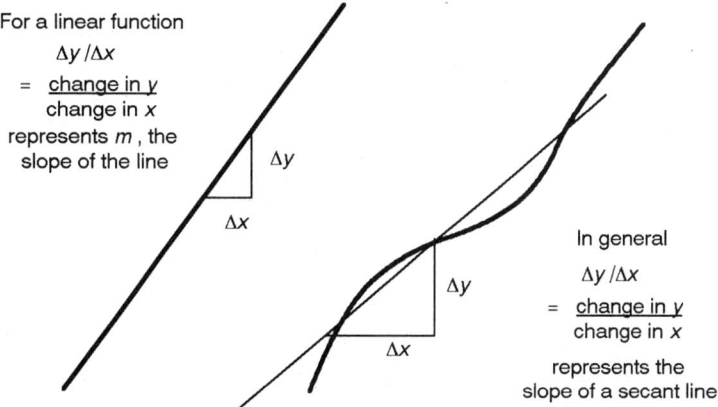

Figure 3.2 Graphical interpretation of the difference quotient.

Figure 3.2 shows the graphical interpretation of the difference quotient. For a linear function, the difference quotient $\Delta y / \Delta x$ simply gives us the slope m, no matter what two different inputs x_1 and x_2 we choose. For a nonlinear function, the value of the difference quotient will depend very much on the two particular inputs we choose. We can interpret the value of the difference quotient of a general function as the slope of a **secant line** to the graph $y = f(x)$ through the points $(x_1, f(x_1))$ and $(x_2, f(x_2))$.

EXAMPLE 7 Suppose $f : x \longmapsto x^2$. Calculate $\dfrac{\Delta f}{\Delta x}$ between $x_1 = 2$ and $x_2 = 4$, and between $x_1 = 3$ and $x_2 = 5$.

Solution Between $x_1 = 2$ and $x_2 = 4$, we have

$$\frac{\Delta f}{\Delta x} = \frac{f(x_2) - f(x_1)}{x_2 - x_1} = \frac{4^2 - 2^2}{4 - 2} = \frac{12}{2} = 6.$$

Between $x_1 = 3$ and $x_2 = 5$, we have

$$\frac{\Delta f}{\Delta x} = \frac{f(x_2) - f(x_1)}{x_2 - x_1} = \frac{5^2 - 3^2}{5 - 3} = \frac{16}{2} = 8.$$

∎

EXAMPLE 8 Suppose $y = \sin x$. Calculate $\dfrac{\Delta y}{\Delta x}$ between $x_1 = \pi/3$ and $x_2 = 0$.

Solution

$$\frac{\Delta y}{\Delta x} = \frac{y_2 - y_1}{x_2 - x_1} = \frac{\sin 0 - \sin \pi/3}{0 - \pi/3} = \frac{-1/2}{-\pi/3} = \frac{3}{2\pi}.$$

Note that interchanging the values of x_1 and x_2 does not affect the value of the difference quotient since both the numerator and the denominator change sign. ∎

Local linearity and piece-wise linear functions

If you have a collection of data points on graph paper, no two of which are in the same vertical line, and connect the points from left to right with straight line segments, the result is called a *line graph* and the function it represents is called a **polygonal** or **continuous piece-wise linear function**.

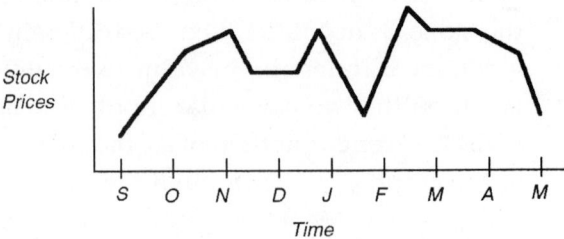

Figure 3.3 A line graph for a continuous piece-wise linear function.

These graphs are used quite often in displaying numerical data gathered over time. Figure 3.3 illustrates a graph representing stock prices over a nine-month period. In essence, line graphs *interpolate* the values between the given data points, and we might use such a graph to approximate output values for inputs falling between two known points. Line graphs are often used to spot trends or patterns in the output values.

If we focused our attention on a small piece of the graph between two consecutive corners, we would be unable to distinguish it from the graph of an ordinary linear function (hence, the name *piece-wise linear*). Indeed, we could even measure the slope of the graph and find an equation for the line that coincides with it. However, a true linear function has a graph with a *global slope* value equal to the difference quotient for any two distinct points on the graph. In contrast, the slope of a piece-wise linear function depends on the particular piece we examine. When we pass a corner, the slope suddenly changes. This leads us naturally to consider the idea of *local slope* and **locally linear** functions.

Definition 2

A function f is **linear over an interval** if the difference quotient

$$\frac{\Delta f}{\Delta x} = \frac{f(x_2) - f(x_1)}{x_2 - x_1}$$

is constant over that interval. That is, $\frac{\Delta f}{\Delta x}$ yields the same value m for any two distinct inputs x_1 and x_2 *in that interval*. This constant of proportionality m is called the **local slope** over the interval. A function is said to be **linear at the input** x_0 if f is linear over some open interval containing x_0. In this case, the slope over this interval is also called the **slope at the input** x_0.

An open interval containing a point x_0 is sometimes called a **neighborhood** of x_0. The importance of a neighborhood being *open* is that this always allows some room (however small) on both sides of the point for lots of "neighbors" —other points that belong to the same neighborhood. For example, if you pick any point x_0 in the open interval $(0, 1)$, there are always infinitely many points to both the left and right of x_0 that also belong to $(0, 1)$. Notice that we cannot say the same thing for either endpoint of a closed interval like $[0, 1]$. Using the language of neighborhoods, we can give a graphical characterization of a locally linear function.

A function f is linear at x_0 if the graph of f is straight in a neighborhood of x_0.

Piece-wise linear functions are the simplest functions that are linear at most of their inputs. They are one step more complicated than globally linear functions and they serve as good illustrations for many of the concepts that we will encounter later in the study of calculus.

Figure 3.4 shows another example of a piece-wise linear function.

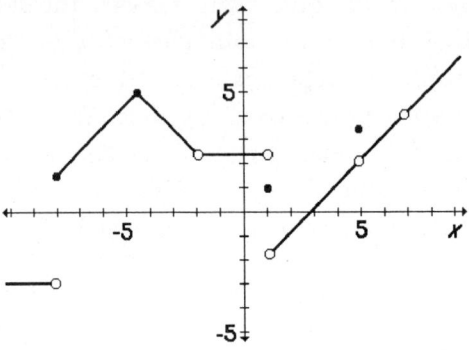

Figure 3.4 Graph of a piece-wise linear function f.

The small open circles represent points not included in the graph. The points at which the function is *not locally linear* are called **vertices**.

EXAMPLE 9 Describe the behavior of the function f illustrated in Figure 3.4.

Solution For $x < -8$, the graph is horizontal and has slope 0.

The graph jumps vertically at $x = -8$ and has *positive* slope for $-8 < x < -4.5$.

At $x = -4.5$, the graph has a *corner* and the slope changes to a *negative* value for $-4.5 < x < -2$.

At $x = -2$, there is a hole in the graph, and the slope is 0 for $-2 < x < 1$.

At $x = 1$, the graph skips to the function value $f(1) = 1$, then jumps down further to resume a positive slope for $1 < x < 5$.

At $x = 5$, the graph skips up momentarily to the function value $f(5) = 3.5$, then continues with the same positive slope for $5 < x < 7$.

At $x = 7$, the graph has another hole, but then continues with a positive slope for $x > 7$. ∎

3.1 LINEAR FUNCTIONS

A vertex corresponds to a hole, a skip, a jump, or a corner in the graph of a piece-wise linear function. A piece-wise linear function is *continuous* at a point a if the graph is unbroken at that point. The vertices are points at which a break could occur. In Figure 3.4, the only vertex at which this piece-wise linear function is continuous is the one at $x = -4.5$. There need not be a function output value at a vertex, as is the case with $x = -2$ and $x = 5$. At $x = 1$, there is a function output, but the corresponding point on the graph matches neither endpoint of the adjacent line segments. Note that the graph has undefined slope at each of the vertex values $x = -8$, -4.5, -2, 1, 5, and 7. The slope is defined for all other x values shown.

Step functions

An important special case of a piece-wise linear function is a *step function*. Step functions are constant between their vertices (with slope 0), and get their name from their graphs, which resemble a sequence of stairsteps. Two step functions worthy of special names are the **floor** and **ceiling** functions, whose graphs are illustrated in Figure 3.5.

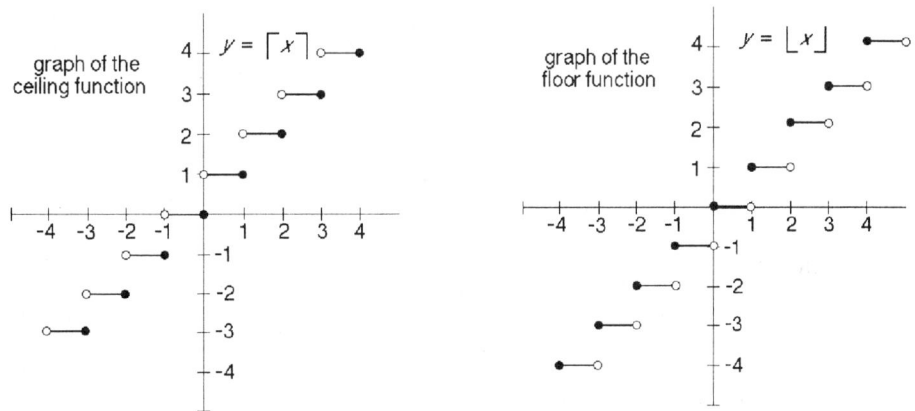

Figure 3.5 Examples of step functions.

EXAMPLE 10 The floor function is denoted

$$\lfloor \ \rfloor : \mathbb{R} \longrightarrow \mathbb{R}$$

where $\lfloor x \rfloor$ is the largest integer n such that $n \leq x$. Thus,

$$\lfloor 1.2 \rfloor = 1, \quad \lfloor 7 \rfloor = 7, \quad \lfloor -3.4 \rfloor = -4, \quad \lfloor -2 \rfloor = -2.$$

For a *positive* input x, the floor function simply truncates (drops) the fractional part of the input. Notice, however, that truncation would not produce the appropriate output for a negative input such as $x = -3.4$. The floor function is sometimes called the **greatest integer function**, since its output $\lfloor x \rfloor$ is the greatest integer less than or equal to the input x. (The older notation for greatest integer function is $[x]$.)

The ceiling function is denoted

$$\lceil \ \rceil : \mathbb{R} \longrightarrow \mathbb{R}$$

where $\lceil x \rceil$ is the smallest integer n such that $n \geq x$. Thus,

$$\lceil 1.2 \rceil = 2, \quad \lceil 7 \rceil = 7, \quad \lceil -3.4 \rceil = -3, \quad \lceil -2 \rceil = -2.$$

As you can see, the $\lfloor x \rfloor = \lceil x \rceil$ if and only if x is an integer. When x is not an integer, we have $\lceil x \rceil = \lfloor x \rfloor + 1$. ∎

The floor and ceiling functions are used in many situations where a real number result is computed, but a whole number result must be used. For example, suppose you determined that 17.54 workers are needed to complete a job by a certain deadline. This indicates that you actually need $18 = \lceil 17.54 \rceil$ workers to complete the job on schedule.

Finding local slopes and Taylor forms

At any input other than a vertex, the graph of a piece-wise linear function looks just like a straight line and has a slope. At a vertex, the slope is *undefined*. For example, step functions have slope 0 except at their vertices. Of course, the slope can change whenever a vertex is crossed.

To find the local slope of a piece-wise linear function at an input x_0 between two consecutive vertices, we first need to find one other input x_1 between the same two vertices. Then we simply find the corresponding outputs y_0 and y_1, and use the usual two-point formula to compute the slope

$$m_0 = \frac{y_1 - y_0}{x_1 - x_0}.$$

We label this slope m_0 with a subscript matching the original input x_0 because this represents the slope at this particular point. If we move to another point between two other vertices, we should expect that the two-point formula may give us a different slope there.

3.1 LINEAR FUNCTIONS

We can use this slope value to find the **Taylor form** of the piece-wise linear function at x_0, namely

$$f(x) = m_0(x - x_0) + f(x_0),$$

but now this evaluation formula is valid only for those inputs x lying between the same two vertices as x_0. In some cases, it may also be valid for one or both of the adjacent vertices.

EXAMPLE 11 Suppose f is the continuous piece-wise linear function corresponding to the table of vertex inputs and graph in Figure 3.6. Find the slope at $x_0 = 1.4$. Then use this information to find $f(1.4)$ and the Taylor form of f at $x_0 = 1.4$.

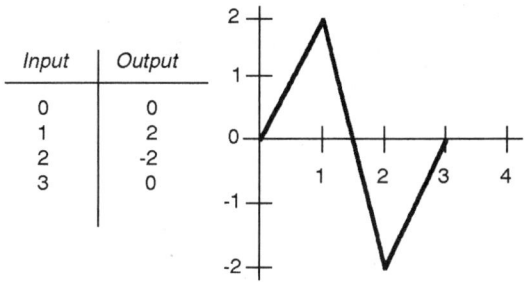

Figure 3.6 Vertices and graph of a continuous piece-wise linear function.

Solution Since 1.4 lies between vertices 1 and 2, the graph of the function in this region is the straight line segment connecting $(1, 2)$ with $(2, -2)$. The slope of the segment is given by

$$m_0 = \frac{f(2) - f(1)}{2 - 1} = \frac{-2 - 2}{1} = -4.$$

Now we can determine $f(1.4)$ by noting that

$$\frac{f(1.4) - 2}{1.4 - 1} = -4,$$

so

$$f(1.4) = (-4)(0.4) + 2 = 0.4.$$

The Taylor form of f at $x_0 = 1.4$ is

$$f(x) = (-4)(x - 1.4) + 0.4$$

and is valid for evaluating $f(x)$ whenever $1 \leq x \leq 2$. ■

Linearity under algebra and compositions

When we combine functions algebraically or through composition, we are interested in what properties the resulting new function inherits from the originals. In the case of linearity, what happens when we combine two linear functions algebraically or by composition? Let's find out directly.

Suppose f and g are both linear with

$$f(x) = m_1 x + b_1 \quad \text{and} \quad g(x) = m_2 x + b_2$$

for all real numbers x.

The formulas for $f + g$ and $f - g$ are

$$(f + g)(x) = (m_1 + m_2)x + (b_1 + b_2)$$
$$(f - g)(x) = (m_1 - m_2)x + (b_1 - b_2).$$

We can see that $f+g$ and $f-g$ are linear, and their graphs have slopes $m_1 + m_2$ and $m_1 - m_2$, respectively.

On the other hand,

$$(fg)(x) = (m_1 m_2)x^2 + (m_1 b_2 + m_2 b_1)x + b_1 b_2$$
$$\frac{f}{g}(x) = \frac{m_1 x + b_1}{m_2 x + b_2}.$$

Now we can see that fg is *not* linear unless at least one of the two functions is constant, and f/g is not linear unless g is a *nonzero* constant function. If $f : x \longmapsto mx + b$ and $g : x \longmapsto c$, then $fg(x) = cmx + cb$, and the slope is cm. If $c \neq 0$, then $(f/g)(x) = (m/c)x + (b/c)$, and the slope is m/c.

For composition, we have

$$(f \circ g)(x) = m_1 m_2 x + (m_1 b_2 + b_1),$$

which shows that $f \circ g$ is also linear and the slope is the *product* $m_1 m_2$.

If we combine piece-wise linear functions algebraically, the results are similar to combining globally linear functions. If f and g are both piece-wise linear, then $f + g$ and $f - g$ will also be piece-wise linear. More generally, if f and g are both locally linear at the same point, then their sum and difference is linear at that point. As for the composition $f \circ g$, if g is linear at x_0, and f is linear at $y_0 = g(x_0)$, then $f \circ g$ will be linear at x_0.

3.1 LINEAR FUNCTIONS

Detecting linearity in data

If we are presented with a function as a table of input and output values, we can test for linearity over an interval by comparing the value of the difference quotient for different pairs of inputs within that interval. If the function is linear, we should always get the same value for the difference quotient.

EXAMPLE 12   Consider Table 3.1 of data values obtained from a certain process. Test for linearity over the interval $(1.27320, 1.27330)$.

Input	Output
1.27319	3.81456
1.27320	3.26714
1.27321	3.30551
1.27322	3.92017
1.27323	4.14405
1.27324	4.66292
1.27325	4.69141
1.27326	4.65674
1.27327	4.61993
1.27328	4.59550
1.27329	4.58799
1.27330	4.52556

Table 3.1 Data values for a process.

Solution   We can compare the change in output to the change in input for different pairs of inputs selected from this interval. In particular, for the inputs 1.27325 and 1.27326, we have

$$\frac{\Delta y}{\Delta x} = \frac{4.65674 - 4.69141}{1.27326 - 1.27325} = \frac{-0.03467}{0.00001} = -3467.$$

Now, if we perform a similar computation for the pair of inputs 1.27324 and 1.27325, we obtain

$$\frac{\Delta y}{\Delta x} = \frac{4.69141 - 4.66292}{1.27325 - 1.27325} = \frac{0.02849}{0.00001} = 2849.$$

These two difference quotients do not match, so we can conclude that the function is *not* linear over this range of inputs. Graphing the input-output pairs from the table of values would also reveal the non-linearity. ∎

What if the data pairs from this example had lined up in a graph? Unless we knew that these were the only possible inputs from the interval $(1.27320, 1.27330)$, we could not be certain that the process was linear. What we would have at least is some supporting evidence of linearity.

In general, we can certainly use graphic and numeric information to show convincingly that a certain process *is not linear*. In contrast, graphic and numeric information can only lend supporting evidence, not proof, that a given process *is linear*.

The fact that linear functions have such a simple structure is made even more important by the large number of natural processes that appear to be *almost* linear over small input intervals. There have been countless experiments in which a researcher has taken a measurement, manipulated an independent variable by a little bit and measured again, manipulated the independent variable by the same amount again and remeasured. Finding that the difference between the first and third measurement is twice the difference between the first and second the researcher proclaims "Aha! Linearity!" Finding a process to be approximately linear means that the process can be analyzed by many standard techniques (like calculus) and it also provides clues to the underlying structure of the process itself.

EXERCISES

Exercises 1-12 refer to two linear functions f and g, where the graph of f has slope 3 and passes through the point $(-4, 1)$. The graph of g passes through the two points $(-1, -2)$ and $(1, -3)$.

1. Find the slope-intercept form of f.

2. Find the Taylor form of f at $x_0 = 2$.

3. Find the slope-intercept form of g.

4. Find the slope-intercept form of the line passing through the origin and parallel to the graph of g.

5. Find the Taylor form of g at $x_0 = -1.5$.

6. Find the Taylor form at $x_0 = 4$ for the line perpendicular to the graph of f at the point $(4, f(4))$.

7. Find the constant function whose graph passes through the intersection of the graphs of f and g.

8. Find the slope-intercept form for g^{-1}.

9. Find the Taylor form of f^{-1} where the graphs of f and f^{-1} intersect.

10. Find the slope-intercept forms for $f+g$, $f-g$, $f \circ g$. Verify that the slopes of these functions are the sum, difference, and product, respectively, of the slopes of f and g.

11. Do $f \circ g$ and $g \circ f$ have the same slope? Are their graphs the same line?

12. What change in input is required to produce a 1 unit change in output for each of the functions f, g, $f+g$, $f-g$, and $f \circ g$?

For each function in exercises 13-18, a) graph the function, b) find the slope-intercept form, c) find the Taylor form at $x_0 = 1$, d) determine the slope directly by calculating the two-point formula using $x_0 = 1$ as one input and using each of the following values as the second input:

$$x_1 = .9, \quad x_2 = .99, \quad x_3 = .999, \quad x_4 = .9999, \quad \text{and } x_5 = .99999.$$

Finally, e) determine which of the inputs $x_1, \ldots, x_5$ gives the most accurate slope value using a calculator.

13. $x \longmapsto \dfrac{(x-5)^2 - (x+3)^2}{9}$

14. $x \longmapsto 1 + 0.12345678(x-1)$

15. $x \longmapsto \dfrac{x^3 + x^2 + x + 1}{3x^2 + 3}$

16. $x \longmapsto \dfrac{2(x + 3(x + 4(x + 5)))}{29}$

17. $x \longmapsto \dfrac{x}{12345}$

18. $x \longmapsto \dfrac{5-x}{7}$

---

*Suppose the graph of the line $y = mx + b$ is at an angle θ (measured counterclockwise) to a horizontal line ($0 \leq \theta < \pi$). This is called the **angle of inclination** for the line. Exercises 19-22 refer to the angle of inclination.*

19. Find $\tan \theta$ in terms of the slope m. (Hint: Look at the triangle on the left in Figure 3.1.)

20. Use exercise 19 to find the angle of inclination of the line $y = \sqrt{3}x - 1$.

21. Use exercise 19 to find the Taylor form of a line passing through the point $(-3, 0)$ and having angle of inclination $3\pi/4$.

22. For what angle of inclination is the slope of a line undefined?

Find the Taylor forms of each of the polynomials in exercises 23-25 at $x = 1$ and $x = -2$.

23. $x^2 + x + 1$

24. $2x^3 - 3x^2 + x/3 - 7$

25. x^4

Here is a table of input-output data pairs. Assume the given inputs are the vertices of a continuous piece-wise linear function f in exercises 26-30.

Input	Output
0	5.32
1	6.11
2	2.20
3	4.07

26. Graph f over the interval $[0,3]$. Is the function linear over this interval?

27. Find the Taylor form for the function at $x_0 = 0.5$. For what interval of values is this evaluation formula valid?

28. Find the Taylor form for the function at $x_0 = 1.5$. For what interval of values is this evaluation formula valid?

29. Find the Taylor form for the function at $x_0 = 2.5$. For what interval of values is this evaluation formula valid?

30. Using the Taylor forms provided by exercises 27-29, produce a split formula valid for evaluating the function over the interval $[0,3]$ and determine $f(1.462)$.

Find all the vertices of the functions in exercises 31-38 by referring to the graphs of the two continuous piece-wise linear functions pictured below.

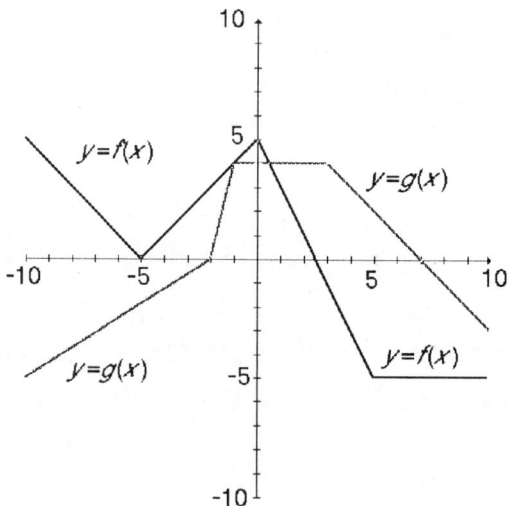

Then find the slopes (if defined) and the Taylor form of these functions at each of the following inputs: $-7.5, -4, -1.5, -0.5, 0, 0.5, 1.5, 4, 7.5$.

31. f
32. g
33. $f + g$
34. $f - g$
35. $f \circ g$
36. $g \circ f$
37. $2f$
38. $g/3$

Now graph fg and f/g for these two functions and answer the questions in exercises 39 and 40.

39. Over what intervals is fg linear?

40. Over what intervals is $\dfrac{f}{g}$ linear?

Graph each of the functions indicated in exercises 41-54 over the interval $[-10, 10]$.

41. $y = \lfloor x \rfloor$
42. $y = \lceil x \rceil$
43. $y = \lfloor x \rfloor^2$
44. $y = \lfloor x^2 \rfloor$
45. $y = (\lfloor x \rfloor + \lceil x \rceil)/2$
46. $y = \lceil x \rceil - \lfloor x \rfloor$
47. $y = \lceil \lfloor x \rfloor \rceil$
48. $y = \lfloor \lceil x \rceil \rfloor$
49. $y = \lfloor x \rfloor \lceil x \rceil$
50. $y = \lceil x \rceil - 1$
51. $y = \lceil x + 1 \rceil$
52. $y = -2\lfloor x \rfloor$
53. $y = \lfloor -2x \rfloor$
54. $y = \dfrac{1}{\lceil x \rceil}$

3.2 WHAT IS A DERIVATIVE?

When two variable quantities are connected in a functional relationship, one central question we want to ask is "How fast does the dependent variable change relative to change in the independent variable?" The *derivative* of a function provides us with a measure of this rate of change.

The word rate may bring to mind a familiar formula from algebra, namely

$$d = rt$$

where d represents *distance* covered, t represents elapsed *time*, and r represents the *rate* or speed. For example, an automobile traveling at a rate of 40 *mph* (miles per hour) for a time of 2 hours will cover a distance of 80 miles.

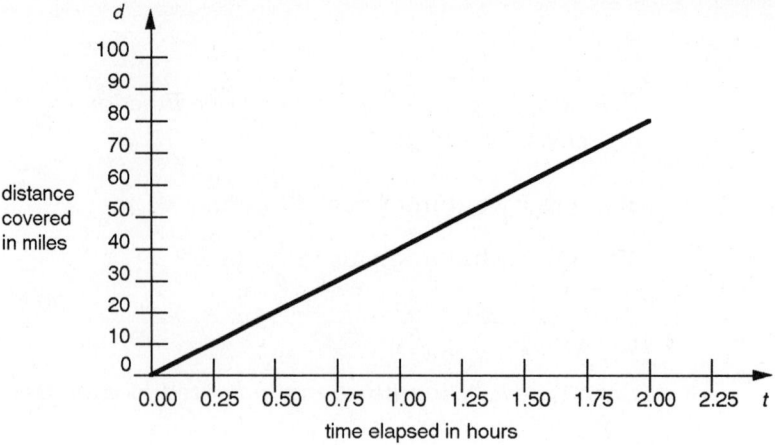

Figure 3.7 Graph of distance vs. time for a car traveling at a constant rate of 40 mph.

Graphically, we can diagram the relationship between d, t, and r with a straight line as in Figure 3.7. Here the vertical axis represents d, the distance covered in miles, and the horizontal axis represents t, the time elapsed in hours.

But what represents r, the rate? The answer is found in the slope of the line. The slope provides the relationship between d and t. We have an increase in the value of d of 40 miles for every unit increase (1 hour) of t. Graphically, r simply represents the *rise/run* of the line. With units attached, we have

$$r = \frac{d}{t} = 40 \frac{\text{miles}}{\text{hour}}.$$

Rates of change— constant, average and instantaneous

The graph in Figure 3.7, however, represents a very unrealistic situation. How often have you been in a car that was able to maintain a speed of exactly 40 mph for two solid hours? Unless the situation was one under artificial control (like an automobile alone on a deserted racetrack), attempting such a feat is not advisable. In Figure 3.8 we have a graph of a more realistic car ride, where the speed of the car fluctuates over the course of the trip. Here, again, the total distance covered by the car over 2 hours is 80 miles, but the graph of distance *versus* time is *not* a straight line.

3.2 WHAT IS A DERIVATIVE?

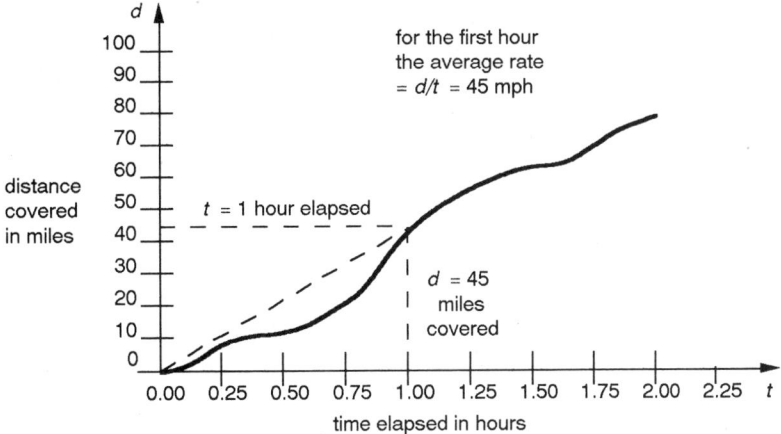

Figure 3.8 The graph of distance vs. time for a car traveling at varying speed.

We can see that the slope of the graph starts out flat, representing that the car began the trip from a stopped position. The slope becomes gradually steeper corresponding to the increase in speed of the car. Subsequent fluctuations in the slope of the graph occur at times when the car is slowing down or speeding up. The flat places in the graph correspond to the car being completely stopped, since the distance traveled is not changing at all in these instances.

The rate or speed of the first car is *constant*— the car travels at 40 mph for the entire duration of the trip. The second car travels the same distance over the same elapsed time, but now we have to speak of 40 mph as being the *average* speed for the trip. If we examine just the first hour of the trip, we note that this car has traveled 45 miles, so for the first hour, this car has an average speed of 45 miles per hour. For the second hour of the trip, this car has an average speed of 35 miles per hour. The formulas $d = rt$ and $r = d/t$ both still hold for the interpretation of r as an average rate. In other words,

$$average\ rate = \frac{total\ distance\ covered}{total\ elapsed\ time}.$$

Take care to note that the average rate will depend on the *specific* time interval under consideration.

The *instantaneous* rate refers to the speed of our car at a particular instant. Graphically, we could think of it as corresponding to the slope of our distance vs. time graph at a single point in time.

Speedometers, odometers, and clocks

Each of the three variables d, t, and r can be thought of as representing instrument readings from inside our car. An *odometer* measures d, the distance covered. A *clock* measures t, the time elapsed. The *speedometer* measures r, the speed of our car at any particular instant.

Together, an odometer and a clock can be used to calculate the *average* speed of our car over any time interval $t_0 \leq t \leq t_1$. If $d(t)$ represents the position of the car at time t, then the average speed over this time interval is given by the difference quotient

$$\frac{d(t_1) - d(t_0)}{t_1 - t_0}.$$

The difference $d(t_1) - d(t_0)$ is the distance covered between our starting position $d(t_0)$ and our final position $d(t_1)$, while the difference $t_1 - t_0$ provides the elapsed time between our initial and end times.

Graphically, we note that this average speed corresponds to the slope of the line connecting the two points $(t_0, d(t_0))$ and $(t_1, d(t_1))$. In contrast, the speedometer reading provides the instantaneous speed of our car at a single point in time, and corresponds to the slope of the graph at a single point.

Measuring instantaneous speed

How could you get a reading of the instantaneous speed of a car without seeing its speedometer? Barring a drastic change in speed while we take the measurements, it would be reasonable to approximate the instantaneous speed of the car by measuring its average speed over a very short time interval.

EXAMPLE 13 Estimate the speedometer reading if your car covered 0.1 miles in 6 seconds.

Solution Your average speed over that time is

$$r = \frac{d}{t} = \frac{0.1 \text{ miles}}{6 \text{ seconds}} = 60 \text{ miles per hour.}$$

This should be a good approximation of what your speedometer would have read during this time. ∎

3.2 WHAT IS A DERIVATIVE?

We are not directly measuring the speed at a particular time, but rather we are measuring the total distance covered over a very small interval of time and calculating the average speed from these measurements. This will be a good approximation if the speed doesn't change abruptly while we are trying to measure it, that is, if the distance covered is very nearly *linear* with respect to time. If we had an extremely accurate odometer (measuring thousandths of a mile) and a stopwatch that could measure tenths of a second, we could possibly obtain an even better estimate of instantaneous speed by using a smaller time interval to calculate average speed.

This is precisely the principle by which police radar determines the speed of a car on the highway. By calculating the time it takes for the radar beam to bounce off a car and return, the radar device can measure the distance to an approaching car. If two beams are bounced off the car separated by a very small (known) time difference, then the device can take the difference in distance measurements and calculate the average speed. Since the time interval is so tiny, the radar reading is essentially identical to the car's speedometer reading.

Visually, if we magnify this graph at a single point until the graph appears linear, the apparent slope may be a good approximation of the slope at that point.

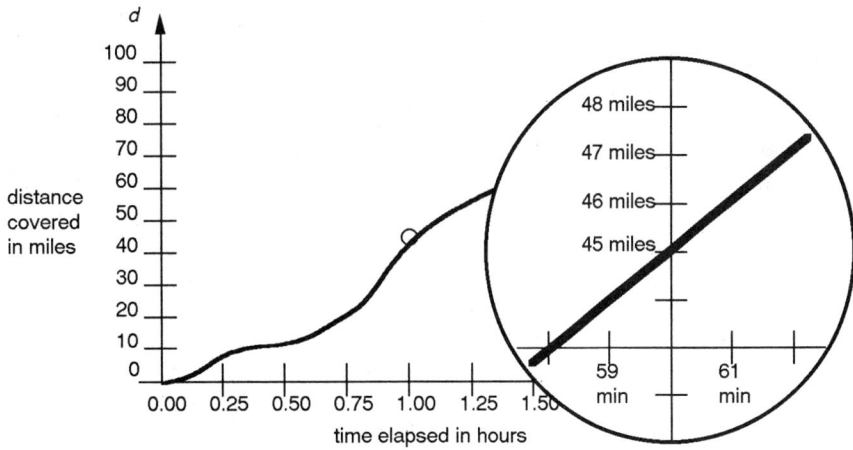

Figure 3.9 The car's speed appears to be 60 mph at t = 1 hour.

EXAMPLE 14 Figure 3.9 shows a magnified view of the graph of our second car at the time $t = 1$ hour. Approximate the slope (speed of the car) at this instant.

Solution The graph looks approximately linear in the close-up view. The rise/run between $t = 60$ minutes (1 hour) and $t = 61$ minutes appears to be 1 mile per minute or 60 miles per hour. ∎

On the other hand, if the speed changes considerably while we are measuring it, our approximation may not be a good one. In particular, if we remeasured over a smaller time interval, we might very well come up with a very different speed. Graphically, a more magnified close-up of the graph might reveal a quite different slope than we first approximated.

What we need is a precise way of calculating the instantaneous rate of change or, equivalently, the slope of the graph at a single point. In a nutshell, calculating the **derivative** gives us this information. In other words, you can think of a derivative as connecting a *speedometer* to a function process, or as providing a *slope machine* for reading the slope of the function graph at any single point. The rest of this chapter is concerned with the mathematical task of finding derivatives in general.

EXERCISES

The illustration below shows a graph of the distance travelled by a car over the first hour of a trip. Exercises 1-8 refer to this graph.

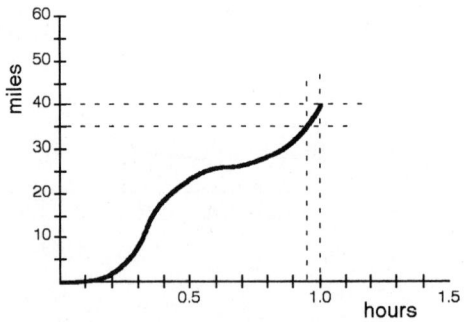

1. Find the average speed of the car for the entire trip.

2. Find the average speed of the car for the time interval represented by the two vertical dashed lines.

3. Is the car speeding up or slowing down over the first half-hour of the trip?

4. Is the car speeding up or slowing down over the second half-hour of the trip?

5. Assuming that the car maintains its speed beyond the last recorded point, estimate graphically when it will cross the 50 mile mark.

6. Assuming that the car maintains its speed beyond the last recorded point, estimate graphically where the car will be at $t = 1.5$ hours.

7. Estimate the instantaneous speed of the car at $t = 0.5$ hours and at $t = 1.0$ hours.

8. Estimate graphically when the instantaneous speed of the car was the same as the average speed for the first hour of the trip.

The illustration below also shows the graph of the distanced travelled by a car over time. Use this graph to answer exercises 9-15.

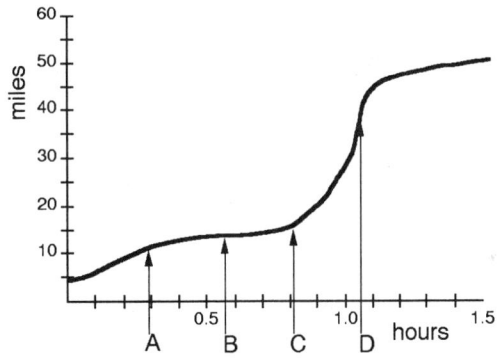

9. At which of the times $A, B, C,$ or D is the car travelling the fastest?

10. At which of the times $A, B, C,$ or D is the car stopped?

11. At which of the times $A, B, C,$ or D is the car gaining speed the fastest?

12. At which of the times $A, B, C,$ or D is the car losing speed the fastest?

13. Find the average speed for the entire trip, the average speed between times A and D, the average speed between B and D, and the average speed between C and D.

14. Estimate the speedometer reading at each of the times $A, B, C,$ and D.

15. At which of the times $A, B, C,$ or D is the car's instantaneous speed the closest to its average speed for the entire trip?

Below is an illustration of an aquarium along with a graph of its water level as a function of time. When the faucet is on, the water level rises at a steady rate. Similarly, when the plug is pulled out, the water level falls at a steady rate (but slower than the faucet's rate). At various times some events happen that affect the water level and/or the rate at which the water level changes. In exercises 16-25 you are asked to identify at exactly what time the given event occurred.

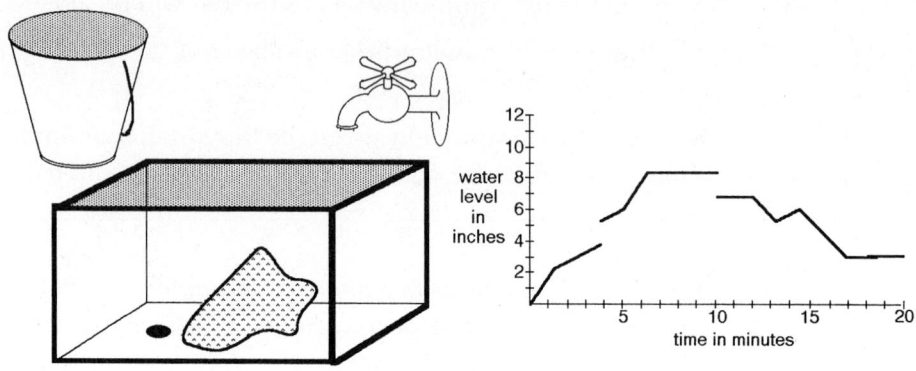

16. The plug is pulled out with the faucet turned off.

17. A large rock is pulled out of the aquarium.

18. The plug is pulled out with the faucet turned on.

19. The plug is put in with the faucet turned off.

20. The plug is put in with the faucet turned on.

21. The faucet is turned on with the plug in.

22. The faucet is turned on with the plug out.

23. A bucket of water is dumped into the aquarium all at once.

24. The faucet is turned off with the plug in.

25. The faucet is turned off with the plug out.

---

26. Now, assume that the rock is placed back in the aquarium at $t = 20$ minutes and the faucet is turned back on. Suppose that the aquarium is 12 inches deep. When will the aquarium overflow?

3.3 COMPUTING AND ESTIMATING DERIVATIVES

Our instantaneous speed measurement problem corresponds to finding the slope of a distance vs. time graph at a single point. We can generalize this problem to finding the slope of any function graph *at a single point*. From algebra, we are already used to measuring the slope of a straight line by taking any two points on the line and dividing the vertical change (the *rise*) by the horizontal change (the *run*). We have seen earlier that we can extend this to finding the slope of a locally linear function in an interval, provided we choose our two points in the interval of linearity.

For general function graphs, our strategy to approximating the instantaneous speed of the car translates to approximating the slope at a single point by picking another point very close to the given point and computing the difference quotient (as if the function were actually linear). As long as the function is approximately linear at this point, and we pick another point close enough, the difference quotient should provide a good estimate of the slope.

Definition of the derivative

If there was a way to refine our measurement techniques sufficiently so that we could measure over arbitrarily small time intervals, and if the measured average speed could be shown to stabilize toward some specific limiting value, then we would be justified in calling this value the true **instantaneous speed**.

Graphically, the average speed is represented by the slope of the secant line passing through two points. The problem of measuring instantaneous speed corresponds to finding a limiting value for these slopes as one point is chosen arbitrarily close to the other. Stated in this way, we can formulate the problem for functions in general: For a function f and a particular input x_0, if the difference quotients

$$\frac{\Delta f}{\Delta x} = \frac{f(x) - f(x_0)}{x - x_0}$$

appear to stabilize to some limiting value for inputs x chosen arbitrarily close to x_0, then we would be justified in calling that value the slope of f at x_0. This is precisely the motivation for the definition of derivative.

Definition 3 | **The derivative of a function at a point.**
Suppose $f : D \longrightarrow \mathbb{R}$ with $f : x \longmapsto f(x)$.
If $x_0 \in D$, then we say f is **differentiable at** x_0 provided that

$$\lim_{x \to x_0} \frac{f(x) - f(x_0)}{x - x_0}$$

exists. If f is differentiable at x_0, we call the value of this limit the **derivative of** f (with respect to the independent variable x) **at** x_0 and denote it by $f'(x_0)$.

Let's analyze this definition. The quantity

$$\frac{f(x) - f(x_0)}{x - x_0}$$

is simply the difference quotient for the function f over the two inputs x and x_0. So it represents the *ratio of the change in output to the change in inputs*. Graphically, it is the slope of the (secant) line passing through the points $(x_0, f(x_0))$ and $(x, f(x))$.

Now, if the sequence of slope values generated by these secant lines approach a single value (Figure 3.10), we call that value the derivative of f at x_0 or $f'(x_0)$ (read usually as "f prime of x_0"), and we can think of it as the slope of the graph of $y = f(x)$ at the single point $(x_0, f(x_0))$.

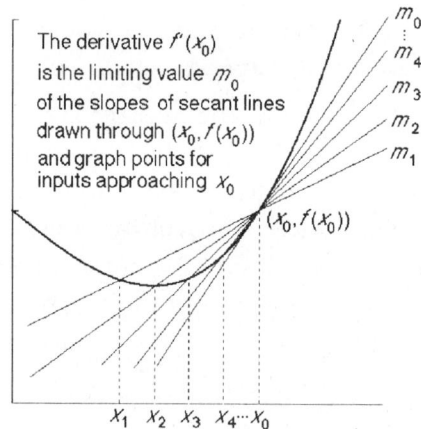

Figure 3.10 The derivative as limiting value of secant line slopes.

3.3 COMPUTING AND ESTIMATING DERIVATIVES

The derivative represents the instantaneous rate of change of output relative to change in input. Since Δx represents the difference $x - x_0$ in the definition of derivative, we must have $\Delta x \longrightarrow 0$ as $x \longrightarrow x_0$. We could write

$$x = x_0 + (x - x_0) = x_0 + \Delta x.$$

This allows us to rewrite the definition of the derivative as

$$f'(x_0) = \lim_{\Delta x \to 0} \frac{f(x_0 + \Delta x) - f(x_0)}{\Delta x}.$$

Yet another way of writing this limit is by using the letter h to represent Δx. This gives us

$$f'(x_0) = \lim_{h \to 0} \frac{f(x_0 + h) - f(x_0)}{h}.$$

However we write it, the meaning is the same: the derivative of a function f at a point x_0 is a limit of difference quotients.

If y is used to denote the dependent variable (so that $y = f(x)$), sometimes $y'(x_0)$ is written instead of $f'(x_0)$, or we could write this as

$$y'(x_0) = \lim_{\Delta x \to 0} \frac{\Delta y}{\Delta x}$$

where it is understood that

$$\Delta y = f(x_0 + \Delta x) - f(x_0).$$

Examples of derivative computations

The words "find the derivative of f at x_0" and "differentiate f at x_0" are synonymous. Computing a derivative of a function at a point requires finding the limit of difference quotients indicated by the definition.

EXAMPLE 15 Find the derivative of $k : x \longmapsto x^2$ at $x_0 = 3$.

Solution If we form the difference quotient for $k : x \longmapsto x^2$ at $x_0 = 3$, we have

$$\frac{k(x) - k(3)}{x - 3} = \frac{x^2 - 3^2}{x - 3} = \frac{(x+3)(x-3)}{(x-3)} = x + 3$$

as long as $x \neq 3$. Now, as $x \to 3$, $x + 3 \to 6$, so we have $k'(3) = 6$. ∎

EXAMPLE 16 Differentiate the function $g : x \longmapsto \sin x$ at $x_0 = 0$.

Solution Taking the appropriate limit of difference quotients, we have

$$g'(0) = \lim_{x \to 0} \frac{\sin x - 0}{x - 0} = \lim_{x \to 0} \frac{\sin x}{x} = 1.$$

We have used the fact verified earlier (in Chapter 2) that $(\sin x)/x \longrightarrow 1$ as $x \to 0$. ∎

EXAMPLE 17 Find $y'(3)$ if $y = \dfrac{1}{x}$.

Solution The difference quotient can be written as

$$\frac{\Delta y}{\Delta x} = \frac{\dfrac{1}{3 + \Delta x} - \dfrac{1}{3}}{\Delta x}$$

$$= \frac{\dfrac{3}{3 \cdot (3 + \Delta x)} - \dfrac{3 + \Delta x}{3 \cdot (3 + \Delta x)}}{\Delta x}$$

$$= \frac{-\Delta x}{\Delta x \cdot 3(3 + \Delta x)}$$

$$= \frac{\Delta x}{\Delta x} \cdot \frac{-1}{3(3 + \Delta x)}.$$

Taking the limit of these difference quotients, we have

$$y'(3) = \lim_{\Delta x \to 0} \frac{\Delta y}{\Delta x} = \lim_{\Delta x \to 0} -\frac{1}{3(3 + \Delta x)} = -\frac{1}{9}.$$

∎

EXAMPLE 18 Find $y'(5)$ if $y = \sqrt{x}$.

Solution The difference quotient can be written as

$$\frac{\sqrt{5 + h} - \sqrt{5}}{h} = \frac{\sqrt{5 + h} - \sqrt{5}}{h} \cdot \frac{\sqrt{5 + h} + \sqrt{5}}{\sqrt{5 + h} + \sqrt{5}}$$

$$= \frac{(\sqrt{5 + h})^2 - (\sqrt{5})^2}{h \cdot (\sqrt{5 + h} + \sqrt{5})}$$

$$= \frac{(5 + h) - 5}{h \cdot (\sqrt{5 + h} + \sqrt{5})}$$

$$= \frac{h}{h} \cdot \frac{1}{\sqrt{5 + h} + \sqrt{5}}.$$

3.3 COMPUTING AND ESTIMATING DERIVATIVES

Taking the limit of these difference quotients, we have

$$y'(5) = \lim_{h \to 0} \frac{1}{\sqrt{5+h} + \sqrt{5}} = \frac{1}{2\sqrt{5}}.$$

∎

EXAMPLE 19 Suppose f is the absolute value function $f : x \longmapsto |x|$. Calculate $f'(2), f'(-3)$ and $f'(0)$ if they exist.

Solution As soon as x is close enough to 2 (but $x \neq 2$) so that we are sure that x is positive, then we have

$$\frac{f(x) - f(2)}{x - 2} = \frac{|x| - |2|}{x - 2} = \frac{x - 2}{x - 2} = 1.$$

Hence

$$f'(2) = \lim_{x \to 2} \frac{f(x) - f(2)}{x - 2} = 1.$$

If x is very close to -3 (but $x \neq -3$), so that we are sure that x is negative, then

$$\frac{f(x) - f(-3)}{x - (-3)} = \frac{|x| - |-3|}{x + 3} = \frac{-x - 3}{x + 3} = -1.$$

Thus,

$$f'(-3) = \lim_{x \to -3} \frac{f(x) - f(-3)}{x - 3} = -1.$$

For values of x close to zero, the value of $|x|$ will depend on whether x is positive or negative:

$$\lim_{x \to 0^+} \frac{|x| - |0|}{x - 0} = \lim_{x \to 0^+} \frac{x - 0}{x - 0} = \lim_{x \to 0^+} \frac{x}{x} = 1$$

$$\lim_{x \to 0^-} \frac{|x| - |0|}{x - 0} = \lim_{x \to 0^+} \frac{-x - 0}{x - 0} = \lim_{x \to 0^+} \frac{-x}{x} = -1.$$

Since the left-hand and right-hand limits do not match, $f'(0)$ doesn't exist and f is *not* differentiable at 0. ∎

Derivatives of locally linear functions

If a function f is locally linear at x_0 with slope m_0, then its graph coincides with a straight line over some neighborhood of x_0. As soon as we choose x close enough to x_0, we will get this slope as the value of the corresponding difference quotient. This means that the derivative value at x_0 will match up with the slope:

$$f'(x_0) = m_0.$$

The absolute value function is an example of a piece-wise linear function, and we have just seen how the derivative gave us the local slope values at $x = 2$ and $x = -3$, but the derivative did not exist at the vertex $x = 0$ where there is no slope. At any point other than a vertex, a piece-wise linear function is locally linear, and the derivative will give us the well-defined slope at that point. Since the slope can vary from point to point, we should think of it as the *local slope*.

In general, when a function has a derivative at a point, it is at least *approximately* locally linear, and we can still talk of there being a well-defined local slope at that point, namely the value of the derivative. The derivative values for a differentiable function may vary from point to point. In fact, a differentiable function could have a different slope value at every single point in its domain!

Estimating derivatives numerically

Because the derivative of a function f at a point x_0 is the limit of the difference quotients, we can often get reasonable estimates of the derivative by simply computing the difference quotient for values of x close to x_0. That is,

$$f'(x_0) \approx \frac{f(x_0 + h) - f(x_0)}{h}$$

for small values of h.

EXAMPLE 20 Approximate $g'(0)$ numerically if $g : x \longmapsto \sin x$ using $h = \pm .01$ (with x and h measured in radians).

Solution For $h = .01$, we have

$$\frac{g(.01) - g(0)}{.01 - 0} = \frac{\sin(.01) - \sin(0)}{.01} \approx \frac{0.00999983 - 0}{.01} = 0.999983.$$

3.3 COMPUTING AND ESTIMATING DERIVATIVES

For $h = -.01$, we have

$$\frac{g(0) - g(-.01)}{0 - .01} = \frac{\sin(0) - \sin(.01)}{-.01} \approx \frac{0 - (-0.00999983)}{-.01} = 0.999983.$$

On the basis of this numerical evidence, we might guess $g'(0) \approx 1$. ■

The numerical approximation of a derivative value corresponds to approximating instantaneous speed by computing average speed over a small time interval.

EXAMPLE 21 Suppose that the actual distance travelled by a car can be modelled by the function

$$d : (0, 2) \longrightarrow \mathbb{R}$$

$$d : t \longmapsto 58t + t^2$$

where t is measured in hours, and $d(t)$ is measured in miles. Estimate the instantaneous speed after exactly one hour $d'(1)$ by calculating average speed over small time intervals, and then check the reasonableness of those estimates by calculating the exact instantaneous speed.

Solution The total distance travelled at $t_0 = 1$ is $d(1) = 59$ miles. The total distance travelled 6 minutes later ($t = 1.1$ hours) is $d(1.1) = 65.01$ miles. Therefore, one estimate for its speed at $t = 1$ is

$$\frac{d(1.1) - d(1)}{1.1 - 1} = \frac{65.01 - 59}{0.1} = 60.1 \text{ mph.}$$

If we use $t = 0.9$ hours as a second sample point, our estimate is

$$\frac{d(0.9) - d(1)}{0.9 - 1} = \frac{53.01 - 59}{-0.1} = 59.9 \text{ mph.}$$

If we use $t = 1.01$ hours as a second sample point, our estimate is

$$\frac{d(1.01) - d(1)}{1.01 - 1} = 60.01 \text{ mph.}$$

If we use $t = 0.99 \, hr$ as a second sample point, our estimate is

$$\frac{d(0.99) - d(1)}{0.99 - 1} = 59.99 \text{ mph.}$$

It appears that these refinements are stabilizing toward 60 mph.

We can check this by computing the *exact* speed $d'(1)$ using the definition of derivative:

$$\frac{d(t) - d(1)}{t - 1} = \frac{(58t + t^2) - (58 \cdot 1 + 1^2)}{t - 1}$$

$$= \frac{58(t - 1) + (t^2 - 1^2)}{t - 1}$$

$$= 58 + (t + 1) \text{ mph}.$$

for $t \neq 1$. The limiting value of this difference quotient as we take t closer and closer to 1 is $58 + 1 + 1 = 60$ mph, confirming that our estimates were reasonable. ∎

Another numerical technique for estimating the value of a derivative $f'(x_0)$ is the use of the **symmetric difference quotient**:

$$f'(x_0) \approx \frac{f(x_0 + h) - f(x_0 - h)}{2h}$$

for small values of h. This is just the average of the two difference quotients

$$\frac{f(x_0 + h) - f(x_0)}{h} \quad \text{and} \quad \frac{f(x_0) - f(x_0 - h)}{h}$$

and can often give a sharper estimate than a *one-sided* difference quotient.

EXAMPLE 22 Approximate $k'(2)$ numerically if $k : x \longmapsto x^3$, using $h = .001$ in the symmetric difference quotient.

Solution We have $k'(2) \approx$

$$\frac{k(2.001) - k(1.999)}{.002} = \frac{8.012006011 - 7.988005999}{.002} = \frac{.024000002}{.002} = 12.000001.$$

Here we have obtained a very good approximation to the exact derivative value $k'(3) = 12$. ∎

There are cases, however, where the symmetric difference quotient can give a very misleading estimate of the derivative.

3.3 COMPUTING AND ESTIMATING DERIVATIVES

EXAMPLE 23 Approximate $f'(0)$ numerically if $f : x \longmapsto |x|$, using $h = .0001$ in the symmetric difference quotient.

Solution   We already know that $f'(0)$ is actually undefined, since its graph has a sharp corner at $x = 0$ and we have no local slope there. But the symmetric difference quotient estimate gives us

$$f'(0) \approx \frac{f(.0001) - f(-.0001)}{.0002} = \frac{.0001 - .0001}{.002} = 0.$$

Moreover, *any* nonzero value of h will give us the same estimate of 0 using the symmetric difference quotient. In this case, the symmetric difference quotient gives us a bad estimate of the derivative. (Indeed, the difficulty in this particular case is due to the absolute value function being an *even* function: $f(h) = f(-h)$ for any value h.)   ∎

Estimating derivatives graphically

The key property of a function that is linear at an input x_0 is that its graph is straight over some neighborhood of x_0. Most functions are not locally linear in this strict sense, but, remarkably, many functions are differentiable at most of their inputs. If we examine the graph of a function at such a point under sufficient magnification, it should look like a straight line.

Using machine graphics, if we zoom in far enough with equal scaling vertically and horizontally on the graph of a function having a derivative at that point, then the graph will appear straight. The slope of this straight line should be reasonably close to the derivative at that point.

To estimate the derivative value $f'(x_0)$ *graphically*, zoom in on the function's graph at the point $(x_0, f(x_0))$ until the graph appears straight. The slope of this line will be an approximation to the value $f'(x_0)$.

EXAMPLE 24 Estimate the local slope of the graph $y = \sin(x)$ at the input $x = 0$.

Solution   Graphing f and zooming in near input 0 we find that the graph looks like a straight line with slope 1 (Figure 3.11).

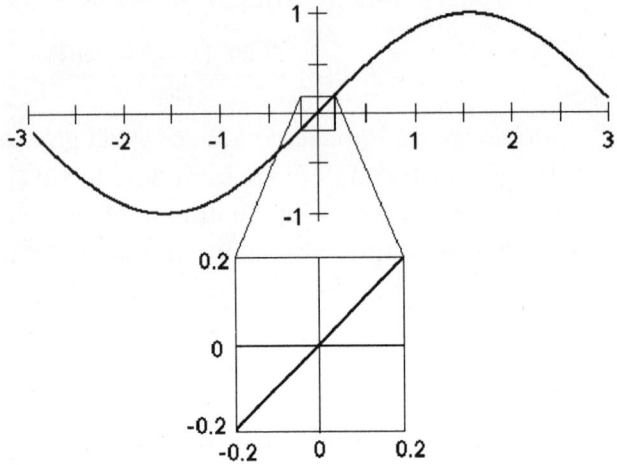

Figure 3.11 Zooming in on the graph of $f : x \longmapsto \sin(x)$ near 0.

Therefore, we would estimate the slope of the graph of $y = \sin(x)$ at $x = 0$ to be 1.   ∎

Realize that the numerical and graphical estimation methods are essentially equivalent—to compute the slope of the function's graph, we must calculate a difference quotient for two points. In a close-up window of the graph, the two points will necessarily be close together.

☞ **With the bounds on the precision of machine computation, keep in mind that two points chosen *too close* together could result in *worse*, not better accuracy in the calculation of the difference quotient.**

EXAMPLE 25 Find the local slope of the absolute value function $f : x \longmapsto |x|$ graphically at the inputs $x_0 = 2, -3,$ and 0.

Solution   Graphing $y = |x|$ we find that the graph looks like a straight line with slope 1 at $x = 2$. Similarly, at $x = -3$, we find that the graph looks like a straight line with slope -1. Therefore, the local slope of f at 2 is 1, and the local slope at -3 is -1 (see Figure 3.12).

3.3 COMPUTING AND ESTIMATING DERIVATIVES

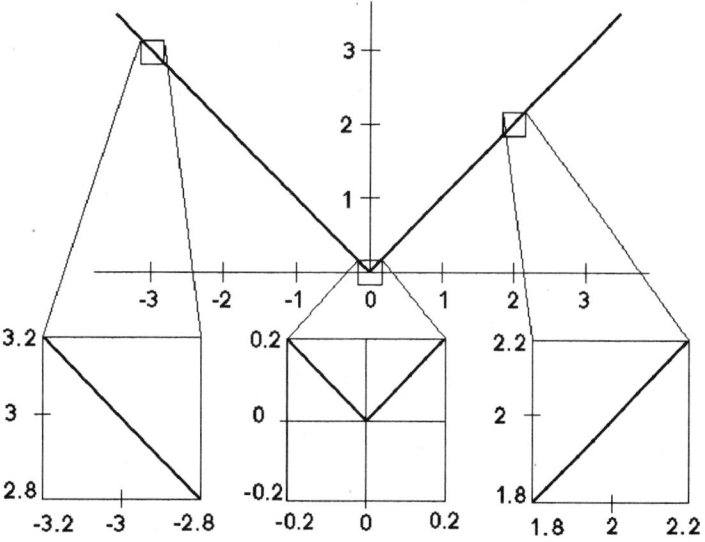

Figure 3.12 Zooming in on the graph of $f : x \longmapsto |x|$.

At $x = 0$, we find that the graph doesn't look like a straight line. Moreover, no matter how much we zoom in, there is always a sharp corner in the graph at the origin. This corresponds to the fact that the derivative $f'(0)$ is *undefined*. ∎

Left and right-hand derivatives

Just as we can speak of left and right limits or left- and right-hand continuity at a point, it is possible to speak of left- and right-hand derivatives.

Definition 4

The **left-hand derivative** of f at $x = x_0$ is denoted

$$f'_-(x_0) = \lim_{x \to x_0^-} \frac{f(x) - f(x_0)}{x - x_0}$$

provided that x_0 is in the domain of f and this left-hand limit exists. The **right-hand derivative** of f at $x = x_0$ is denoted

$$f'_+(x_0) = \lim_{x \to x_0^+} \frac{f(x) - f(x_0)}{x - x_0}$$

provided that x_0 is in the domain of f and this right-hand limit exists.

We can see from the definition that the derivative $f'(x_0)$ exists if and only if both $f'_-(x_0)$ and $f'_+(x_0)$ exist and $f'_-(x_0) = f'_+(x_0)$.

EXAMPLE 26 Find $f'_-(0)$ and $f'_+(0)$ (if they exist) for $f : x \longmapsto |x|$.

Solution   For the left-hand derivative:

$$f'_-(0) = \lim_{x \to 0^-} \frac{f(x) - f(0)}{x - 0} = \lim_{x \to 0^-} \frac{|x| - |0|}{x - 0}$$

$$= \lim_{x \to 0^-} \frac{-x}{x} \quad (|x| = -x \text{ since } x < 0)$$

$$= -1.$$

For the right-hand derivative:

$$f'_+(0) = \lim_{x \to 0^+} \frac{f(x) - f(0)}{x - 0} = \lim_{x \to 0^+} \frac{|x| - |0|}{x - 0}$$

$$= \lim_{x \to 0^+} \frac{x}{x} \quad (|x| = x \text{ since } x > 0)$$

$$= 1.$$

Both the left- and right-hand derivative values at $x = 0$ exist, but they do not match. The derivative $f'(x_0)$ is undefined. ∎

EXAMPLE 27 Find $g'_-(2)$ and $g'_+(2)$ (if they exist) for $g : x \longmapsto (x - 2)^{3/2}$.

Solution   Since $g(x) = (x - 2)^{3/2} = \sqrt{(x - 2)^3}$ is undefined when $x < 2$, the left-hand derivative $g'_-(2)$ does not exist (we cannot even approach $x = 2$ from the left-hand side). On the other hand,

$$g'_+(2) = \lim_{x \to 2^+} \frac{g(x) - g(2)}{x - 2} = \lim_{x \to 2^+} \frac{(x - 2)^{3/2} - 0}{x - 2}$$

$$= \lim_{x \to 2^+} (x - 2)^{1/2}$$

$$= 0.$$

Since there is no left-hand derivative, the derivative $g'(2)$ does not exist. ∎

EXERCISES

In exercises 1-8, calculate the derivative value indicated by computing the appropriate limit of difference quotients.

1. $f'(1.25)$ where $f : x \longmapsto 4x + 1$
2. $f'(-2)$ where $f : x \longmapsto x^2 + x + 1$
3. $f'(-1.5)$ where $f : x \longmapsto \dfrac{1}{2x + 1}$
4. $f'(3)$ where $f : x \longmapsto \sqrt{2x + 3}$
5. $f'(2.9)$ where $f : x \longmapsto -3.5$
6. $f'(1.5)$ where $f : x \longmapsto \dfrac{|-2x + 3|}{5}$

3.3 COMPUTING AND ESTIMATING DERIVATIVES

7. $f'(6)$ where $f : x \longmapsto \dfrac{x-1}{4}$ **8.** $f'(1)$ where $f : x \longmapsto x^3$

In exercises 9-16:
a) Estimate the indicated derivative numerically by using $h = \pm.001$ in the difference quotient.
b) Estimate the indicated derivative numerically by using $h = .001$ in the symmetric difference quotient.
c) Compare the accuracy of each of the three numerical estimates with the actual derivative value computed in exercises 1-8.

9. $f'(1.25)$ where $f : x \longmapsto 4x + 1$ **10.** $f'(-2)$ where $f : x \longmapsto x^2 + x + 1$
11. $f'(-1.5)$ where $f : x \longmapsto \dfrac{1}{2x+1}$ **12.** $f'(3)$ where $f : x \longmapsto \sqrt{2x+3}$
13. $f'(2.9)$ where $f : x \longmapsto -3.5$ **14.** $f'(1.5)$ where $f : x \longmapsto \dfrac{|-2x+3|}{5}$
15. $f'(6)$ where $f : x \longmapsto \dfrac{x-1}{4}$ **16.** $f'(1)$ where $f : x \longmapsto x^3$

In exercises 17-26, zoom in on the graph of the indicated function at the point $(1,1)$ to estimate its derivative at $x = 1$ graphically.

17. $x \longmapsto x^{2/3}$ **18.** $x \longmapsto x^{3/2}$
19. $x \longmapsto x^{4/3}$ **20.** $x \longmapsto x^{3/4}$
21. $x \longmapsto x^{-1/2}$ **22.** $x \longmapsto x^{-3}$
23. $x \longmapsto \arctan x$ **24.** $x \longmapsto \text{arccot } x$
25. $x \longmapsto 2^x$ **26.** $x \longmapsto \log_{10} x$

In exercises 27-36, zoom in on the graph of the indicated function to estimate its derivative graphically at $x = \pi/4$ and $x = 2\pi/3$.

27. $x \longmapsto \sin x$ **28.** $x \longmapsto \cos x$
29. $x \longmapsto \tan x$ **30.** $x \longmapsto \sec x$
31. $x \longmapsto \csc x$ **32.** $x \longmapsto \cot x$
33. $x \longmapsto \sin x^2$ **34.** $x \longmapsto \sin^2 x$
35. $x \longmapsto \sin^2 x + \cos^2 x$ **36.** $x \longmapsto \sin 6x$

Find left-hand and right-hand derivatives $f'_-(x_0)$ and $f'_+(x_0)$, if they exist, at the indicated points for the functions f in exercises 37-40.

37. $f : x \longmapsto \dfrac{|-2x+3|}{5}$; $x_0 = 1.5$ **38.** $f : x \longmapsto \lfloor x \rfloor$; $x_0 = 0$
39. $f : x \longmapsto |\sin x|$; $x_0 = \pi$ **40.** $f : x \longmapsto \sqrt{1-x^2}$; $x_0 = 1$

A falling object released from a height of 15 meters has the function

$$H : t \longmapsto (15 - 6.729 \cdot t^2)$$

describing its height as a function of elapsed time. Here t is measured in seconds, and $H(t)$ is measured in meters. Exercises 41-55 refer to this falling object.

41. Find $H(0)$. (How high is the object when it is released?)

42. Find when the object is exactly two meters off the ground.

43. Find how high the object is at $t = 0.3$ seconds.

44. Find when the object hits the ground. (For what value of t is $H(t) = 0$ meters?)

45. How far off the ground is the object exactly 0.5 seconds before it hits the ground?

46. Graph $y = H(t)$ for the time interval from the object's initial release to the time it hits the ground.

47. Find the average speed (in meters per second) of the object over the time from release until it hits the ground.

48. Find the average speed of the object between $t = 0.3$ seconds and $t = 0.4$ seconds.

49. Find the average speed of the object during the last 0.5 seconds before it hits the ground.

50. Find the average speed of the object between the time of its release and the time it is two meters off the ground.

51. Compute the instantaneous speed (in meters per second) of the object (by computing a limit of difference quotients) at $t = 0.3$ seconds.

52. Compute the instantaneous speed of the object when it is exactly two meters off the ground.

53. Compute the instantaneous speed of the object at the instant it hits the ground.

54. Compute the instantaneous speed of the object exactly 0.5 seconds before it hits the ground.

55. Compute the instantaneous speed of the object at its moment of release.

3.4 DERIVATIVE FORMULAS, NOTATIONS, AND PROPERTIES

So far, we have concentrated on calculating the derivative of a given function f at a particular point x. Now, imagine performing this process at every possible input x for which the limit

$$\lim_{h \to 0} \frac{f(x+h) - f(x)}{h}$$

exists.

What we have just described is a new *function* "derived" from our original function f. We could illustrate the process as a slope machine (or speedometer) as in Figure 3.13. Given any input x, the machine outputs the instantaneous rate of change $f'(x)$, which provides the slope of the graph of the original function at $(x, f(x))$.

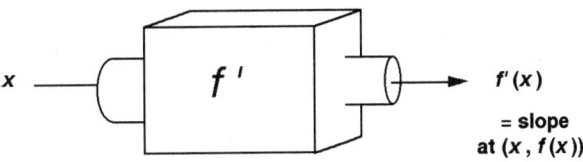

Figure 3.13 The derivative as a slope machine.

Sketching the graph of the derivative function

By applying the idea of a local slope to several points along the graph of a function, we can sketch approximately how the graph of the derivative function must look. The idea is to estimate the slope at each point on the graph of the original function, and then to plot *this value* as the output of the derivative function.

EXAMPLE 28 Sketch the graph of the derivative of the function $f : x \mapsto x^2$.

Solution Figure 3.14 shows the graph of $y = x^2$ with close-ups of the graph for the inputs $x = -3, -2, -1, 0, 1, 2,$ and 3. The slopes at each of these points, respectively, appear to be approximately $-6, -4, -2, 0, 2, 4,$ and 6.

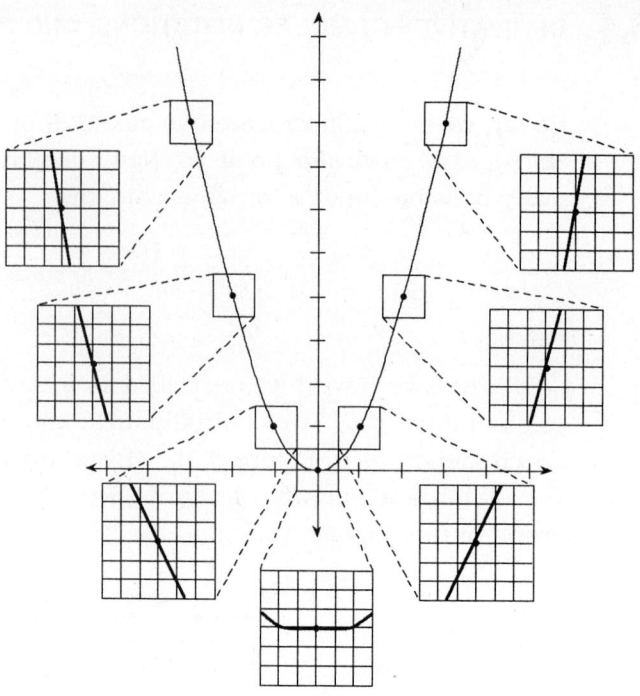

Figure 3.14 Estimating slope values from the graph $y = x^2$.

If we pair each input with the corresponding *slope value*, and then plot these ordered pairs, we can get a good sketch of how the graph of the derivative should look.

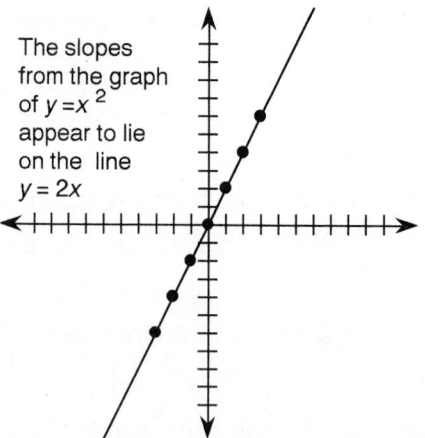

Figure 3.15 Sketch of the derivative f' of $f : x \longmapsto x^2$.

In this case, the graph of the derivative function appears to line up with the graph of $y = 2x$ (Figure 3.15). ∎

3.4 DERIVATIVE FORMULAS, NOTATIONS, AND PROPERTIES

Calculating the derivative function

To obtain a precise derivative function means we must obtain the precise derivative value at each point in the domain at which the limit of the difference quotients exists. Certainly it is neither practical nor feasible to carry out a separate limit calculation at every possible input. But, if we have a simple formula for the function f, we can often encapsulate all the derivative computations into a single formula. This shortcuts the work of re-computing the limiting difference quotient at each point.

EXAMPLE 29 Find the formula for the derivative f' of $f : x \longmapsto x^2$.

Solution The general difference quotient for f is

$$\frac{f(x+h) - f(x)}{h} = \frac{(x+h)^2 - x^2}{h}$$

$$= \frac{(x^2 + 2xh + h^2) - x^2}{h}$$

$$= \frac{2xh + h^2}{h}$$

$$= \frac{h}{h} \cdot (2x + h).$$

For $h \neq 0$, the general difference quotient has the value $2x + h$, so

$$f'(x) = \lim_{h \to 0} 2x + h = 2x.$$

∎

Our sketch from the previous example is vindicated. We now have a formula giving us a readout of the slope of the function $f : x \longmapsto x^2$ at any point x_0.

Fortunately, almost all of our standard functions have simple formulas for their derivative functions. In the remainder of this section we work out the formulas for the derivatives of a few of the most commonly used functions, and then we'll note some alternative notations and basic properties of derivatives.

To fix our notation, if $f : x \longmapsto f(x)$ is a function with domain D, then we will denote the derivative function by

$$f' : x \longmapsto f'(x).$$

The domain D' of the derivative function is

$$D' = \{x \in D : f \text{ is differentiable at } x\}$$

(The acceptable inputs for f' are those inputs for f where the derivative exists.) The output value $f'(x)$ at any $x \in D'$ is the derivative of the original function f at x.

To find the limit of a difference quotient at a general input value x, we'll find it most convenient to write it as

$$f'(x) = \lim_{h \to 0} \frac{f(x+h) - f(x)}{h}.$$

Our goal is to determine the formulas for f' when f is one of the basic functions.

Linear functions

First on our list of basic functions should be the linear ones, since the derivative is so easy to compute. The derivative in this case is independent of the input and its value is simply the *slope* of the original function.

If $f : x \longmapsto mx + b$ where m and b are constants, then the derivative of f is simply $f' : x \longmapsto m$.

EXAMPLE 30   Find a formula for f' when f is the function $f : x \longmapsto 3x - 17$.

Solution   Since f is a linear function, it has the same local slope everywhere, $m = 3$. Therefore $f' : x \longmapsto 3$. ∎

This derivative formula for a linear function $f : x \longmapsto mx + b$ follows from noting that the difference quotient $\dfrac{f(x+h) - f(x)}{h} = m$ for any $h \neq 0$.

Constant functions

Constant functions are just a special case of linear functions, but it is worthwhile drawing attention to them.

If $f : x \longmapsto c$ is a constant function, then $f' : x \longmapsto 0$.

3.4 DERIVATIVE FORMULAS, NOTATIONS, AND PROPERTIES

EXAMPLE 31 Find g' when $g : x \longmapsto 3$.

Solution Since g is a constant function, $g' : x \longmapsto 0$. This simply says that the instantaneous rate of change of a constant function is always 0. ∎

The monomial functions $x \longmapsto x^n$

The particular monomial functions $x \longmapsto x^n$ (n a nonnegative integer) can be used as building blocks for polynomial and rational functions. Once we know their derivatives, this information will help us find the derivatives of any polynomial or rational function.

We've already seen how to compute these derivatives for the cases $n = 0$ (the constant function $x \longmapsto 1$), $n = 1$ (the identity function $x \longmapsto x$), and $n = 2$ (the squaring function $x \longmapsto x^2$).

When $f : x \longmapsto x^0$, then $f' : x \longmapsto 0$.

When $f : x \longmapsto x^1$, then $f' : x \longmapsto 1$.

When $f : x \longmapsto x^2$, then $f' : x \longmapsto 2x$.

Now let's try the case $n = 3$.

EXAMPLE 32 Find the formula for f', the derivative of $f : x \longmapsto x^3$.

Solution The general difference quotient for f is

$$\frac{f(x+h) - f(x)}{h} = \frac{(x+h)^3 - (x^3)}{h}$$

$$= \frac{(x^3 + 3x^2h + 3xh^2 + h^3) - x^3}{h}$$

$$= \frac{3x^2h + 3xh^2 + h^3}{h}$$

$$= \frac{h}{h} \cdot (3x^2 + 3xh + h^2).$$

For $h \neq 0$, this general difference quotient has the value $3x^2 + 3xh + h^2$, which approaches $3x^2$ as $h \to 0$. ∎

When $f : x \longmapsto x^3$, then $f' : x \longmapsto 3x^2$.

Do you see a pattern emerging? In every case so far ($n = 0, 1, 2, 3$):

☞ **If $f : x \longmapsto x^n$, then** $f' : x \longmapsto nx^{n-1}$.

Does this pattern hold for all nonnegative integers n? The answer is yes! To see why, we need to note that when expanded

$$(x+h)^n = x^n + nhx^{n-1} + h^2 \cdot \text{(other terms)}.$$

When we form the difference quotient for $f : x \longmapsto x^n$ we obtain

$$\frac{(x+h)^n - x^n}{h} = \frac{x^n + nhx^{n-1} + h^2 \cdot \text{(other terms)} - x^n}{h}.$$

After cancelling $x^n - x^n$ in the numerator and factoring out h/h, we are left with

$$\frac{h}{h} \cdot (nx^{n-1} + h \cdot \text{(other terms)}).$$

For $h \neq 0$, the factor $h/h = 1$, and as $h \to 0$, $h \cdot$(other terms) vanish, leaving only $nx^{n-1} = f'(x)$.

EXAMPLE 33 Find f' when $f : x \longmapsto x^{1000}$.

Solution $f' : x \longmapsto 1000x^{999}$. ∎

The sine function

Before we start with calculating the derivative of the sine function through the limit of a difference quotient, let's take a look at a graph of the sine function. If we look close-up at the graph at several points, and sketch the graph of the derivative, we obtain a graph which looks remarkably like that of $y = \cos x$.

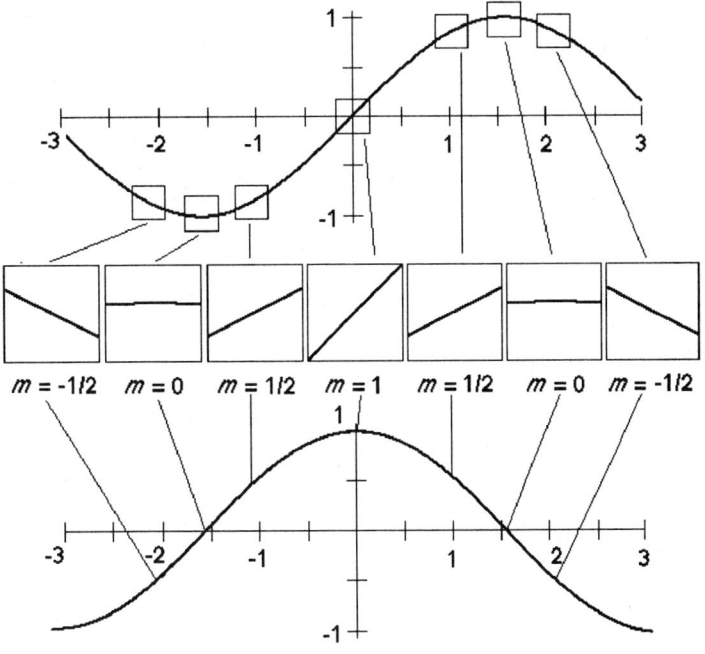

Figure 3.16 Sketching the graph of the derivative of $\sin x$.

Indeed, this is precisely the case.

☞ **If** $f : x \longmapsto \sin x$, **then** $f' : x \longmapsto \cos x$.

We can verify this by using the addition formula for sine and some algebra. The difference quotient is

$$\frac{\sin(x+h) - \sin x}{h} = \frac{\sin x \cos h + \cos x \sin h - \sin x}{h}$$

$$= \sin x \cdot \frac{\cos h - 1}{h} + \cos x \cdot \frac{\sin h}{h}.$$

In section 2.4 (and the exercises following it) we noted that

$$\lim_{\theta \to 0} \frac{\sin \theta}{\theta} = 1 \quad \text{and} \quad \lim_{\theta \to 0} \frac{\cos \theta - 1}{\theta} = 0.$$

In the difference quotient we see these same limits appearing with h in place of θ. So, as $h \to 0$,

$$\sin x \cdot \frac{\cos h - 1}{h} + \cos x \cdot \frac{\sin h}{h} \longrightarrow \sin x \cdot 0 + \cos x \cdot 1 = \cos x.$$

The cosine function

Given that the derivative of the sine function is the cosine function, one might be tempted to guess that the opposite also holds. However, if we check the graph of $y = \cos x$, then we can see that as we move to the right of 0, the slope is *negative*, while the value of $\sin x$ for the same inputs is *positive*. It turns out that simply taking the negative of the sine function does the trick.

If $f : x \longmapsto \cos x$, **then** $f' : x \longmapsto -\sin x$.

Now the difference quotient is

$$\frac{\cos(x+h) - \cos x}{h} = \frac{\cos x \cos h - \sin x \sin h - \cos x}{h}$$

$$= \cos x \cdot \frac{\cos h - 1}{h} - \sin x \cdot \frac{\sin h}{h}.$$

Using the same limit results as before, as $h \to 0$,

$$\cos x \cdot \frac{\cos h - 1}{h} - \sin x \cdot \frac{\sin h}{h} \longrightarrow \cos x \cdot 0 - \sin x \cdot 1 = -\sin x.$$

As you can guess, it's easy to mix up which of these two derivative formulas requires the negative sign. If you forget, our suggestion is to visualize what the graph of the derivative should look like and let that guide you.

Alternative notations for the derivative

Another notation for derivative that is used to emphasize the connection to difference quotients is

$$\frac{dy}{dx} = \lim_{\Delta x \to 0} \frac{\Delta y}{\Delta x}.$$

The symbol dy/dx is read simply as "dee-why-dee-ex" and is called the **Leibniz** notation for derivatives, named after Gottfried Leibniz (1646-1716). Along with Isaac Newton (1642-1727), Leibniz is generally credited as one of the inventors or discoverers of calculus.

3.4 DERIVATIVE FORMULAS, NOTATIONS, AND PROPERTIES

 The Leibniz notation dy/dx does NOT represent the ratio of two *separate* quantities dy and dx.

The *single* quantity dy/dx is the limit of the ratios $\Delta y/\Delta x$ as $\Delta x \to 0$. To be sure, a much better way of thinking of the Leibniz notation is in terms of an *operator* acting on the function represented by y. That is, think of dy/dx as

$$\frac{dy}{dx} = \frac{d}{dx}(y),$$

where d/dx is read as "the derivative with respect to x of" and operates on the function expressed by y to produce the derivative y'.

(This does not mean that you will never encounter the symbols dy and dx separately. Some people formally define $dy = y' \cdot dx$ so that the arithmetic $dy/dx = y'$ comes out correctly. Written this way, dy and dx are called **differentials**. We will encounter dy and dx later, but in a different context with a different interpretation attached to these symbols.)

The prime notation makes it easy to distinguish between a derivative *function* f' and the specific numerical *value* of the derivative at a particular point $f'(x_0)$. The symbol dy/dx by itself does not tell us at which point x_0 we are measuring the slope or rate of change. To remedy this, we can use a long vertical bar with a subscript specifying this point:

$$\left.\frac{dy}{dx}\right|_{x=x_0} = y'(x_0).$$

Now all the essential information to communicate a derivative value is present: the original dependent variable y, the derivative operator d/dx (which also tells us the independent variable), and the specific point x_0. The Leibniz notation

$$\frac{dy}{dx}$$

should be used to refer to the derivative *function*, while

$$\left.\frac{dy}{dx}\right|_{x=x_0}$$

should be used for the derivative's *value* at $x = x_0$. This notation using the vertical bar is read "the value of dy/dx where $x = x_0$."

EXAMPLE 34 If $y = -3x^5$, find $\dfrac{dy}{dx}$ and $\left.\dfrac{dy}{dx}\right|_{x=2}$

Solution   The derivative function is

$$\dfrac{dy}{dx} = -15x^4.$$

The value of this function at $x = 2$ is

$$\left.\dfrac{dy}{dx}\right|_{x=2} = -15(2)^4 = -15 \cdot 16 = -240.$$

∎

We will see that the Leibniz notation for derivatives has a few advantages. Many (but not all) properties and rules about derivatives look like simple algebraic facts when written in Leibniz notation. That can help us as a memory aid to remembering these derivative facts. If a particular function name or dependent variable label is not important, then the Leibniz notation can be used with the function formula itself.

EXAMPLE 35 Write the derivative of $x \longmapsto x^3$ at $x = 5$ in Leibniz notation.

Solution   In Leibniz notation, we could write $\left.\dfrac{d}{dx}(x^3)\right|_{x=5}$. ∎

Letters can be used for a variety of symbolic quantities—variables, constants, parameters, functions, etc. When several letters appear in a single functional expression, then the question "What is the derivative?" makes no sense unless we know exactly which letter represents the independent variable. Certainly a machine cannot be expected to guess this information correctly unless we specify the independent variable.

For example, suppose

$$y = 3x^2 + 2t^4$$

Is y a function of variable x (with t constant), a function of variable t (with x constant) or possibly a function of both variables x and t? We really have no way of knowing the right interpretation of the letters x and t without some context. The denominator of the Leibniz notation for derivative specifies the independent variable.

3.4 DERIVATIVE FORMULAS, NOTATIONS, AND PROPERTIES

EXAMPLE 36 Find $\dfrac{dy}{dx}$ and $\dfrac{dy}{dt}$ for $y = 3x^2 + 2t^4$.

Solution Letters other than the specified independent variable are assumed to represent constants.

$$\frac{dy}{dx} = \text{derivative of } y \text{ with respect to } x$$

$$= 6x \text{ (since we assume } t \text{ is constant)}$$

$$\frac{dy}{dt} = \text{derivative of } y \text{ with respect to } t$$

$$= 8t^3 \text{ (since we assume } x \text{ is constant)}$$

∎

In the prime notation, if there is any doubt as to what the independent variable is, we should include mention of it.

There are a couple of other notations for derivative that enjoy quite a bit of use, particulary in the study of differential equations. One of these is called **operator notation** and is denoted by a D prefixed to the function name. In operator notation, we can write

$$Df \quad \text{or} \quad Dy$$

to denote the derivative of a function $y = f(x)$.

We could add a subscript to D to indicate the independent variable. The notation $D_x y$ means derivative of y with respect to x, and $D_t f$ means derivative of f with respect to t.

Newtonian notation for the derivative (named after Newton) employs a dot over the function name instead of the prime symbol. In Newtonian notation, we can write

$$\dot{f} \quad \text{or} \quad \dot{y}$$

to denote the derivative of a function $y = f(x)$.

The fact that there are so many different notations for derivatives is testimony to their widespread use. The preferred notation depends on the context, but in this book, we will use the prime notation and the Leibniz notation almost exclusively.

Linearity properties of the derivative

When we add two linear functions together, the slope of the sum function is the sum of the slopes. Similarly, the slope of the difference of two linear functions is the difference in their slopes. The exact same relationships hold for the sum and difference of two differentiable functions f and g:

$$(f+g)' = f' + g' \qquad (f-g)' = f' - g'.$$

If we use u to represent $f(x)$ and v to represent $g(x)$, then in Leibniz notation the sum and difference properties are:

$$\frac{d(u+v)}{dx} = \frac{du}{dx} + \frac{dv}{dx} \qquad \frac{d(u-v)}{dx} = \frac{du}{dx} - \frac{dv}{dx}.$$

The sum and difference properties for derivatives follow from the algebra of limits applied to the difference quotients. For the sum rule, the difference quotient

$$\frac{(f+g)(x+h) - (f+g)(x)}{h} = \frac{[f(x+h) + g(x+h)] - [f(x) + g(x)]}{h}$$

$$= \frac{[f(x+h) - f(x)] + [g(x+h) - g(x)]}{h}$$

$$= \frac{f(x+h) - f(x)}{h} + \frac{g(x+h) - g(x)}{h}.$$

We recognize this as the sum of the difference quotients we use to calculate the derivatives of f and g separately. When we take the limit of both sides of the equation as $h \to 0$, we get

$$(f+g)'(x) = f'(x) + g'(x).$$

The difference property is verified in a similar way.

EXAMPLE 37 $\quad \dfrac{d}{dx}(x^3 - \sin x) = 3x^2 - \cos x.$ ∎

If you multiply a linear function by a constant, the result is a new linear function whose slope is that constant times the original slope. The same is true for the derivative. If f is a differentiable function and c is a constant, then

$$(cf)' = cf'.$$

In Leibniz notation:

$$\frac{d(cy)}{dx} = c\frac{dy}{dx}.$$

3.4 DERIVATIVE FORMULAS, NOTATIONS, AND PROPERTIES

 This simple rule for multiplication works *only* when the multiplier is a constant. The derivative of the product of functions is somewhat more complicated and will be covered in the next section.

EXAMPLE 38 Find $\dfrac{d}{dx}(5\sin x)$.

Solution $\dfrac{d}{dx}(5\sin(x)) = \dfrac{5}{-}ddx(\sin(x)) = 5\cos(x)$.

The constant multiple property holds because we can factor the constant out of the difference quotient.

$$\frac{cf(x+h) - cf(x)}{h} = c\left[\frac{f(x+h) - f(x)}{h}\right].$$

When we take the limit as $h \to 0$ we obtain

$$(cf)'(x) = cf'(x).$$

(Notice that the difference property could be obtained using the sum and constant multiple property, since $f - g = f + (-1)g$.)

Together, these properties of the derivative are called *linearity properties*. In summary,

 If f and g are differentiable, and a and b are constants, then

$$(af + bg)' = af' + bg'.$$

The linearity properties allow us easily to take the derivative of any polynomial function.

EXAMPLE 39 Find f' if $f : x \longmapsto 3x^4 - 5x^2 + \dfrac{x}{3} - \sqrt{2}$.

Solution $f' : x \longmapsto 12x^3 - 10x + \dfrac{1}{3}$.

EXAMPLE 40 Find $\dfrac{dy}{dx}$ when $y = \dfrac{4x - 2x^5}{7}$.

Solution Since $\dfrac{4x - 2x^5}{7} = \dfrac{4}{7}x - \dfrac{2}{7}x^5$, we have $\dfrac{dy}{dx} = \dfrac{4}{7} - \dfrac{10}{7}x^4$. ∎

EXERCISES

In exercises 1-10, find the derivative of the given function.

1. $x \longmapsto x^3 + x^2 + x + 1$
2. $x \longmapsto 2x^3 - 3x^2 + 6x - 12$
3. $x \longmapsto \dfrac{5x^2}{3} - \dfrac{7x^5}{4}$
4. $x \longmapsto (2x - 3)^3$ (Expand first.)
5. $x \longmapsto \dfrac{9 - x^2}{6}$
6. $x \longmapsto 2\sin x + 3\cos x$
7. $x \longmapsto \dfrac{\cos x - \sin x}{7}$
8. $x \longmapsto \dfrac{x^4}{4} + \dfrac{x^3}{3} + \dfrac{x^2}{2}$
9. $x \longmapsto (4x^3 + 2x - 5)(8 - 9x^3)$
10. $x \longmapsto \dfrac{x^{1001}}{77}$

In each of exercises 11-20, graph the function over the indicated interval, and then sketch its derivative function over the same interval.

11. $x \longmapsto x^{3/2}$; $[0, 9]$
12. $x \longmapsto x^{2/3}$; $[-8, 8]$
13. $x \longmapsto x^{-2}$; $[-3, 3]$
14. $x \longmapsto x^{1/3}$; $[-8, 8]$
15. $x \longmapsto \sin x$; $[-\pi, 3\pi]$
16. $x \longmapsto \cos x$; $[-\pi, 3\pi]$
17. $x \longmapsto \tan x$; $(-\pi, \pi)$
18. $x \longmapsto \csc x$; $(0, 2\pi)$
19. $x \longmapsto \sec x$; $(-\pi, \pi)$
20. $x \longmapsto \cot x$; $(0, 2\pi)$

In exercises 21-28, you are given the graph of a function against a grid. Assuming that the grid lines are spaced 1 unit apart both horizontally and vertically, sketch the graph of the derivative of each function over the same interval.

3.4 DERIVATIVE FORMULAS, NOTATIONS, AND PROPERTIES

21.
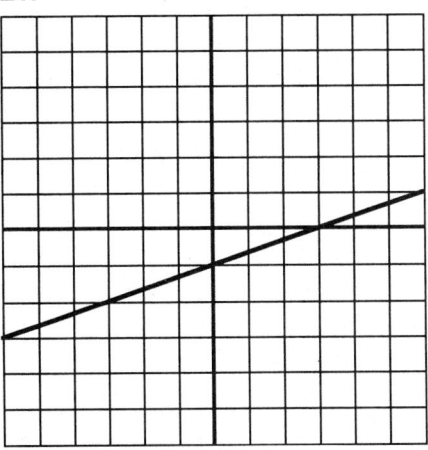

22.
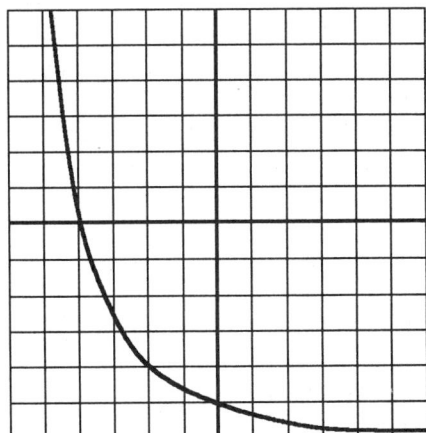

23.
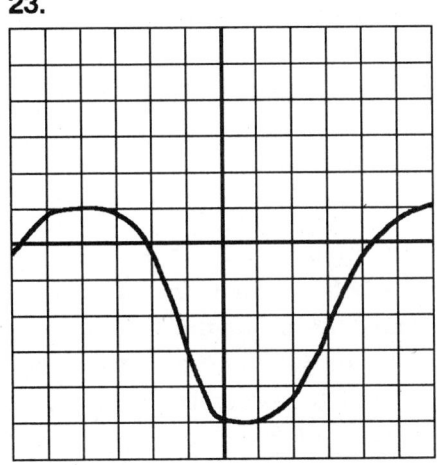

24.
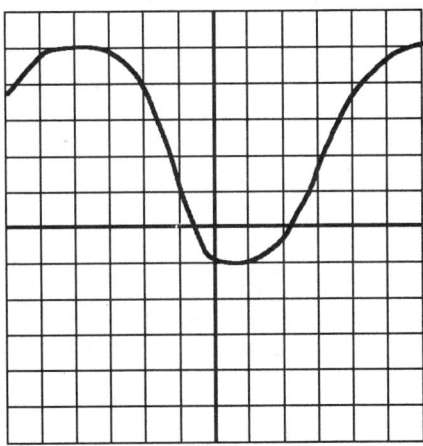

25.
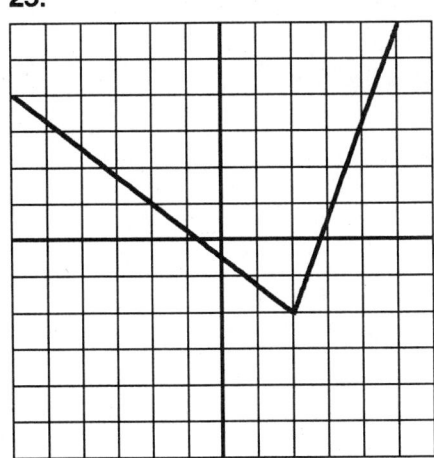

26.
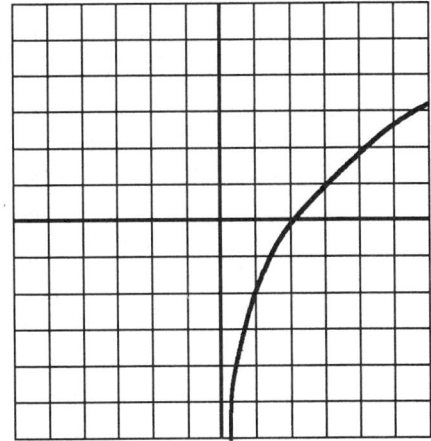

27.

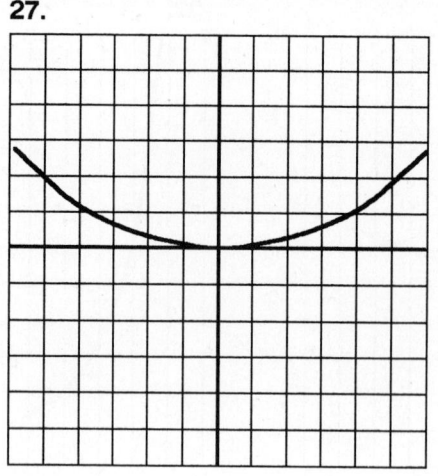

28.

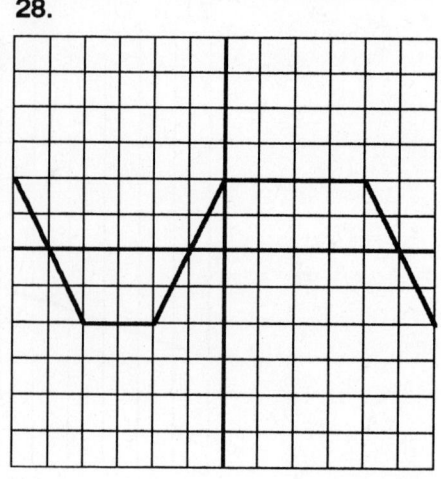

3.5 NEW DERIVATIVES FROM OLD

In this section we will derive more of the basic rules which govern how the derivatives of functions behave when the functions are combined algebraically or through composition. In the last section we saw that the derivative of a sum or difference of two functions is simply the sum or difference of the derivatives, and the derivative of a constant multiple of a function is the same constant multiple of the derivative.

For products, quotients, and compositions of functions, the rules for finding derivatives are slightly more complicated. Once we have these rules, then we will be able to compute the derivative of almost any function built up from the basic functions.

The product rule

The linearity properties for derivatives are very natural. At first glance, the derivative rule for a product is surprising. While it would be easy to remember, let's make it clear:

  WARNING: The derivative of a product is NOT the product of the derivatives.

3.5 NEW DERIVATIVES FROM OLD

This really should be expected, since the product of two linear functions does not have to be a linear function (think of $x \cdot x = x^2$). Let's look at an example to reason what the product rule for derivatives should be.

EXAMPLE 41 Suppose at a particular instant, the width of a rectangle is w and the rate at which the width is changing is w'. At this same instant, suppose that the height is h and the instantaneous rate of change of the height is h'. How fast is the *area* of the rectangle changing at this instant?

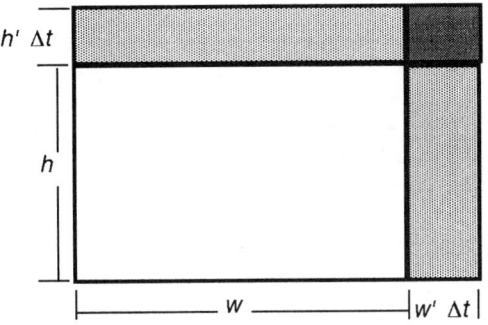

Figure 3.17 Rate of area change for a growing rectangle.

Solution The illustration in Figure 3.17 depicts what our rectangle might look like at this instant and at a short time later.

Now, for a very short time interval, we can use the derivatives to get a good prediction of the new width and height. If we use Δt to represent the elapsed time and multiply by the rates, then

$$\text{new width} \approx w + w' \Delta t \quad \text{and} \quad \text{new height} \approx h + h' \Delta t.$$

This means that the change in area ΔA is the sum of the areas of the three shaded rectangles in the illustration:

$$\Delta A \approx h \cdot w' \Delta t + w \cdot h' \Delta t + (w' \Delta t) \cdot (h' \Delta t).$$

If we form the difference quotient $\Delta A / \Delta t$, then

$$\frac{\Delta A}{\Delta t} \approx h \cdot w' + w \cdot h' + w'h' \Delta t.$$

As $\Delta t \to 0$, the better the individual approximations become. Furthermore, the third term $w'h' \Delta t \to 0$ as $\Delta t \to 0$. These approximations appear to converge on a derivative value

$$A' = w'h + wh'.$$

∎

This is precisely the result known as the product rule.

 If f and g are differentiable functions, then

$$(fg)' = f'g + fg'.$$

The product rule can be formally verified by using a little clever algebra on the difference quotient for the product:

$$\frac{(fg)(x+h) - (fg)(x)}{h} = \frac{f(x+h)g(x+h) - f(x)g(x)}{h}$$

$$= \frac{f(x+h)g(x+h) - f(x)g(x+h) + f(x)g(x+h) - f(x)g(x)}{h}$$

Notice that we've subtracted the new expression $f(x)g(x+h)$ in the numerator and then added it right back, resulting in no change in the value of the quotient. Now we can split the difference quotient up into

$$\frac{f(x+h) - f(x)}{h} \cdot g(x+h) + f(x) \cdot \frac{g(x+h) - g(x)}{h}.$$

Look! The usual difference quotient we use for computing $f'(x)$ is a factor of the first term and the difference quotient for computing $g'(x)$ is a factor of the second term. Since f and g are differentiable, as $h \to 0$, we have

$$\frac{(fg)(x+h) - (fg)(x)}{h} \longmapsto f'(x)g(x) + f(x)g'(x).$$

EXAMPLE 42 Find $\dfrac{dy}{dx}$ when $y = (5x^2 - 3)(2x^3 + 7x)$ using the product rule.

Solution We can consider y to be the product of

$$u = 5x^2 - 3 \quad \text{and} \quad v = 2x^3 + 7x.$$

Using the product rule, we have

$$\frac{dy}{dx} = \frac{d}{dx}(uv) = \frac{du}{dx} \cdot v + u \cdot \frac{dv}{dx}$$

$$= (10x)(2x^3 + 7x) + (5x^2 - 3)(6x^2 + 7)$$

$$= 20x^4 + 70x^2 + 30x^4 - 18x^2 + 35x^2 - 21$$

$$= 50x^4 + 87x^2 - 21.$$

■

3.5 NEW DERIVATIVES FROM OLD

EXAMPLE 43 Find $\dfrac{dy}{dx}$ when $y = (5x^2 - 3)(2x^3 + 7x)$ by first expanding the product and then taking the derivative.

Solution Since $y = (5x^2 - 3)(2x^3 + 7x) = 10x^5 + 29x^3 - 21x$, we have $\dfrac{dy}{dx} = 50x^4 + 87x^2 - 21$ as before. ∎

When taking the derivative of a product of polynomials, we can see that we now have an option. For products of other functions, we may have no choice but to use the product rule.

EXAMPLE 44 Find $\dfrac{df}{dx}$ when $f : x \longmapsto x^3 \sin x$.

Solution $\dfrac{df}{dx} = \dfrac{d}{dx}(x^3) \cdot \sin(x) + x^3 \cdot \dfrac{d}{dx}(\sin(x)) = 3x^2 \sin(x) + x^3 \cos(x)$. ∎

For a product of three or more functions, we need to apply the product rule more than once.

EXAMPLE 45 Find y' when $y = 4x^2 \sin x \cos x$.

Solution We can see that y is the product of three functions:

$$u = 4x^2 \qquad v = \sin x \qquad w = \cos x.$$

To use the product rule on $y = uvw$, we can think of y first as the product $u \cdot (vw)$ so that

$$\dfrac{dy}{dx} = \dfrac{d(4x^2)}{dx}(\sin x \cos x) + 4x^2 \dfrac{d(\sin x \cos x)}{dx}$$

$$= 8x \sin x \cos x + 4x^2 \cdot \left[\dfrac{d(\sin x)}{dx} \cos x + \sin x \dfrac{d(\cos x)}{dx} \right]$$

$$= 8x \sin x \cos x + 4x^2 \cos^2 x - 4x^2 \sin^2 x.$$

∎

Notice that we had to apply the product rule a second time in this example. In general, if $y = uvw$, then

$$y' = u'vw + uv'w + uvw'.$$

The constant multiple rule is consistent with the product rule. If we think of cf as the product of a constant function $x \longmapsto c$ and the function $x \longmapsto f(x)$, then applying the product rule we have

$$(cf)' = 0 \cdot f + c \cdot f' = cf'.$$

Quotient rule

Suppose k, g, and f are differentiable functions such that

$$f = kg.$$

If we take the derivative of f using the product rule, then

$$f' = k'g + kg'.$$

If we substitute $k = f/g$ and solve for k' we get the *quotient rule*:

$$k' = \left(\frac{f}{g}\right)' = \frac{f'g - fg'}{g^2}$$

which is valid for any value x at which $g(x) \neq 0$. In Leibniz notation, if $y = u/v$, where u and v are functions of x, then

$$\frac{dy}{dx} = \frac{(du/dx) \cdot v - u \cdot (dv/dx)}{v^2}.$$

The quotient rule allows us to take the derivative of any rational function (quotient of two polynomial functions).

EXAMPLE 46 Find $\dfrac{dy}{dx}$ when $y = \dfrac{2x^3 - 5x + 7}{x^2 - 3x}$.

Solution Letting $u = 2x^3 - 5x + 7$ and $v = x^2 - 3x$, we have

$$\frac{dy}{dx} = \frac{(du/dx) \cdot v - u \cdot (dv/dx)}{v^2}$$

$$= \frac{(6x^2 - 5)(x^2 - 3x) - (2x^3 - 5x + 7)(2x - 3)}{(x^2 - 3x)^2}$$

$$= \frac{6x^4 - 18x^3 - 5x^2 + 15x - 4x^4 + 6x^3 + 10x^2 - 15x - 14x + 21}{x^4 - 6x^3 + 9x^2}$$

$$= \frac{2x^4 - 12x^3 + 5x^2 - 14x + 21}{x^4 - 6x^3 + 9x^2}.$$

∎

The quotient rule also allows us to find the derivatives of the other four trigonometric functions.

3.5 NEW DERIVATIVES FROM OLD

EXAMPLE 47 Find the derivative formula for $\frac{d}{dx}(\tan(x))$.

Solution Since $\tan(x) = \frac{\sin(x)}{\cos(x)}$,

$$\frac{d}{dx}(\tan(x)) = \frac{\frac{d}{dx}(\sin(x)) \cdot \cos(x) - \sin(x)\frac{d}{dx}(\cos(x))}{\cos^2(x)}$$

$$= \frac{\cos^2(x) + \sin^2(x)}{\cos^2(x)} = \frac{1}{\cos^2(x)} = \sec^2(x).$$

In the second line we've used $\cos^2(x) + \sin^2(x) = 1$ and $\sec(x) = \frac{1}{\cos(x)}$. ∎

EXAMPLE 48 Find the derivative formula for $\frac{d}{dx}(\csc(x))$.

Solution Since $\csc(x) = \frac{1}{\sin(x)}$, we have

$$\frac{d}{dx}(\csc(x)) = \frac{0 \cdot \sin(x) - 1 \cdot \cos(x)}{\sin^2(x)}$$

$$= \frac{-\cos(x)}{\sin^2(x)} = \frac{-1}{\sin(x)}\frac{\cos(x)}{\sin(x)} = -\csc(x)\cot(x).$$

∎

In the exercises you will be asked to verify the other two trigonometric derivative formulas

$$\frac{d}{dx}(\sec(x)) = \sec(x)\tan(x) \qquad \frac{d}{dx}(\cot(x)) = -\csc^2(x).$$

We can extend our derivative formula for monomial functions to now include negative integer exponents. In other words, if n is a positive integer and

$$f : x \longmapsto x^{-n}$$

then

$$f' : x \longmapsto -nx^{-n-1}.$$

We can verify this using the quotient rule:

$$\frac{d(x^{-n})}{dx} = \frac{d(1/x^n)}{dx}$$

$$= \frac{d(1)/dx \cdot x^n - 1 \cdot d(x^n)/dx}{(x^n)^2}$$

$$= \frac{0 \cdot x^n - nx^{n-1}}{x^{2n}}$$

$$= -nx^{n-1-2n} = -nx^{-n-1}.$$

EXAMPLE 49   Find f' when $f : x \longmapsto 1/x^3$.

Solution   Since $1/x^3 = 1x^{-3}$, we have $f' : x \longmapsto -3x^{-4} = -3/x^4$. ∎

The chain rule

The **chain rule** describes the relationship between the derivative of the *composition* of two functions and the derivatives of the individual functions. In considering the composition of two *linear* functions we found that the slope of the composition was the *product* of the two slopes. The same was true for piece-wise linear functions, but we needed to take care of *where* we measured the slope. Let's review this locally linear situation:

If g is linear at the particular point x_0 (with slope m_1 there), and f is linear at the particular point $y_0 = g(x_0)$ (with slope m_2 there), then $f \circ g$ is linear at x_0, with slope $m_1 m_2$.

The chain rule says that the same situation holds for derivatives:

If g is differentiable at x_0 (with derivative $g'(x_0)$) and f is differentiable at $y_0 = g(x_0)$ (with derivative $f'(y_0)$), then $f \circ g$ is differentiable at x_0 with derivative

$$(f \circ g)'(x_0) = f'(y_0)g'(x_0) = f'(g(x_0))g'(x_0).$$

As a physical illustration of the chain rule, consider two gears A and B meshed together. As gear A rotates, it causes gear B to rotate also. Now suppose gear B rotates 6 times for every rotation of gear A, and suppose gear A rotates 5 times per minute. How fast does gear B rotate relative to time? The answer is clearly 30 times per minute, the *product* of B's rate relative to A and A's rate relative to time.

3.5 NEW DERIVATIVES FROM OLD

Thought of in this way, the chain rule seems obvious: If y is a function of u, and u is a function of x, then the rate of change of y with respect to x is the product of the rate of change of y with respect to u and the rate of change of u with respect to x. The formal proof of the chain rule, however, has some subtleties we will not go into here.

The chain rule takes on a particularly nice form when written in Leibniz notation:

$$\frac{dy}{dx} = \frac{dy}{du}\frac{du}{dx}.$$

This symbolic appearance suggests a simple cancellation in fraction multiplication, but it is really a statement of the relationship between the derivatives of two functions composed together. Note also that this Leibniz form for the chain rule lacks the crucial information regarding where the various derivatives are to be evaluated:

$$\left.\frac{dy}{dx}\right|_{x=x_0} = \left.\frac{dy}{du}\right|_{u=u(x_0)} \left.\frac{du}{dx}\right|_{x=x_0}.$$

EXAMPLE 50 If $f : x \longmapsto \sin^3 x$, find f'.

Solution Let $u = \sin x$ and $y = u^3$. Considered as a function of x, we have

$$y = u^3 = (\sin x)^3 = \sin^3 x = f(x).$$

Using the chain rule, we have

$$f'(x) = \frac{dy}{dx} = \frac{dy}{du}\frac{du}{dx} = 3u^2 \cdot \cos x.$$

Substituting $u = \sin x$ back in to get this expression entirely in terms of x:

$$f' : x \longmapsto 3\sin^2 x \cos x.$$

■

Forgetting to apply the chain rule in this way is a very common error. This particular application of the chain rule occurs quite often. If y is a differentiable function of x and n is a positive integer, then

$$\frac{d(y^n)}{dx} = ny^{n-1} \cdot \frac{dy}{dx}.$$

This particular use of the chain rule is sometimes called the *power rule*.

EXAMPLE 51 If $g : x \longmapsto \sin(x^3)$, find g'.

Solution This time, let $u = x^3$ and $y = \sin u$. Then $y = \sin(x^3)$ and

$$g'(x) = \frac{dy}{dx} = \frac{dy}{du}\frac{du}{dx} = \cos u \cdot (3x^2) = \cos(x^3) \cdot (3x^2).$$

∎

Rational power functions

Suppose q is a positive integer and x is such that $(x^{1/q})^q = x$ (can this not happen?). Now, if we write $y = x^{1/q}$, then

$$y^q = x,$$

and the power rule gives us

$$qy^{q-1}y' = 1.$$

If we solve for y' we have

$$y' = \frac{1}{qy^{q-1}} = \frac{1}{qx^{1-1/q}} = \frac{1}{q} \cdot x^{1/q-1}.$$

Once we know y', we could differentiate y^p for any integer power p:

$$(y^p)' = py^{p-1}y' = p(x^{1/q})^{(p-1)} \cdot \frac{1}{q} \cdot x^{1/q-1} = \frac{p}{q}x^{(p/q-1)}.$$

We can state this now as a differentiation rule for rational power functions.

If $f : x \longmapsto x^{p/q}$ where p, q are integers and $q > 0$, then

$$f' : x \longmapsto \frac{p}{q}x^{(p/q-1)}$$

for any value x in the domain of f satisfying $(\sqrt[q]{x})^q = x$.

The derivative formula for power functions $x \longmapsto x^r$ is wonderfully consistent. So far, we have seen that if r is any rational number (including positive and negative integers), then

$$\frac{d}{dx}(x^r) = rx^{r-1}.$$

To take advantage of this, function formulas involving radicals should be rewritten with rational exponents.

3.5 NEW DERIVATIVES FROM OLD

EXAMPLE 52 Find $\dfrac{d}{dx}(\sqrt[3]{x^2})$.

Solution $\dfrac{d}{dx}(\sqrt[3]{x^2}) = \dfrac{d}{dx}(x^{2/3}) = \dfrac{2}{3}\cdot x^{-1/3} = \dfrac{2}{3\sqrt[3]{x}}$. ∎

In fact, the formula $\dfrac{d}{dx}(x^r) = rx^{r-1}$ holds for *any* real number r, as we'll verify in a later chapter.

Dictionary of derivative formulas and rules

The derivatives of all the basic functions mentioned in Chapter 1 are summarized below. We have included for completeness the derivatives of the exponential, logarithmic, and inverse trigonometric functions. We will verify these formulas in later chapters.

DERIVATIVE FORMULAS

$f : x \longmapsto x^r$ $\qquad\qquad f' : x \longmapsto rx^{r-1}$

$f : x \longmapsto a^x \,(a > 0,\ a \neq 1)$ $\qquad\qquad f' : x \longmapsto \ln(a)\cdot a^x$

$f : x \longmapsto \log_a(x) \,(a > 0,\ a \neq 1)$ $\qquad\qquad f' : x \longmapsto \dfrac{1}{x\ln(a)}$

$f : x \longmapsto \sin(x)$ $\qquad\qquad f' : x \longmapsto \cos(x)$

$f : x \longmapsto \cos(x)$ $\qquad\qquad f' : x \longmapsto -\sin(x)$

$f : x \longmapsto \tan(x)$ $\qquad\qquad f' : x \longmapsto \sec^2(x)$

$f : x \longmapsto \csc(x)$ $\qquad\qquad f' : x \longmapsto -\csc(x)\cot(x)$

$f : x \longmapsto \sec(x)$ $\qquad\qquad f' : x \longmapsto \sec(x)\tan(x)$

$f : x \longmapsto \cot(x)$ $\qquad\qquad f' : x \longmapsto -\csc^2(x)$

$f : x \longmapsto \arcsin(x)$ $\qquad\qquad f' : x \longmapsto \dfrac{1}{\sqrt{1-x^2}}$

$f : x \longmapsto \arccos(x)$ $\qquad\qquad f' : x \longmapsto \dfrac{-1}{\sqrt{1-x^2}}$

$f : x \longmapsto \arctan(x)$ $\qquad\qquad f' : x \longmapsto \dfrac{1}{1+x^2}$

$f : x \longmapsto \text{arccsc}(x)$ $\qquad\qquad f' : x \longmapsto \dfrac{-1}{\sqrt{x^2-1}}$

$f : x \longmapsto \text{arcsec}(x)$ $\qquad\qquad f' : x \longmapsto \dfrac{1}{\sqrt{x^2-1}}$

$f : x \longmapsto \text{arccot}(x)$ $\qquad\qquad f' : x \longmapsto \dfrac{-1}{1+x^2}$

DERIVATIVE RULES

If f, g are differentiable, c is a constant and x_0 is a specific input:

Linearity Properties
$$(cf)'(x_0) = c \cdot f'(x_0)$$
$$(f+g)'(x_0) = f'(x_0) + g'(x_0)$$
$$(f-g)'(x_0) = f'(x_0) - g'(x_0)$$

Product Rule
$$(fg)'(x_0) = f'(x_0)g(x_0) + f(x_0)g'(x_0)$$

Quotient Rule
$$(f/g)'(x_0) = \frac{f'(x_0)g(x_0) - f(x_0)g'(x_0)}{g^2(x_0)}$$

Chain Rule
$$(f \circ g)'(x_0) = f'(g(x_0)) \cdot g'(x_0)$$

Power Rule
$$(f^n)'(x_0) = nf^{n-1}(x_0) \cdot f'(x_0)$$

The basic formulas and rules allow us to differentiate very complicated functions. Indeed, you can see how a machine with the basic formulas memorized and the basic rules programmed can automate the differentiation process, at least for functions built out of the basic ones. For functions not built out of these basic functions, we can always resort back to the definition of derivative as a limit of difference quotients.

EXAMPLE 53 Find $\dfrac{dy}{dx}$ when $y = x^2 \sin\left(\dfrac{5x^3 - 4}{12x + \pi}\right) - \tan^2(\sqrt[5]{x^2})$.

Solution For a function this complicated, it can be very useful to make some intermediate substitutions. Here, let $u = \dfrac{5x^3 - 4}{12x + \pi}$ and $v = \sqrt[5]{x^2} = x^{\frac{2}{5}}$.

Then $y = x^2 \sin u - \tan^2 v$

and

$$\frac{dy}{dx} = \left[2x \sin u + x^2 \cos u \cdot \frac{du}{dx}\right] - 2 \tan v \cdot \sec^2 v \frac{dv}{dx}.$$

Now we can compute $\dfrac{du}{dx}$ and $\dfrac{dv}{dx}$ and substitute into this expression.

$$\frac{du}{dx} = \frac{15x^2(12x + \pi) - (5x^3 - 4) \cdot 12}{(12x + \pi)^2}$$

$$\frac{dv}{dx} = \frac{2}{5}x^{-\frac{3}{5}} = \frac{2}{5\sqrt[5]{x^3}}.$$

3.5 NEW DERIVATIVES FROM OLD

Our final expression for $\dfrac{dy}{dx}$ is

$$\dfrac{dy}{dx} = 2x \sin\left(\dfrac{5x^3 - 4}{12x + \pi}\right)$$
$$+ x^2 \cdot \cos\left(\dfrac{5x^3 - 4}{12x + \pi}\right) \cdot \left(\dfrac{15x^2(12x + \pi) - (5x^3 - 4) \cdot 12}{(12x + \pi)^2}\right)$$
$$- 2\tan(\sqrt[5]{x^2}) \cdot \sec^2(\sqrt[5]{x^2}) \cdot \dfrac{2}{5\sqrt[5]{x^3}}.$$

∎

The point to be made is that when applied to compositions and algebraic combinations of basic functions, differentiation is a quite mechanical (although sometimes tedious) process.

Far more important activities for the human user of calculus will be finding the *right* function that describes a real world process and then analyzing and interpreting the information that the derivative of that function can provide.

EXERCISES

Below are shown the outputs that the functions $f, g,$ and h and their derivatives assign to certain inputs. Use this information to compute the derivatives indicated in exercises 1-15.

f		f'		g		g'	
$1 \xrightarrow{f} 4$		$1 \xrightarrow{f'} -7.2$		$1 \xrightarrow{g} 3$		$1 \xrightarrow{g'} -0.7$	
$2 \xrightarrow{f} 3$		$2 \xrightarrow{f'} 0.5$		$2 \xrightarrow{g} 2$		$2 \xrightarrow{g'} -0.3$	
$3 \xrightarrow{f} 2$		$3 \xrightarrow{f'} -0.5$		$3 \xrightarrow{g} 1.9$		$3 \xrightarrow{g'} -0.85$	
$4 \xrightarrow{f} 5$		$4 \xrightarrow{f'} -4$		$4 \xrightarrow{g} 1.3$		$4 \xrightarrow{g'} -0.6$	
$5 \xrightarrow{f} 1$		$5 \xrightarrow{f'} 1$		$5 \xrightarrow{g} 1$		$5 \xrightarrow{g'} -0.23$	

$1 \xrightarrow{h} 1$	$1 \xrightarrow{h'} 5$
$2 \xrightarrow{h} 4$	$2 \xrightarrow{h'} 6$
$3 \xrightarrow{h} 7$	$3 \xrightarrow{h'} 7$
$4 \xrightarrow{h} 12$	$4 \xrightarrow{h'} 8$
$5 \xrightarrow{h} 18$	$5 \xrightarrow{h'} 9$

1. $F'(3)$; $F: x \longmapsto 7.1 \cdot f(x)$
2. $F'(1)$; $F: x \longmapsto g(x)/2$
3. $F'(2)$; $F: x \longmapsto 2.3 + h(x)$
4. $F'(1)$; $F: x \longmapsto (0.6)f(x) + 5g(x)$
5. $F'(6)$; $F: x \longmapsto h(3x - 16)$
6. $F'(2)$; $F: x \longmapsto f(x^2)$
7. $F'(1)$; $F: x \longmapsto g(x) - f(x)$
8. $F'(3)$; $F: x \longmapsto \dfrac{5}{h(x)}$
9. $F'(2)$; $F: x \longmapsto (2f(x) + 3h(x))^2$
10. $F'(1)$; $F: x \longmapsto -f(g(x))$
11. $F'(3)$; $F: x \longmapsto h(x - f(x))$
12. $F'(1)$; $F: x \longmapsto h(h(h(x)))$
13. $F'(2)$; $F: x \longmapsto f((g(x))^2)$
14. $F'(1)$; $F: x \longmapsto \dfrac{f(x)}{f(g(x))}$
15. $F'(2)$; $F: x \longmapsto f(x)g(x)h(x)$

Using f and g given by the graphs below, compute the derivatives indicated in exercises 16-21.

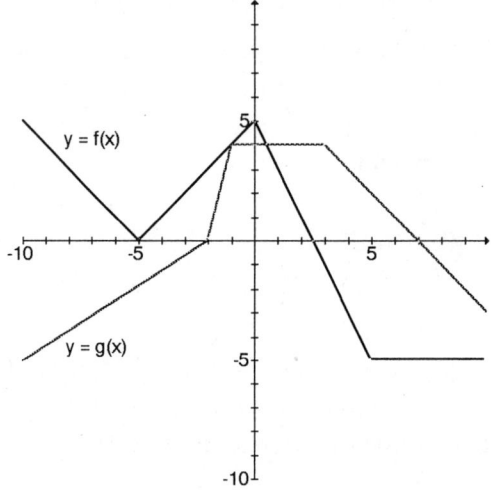

16. $F'(4)$; $F: x \longmapsto f(x) + g(x)$
17. $F'(-4)$; $F: x \longmapsto f(x) - g(x)$
18. $F'(1)$; $F: x \longmapsto f(x) \cdot g(x)$
19. $F'(1)$; $F: x \longmapsto \dfrac{f(x)}{g(x)}$
20. $F'(-3)$; $F: x \longmapsto f(g(x))$
21. $F'(1)$; $F: x \longmapsto g(f(x))$

For each of the functions f indicated in exercises 22-38 find f'.

22. $f: x \longmapsto (x^2 - x - 2)(x^2 + 2x - 8)$
23. $f: x \longmapsto \dfrac{x^2 - 3x + 2}{x^3 + 8}$
24. $f: x \longmapsto (4x^3 + 2x - 5)(8 - 9x^3)$
25. $f: x \longmapsto \dfrac{2x^2 - x - 3}{x^2 + 3x + 2}$
26. $f: x \longmapsto (x^2 - 6x + 5)^4$
27. $f: x \longmapsto \csc x$
28. $f: x \longmapsto \cot x$
29. $f: x \longmapsto (x^3 + 27)\sin(x^2 + 2x - 3)$
30. $f: x \longmapsto \cos(x^2 - 3x + 4)$
31. $f: x \longmapsto \tan(2x^2 - 5x + 1)$
32. $f: x \longmapsto \sec(x^2 + 6x + 8)$
33. $f: x \longmapsto \sin^2(6 - 4x - 2x^2)$

3.5 NEW DERIVATIVES FROM OLD

34. $f : x \longmapsto (2x^2 + 3x - 2)^{2/3}$
35. $f : x \longmapsto \sqrt[4]{7 + 4x - 2x^2}$
36. $f : x \longmapsto \dfrac{\sqrt{x^2 - 25}}{x^2 + 6x + 5}$
37. $f : x \longmapsto \dfrac{(11x^2 - 10x + 1)^{3/2}}{(3x^2 + 5x + 7)^{4/3}}$
38. $f : x \longmapsto (5x^2 + 3x - 10)(2x^2 - 6x + 8)(x^2 - 5x + 4)$

39. Find $\dfrac{d}{dx}((x^2 + 1)^4)$.
40. Find $\dfrac{d}{dx}(\sin^3(x))$.
41. Find $\dfrac{d}{dx}(\sqrt[3]{5x^2 - 17x})$.
42. Find $\dfrac{d}{dx}((2x)^{-2})$.
43. Find $\dfrac{d}{dx}(\sin(\sin(x)))$.
44. Find $\dfrac{d}{dx}(1/\cos^3(x))$.
45. Find $\dfrac{d}{dx}(\sin^2 x + \cos^2 x)$.
46. Find $\dfrac{d}{dx}(\sin(x))$ when x is measured in degrees, given that

$$\sin(x°) = \sin(x \cdot \dfrac{\pi}{180} \; radians).$$

For exercises 47-56, determine the numerical value of the indicated derivative, if it exists.

47. $f'(27)$ where $f : x \longmapsto \sqrt[3]{x}$
48. $g'(8)$ where $g : x \longmapsto \sqrt{1 + \sqrt{1 + x}}$
49. $k'(\sqrt{\pi})$ where $k : x \longmapsto \sin(x^2)$
50. $D_x y(1)$ where $y = \sqrt{\dfrac{x+3}{x+5}}$
51. $\dot{f}(\pi/6)$ where $f : x \longmapsto 1 - (\sin(x))^2$
52. $\left.\dfrac{dy}{dx}\right|_{x=-36}$ where $y = \sqrt{x^2}$
53. $\left.\dfrac{d(x^{-5})}{dx}\right|_{x=1}$

CHAPTER 4

Derivative as Measurement Tool

In the last chapter we noted several interpretations of the derivative. While all the interpretations are closely related, each has its own unique flavor to add to our understanding of derivatives.

Symbolically, the derivative f' may have a formula that can be *derived* directly from the formula for the original function f.

Graphically, f' provides a *slope* machine for the graph of the function f.

Numerically, f' provides a *limiting value* of difference quotients of the function f at each input.

Physically, f' acts as a *speedometer* for the function f, providing a reading of the instantaneous rate of change of the dependent variable with respect to the independent variable.

Is one interpretation more useful than another? It depends very much on the context and our specific use of the derivative. Whenever it is appropriate, assigning multiple interpretations to the derivative can help you check the reasonableness of your computations as well as shed new light on what those computations mean. In this chapter, we take a closer look at how the derivative can be used as a measurement tool.

First, we'll take a closer look at the idea of derivative measuring rate of change through the use of some familiar geometric examples. The physical example of an object's position and velocity as functions of time will serve us well. A differentiable function is *approximately* locally linear. As a consequence, we can use the derivative to find a *best* linear approximation to the function that can be used to make short-term predictions about the output of the function.

The derivative provides a powerful tool for analyzing the behavior of a function. We'll examine in detail how we can use the derivative to analyze the increasing and decreasing behavior of a function, find the extreme output values, and even how to find how *curved* the graph of a function is. Along the way we'll meet *higher order* derivatives (that is, derivatives of derivatives) and interpret the information they can provide both graphically and physically.

4.1 PHYSICAL INTERPRETATION OF DERIVATIVE— RATE OF CHANGE

In terms of units, we can interpret the derivative as measuring the number of units change in the dependent variable per unit change in the independent variable. For example, if $y = f(x)$ is measured in centimeters (cm) and x is measured in grams (g), then the values of the derivative dy/dx will have units of centimeters per gram (cm/g).

Examples from geometry

Let's look at a few familiar formulas from geometry to get a feel for the idea of a derivative as measuring rate of change. In each of the examples below, we'll examine the reasonableness of this idea by comparing the derivative value (instantaneous rate) to the difference quotient value (average rate) over a very small change in the independent variable.

EXAMPLE 1 As a function of its side length x, the area of a square is given by the formula

$$A = x^2.$$

Calculate dA/dx, and interpret it as a rate of change of A with respect to its side length value x when $x = 5$ meters.

Solution   We calculate $dA/dx = 2x$. Since x is measured in *meters*, A is measured in *square meters*. The units of dA/dx are *square meters per meter*. When $x = 5$ meters, the rate of change of A with respect to x is

$$\left.\frac{dA}{dx}\right|_{x=5} = 10 \; \frac{\mathrm{m}^2}{\mathrm{m}}.$$

In this example, it's tempting to reduce the units m²/m to m, but that disguises the true nature of dA/dx as a rate of change of one quantity relative to another. At the instant that the side of a square is 5 meters long, the rate of change of its area is 10 square meters increase per meter increase in side length. ■

The reasonableness of the derivative formula $dA/dx = 2x$ can be seen geometrically if we approximate the difference quotient $\Delta A/\Delta x$ for a small change in side length Δx. Figure 4.1 illustrates that when we add a small amount Δx to the side length, most of the change in area ΔA is accounted for by the sum of the areas of two thin rectangles $2x\Delta x$.

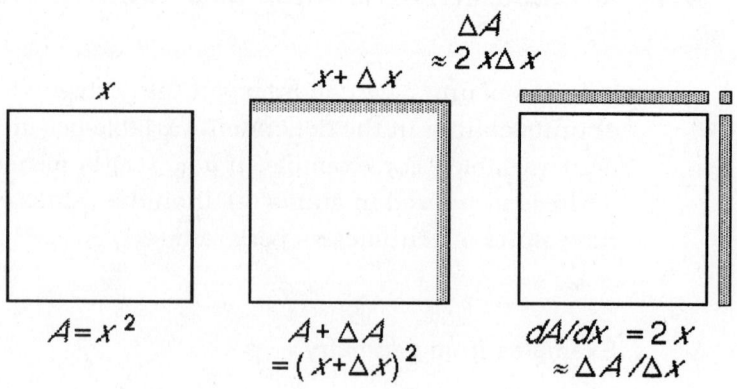

Figure 4.1 Rate of change in area of the square.

If we use this approximation of ΔA to compute the difference quotient, we have

$$\frac{\Delta A}{\Delta x} \approx \frac{2x\Delta x}{\Delta x} = 2x = \frac{dA}{dx}.$$

The error between this approximate difference quotient and the true difference quotient is simply $(\Delta x)^2/\Delta x = \Delta x$ (we're missing the area of the small square in the numerator ΔA). As $\Delta x \to 0$, this error vanishes.

EXAMPLE 2 As a function of its side length x, the volume of a cube is given by the formula

$$V = x^3.$$

Calculate dV/dx and interpret it as a rate of change when $x = 2$ inches.

Solution We calculate $dV/dx = 3x^2$. Since x is measured in *inches*, V is measured in *cubic inches*. The units of dV/dx are cubic inches per inch. When $x = 2$ inches the rate of change of V with respect to x is

$$\left.\frac{dV}{dx}\right|_{x=2} = 12\,\frac{\text{in}^3}{\text{in}}.$$

∎

Figure 4.2 illustrates that most of the change in volume ΔV after a small increase in side length Δx is accounted for by the three flat square "slabs," whose total volume is $3x^2 \Delta x$ (using *length · width · thickness* and adding the results).

4.1 PHYSICAL INTERPRETATION OF DERIVATIVE— RATE OF CHANGE

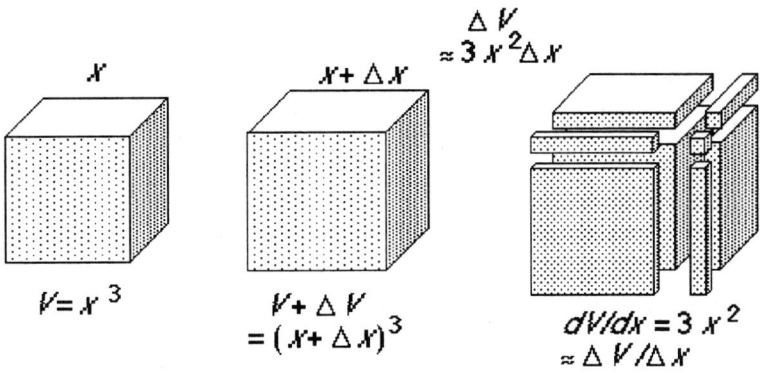

Figure 4.2 Rate of change in volume of the cube.

Using this as an approximation for ΔV gives us

$$\frac{\Delta V}{\Delta x} \approx \frac{3x^2 \Delta x}{\Delta x} = 3x^2 = \frac{dV}{dx}.$$

The error in our approximation to $\Delta V/\Delta x$ is

$$\frac{3x(\Delta x)^2 + (\Delta x)^3}{\Delta x} = 3x\Delta x + (\Delta x)^2.$$

Again, the smaller Δx is, the more insignificant this error is.

EXAMPLE 3 As a function of its radius r, the area of a circle is given by the formula

$$A = \pi r^2.$$

Calculate dA/dr, and interpret it as a rate of change of A with respect to the radius r when $r = 7$ cm.

Solution We calculate $dA/dr = 2\pi r$. (Remember, π is a *constant*.) Since r is measured in cm, A is measured in cm^2. The units of dA/dr are cm^2 per cm. When $r = 7$ cm, the rate of change of A with respect to r is

$$\left.\frac{dA}{dr}\right|_{r=7} = 14\pi \ \frac{\text{cm}^2}{\text{cm}}.$$

∎

The change in area ΔA corresponding to a small change in radius Δr corresponds to the area of a thin *annulus* or band as shown in Figure 4.3.

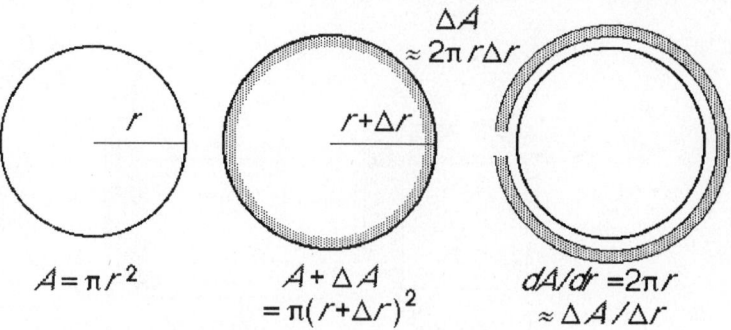

Figure 4.3 Rate of change in area of the circle.

If we imagine "snipping" the band and straightening it out to a long strip, we can see that the area of the band should be reasonably close to the product of its length ($\approx 2\pi r$, the circumference of the original circle) and its thickness Δr. Using this as our approximation for ΔA gives us

$$\frac{\Delta A}{\Delta r} \approx \frac{2\pi r \Delta r}{\Delta r} = 2\pi r = \frac{dA}{dr}.$$

Position and velocity

One of the early uses of calculus was in the study of motion and force in physics. Indeed, we used the motion of a car as our motivating illustration of a derivative value as a speedometer reading. Let's resume that discussion now.

To avoid confusion with the Leibniz notation's d, we'll use s to denote the *position of the car as a function of time t* . Now, if s is a differentiable function of t, then its derivative gives us the **velocity** or instantaneous rate of change of position with respect to t. We write v for velocity, so that

$$v(t) = \frac{ds}{dt} = s'(t).$$

In our initial discussion of the car illustration, we used the term *speed* instead of *velocity* . Velocity is really a more accurate term for our purposes, since speed refers to how fast the car is moving, while velocity also tells us the *direction of motion*. If our car travels in a straight line we can interpret a *positive* velocity $v = 40$ mph as *forward* motion, while a *negative* velocity $v = -40$ mph should be interpreted as *reverse* motion at the same speed.

In other words, speed gives us the *magnitude* (absolute value) of velocity, or

$$speed = |velocity|.$$

4.1 PHYSICAL INTERPRETATION OF DERIVATIVE— RATE OF CHANGE

For vertical motion, positive and negative velocity will correspond to upward and downward movement, but exactly which will depend on how we choose to measure the position. In one setting, we might measure the depth of a canyon as a positive distance. In another setting, we might choose to measure height above sea level, so that the same canyon floor might be considered to be at a *negative* height.

EXAMPLE 4 The vertical height (in feet) of a ball thrown upwards from a tall building is described by the position function

$$s : [0, 10] \longrightarrow \mathbb{R}$$

$$s : t \longmapsto -16t^2 + 96t + 640$$

where t is measured in seconds.

a) Find its initial velocity (at time $t = 0$).

b) Find its height when the velocity is 0.

c) Find its velocity when the ball returns to earth ($s = 0$).

Solution The velocity is

$$v(t) = \frac{ds}{dt} = s'(t) = -32t + 96.$$

a) The initial velocity is

$$v(0) = f'(0) = -32 \cdot 0 + 96 = 96 \text{ feet per sec}.$$

b) The velocity is 0 when $-32t + 96 = 0$. Solving for t we have $t = 3$ seconds. Now, substituting $t = 3$ back into the original position function to find the ball's height, we have

$$s(3) = -16 \cdot 3^2 + 96 \cdot 3 + 640 \text{ feet}$$

$$= 784 \text{ feet}.$$

c) $s = 0$ when $-16t^2 + 96t + 640 = 0$. Dividing both sides by -16 yields

$$t^2 - 6t - 40 = 0.$$

The solutions to this quadratic equation are

$$t = 10 \text{ sec} \quad \text{and} \quad t = -4 \text{ sec}.$$

The second solution satisfies the quadratic equation, but does not make sense in this particular physical setting (we assume we cannot go back in time). Using $t = 10$ in our expression for velocity gives us

$$v(10) = -32 \cdot 10 + 96 = -224 \text{ feet per sec.}$$

The negative value for the velocity indicates that the ball was heading downward (naturally) when it hit the ground. ■

EXERCISES

Below is an illustration of a sphere of radius r and a sphere of slightly larger radius $r + \Delta r$. The change in volume is the volume of the shell shown. Exercises 1-6 refer to these spheres.

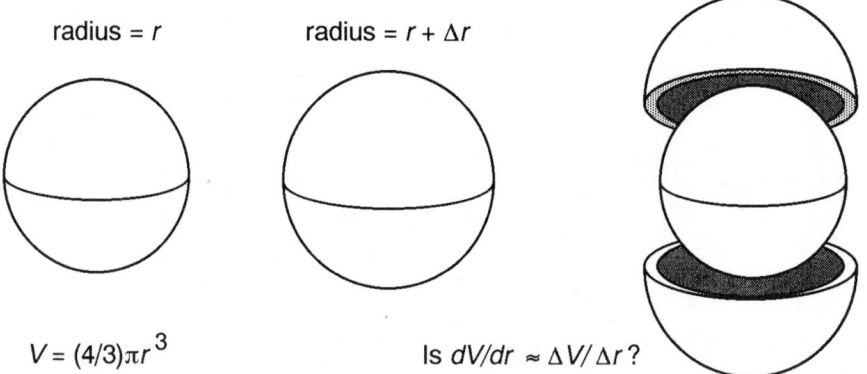

radius = r

radius = $r + \Delta r$

$V = (4/3)\pi r^3$

Is $dV/dr \approx \Delta V/\Delta r$?

1. Approximate the change in volume by taking the product of the surface area of the original sphere and the thickness of the shell.

2. Use this approximation of change in volume to approximate the difference quotient $\Delta V/\Delta r$.

3. As a function of its side length r, the volume of a sphere is given by the formula

$$V = \frac{4}{3}\pi r^3.$$

Calculate dV/dr and interpret it as a rate of change when $x = 3$ feet.

4. Compare dV/dr to the approximate difference quotient in exercise 2.

5. Calculate the true volume of the shell (compute the difference in the volume of the two spheres).

4.1 PHYSICAL INTERPRETATION OF DERIVATIVE— RATE OF CHANGE

6. What is the error in the approximate difference quotient found in exercise 2? What happens to this error quantity as $\Delta r \to 0$?

In exercises 7-16, a geometrical quantity is given that can be expressed as a function of another variable measured in units of length. For each, find the appropriate functional relationship, calculate the derivative, and interpret the derivative as a rate of change when the independent variable has the value 3.6 cm.

7. The perimeter P of a square is a function of its side length x.

8. The surface area S of a cube is a function of its side length x.

9. The diagonal y of a square is a function of its side length x.

10. The diagonal z of a cube is a function of its side length x.

11. The volume V of a cube is a function of its diagonal length z.

12. The circumference C of a circle is a function of its radius r.

13. The circumference C of a circle is a function of its diameter d.

14. The area A of a circle is a function of its diameter d.

15. The area A of a circle is a function of its circumference C.

16. The surface area S of a sphere is a function of its radius r.

In each of exercises 17-20, you are given the formula for the position function $s : t \longmapsto s(t)$ of some object over the time interval $0 \leq t \leq 60$. (s is measured in feet and t is measured in seconds.) For each function s, find the following:
a) *its initial velocity (at time $t = 0$)*
b) *the specific times in the interval $[0, 60]$ for which the velocity is 0*
c) *the specific times in the interval $[0, 60]$ for which the object is back at its original position*
d) *the specific positions (values of s) for which the velocity is 0*
e) *the graph of the position $y = s(t)$ over the interval $[0, 60]$*
f) *the graph of the velocity $y = v(t)$ over the interval $[0, 60]$.*

17. $s : t \longmapsto -16t^2 + 64000$

18. $s : t \longmapsto .002t^3 - 0.27t^2 + 12t + 10$

19. $s : t \longmapsto |90 - t^2/30|$

20. $s : t \longmapsto 2 \sin \pi t/20$

Exercises 21-25 refer to an open cone made by cutting a sector (piece of pie) of angle θ out of a circle of radius 6 inches and then connecting the edges.

21. Express the area A of the cut-out sector of the circle as a function of the sector angle θ. Calculate $dA/d\theta$ and interpret it as a rate of change when θ is $120°$ and $3\pi/2$ *radians*, respectively.

22. Express the circumference of the opening of the cone as a function of θ and find its derivative with respect to θ. Interpret it as a rate of change when θ is $120°$ and $3\pi/2$ *radians*, respectively.

23. Express the surface area of the open cone as a function of θ and find its derivative with respect to θ. Interpret it as a rate of change when θ is $120°$ and $3\pi/2$ *radians*, respectively.

24. Express the volume of the cone as a function of θ and find its derivative with respect to θ. Interpret it as a rate of change when θ is $120°$ and $3\pi/2$ *radians*, respectively.

25. Express the height of the cone as a function of θ and find its derivative with respect to θ. Interpret it as a rate of change when θ is $120°$ and $3\pi/2$ *radians*, respectively.

4.2 BEST LINEAR APPROXIMATIONS

When it is possible to find the derivative f' directly, then we can use it to *predict* the output of the function f. In the case of a moving car, for example, we could use the speedometer reading together with the car's position to predict where the car will be a short time in the future.

EXAMPLE 5 Suppose that the time it takes you to get your foot off the gas pedal when you see someone's brake lights come on is 0.5 seconds. What is a safe distance to leave between your car and the car in front of you at 60 mph?

Solution Assuming that your brakes work at least as well as those of the car in front of you, you won't collide as long as your car doesn't cover the entire distance separating your cars in the 0.5 seconds it takes you to react.

4.2 BEST LINEAR APPROXIMATIONS

If you are travelling at approximately constant velocity during this short time interval, then the car will travel approximately

$$60 \frac{\text{miles}}{\text{hours}} \cdot (0.5 \text{ sec}) \cdot \frac{1 \text{ hour}}{60 \cdot 60 \text{ sec}} = \frac{1}{120} \text{ miles} = 44 \text{ feet}.$$

So, if you leave more than 44 feet between the cars, you should avoid a collision. ∎

Figure 4.4 illustrates the positions of the cars in this example.

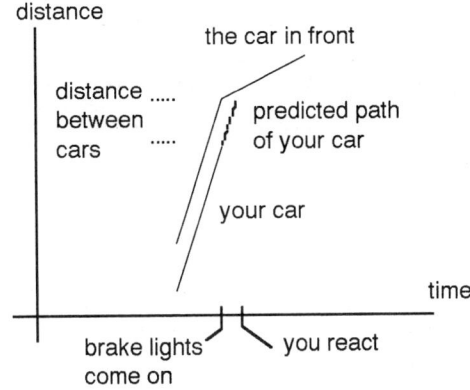

Figure 4.4 Reaction time and distance between cars.

Assuming that the velocity is approximately constant during the reaction time is equivalent to assuming that the position function of the car is approximately a *linear* function of time during this interval. The slope of this linear approximation function is the velocity 60 mph. A linear function is determined by the slope of its graph and the output value at one particular input. Here we used the instantaneous velocity (speedometer reading) as the slope. The original position of the car (odometer reading) could be considered the output value at the specific input time (clock reading) we see the brakes come on.

We can use this strategy to find a linear approximation function for any differentiable function at a particular point. That is, for a differentiable function f and an input x_0, we can use the derivative $f'(x_0)$ as the slope of a line passing through the point $(x_0, f(x_0))$. The function g whose graph is this line is a linear approximation function for the original function f. In fact, this linear approximation function g is called the *best* linear approximation function.

Definition 1 The **best linear approximation** to a differentiable function f at the point x_0 is a linear function g of the form

$$g : x \longmapsto f(x_0) + m(x - x_0)$$

where the slope is $m = f'(x_0)$. Note that $g(x_0) = f(x_0)$.

This linear function g can be thought of as a *local* approximation for f: for values x close to x_0, we have

$$f(x) \approx g(x) = f(x_0) + m(x - x_0).$$

The anatomy of this best linear approximation can be dissected in terms of the instrument readings in our car (Figure 4.5).

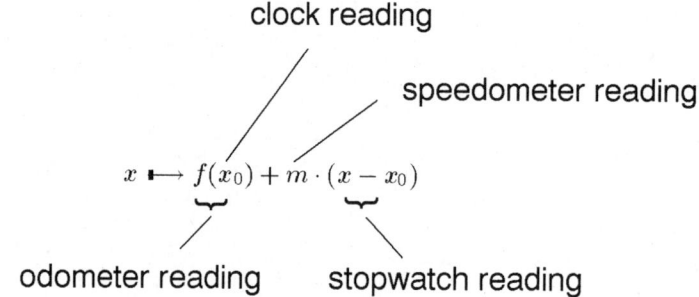

Figure 4.5 Anatomy of the best linear approximation g.

Since both the odometer reading and the speedometer reading will depend on the point at which the readings are taken, the best linear approximation only makes sense as the best linear approximation *at some particular point*.

EXAMPLE 6 Suppose that the actual distance travelled by a car could be modelled by the function

$$s : (0, 2) \longrightarrow \mathbb{R}$$

$$s : t \longmapsto 58t + t^2$$

where t is measured in hours, and s is measured in miles. Find the best linear approximation function for the distance traveled at the instant $t = 1$.

4.2 BEST LINEAR APPROXIMATIONS

Solution The instantaneous velocity of the car at time t is

$$s'(t) = 58 + 2t$$

(units of mph). At $t = 1$ we have $s'(1) = 60$. We use this as the slope m of our linear approximation. The actual position of the car at $t = 1$ would be $s(1) = 59$, so we can write the linear approximation as

$$s(t) \approx 59 + 60(t - 1).$$

■

In general, the relative accuracy of the best linear approximation gets better the closer we are to the point at which we computed the slope (speedometer reading). Exactly what do we mean by "best" here? Of all the *linear* functions we might choose to approximate the actual distance covered, this one is the best in the sense that not only does the prediction become more accurate as we focus our attention on smaller elapsed times, but also *it becomes more accurate faster than any other possible choice*.

EXAMPLE 7 Compare the error between the actual position function and its best linear approximation at $t = 1$ for the times $t = 1.5$ hours and $t = 1$ hour and 0.5 seconds.

Solution For $t = 1.5$ the actual position function gives us (in miles)

$$s(1.5) = 58(1.5) + (1.5)^2 = 89.25.$$

Using the best linear approximation we would have (in miles)

$$s(1.5) \approx 59 + 60(1.5 - 1) = 59 + 30 = 89.$$

Hence, the error is -0.25 miles. For $t = 1$ hour and 0.5 seconds (≈ 1.000014 hours) we find that the error is about -0.0075 miles. ■

In the case of a polynomial function, we can find its best linear approximation at a point by simply writing it in Taylor form and dropping all but the linear portion. The slope of this line will be the derivative of the polynomial function at this point.

EXAMPLE 8 Find the best linear approximation and the derivative of f at $x = 5.7$ if $f : x \longmapsto x^2 - 3$.

Solution We can put f in Taylor form at 5.7 by substituting $(x - 5.7) + 5.7$ for x and then expanding and collecting terms:

$$f(x) = ((x - 5.7) + 5.7)^2 - 3$$
$$= (x - 5.7)^2 + 2(5.7)(x - 5.7) + (5.7)^2 - 3$$
$$= 29.49 + 11.4(x - 5.7) + (x - 5.7)^2.$$

We can think of this as a linear approximation $29.49 + 11.4(x - 5.7)$ plus an *error term* $(x - 5.7)^2$. When x is very close to 5.7, the quantity $(x - 5.7)^2$ will be *much* smaller than the remaining linear part

$$29.49 + 11.4(x - 5.7)$$

which will be the *best* linear approximation. Since the slope of this line is 11.4, we have $f'(5.7) = 11.4$.

You can check that this is indeed the correct derivative value by differentiating f and substituting $x = 5.7$. ∎

Tangent and normal lines

Once we have found the best linear approximation to a function f at a particular input x_0, we can graph it as the line with slope $f'(x_0)$ and passing through the point $(x_0, f(x_0))$.

Geometrically, this line will often appear to "kiss" the graph of the original function at the point $(x_0, f(x_0))$, much like a tangent line to a circle. For this reason, the **tangent line** to the graph of a function f at the point $(x_0, f(x_0))$ is *defined* to be the graph of this best linear approximation.

Unlike the tangent line to a circle, the tangent line to a graph could well intersect the graph at other points (see Figure 4.6). If you graphed both the tangent line and the original function $y = f(x)$ using machine graphics, and then zoomed in on the point of tangency far enough, it would become impossible to distinguish the two graphs from one another (unless you zoomed in so far as to have round-off precision errors take effect). This suggests a way to check visually the reasonableness of a tangent line equation.

4.2 BEST LINEAR APPROXIMATIONS

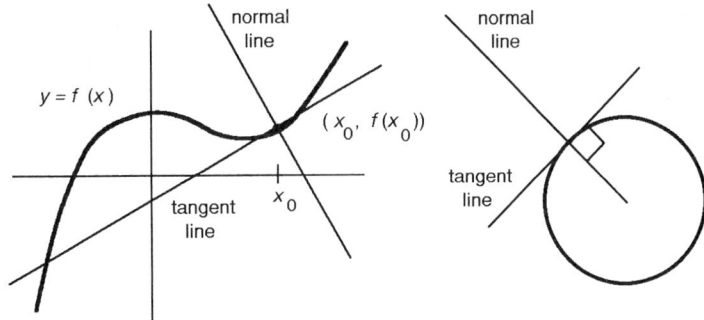

Figure 4.6 Tangent and normal lines to a graph and a circle.

A **normal line** to the graph of a function f at the point $(x_0, f(x_0))$ is defined to be the *line perpendicular to the tangent line at that point.* (If $m \neq 0$ is the slope of a line, then $-1/m$ is the slope of a line perpendicular to it.) A well-known property of the normal line to a circle is that it must pass through the center of the circle (see Figure 4.6). As related to the graph of a function $y = f(x)$, we can think of the normal line as providing the direction to move if we want to move away from the graph as quickly as possible.

EXAMPLE 9 Find the equation of the tangent line to the graph of $y = x^2$ at the point $(3, 9)$.

Solution Since $\dfrac{dy}{dx} = 2x$, the slope of the tangent line is given by

$$\left.\dfrac{dy}{dx}\right|_{x=3} = 6$$

and the equation of the tangent line at $(3, 9)$ is $y = 6(x - 3) + 9$. ∎

EXAMPLE 10 Suppose $f : x \longmapsto 2/x^3$. Find the equation of the tangent line to the graph of $y = f(x)$ at the input $x_0 = -2$.

Solution Since $f(x) = 2x^{-3}$, we have $f'(x) = -6x^{-4}$. The tangent line must pass through the point

$$(x_0, f(x_0)) = (-2, 2/(-2)^3) = (-2, -1/4)$$

and the slope of the tangent line must be

$$m = f'(-2) = -6(-2)^{-4} = -6/16 = -3/8.$$

Hence, the equation of the tangent line is

$$y = -\frac{3}{8}(x+2) - \frac{1}{4}.$$

∎

EXAMPLE 11 Find the equation of the normal line to the graph of $y = x^{2/3}$ at the point $(-8, 4)$.

Solution Since $\dfrac{dy}{dx} = \dfrac{2}{3}x^{-1/3}$, the slope of the tangent line is given by

$$m = \left.\frac{dy}{dx}\right|_{x=-8} = \frac{2}{3} \cdot \frac{1}{(-8)^{1/3}} = -\frac{1}{3}.$$

The slope of the normal line would be the *negative reciprocal*

$$\frac{-1}{m} = \frac{-1}{-1/3} = 3.$$

Hence the equation of the normal line at $(-8, 4)$ is $y = 3(x+8) + 4$. ∎

EXAMPLE 12 Suppose $g : x \longmapsto \cos(x)$. Find the equation of the normal line to the graph of $y = g(x)$ at $x_0 = 0$.

Solution Our line must pass through the point $(x_0, g(x_0)) = (0, \cos(0)) = (0, 1)$. Since $g(x) = \cos(x)$, we have

$$g'(x) = -\sin(x)$$

so the slope of the tangent line at $x_0 = 0$ is

$$m = -\sin(0) = 0.$$

This means the tangent line is *horizontal*, requiring the normal line to be *vertical*. Since the normal line passes through $(0, 1)$, we conclude that its equation is

$$x = 0.$$

Conversely, a normal line will be horizontal at point where the tangent line is vertical.

4.3 USING THE DERIVATIVE TO ANALYZE FUNCTION BEHAVIOR

EXAMPLE 13 The normal line to the graph of $y = x^{1/3}$ at $x_0 = 0$ is the horizontal line $y = 0$. (Graph $y = x^{1/3}$.) ∎

EXERCISES

In exercises 1-10, find the best linear approximation of the given function at the specified point x_0. Use the best linear approximation to predict the function's output value at $x_0 + 1$, $x_0 - 1$, $x_0 + .001$, and $x_0 - .001$. Compare the predicted values with the actual function value at each of these points and calculate the error. Finally, graph both the original function and its best linear approximation over the interval $[x_0 - .1, x_0 + .1]$.

1. $f : x \longmapsto x^2$; $x_0 = -3$
2. $f : x \longmapsto \sqrt{x}$; $x_0 = 9$
3. $f : x \longmapsto x^{2/3}$; $x_0 = -8$
4. $f : x \longmapsto x^3 - 4x$; $x_0 = -1.5$
5. $f : x \longmapsto \sin x$; $x_0 = \pi/3$
6. $f : x \longmapsto \cos x$; $x_0 = 0$
7. $f : x \longmapsto \tan x$; $x_0 = \pi/4$
8. $f : x \longmapsto 1/x$; $x_0 = -2$
9. $f : x \longmapsto x$; $x_0 = 0$
10. $f : x \longmapsto 5.42$; $x_0 = 132.78$

In exercises 11-16, find the best linear approximation of the given polynomial function p at the specified point x_0 by using the algebraic technique illustrated in Example 8. Find the derivative $p'(x_0)$ directly, and compare it to the slope of your best linear approximation (they should match exactly).

11. $p : x \longmapsto x^2 + x + 1$; $x_0 = 1.6$
12. $p : x \longmapsto 2x^2 - 7x + 3$; $x_0 = 3/2$
13. $p : x \longmapsto x^3 - 12x$; $x_0 = -2$
14. $p : x \longmapsto \dfrac{3x - 17}{5}$; $x_0 = 4$
15. $p : x \longmapsto x^5$; $x_0 = 1$
16. $p : x \longmapsto 3$; $x_0 = 2$

In exercises 17-26, find the equations of the tangent and normal lines to the given functions at the specified point.

17. $p : x \longmapsto x^2 + x + 1$; $x_0 = 1.6$
18. $p : x \longmapsto 2x^2 - 7x + 3$; $x_0 = 3/2$
19. $p : x \longmapsto x^3 - 12x$; $x_0 = -2$
20. $p : x \longmapsto \dfrac{3x - 17}{5}$; $x_0 = 4$
21. $p : x \longmapsto x^5$; $x_0 = 1$
22. $p : x \longmapsto 3$; $x_0 = 2$
23. $f : x \longmapsto \sin^2 x$; $x_0 = \pi/3$
24. $f : x \longmapsto \tan x$; $x_0 = 3\pi/4$
25. $f : x \longmapsto \sin x^2$; $x_0 = 0$
26. $f : x \longmapsto \sqrt{a^2 - x^2}$; $x_0 = b$

(Note: In exercise 26, a and b are positive constants, with $-a < b < a$.)

4.3 USING THE DERIVATIVE TO ANALYZE FUNCTION BEHAVIOR

The old saying that "a picture is worth a thousand words" is truly an understatement in mathematics. Graphs are extremely powerful tools for analyzing a function, and many types of function behavior translate directly to easily visualized characteristics of the graph of the function. In this section we'll see how the interpretation of derivative as slope can be exploited in analyzing a function's behavior.

Differentiability means smooth behavior

Knowing that a function f is *continuous* at a point a guarantees that the graph of f can have no hole or break at that point. Knowing that a function f is *differentiable* at a point a automatically means that f is continuous at $x = a$, because the difference quotients

$$\frac{f(x) - f(a)}{x - a}$$

could not possibly have a limiting value unless $f(x) \to f(a)$ as $x \to a$. So a hole, skip, jump, pole or other discontinuous behavior cannot occur at a point where the function has a derivative.

But differentiability at a point means more—the graph of the function cannot have a sharp corner or cusp at that point. Figure 4.7 shows the graph of $g : x \longmapsto x^{2/3}$. The slope is undefined at $x = 0$ ($g'(x) = \frac{2}{3}x^{-1/3}$) and the nondifferentiability there is evidenced by a sharp *cusp* in the graph. Similarly, the absolute value function is not differentiable at $x = 0$, as evidenced by the sharp corner in its graph at $x = 0$. Both of these functions, however, are continuous at $x = 0$.

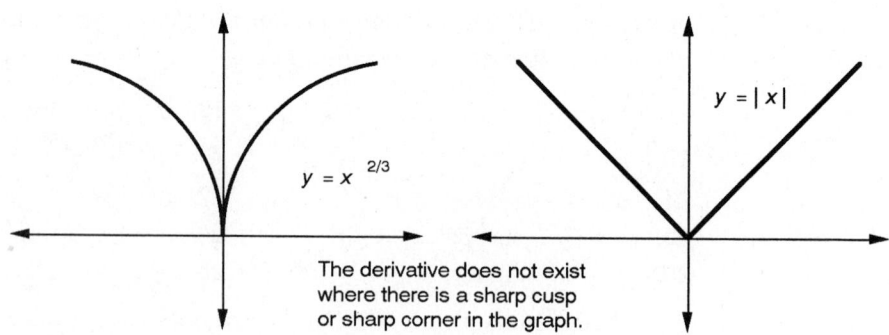

Figure 4.7 Derivative does not exist at a cusp or sharp corner.

4.3 USING THE DERIVATIVE TO ANALYZE FUNCTION BEHAVIOR

Could a function have a smooth graph at a point where the function is not differentiable? The graph of the function $f : x \longmapsto x^{1/3}$ is illustrated in Figure 4.8. The graph certainly looks smooth at $x = 0$, but the tangent line there is vertical and the graph has undefined slope at $x = 0$. (Note that $f'(x) = \frac{1}{3}x^{-2/3}$ is undefined when $x = 0$.)

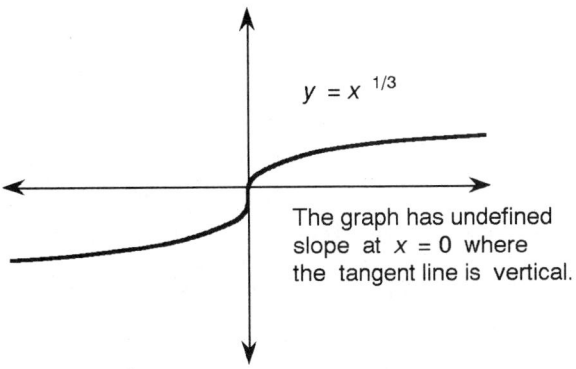

Figure 4.8 The derivative is undefined when the tangent line is vertical.

Summarizing what we have noted so far, the existence of $f'(x_0)$ requires that the graph of $y = f(x)$ can have no hole, skip, jump, pole, sharp corner, cusp, or vertical tangent at $(x_0, f(x_0))$.

Increasing and decreasing behavior

In general, a function is said to be *increasing* or *decreasing* depending on whether its outputs increase or decrease as the inputs increase. Let's make this more precise mathematically.

Definition 2

> Suppose the function f is defined on an interval (a, b). Then we say
> f is **increasing** if whenever $a \leq x_1 < x_2 \leq b$, then $f(x_1) \leq f(x_2)$.
> f is **strictly increasing** if whenever $a \leq x_1 < x_2 \leq b$, then $f(x_1) < f(x_2)$.
> f is **decreasing** if whenever $a \leq x_1 < x_2 \leq b$, then $f(x_1) \geq f(x_2)$.
> f is **strictly decreasing** if whenever $a \leq x_1 < x_2 \leq b$, then $f(x_1) > f(x_2)$.
> If f is either increasing or decreasing over an interval $[a, b]$, we say f is **monotonic**. If f is either strictly increasing or strictly decreasing on an interval $[a, b]$, then we can say that f is **strictly monotonic**.

Note carefully the distinction between the terms *increasing* and *strictly increasing*. It might be more appropriate to call an *increasing* function *non-decreasing* (as the inputs get larger, the outputs cannot get smaller). In contrast, the outputs of a *strictly increasing* function get strictly bigger as the inputs get bigger. A similar distinction must be made between *decreasing* and *strictly decreasing* functions. An increasing function has a graph that either rises (from left to right) or stays level over the given interval, while a strictly increasing function has a graph that strictly rises over the given interval. A decreasing function has a graph that either falls (from left to right) or stays level over the given interval, while a strictly decreasing function has a graph that strictly falls over the given interval. No function can be both strictly increasing and decreasing, but a constant function is technically both increasing and decreasing!

The slope of a linear or a piece-wise linear function's graph tells us instantly whether the function is strictly increasing (positive slope), strictly decreasing (negative slope), or constant (zero slope). The derivative of a differentiable function gives us similar information.

 If $f'(x) > 0$ on an interval (a, b), then the original function f is strictly *increasing* on that interval.

 If $f'(x) < 0$ on an interval (a, b), then the original function f is strictly *decreasing* on that interval.

Critical points

Now, what happens at a point x_0 (in the domain of f) where $f'(x_0)$ is neither positive nor negative? There are only two possibilities—either $f'(x_0) = 0$ or $f'(x_0)$ is not even defined. In either case, we call x_0 a *critical point* for the function f.

Definition 3

A **critical point** x_0 for a function f is a point in the domain of f such that either $f'(x_0) = 0$ or $f'(x_0)$ is *undefined*.

Geometrically, a critical point occurs where the graph of f is either flat (slope 0), or has a sharp corner, cusp, vertical tangent or other such behavior making the slope undefined.

4.3 USING THE DERIVATIVE TO ANALYZE FUNCTION BEHAVIOR

EXAMPLE 14 The input $x = 0$ is a critical point for each of the functions $x \longmapsto x^{2/3}$, $x \longmapsto x^{1/3}$, and $x \longmapsto |x|$, since their derivatives are undefined for $x = 0$. ∎

Figure 4.9 illustrates the graph of a function with nine different critical points indicated ($x_1, x_2, \ldots, x_9$).

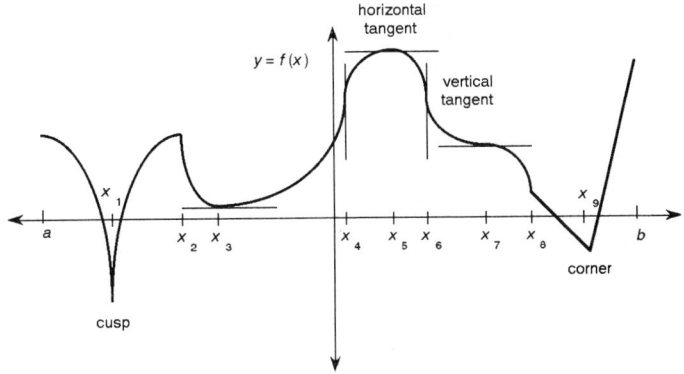

Figure 4.9 Critical points occur where $f'(x)$ is zero or undefined.

The points x_1, x_2, x_8, and x_9 are critical because a sharp change in the direction of the graph occurs (the derivative value is undefined at these points). The points x_4 and x_6 are critical because the tangent to the graph at those points is vertical (again, the derivative value is undefined). The points x_3, x_5, and x_7 are critical because the tangent is horizontal (the derivative value is 0).

EXAMPLE 15 Find the critical points of the function $f : x \longmapsto \dfrac{1}{x} + x^2$.

Solution The derivative of f is

$$f' : x \longmapsto -\frac{1}{x^2} + 2x.$$

The derivative f' is undefined at $x = 0$, but this is *not* a critical point because 0 is not in the domain of the original function. If we set $f'(x) = 0$:

$$-\frac{1}{x^2} + 2x = 0$$

and solve for nonzero values x, we obtain

$$-1 + 2x^3 = 0$$

(multiply both sides of the equation by x^2). We conclude that $x = 1/\sqrt[3]{2}$ is a critical point of f. ∎

Local extrema and the Critical Point Theorem

Critical points deserve special attention in the analysis of a function, for they are potential locations for extreme values of the function. Here we mean extreme in a *local* sense. Let's make that more clear with the following definition.

Definition 4

> A **local maximum** of a function f occurs at a point x_0, if it is true that $f(x_0) \geq f(x)$ for all x in some neighborhood of x_0.
> A **local minimum** of a function f occurs at a point x_0, if it is true that $f(x_0) \leq f(x)$ for all x in some neighborhood of x_0.

The point x_0 is a local maximum (or local minimum) if there is an open interval containing x_0 such that $f(x_0)$ is the maximum (or the minimum) output value over that interval. If either a local minimum or local maximum for f occurs at x_0, we say that a **local extremum** occurs there. Some books use the term **relative** in place of the term **local** as we have used here.

 Be careful to distinguish between a *local extremum* **and its** *location*. **A local minimum or local maximum refers to the output $f(x_0)$ of the function. On the other hand, the location is the** *input $x = x_0$.*

Figure 4.10 illustrates the locations of the local extrema for our critical point example.

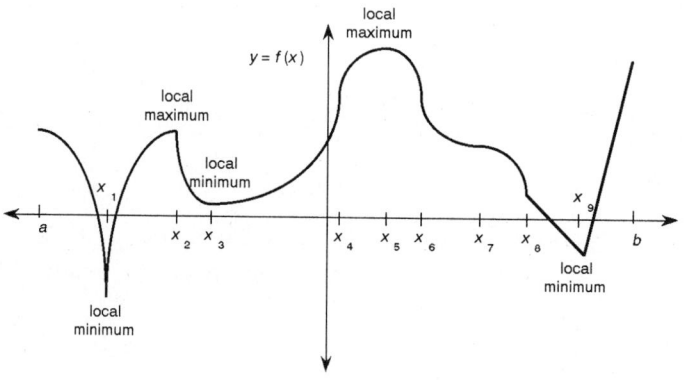

Figure 4.10 Local extrema occur at critical points.

4.3 USING THE DERIVATIVE TO ANALYZE FUNCTION BEHAVIOR

Notice that every local minimum and local maximum occurred at a critical point. That is the conclusion of the *Critical Point Theorem*.

Theorem 4.1

Critical Point Theorem
Hypothesis: f has a local extremum at x_0.
Conclusion: Either $f'(x_0) = 0$ or $f'(x_0)$ is undefined.

Reasoning Let's think about the situation of a local maximum first. (Think of x_0 as being x_2 or x_5 in Figure 4.10 as an illustration.) In some neighborhood of x_0, $f(x_0) \geq f(x)$ for all the other values of x in the neighborhood.

That means that if $x < x_0$ then

$$\frac{f(x) - f(x_0)}{x - x_0} \geq 0.$$

If $x > x_0$ then

$$\frac{f(x) - f(x_0)}{x - x_0} \leq 0.$$

So, as $x \to x_0^-$ all the difference quotients are greater than or equal to 0, and as $x \to x_0^+$ all the difference quotients are less than or equal to 0. If there is a *single limiting value* to these difference quotients, it *must be* 0. If there is *not* a single limiting value, then the derivative doesn't even exist at that point. In either case, we would conclude that x_0 is a critical point of f.

The situation of a local minimum is similar. (Think of x_0 as being x_1, x_3, or x_9 in Figure 4.10 as an illustration.) Now the difference quotients from the left are less than or equal to 0, and the difference quotients from the right are greater than or equal to 0. If f has a derivative value at the local minimum, then $f'(x_0) = 0$. Otherwise $f'(x_0)$ is undefined. □

Be careful how you read this theorem, for its *converse* is not true. The theorem says that if f has any local extrema, they must occur at critical points. It *does not say* that if f has a critical point, then an extremum must occur there. In Figure 4.10, we can see that x_4, x_6, x_7, and x_8 are all critical points, but not a single one is the location of a local extremum.

Finding absolute extrema—first derivative test for extrema

If the derivative of a continuous function is defined in a neighborhood of a critical point (except possibly at the critical point itself), then we can use it to determine whether the critical point is the location of a local minimum, a local maximum, or neither. This **derivative test** for local extrema is very simple, and we can reason graphically as follows:

As we move past the location x_0 of a local maximum from left to right, the function's graph must change from rising *uphill* to the peak to falling *downhill* past the peak. More precisely, if $f'(x)$ is positive on an interval (a, x_0), and $f'(x)$ is negative on an interval (x_0, b), then $f(x_0)$ is a local maximum.

Similarly, if we move past the location of a local minimum from left to right, the function's graph must change from falling downhill to rising uphill. Or, in terms of the derivative, if $f'(x)$ is negative on an interval (a, x_0), and $f'(x)$ is positive on an interval (x_0, b), then $f(x_0)$ is a local minimum.

Let's summarize these observations:

The Derivative Test for Local Extrema

☞ **If $f'(x)$ changes sign from positive to negative at x_0, then a local maximum occurs at x_0.**

☞ **If $f'(x)$ changes sign from negative to positive at x_0, then a local minimum occurs at x_0.**

If $f'(x)$ does not change sign at x_0, then the critical point is the location of neither a local minimum nor a local maximum. (Check this derivative test for yourself on each of the nine critical points in Figure 4.11 below.) By an **absolute maximum** or **absolute minimum**, we mean THE highest or THE lowest output value achieved by the function.

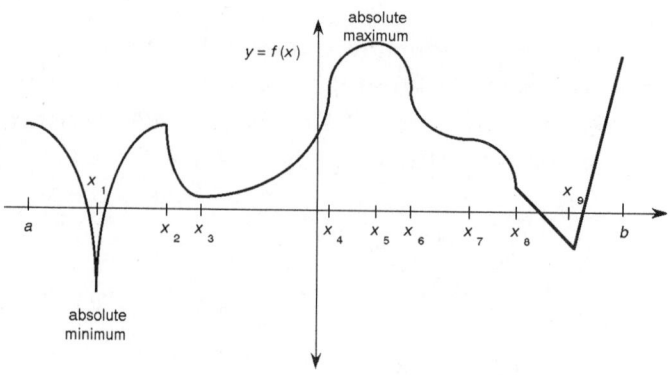

Figure 4.11 Absolute extrema of f over $[a, b]$.

4.3 USING THE DERIVATIVE TO ANALYZE FUNCTION BEHAVIOR

Certainly, a function does not necessarily need to have either an absolute maximum or absolute minimum output value. For example, the function $x \longmapsto 1/x$ takes on *arbitrarily* high and low output values for inputs near $x = 0$. However, if a continuous function is defined on a closed interval $[a, b]$, then we must have absolute maximum and minimum output values occur over that interval (by the Extreme Value Theorem). These extreme values could occur either at the critical points contained in that interval, or possibly at one or more of the endpoints.

In any case, we can outline a step-by-step strategy for locating the absolute extrema of a continuous function f over a closed interval $[a, b]$:

Strategy for Finding Absolute Extrema

Step 1. *Find the derivative f'.*

Step 2. *Find the critical points of f that lie in the interval $[a, b]$.*

This requires finding each of those values $a < x_0 < b$ such that $f'(x_0) = 0$ or $f'(x_0)$ is undefined.

Step 3. *Identify which critical points are local minima and which are local maxima.*

Step 4. *Evaluate $f(x_0)$ for each local extremum and evaluate $f(a)$ and $f(b)$ for comparison to identify the absolute maximum and minimum output values.*

For each critical point x_0 in the interval, we can use the derivative test to identify x_0 as the location of a local minimum, maximum, or neither. We examine the sign of $f'(x)$ to the left and to the right of the critical point x_0. If $f'(x)$ changes sign from positive (+) to negative (−), then $f(x_0)$ is a local maximum value. If $f'(x)$ changes sign from negative (−) to positive (+), then $f(x_0)$ is a local minimum value. (If $f'(x)$ does not change sign, then $f'(x_0)$ is neither a maximum nor minimum.) It is easiest to keep these criteria straight if you visualize the graphical behavior associated with the slope: as we move from left to right, a change from uphill to downhill represents a peak (local maximum) and a change from downhill to uphill represents a valley (local minimum).

Don't forget to compare the function outputs at the critical points to the function outputs $f(a)$ and $f(b)$ at the endpoints of the interval.

EXAMPLE 16 Find the absolute maximum and minimum of $f : x \longmapsto x^3 - 12x$ on the closed interval $[-1, 3]$.

Solution *Step 1.* The derivative is $f' : x \longmapsto 3x^2 - 12$.

Step 2. The derivative is defined over the entire interval. $f'(x) = 3x^2 - 12 = 0$ when $x^2 = 4$ or $x = \pm 2$. Only one of these two critical points is in the given interval $[-1, 3]$: $x = 2$.

Step 3. $3x^2 - 12 < 0$ for $x < 2$ (provided x is reasonably close to 2) and $3x^2 - 12 > 0$ for $x > 2$. This means that f has a *local minimum* at $x = 2$.

Step 4. Finally, we check the actual output values at the one local minimum we found ($x = 2$) and at the endpoints ($x = -1$ and $x = 3$):

$$f(2) = 2^3 - 12 \cdot 2 = -16, \quad f(-1) = (-1)^3 - 12 \cdot (-1) = 11, \quad f(3) = 3^3 - 12 \cdot 3 = -9.$$

From these values, we can see that $f(2) = -16$ is the absolute minimum of f over $[-1, 3]$, and $f(-1) = 11$ is the absolute maximum. ■

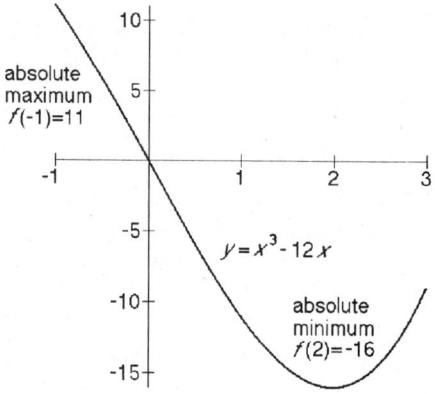

Figure 4.12 Setting the vertical range using the absolute extrema.

We note that knowing this information allows us to specify a vertical range for graphing the function over this interval so that the entire graph fits in our viewing window. Figure 4.12 shows a graph of $y = x^3 - 12x$ using $[-1, 3]$ as the horizontal range and $[-16, 11]$ as the vertical range.

Of course, with machine graphics capability, we can often get a very good estimate of the absolute minimum and maximum function outputs over an interval by simply graphing the function over that interval. We may need to experiment awhile to find a vertical scaling that shows us approximately the locations of the maximum and minimum. Even if our grapher has *autoscaling* (meaning that the machine automatically tries out some inputs from the interval first to determine a reasonable vertical scaling for graphing), we still may need to adjust the scaling to see the approximate value of a local minimum with one view, and the approximate maximum value with another view.

Because a machine grapher ultimately graphs only finitely many values, it's possible that important function behavior is missed between the plotted points. You can see how analyzing the function with the derivative gives us a more definitive way of locating the extrema. We could even use the grapher to aid in our analysis of the derivative. For example, if we graphed *the derivative* over the same interval, then the locations where the derivative's graph crosses over the x-axis indicate local minima or local maxima of the original function.

The Mean Value Theorem for derivatives

The Mean Value Theorem is one of the most important results regarding the derivative of a function. It is a wellspring of information, even though its main idea is fairly simple.

The word *mean* here is used in the sense of *average* and we can think of the mean value theorem as pointing out an important connection between average and instantaneous rate of change. To illustrate the idea of the theorem, let's go back to our example of the car traveling at varying speeds.

Suppose you are driving in a car and over a certain two-hour time interval you cover exactly 80 miles. We already know that this means the *average* velocity for the trip is

$$\frac{total\ distance}{total\ time} = \frac{80\ \text{miles}}{2\ \text{hours}} = 40\ \text{mph}.$$

Now, the question: Did there have to be some instant during the two hours when your instantaneous velocity (speedometer reading) was exactly 40 mph?

The answer is YES. To understand why, suppose for the sake of argument that your speedometer *never* read 40 mph for even a single instant. Then either your car's velocity was always less than 40 mph or always over 40 mph for the entire trip. In the first case, if you always travel at a velocity less than 40 mph for two hours, you cannot possibly cover the entire distance of 80 miles. In the second case, if you always travel at a velocity of more than 40 mph for two hours, you will undoubtedly cover *more* than 80 miles. Since upi covered *exactly* 80 miles on the trip, by process of elimination we must conclude that there is at least one instant at which you are traveling at a velocity of *exactly* 40 mph.

This is the principle by which a toll booth police officer may write a speeding ticket. If your toll ticket has a time stamp, it allows your average speed for the trip between toll booths to be computed. If this average speed is in excess of the speed limit, then the police officer may conclude that you broke the speed limit at least once during the trip.

Now we'll turn to a graphical interpretation of this theorem and the reasoning behind it. First, we'll look at a closely related theorem.

Theorem 4.2

Rolle's Theorem.
Hypothesis 1: f is continuous on $[a, b]$.
Hypothesis 2: f is differentiable on (a, b).
Hypothesis 3: $f(a) = f(b)$.
Conclusion: There is at least one point $a < x_0 < b$ such that $f'(x_0) = 0$.

Reasoning

Geometrically, the hypotheses require the graph of f to connect two points with the same y-coordinate (Hypothesis 3) with an unbroken, smooth curve (Hypotheses 1 and 2). The illustration of such a situation is shown in Figure 4.13 below.

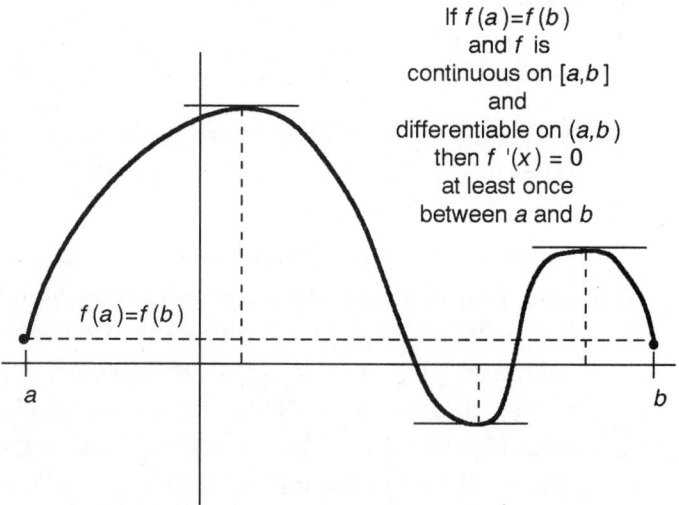

Figure 4.13 Illustration of Rolle's Theorem.

The conclusion of the theorem states that there must be at least one point between a and b where the slope of the graph is $m = 0$. In the particular illustration above, we actually see three such points.

Recall the Extreme Value Theorem: *a continuous function must have definite minimum and maximum output values over a closed interval*. So, Hypothesis 1 guarantees that f must have definite minimum and maximum values on $[a, b]$. If *both* the minimum and maximum happen to occur at the two endpoints, then f must actually be a *constant* function (since $f(a) = f(b)$ by Hypothesis 3). In that case, the conclusion of Rolle's Theorem is easily satisfied, since $f'(x_0) = 0$ for *any* choice of x_0 between a and b (the graph is horizontal between a and b).

If f is not constant, then either the minimum or maximum (or both) occur somewhere strictly between a and b. If x_0 is a point where an extremum occurs, then the Critical Point Theorem (4.1) requires that x_0 be a critical point, which means that $f'(x_0) = 0$ or is undefined. Now Hypothesis 2 is important. Since the derivative is defined for all such values between a and b, we must have $f'(x_0) = 0$. □

Rolle's Theorem can be used to prove the Mean Value Theorem, which we now state formally.

Theorem 4.3

Mean Value Theorem.
Hypothesis 1: f is continuous on $[a, b]$.
Hypothesis 2: f is differentiable on (a, b).
Conclusion: $f'(x_0) = \dfrac{f(b) - f(a)}{b - a}$ for at least one point $a < x_0 < b$.

Notice that the hypotheses are the same as Rolle's Theorem except that we've dropped the hypothesis requiring $f(a) = f(b)$. The function f may take on different values at the endpoints of the interval. Geometrically, the conclusion states that there must be at least one point x_0 between a and b where the slope $f'(x_0)$ matches the slope of the secant line connecting the two points $(a, f(a))$ and $(b, f(b))$.

The illustration of such a situation is shown in Figure 4.14. Note that this illustration looks like a "tilted" version of our illustration for Rolle's Theorem. If $f(a) = f(b)$, the secant line is horizontal and has slope 0. In this case, the conclusion of the Mean Value Theorem is the same as Rolle's Theorem: there is at least one point between a and b where the slope is 0. In fact, we can use Rolle's Theorem to prove the Mean Value Theorem.

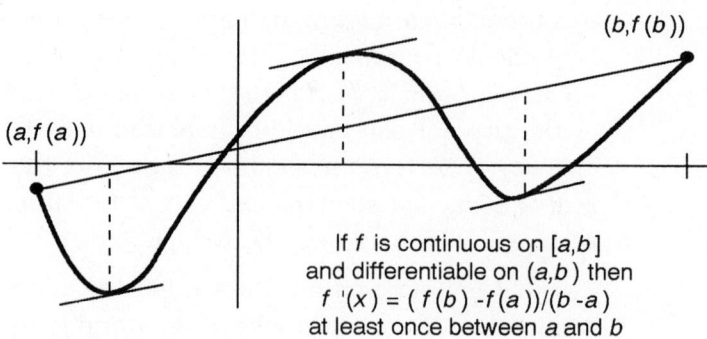

Figure 4.14 Illustration of the Mean Value Theorem.

Reasoning Suppose we let g represent the linear function whose graph is the secant line. Then

$$g(a) = f(a) \quad \text{and} \quad g(b) = f(b)$$

and the slope of the graph of g could be found by using the two-point formula with $(a, f(a))$ and $(b, f(b))$:

$$m = \frac{f(b) - f(a)}{b - a}.$$

Now we can apply Rolle's Theorem to the function difference $f - g$, because $f - g$ still satisfies our first two hypotheses, and we have

$$f(a) - g(a) = 0 = f(b) - g(b).$$

There must be at least one point x_0 such that

$$f'(x_0) - g'(x_0) = 0 \quad \text{or} \quad f'(x_0) = g'(x_0).$$

But g is a linear function, and so its derivative $g'(x_0)$ is just the slope m of the line passing through the two points $(a, f(a))$ and $(b, f(b))$. We can conclude that

$$f'(x_0) = \frac{f(b) - f(a)}{b - a}.$$

$\square$

Rolle's Theorem and the Mean Value Theorem (as well as the Intermediate Value Theorem and the Extreme Value Theorem) can be classified as *existence* theorems. The conclusion of each theorem guarantees the *existence* of one or more real numbers satisfying a particular property or equation. They do not, however, provide us with a recipe or procedure for actually finding these numbers. For that, we are left to our own devices (for example, the Critical Point Theorem tells us exactly *where* to look for local extrema of a function.) Existence theorems are important results in

4.3 USING THE DERIVATIVE TO ANALYZE FUNCTION BEHAVIOR

a couple of ways. For one, they could tell us whether or not we are wasting our time in looking for a solution to an equation. For another, simply knowing that a solution to an equation exists may be far more important information than the value of the actual solution.

Equal derivatives mean parallel graphs

Suppose you know that for all x such that $a < x < b$,

$$f'(x) = 0.$$

Must f be *constant* over the interval (a, b)?

Surely, the derivative of a constant function $f : x \longmapsto c$ will be the constant zero function $f' : x \longmapsto 0$, but we're asking whether or not the *converse* must be true. In other words, could it be possible for the graph of a function to be *flat* at every point over an interval without the graph simply being a continuous flat line over the interval?

We can use the Mean Value Theorem to argue that f must indeed be a constant function over (a, b) by showing that any two inputs x_1 and x_2 in the interval must produce equal function output values $f(x_1) = f(x_2)$. First, note that the slope of the line through the two points $(x_1, f(x_1))$ and $(x_2, f(x_2))$ is

$$m = \frac{f(x_2) - f(x_1)}{x_2 - x_1}.$$

If we apply the Mean Value Theorem to our function f over the interval $[x_1, x_2]$, then there exists at least one point x_0 between x_1 and x_2 such that $f'(x_0) = m$. But we know the derivative $f'(x) = 0$ over the whole interval, so $m = 0$, so the numerator $f(x_2) - f(x_1) = 0$. This means that $f(x_2) = f(x_1)$. We will highlight a generalization of this conclusion as a theorem.

Theorem 4.4

Hypothesis: Suppose two functions f and g have the same derivative over (a, b). That is, $f'(x) = g'(x)$ for $a < x < b$.
Conclusion: f and g differ only by a constant function over (a, b).
In other words, there is some constant C such that $f(x) = g(x) + C$ for $a < x < b$.

Reasoning Since the derivative of $f - g$ is $f' - g'$, the hypothesis means that the derivative of $f - g$ is zero (0) over the whole interval (a, b). From our discussion above, we can conclude that $f - g$ is a constant function $x \longmapsto C$ for some real number C, or

$$f(x) - g(x) = C \quad for \quad a < x < b.$$

Adding $g(x)$ to both sides of the equation gives us the conclusion of the theorem. □

Graphically, just as two lines with the same slope are parallel, two functions with the same derivative have "parallel" graphs, in the sense that the vertical distance (the constant C) between them is always the same.

EXERCISES

For each function given in exercises 1-10:

a) find all the critical points, and classify each as a local minimum, local maximum, or neither

b) identify the intervals over which the function is strictly increasing

c) identify the intervals over which the function is strictly decreasing

d) graph the function and its derivative over the interval $[-10, 10]$ to verify that the original function is strictly increasing when the derivative is positive, and strictly decreasing when the derivative is negative.

1. $x \longmapsto x^3 - x$
2. $x \longmapsto \frac{1}{x}$
3. $x \longmapsto -\frac{4}{(x+1)^2}$
4. $x \longmapsto 3$
5. $x \longmapsto \tan x$
6. $x \longmapsto \csc x$
7. $x \longmapsto \sec x$
8. $x \longmapsto \cot x$
9. $x \longmapsto |2x - 3|$
10. $x \longmapsto \sin^2(x) + \cos^2(x)$

Find the absolute maximum and absolute minimum of each function in exercises 11-18 on the indicated closed interval.

11. $f : x \longmapsto 2x^2 + 3x - 1$ $[-2, 1]$
12. $f : x \longmapsto -4x^3 + 5$ $[1, -1]$
13. $f : x \longmapsto -x^2 - 3x + 6$ $[1, 2]$
14. $f : x \longmapsto x^2 - 3x + 1$ $[-4, -2]$
15. $f : x \longmapsto -2x^2 + 3x - 1$ $[-1, 4]$
16. $f : x \longmapsto |\sin(x/3)|$ $[\pi/4, \pi/3]$

4.3 USING THE DERIVATIVE TO ANALYZE FUNCTION BEHAVIOR

In each of exercises 17-20, you are given the formula for the position function $s : t \longmapsto s(t)$ of some object over the time interval $0 \leq t \leq 60$. (s is measured in feet and t is measured in seconds.) For each function s, find the absolute maximum value of the position over the interval $[0, 60]$ and the time(s) t that this value occurs.

17. $s : t \longmapsto -16t^2 + 64000$

18. $s : t \longmapsto .002t^3 - 0.27t^2 + 12t + 10$

19. $s : t \longmapsto |90 - t^2/30|$

20. $s : t \longmapsto 2 \sin \pi t/20$

---

*In exercises 21-28, you are given the graphs of the **derivatives** of eight functions. For each, indicate the locations of the local maxima and local minima (if any) of the original function.*

21.

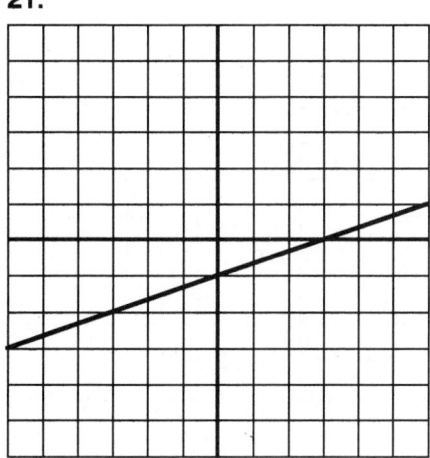

22.

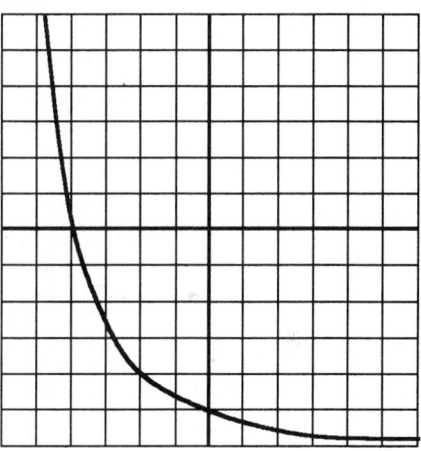

23.

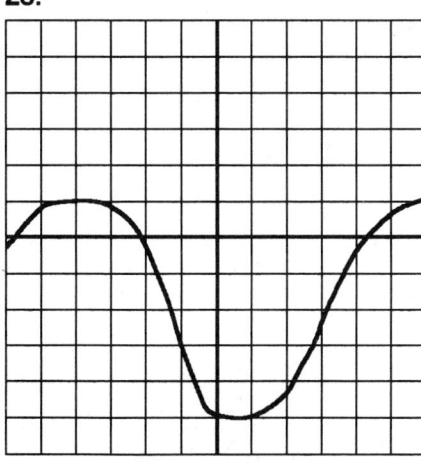

24.

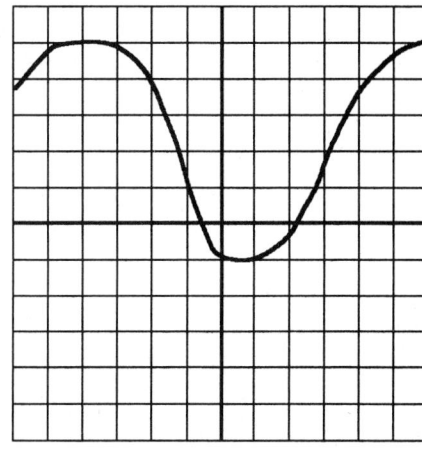

25.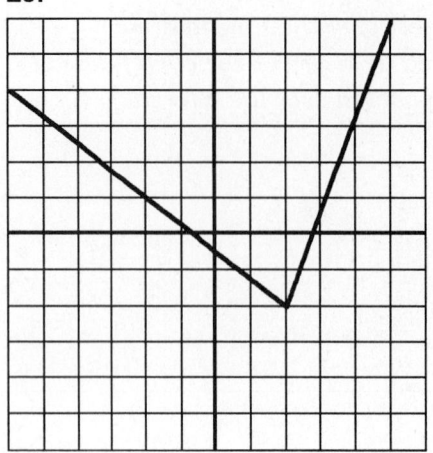

26.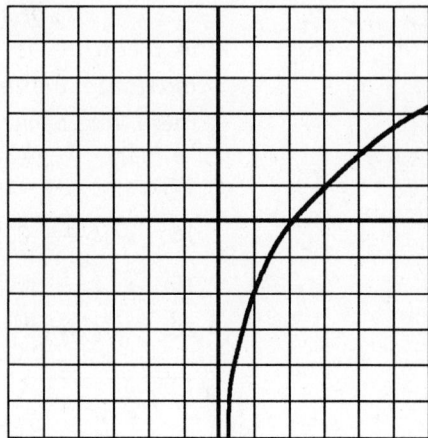

27.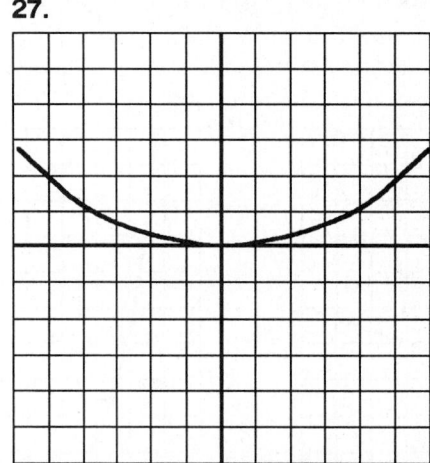

28.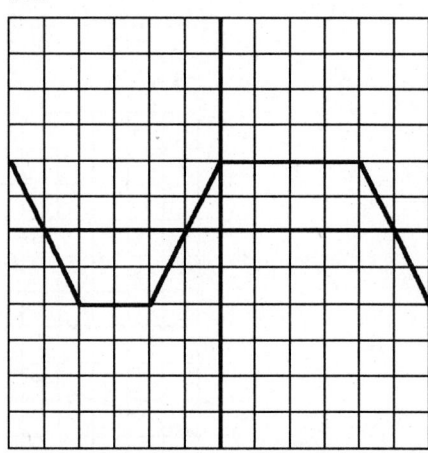

29. For each of the graphs shown in exercises 21-28, approximate the slope of the secant line connecting the leftmost and rightmost points on the graph. Locate visually (if possible) at least one point at which the slope of the tangent line is the same as the slope of this secant line.

30. For which two of the graphs was this impossible? Why doesn't this contradict the Mean Value Theorem?

31. Verify that $f : x \longmapsto (\sin(x) + \cos(x))^2$ and $g : x \longmapsto \sin(2x)$ have the same derivative. Graph both functions to see that the graphs are parallel. What is the constant difference C in this case?

4.4 HIGHER ORDER DERIVATIVES

The derivative of a function f is a new function f'. We could, in turn, find its derivative $(f')'$ or simply f''. This is known as the *second* derivative of the original function f.

EXAMPLE 17 Find the second derivative of $f : x \longmapsto x^3 - 5x^2 + 4x - 1$.

Solution The first derivative is $f' : x \longmapsto 3x^2 - 10x + 4$, and the second derivative is $f'' : x \longmapsto 6x - 10$. ∎

If y is used to denote the output of f, then other notations for the second derivative include

$$D^2 f \qquad D^2 y \qquad \frac{d^2 y}{dx^2} \qquad \frac{d^2 f}{dx^2} \qquad y'' \qquad \ddot{y}.$$

If you are wondering about the placement of the "exponents" on the Leibniz notation for derivative, remember that we think of d/dx as an *operator*, so

$$\frac{d^2 y}{dx^2} \qquad \text{represents} \qquad \frac{d}{dx}\left(\frac{dy}{dx}\right)$$

and we can see the logic behind the superscript locations.

We can continue this process to find higher order derivatives. The third derivative (derivative of the second derivative) is denoted f''' or

$$D^3 f \qquad D^3 y \qquad \frac{d^3 y}{dx^3} \qquad \frac{d^3 f}{dx^3} \qquad y''' \qquad \dddot{y}.$$

For fourth and higher derivatives, the prime notation becomes unwieldy. We replace the prime marks by a superscript in parentheses:

$$f^{(4)}, \quad f^{(5)}, \quad f^{(6)}, \ldots$$

represent the fourth, fifth, sixth, and so on derivatives of f. The parentheses serve to distinguish between the derivative and a power of the function. Note the distinction:

$$y^4 = y \cdot y \cdot y \cdot y \qquad \text{and} \qquad y^{(4)} = y''''.$$

Physical interpretation of the second derivative

What exactly does the second derivative y'' tell us? First, since it is the derivative of y', it tells us how fast the first derivative is changing with respect to change in the independent variable. If we think back to our automobile example, the first derivative of distance with respect to time ds/dt was the velocity of the car. How fast this velocity changes is referred to as the *acceleration*. For example, when someone says a car can go from 0 to 60 mph in 10 seconds, we have been given the data needed to compute the *average acceleration* of the car.

$$\text{average acceleration} = \frac{\text{total change in speed}}{\text{total elapsed time}} = \frac{60 \text{ miles/hour}}{10 \text{ seconds}} = 8.8 \text{ ft/sec}^2.$$

What exactly is a "square second" $\sec^2$? Notice that in this context, the shorthand abbreviation for the units ft/sec^2 is best interpreted as $\frac{\text{ft/sec}}{\sec}$ (feet per second per second).

Just as the instantaneous velocity is given by the first derivative of position, the *instantaneous acceleration* is given by the first derivative of velocity, or equivalently, the second derivative of position:

$$a(t) = \frac{dv}{dt} = \frac{d^2 s}{dt^2}.$$

EXAMPLE 18 The vertical position of an object thrown from the top of a building is given by the function

$$s : [0, 10] \longmapsto \mathbb{R}$$

$$s(t) = -16t^2 + 96t + 640 \text{ feet}.$$

Find the object's position, velocity, and acceleration at $t = 5$ seconds.

Solution $s(5) = -16 \cdot 5^2 + 96 \cdot 5 + 640 = 240$ feet above the ground.

Since the velocity $v(t) = \dfrac{ds}{dt} = -32t + 96$, we have

$$v(5) = -32 \cdot 5 + 96 = -64 \; \frac{\text{ft}}{\sec} \text{ (feet per second).}$$

The negative sign means the object is heading toward the ground at this instant.

Finally, the acceleration $a(t) = \dfrac{dv}{dt} = \dfrac{d^2 s}{dt^2} = -32$, so

$$a(5) = -32 \; \frac{\text{ft}}{\sec^2} \text{ (feet per second per second).}$$

Notice that for this example, the acceleration function is constant. In fact, this is exactly the acceleration due to the force of gravity. The negative sign indicates that the acceleration is also directed downward. ∎

Concavity and inflection points

While the second derivative f'' can be interpreted in a physical sense as acceleration, its graphical interpretations relate to both the graph of the original function and the graph of the first derivative.

Certainly, we can use the second derivative f'' to analyze the behavior of the first derivative f' in exactly the same way we use the first derivative to analyze the behavior of the original function f. In other words,

☞ **If $f''(x) > 0$ on an interval (a,b), then $f'(x)$ is strictly** *increasing* **on that interval.**

☞ **If $f''(x) < 0$ on an interval (a,b), then $f'(x)$ is strictly** *decreasing* **on that interval.**

Definition 5
> When the *slope* of a graph is strictly increasing, we say the graph is **concave up**. When the *slope* of a graph is strictly decreasing, we say the graph is **concave down**.

In terms of the original function's graph, the sign of the second derivative is evidenced by the *concavity*. Figure 4.15 illustrates the visual appearance of concavity in the graph of a function $y = f(x)$. As suggested by this picture, you can use the visual idea of a cup turned up or down to remind you of the graphical meanings of *concave up* and *concave down*.

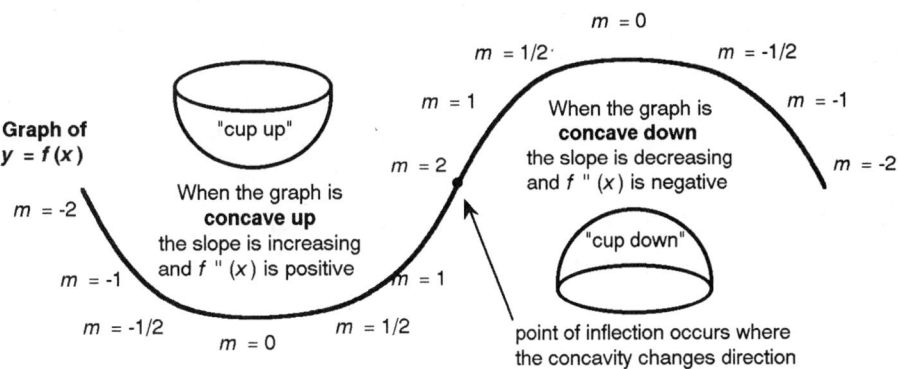

Figure 4.15 Graphical interpretation of concavity.

A point on the graph where the curve changes concavity is called a *point of inflection*. A point of inflection for the curve in Figure 4.15 is indicated. We could define a point of inflection formally using the second derivative.

Definition 6 | A point $(x_0, f(x_0))$ is called a **point of inflection** for the graph of f if there is a neighborhood of x_0 over which $f''(x)$ is positive on one side of x_0 and negative on the other side (f'' changes sign at x_0).

To find the locations of these points, we need to *check* those values x_0 in the domain of the function such that $f''(x_0) = 0$ or is *undefined*. Keep in mind that these are *potential* locations of points of inflection, just as critical points are potential locations of maxima and minima. The true test determining an inflection point is whether the concavity of the graph changes direction.

☞ **At a point of inflection, the second derivative must change sign.**

Figure 4.16 illustrates the intervals over which the graph of f is concave up, concave down, or neither.

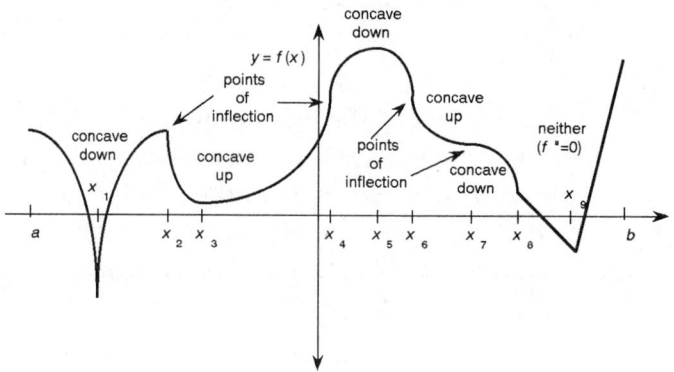

Figure 4.16 Concavity and points of inflection.

EXAMPLE 19 Using Figure 4.16, describe the behavior of the second derivative f'' over the interval (a, b).

Solution For $a < x < x_1$ and $x_1 < x < x_2$ the second derivative value $f''(x)$ is *negative*. The second derivative value $f''(x)$ is also negative for $x_4 < x < x_6$ and $x_7 < x < x_8$.

For $x_2 < x < x_4$ and $x_6 < x < x_7$, $f''(x)$ is *positive*.

The function appears to be linear for $x_8 < x < x_9$ and for $x_9 < x < b$. Since the slope is *constant* over these intervals, we have $f''(x) = 0$ over these intervals.

We note that at $x = x_1, x_2, x_4, x_6, x_8$, and x_9, the second derivative is definitely undefined (since even the first derivative is undefined at these points). The points of inflection for the graph in Figure 4.16 are $(x_2, f(x_2))$, $(x_4, f(x_4))$, $(x_6, f(x_6))$, $(x_7, f(x_7))$. Note that f'' is undefined at $x = x_2$, x_4, and x_6 in Figure 4.16. As for $f''(x_7)$, we cannot tell visually whether or not it is zero or undefined. ∎

Second derivative test for extrema

If $f(x_0)$ is either a local maximum or local minimum, then we have already seen that the first derivative f' must change sign at $x = x_0$. If $f'(x_0)$ is defined, then the only possibility is $f'(x_0) = 0$. If $f''(x_0)$ is also defined, then we can sometimes use its value to identify whether $f(x_0)$ is a local maximum or local minimum.

To illustrate, let's examine the local minimum at $x = x_3$ and the local maximum at $x = x_5$ in Figure 4.16. We have $f'(x_3) = 0$ and $f''(x_3)$ is *positive* (concave up graph), so $(x, f(x_3))$ is the bottom point of the "cup." In contrast, $f'(x_5) = 0$ also, but now $f''(x_5)$ is *negative* (concave down graph), so $(x_5, f(x_5))$ is the point at the top of an overturned cup. We have essentially described what is known as the *second derivative test for extrema*.

The Second Derivative Test for Extrema

If $f'(x_0) = 0$ and $f''(x_0)$ is *positive*, then $f(x_0)$ is a local minimum.

If $f'(x_0) = 0$ and $f''(x_0)$ is *negative*, then $f(x_0)$ is a local maximum.

The second derivative test for extrema is really based on the first derivative test for extrema. The second derivative tells us the *rate of change of the slope*. Hence, if the slope of the graph is zero at a point, but the slope is increasing, then the slope must be changing from negative to positive (a local minimum). If the slope is decreasing at that point, then the slope must be changing from positive to negative (a local maximum).

However, the second derivative test is not as general as the first derivative test. It gives us no information if x_0 is a critical point such that $f'(x_0)$ is undefined (like at a cusp or corner in the graph). Also, note that if $f'(x_0) = 0$ and $f''(x_0) = 0$ or is *undefined*, then $f(x_0)$ may be a local maximum, a local minimum, or neither. In these cases, we must check whether or not the *first* derivative f' changes sign at x_0. In other words, if the second derivative test fails to give us any information, we can always fall back on the first derivative test.

EXAMPLE 20 Use the second derivative test to determine the local extrema of the function $f : x \longmapsto x^3 - 12x$.

Solution The first derivative is $f' : x \longmapsto 3x^2 - 12$, which is defined for all real values. The critical points of f are the zeroes of f':

$$x = \pm 2.$$

The second derivative is $f'' : x \longmapsto 6x$. Since $f''(-2) = -12$ is negative (graph concave down), we conclude that $f(-2) = 16$ is a local *maximum*. Since $f''(2) = 12$ is positive (graph concave up), we conclude that $f(2) = -16$ is a local *minimum*. ∎

Best quadratic approximations

Just as the first derivative enables us to find the best linear approximation to a function at a point, the second derivative enables us to find the best quadratic (second degree) approximation.

The best linear approximation to a function f at a point x_0 will have a graph passing through the point $(x_0, f(x_0))$ and has a slope equal to the derivative value $f'(x_0)$. If we called this linear approximation function g, then we could state these requirements as

$$g(x_0) = f(x_0) \qquad \text{and} \qquad g'(x_0) = f'(x_0).$$

These two requirements completely determine the linear function g.

Definition 7

> The **best quadratic approximation** p to a function f at x_0 is the second degree polynomial function
>
> $$p : x \longmapsto ax^2 + bx + c$$
>
> which satisfies the *three* requirements:
>
> $$p(x_0) = f(x_0) \qquad p'(x_0) = f'(x_0) \qquad p''(x_0) = f''(x_0).$$

These three requirements completely determine the three coefficients a, b, and c, just as the two requirements for the best linear approximation completely determine its slope and y-intercept.

4.4 HIGHER ORDER DERIVATIVES

EXAMPLE 21 Find the best quadratic approximation p to the function $f : x \longmapsto \cos x$ at $x = 0$.

Solution First, we'll calculate

$$f'(x) = -\sin x \quad \text{and} \quad f''(x) = -\cos x.$$

The first two derivatives of a general quadratic function $p : x \longmapsto ax^2 + bx + c$ are

$$p'(x) = 2ax + b \quad \text{and} \quad p''(x) = 2a.$$

The requirements for the *best* quadratic approximation function for f at x_0 are:

$$c = a \cdot 0^2 + b \cdot 0 + c = p(0) = f(0) = \cos 0 = 1$$
$$b = 2a \cdot 0 + b = p'(0) = f'(0) = -\sin 0 = 0$$
$$2a = p''(0) = f''(0) = -\cos 0 = -1.$$

From these we can see that

$$c = 1 \quad b = 0 \quad a = -\frac{1}{2}$$

so that $p(x) = -x^2/2 + 1$. ∎

In Figure 4.17 we have graphed both $y = \cos(x)$ and $y = -x^2/2 + 1$, and we can see how "snug" the fit is between the two graphs near $x = 0$.

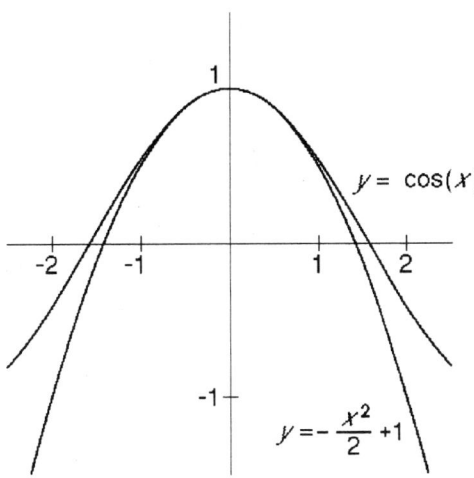

Figure 4.17 The graphs of $y = \cos(x)$ and its best quadratic approximation at $x = 0$.

Cubic splines

In the previous chapter we used piecewise linear functions to motivate the idea of local linearity. When data is gathered in the form of ordered pairs of inputs and outputs, a first step in "fitting a curve" to the data is to plot the ordered pairs and connect them with straight line segments. The result is the graph of a piecewise linear function. The nice properties that this approximation has are:

1) its graph passes through all the data points, and

2) it is a continuous function.

Now, suppose we also know the instantaneous rate of change at each data input. Graphically, this means we know the local slope at each of the points we plot. Can we fit a curve to the data so that it matches both the output value *and* the local slope at each input?

We cannot get by very well with piecewise linear approximations, because the resulting function would not, in general, be continuous, and even if it were continuous, the first derivative might not be. Figure 4.18 illustrates the problems we run into trying to fit a piecewise linear function to the data.

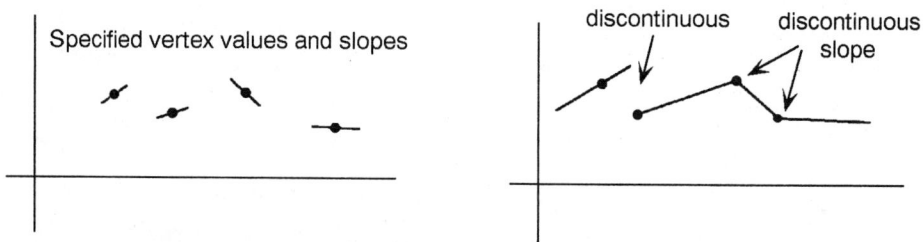

Figure 4.18 Fitting piecewise linear functions to both slopes and outputs.

Our goal, then, is to create a continuous function such that

1) its graph passes through all the data points,

2) the local slopes are correct at each data point, and

3) its first derivative is continuous.

Given a finite set of vertices, we can accomplish this goal by using piecewise *cubic* rather than linear functions. Such a function is called a piecewise cubic approximation or **cubic spline**. Cubic splines are used in some computer graphics programs to produce smooth curves through a set of given points. Also, since polynomials are easy to work with computationally, cubic splines may be used in place of transcendental functions like trigonometric or exponential functions.

4.4 HIGHER ORDER DERIVATIVES

Let's examine how one determines a cubic spline approximation function. If we have a given set of input points (vertices), then between each pair of consecutive vertices x_0 and x_1, our cubic spline function f will have some evaluation formula

$$f : x \longmapsto ax^3 + bx^2 + cx + d$$

for the right choices of the parameters a, b, c, and d. Between another pair of vertices, the choices of a, b, c, and d may change.

Now, we assume that we know the outputs y_0 and y_1 as well as the local slopes y'_0 and y'_1 corresponding to the two inputs x_0 and x_1. Our task is to find a cubic function f that matches the outputs

$$y_0 = f(x_0) \quad \text{and} \quad y_1 = f(x_1),$$

as well as the first derivative values

$$y'_0 = f'(x_0) \quad \text{and} \quad y'_1 = f'(x_1).$$

How can we determine the parameters a, b, c, and d? Note that we have four separate pieces of information (the two outputs and two slopes) that we can use to solve for these four unknown parameters.

Let's look at an example to illustrate.

EXAMPLE 22 Find a cubic function $f : x \longmapsto ax^3 + bx^2 + cx + d$ whose values and slopes agree with $x \longmapsto \sin(x)$ at $x = 0$ and $x = 1$. Compare its graph with that of $x \longmapsto \sin(x)$ on the interval $[0, 2]$.

Solution First, let's gather our pieces of information together.

$$\frac{d(\sin(x))}{dx} = \cos(x),$$

so for $x_0 = 0$, we have

$$y_0 = \sin(0) = 0 \quad \text{and} \quad y'_0 = \cos(0) = 1$$

and for $x_1 = 1$, we have

$$y_1 = \sin(1) \approx 0.84147 \quad \text{and} \quad y'_1 = \cos(1) = \approx 0.54030.$$

Our cubic function f has derivative $f' : x \longmapsto 3ax^2 + 2bx + c$, so

$$f(0) = d = y_0 = 0 \quad \text{and} \quad f'(0) = c = y'_0 = 1$$

from which we can see that $d = 0$ and $c = 1$.

Now at $x_1 = 1$, f must satisfy

$$f(1) = a + b + 1 = y_1 \approx 0.84147 \quad \text{and} \quad f'(1) = 3a + 2b + 1 \approx 0.54030.$$

We can solve for a and b to obtain $a \approx -0.14264$ and $b \approx -0.0.015889$. With all the parameters determined, we have our cubic spline function

$$f(x) \approx (-0.14264)x^3 - (0.015889)x^2 + x.$$

The graphs of f and $x \longmapsto \sin(x)$ over the interval $[0, 2]$ are shown in Figure 4.19. Notice that the two curves fit quite closely over the whole interval $[0, 1]$, but they begin diverging outside this interval. ∎

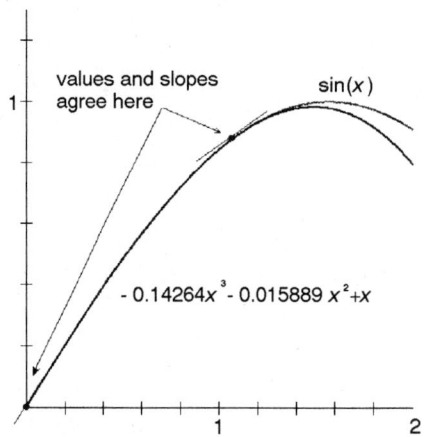

Figure 4.19 Graphs of $x \longmapsto \sin(x)$ and a cubic spline.

With this example as a guide, we can generalize the method to find a cubic spline approximation over any interval $[x_0, x_1]$ where the outputs y_0 and y_1 and the slopes y'_0 and y'_1 are known at the endpoints. The two pairs of requirements that must be met are:

slope requirements:

$$f'(x_0) = 3ax_0^2 + 2bx_0 + c = y'_0 \qquad f'(x_1) = 3ax_1^2 + 2bx_1 + c = y'_1$$

output requirements:

$$f(x_0) = ax_0^3 + bx_0^2 + cx_0 + d = y_0 \qquad f(x_1) = ax_1^3 + bx_1^2 + cx_1 + d = y_1.$$

4.4 HIGHER ORDER DERIVATIVES

While in general $x_0 \neq 0$ and $y_0 \neq 0$, we can simplify the process by first considering the special case where $x_0 = 0$ and $y_0 = 0$. In other words, we'll assume that the first vertex is the origin of our coordinate system. Once we have determined the solution here, we can use what we know about translating function graphs to find the solution for arbitrary x_0 and y_0 (see Figure 4.20).

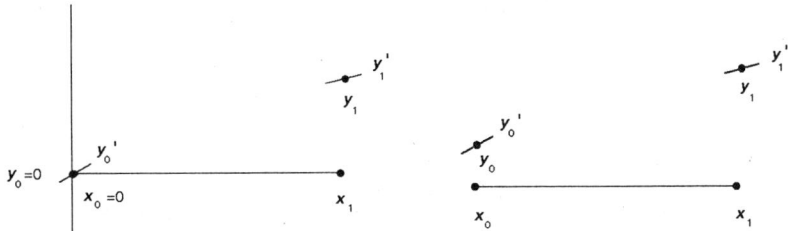

Figure 4.20 Solution for $x_0 = 0$ and $y_0 = 0$ can be translated to other situations.

Since $f(0) = d = 0$, and $f'(0) = c = y_0'$, we must have

$$f(x) = ax^3 + bx^2 + y_0'x.$$

Next, consider the requirement:

$$y_1' = f'(x_1) = 3ax_1^2 + 2bx_1 + y_0'.$$

Solving for b (note that x_1 can't be 0 since $x_0 = 0$ and $x_1 \neq x_0$), we have

$$b = \frac{y_1' - y_0'}{2x_1} - \frac{3}{2}ax_1.$$

This gives us the coefficient b in terms of known quantities and a.

To find a, we consider the last requirement $f(x_1) = ax_1^3 + bx_1^2 + y_0'x_1 = y_1$. If we substitute for b and solve for a, we find

$$a = \frac{y_1' + y_0'}{x_1^2} - \frac{2y_1}{x_1^3}.$$

Substituting this back into the formula for b, we have

$$b = \frac{3y_1}{x_1^2} - \frac{2y_0' + y_1'}{x_1}$$

and we now have formulas giving a, b, and c in terms of y_0', x_1, y_1, and y_1' for the case where $x_0 = 0$ and $y_0 = 0$.

To find the parameters for the cubic approximation when the first vertex is not the origin, we need to recall how we can translate function graphs vertically and horizontally.

To translate the graph of this function vertically by y_0 units and horizontally by x_0 units, we need to replace y and y_1 by $y - y_0$ and $y_1 - y_0$, respectively, and replace x and x_1 by $x - x_0$ and $x_1 - x_0$, respectively.

This will translate the left endpoint of our cubic segment from the origin to the point (x_0, y_0), giving us the formula

$$f : x \mapsto a(x - x_0)^3 + b(x - x_0)^2 + c(x - x_0) + d$$

where $d = y_0$, $c = y_0'$,

$$b = \frac{3(y_1 - y_0)}{(x_1 - x_0)^2} - \frac{2y_0' + y_1'}{x_1 - x_0}, \quad \text{and} \quad a = \frac{y_1' + y_0'}{(x_1 - x_0)^2} - \frac{2(y_1 - y_0)}{(x_1 - x_0)^3}.$$

EXAMPLE 23 Find the piecewise cubic approximation, f, to $x \mapsto \exp(x/2)$ on the interval $[0, 3]$ with vertices 0, 1, 2, and 3.

Solution Our approximation will have a split definition with three regions, $[0, 1]$, $[1, 2]$, and $[2, 3]$. On each of these regions we can use the formulas we derived above.

On the first region, with $x_0 = 0$ and $x_1 = 1$, we have

$$y_0 = \exp(0) \quad y_1 = \exp(1/2) \quad y_0' = \frac{1}{2}\exp(0) \quad y_1' = \frac{1}{2}\exp(1/2)$$

so that over $[0, 1]$

$$f(x) \approx 0.026918(x - 0)^3 + 0.12180(x - 0)^2 + 0.5x + 1.$$

In the second region, with $x_0 = 1$ and $x_1 = 2$, we have

$$y_0 = \exp(1/2) \quad y_1 = \exp(1) \quad y_0' = \frac{1}{2}\exp(1/2) \quad y_1' = \frac{1}{2}\exp(1)$$

so that over $[1, 2]$

$$f(x) \approx 0.044380(x - 1)^3 + 0.20082(x - 1)^2 + 0.82436(x - 1) + 1.648721.$$

In the last region, with $x_0 = 2$ and $x_1 = 3$, we have

$$y_0 = \exp(1) \quad y_1 = \exp(3/2) \quad y_0' = \frac{1}{2}\exp(1) \quad y_1' = \frac{1}{2}\exp(3/2)$$

so that over $[2, 3]$

$$f(x) \approx 0.073171(x - 2)^3 + 0.33110(x - 2)^2 + 1.3591(x - 2) + 2.71828.$$

4.4 HIGHER ORDER DERIVATIVES

Altogether we can describe the function f as

$$f : x \longmapsto \begin{cases} 0.026918(x-0)^3 + 0.12180(x-0)^2 + 0.5x + 1 \\ \qquad\qquad\qquad\qquad\qquad \text{for } 0 \leq x \leq 1 \\ 0.044380(x-1)^3 + 0.20082(x-1)^2 + 0.82436(x-1) + 1.648721 \\ \qquad\qquad\qquad\qquad\qquad \text{for } 1 \leq x \leq 2 \\ 0.073171(x-2)^3 + 0.33110(x-2)^2 + 1.3591(x-2) + 2.71828 \\ \qquad\qquad\qquad\qquad\qquad \text{for } 1 \leq x \leq 2 \\ \text{undefined} \quad \text{otherwise.} \end{cases}$$

■

EXERCISES

For each of the functions in exercises 1-8:
a) find all critical point(s) of f
b) apply the second derivative test to each critical point to determine whether a local minimum or local maximum occurs there (if the second derivative test fails, use the first derivative test)
c) find the interval(s) where the graph of f is concave up
d) find the interval(s) where the graph of f is concave down
e) find the point(s) of inflection for the graph of f.

1. $f : x \longmapsto -3x^4 + 4x^3$
2. $f : x \longmapsto x^3 + 3x^2 - 9x - 2$
3. $f : x \longmapsto -x^4 + 8x^3 - 18x^2 + 1$
4. $f : x \longmapsto 5x^3 - x^2 - x + 2$
5. $f : x \longmapsto 2x^3 + 2x^2 - 2x - 1$
6. $f : x \longmapsto -3x^4 - 8x^3 - 6x^2 + 3$
7. $f : x \longmapsto \dfrac{x^2}{x^2 - 4}$
8. $f : x \longmapsto \dfrac{1}{x^3 - x}$

Since each of the trigonometric functions is periodic, any points of inflection for their graphs will be spaced evenly apart. Where do the points of inflection occur for each of the trigonometric functions in exercises 9-14?

9. $x \longmapsto \sin(x)$
10. $x \longmapsto \cos(x)$
11. $x \longmapsto \tan(x)$
12. $x \longmapsto \cot(x)$
13. $x \longmapsto \csc(x)$
14. $x \longmapsto \sec(x)$

Each of the functions in exercises 15-20 is either strictly increasing or strictly decreasing over its entire domain. Identify which and indicate the points of inflection (if any) and over what intervals the graph of the function is concave up or concave down.

15. $x \longmapsto \arcsin(x)$
16. $x \longmapsto \arccos(x)$
17. $x \longmapsto \arctan(x)$
18. $x \longmapsto \log_{10}(x)$
19. $x \longmapsto 2^x$
20. $x \longmapsto 3^{-x}$

In exercises 21-28, you are given the graphs of the **derivatives** of eight functions. For each, indicate the locations of the points of inflection and over what intervals the graph of the original function is concave up or concave down.

21.

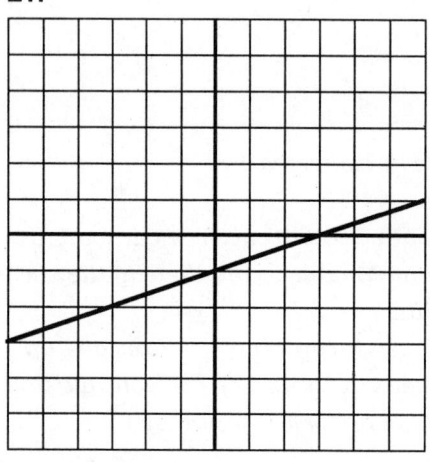

22.

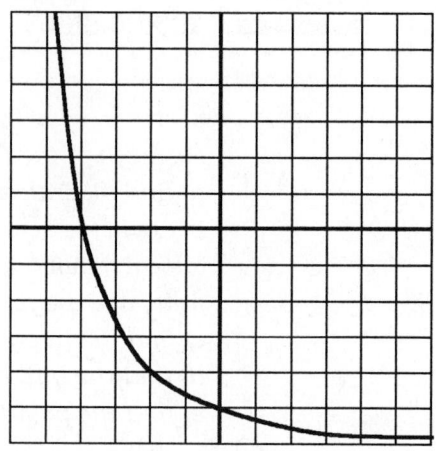

23.

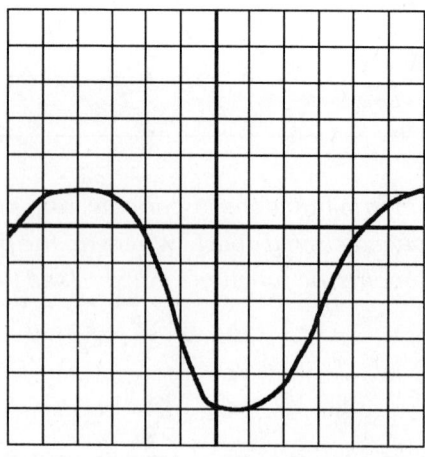

24.

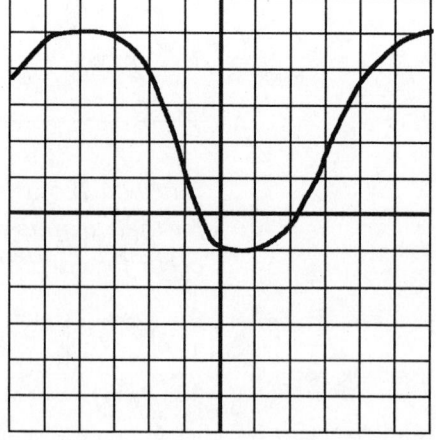

4.4 HIGHER ORDER DERIVATIVES

25.
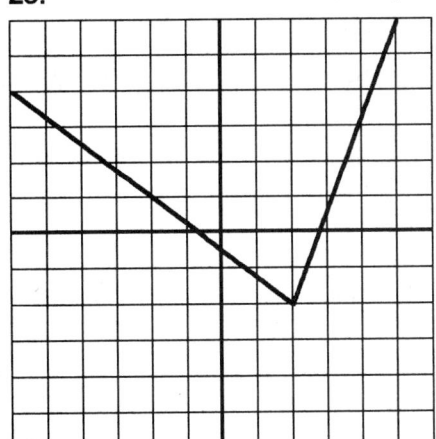

26.
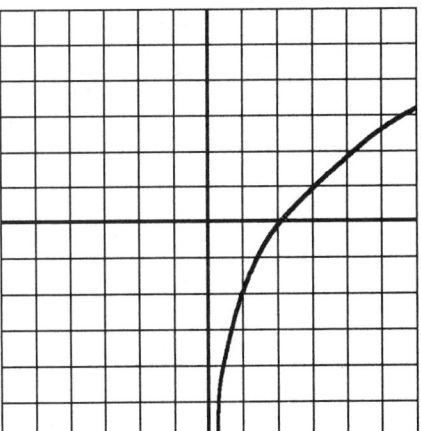

27.
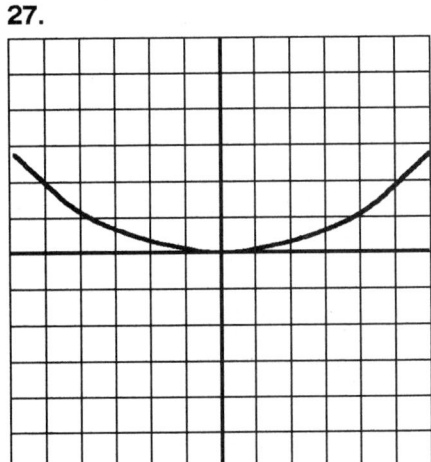

28.
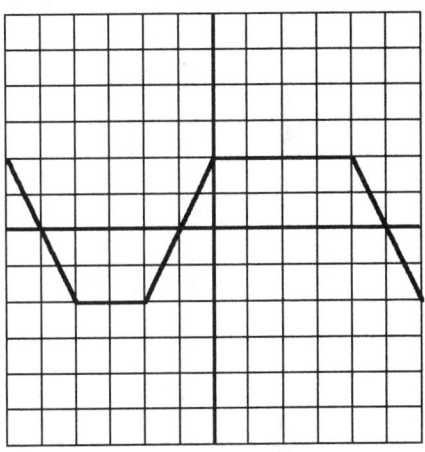

Rational power functions provide a good set of examples of the different kinds of behavior a function can have at a point. Graph each of the functions given in exercises 29-38 over the interval $[-5, 5]$. Each of your graphs should pass through the origin $(0,0)$. Using your graphs and analyzing the first and second derivatives, answer the questions in exercises 39-48.

29. $f : x \longmapsto x^{1/2}$ **30.** $f : x \longmapsto x^{1/3}$

31. $f : x \longmapsto x^{3/2}$ **32.** $f : x \longmapsto x^{2/3}$

33. $f : x \longmapsto x^{5/2}$ **34.** $f : x \longmapsto x^{5/3}$

35. $f : x \longmapsto x^{4/3}$ **36.** $f : x \longmapsto x^{7/3}$

37. $f : x \longmapsto x^{4}$ **38.** $f : x \longmapsto x^{3}$

39. Which functions have a critical point at $x = 0$?

40. Which of the graphs have a point of inflection at $(0,0)$?

41. Which of the graphs have a horizontal tangent at $x = 0$?

42. Which of the graphs have a vertical tangent at $x = 0$?

43. Which of the graphs have a cusp at $x = 0$?

44. For which of these functions does $f'(0) = 0$, yet f has neither a local minimum nor a local maximum at $x = 0$?

45. For which of these functions does $f''(0) = 0$, yet f does not have $(0,0)$ as a point of inflection?

46. For which of these functions is $f'(0)$ undefined, yet f has neither a local minimum nor a local maximum at $x = 0$?

47. For which of these functions is $f'(0) = 0$, but $f''(0)$ is undefined?

48. For which of these functions is $f''(0)$ undefined, yet f does not have $(0,0)$ as a point of inflection?

In each of exercises 49-52, you are given the formula for the position function $s: t \longmapsto s(t)$ of some object over the time interval $0 \leq t \leq 60$. (s is measured in feet and t is measured in seconds.) For each function s, find the following:
a) its initial acceleration (at time $t = 0$)
b) the specific times in the interval $[0, 60]$ for which the acceleration is 0
c) the specific positions (values of s) for which the velocity is at its greatest
d) the specific times in the interval $[0, 60]$ for which the acceleration is at its greatest
e) the graph of the acceleration $y = a(t)$ over the interval $[0, 60]$.

49. $s: t \longmapsto -16t^2 + 64000$

50. $s: t \longmapsto .002t^3 - 0.27t^2 + 12t + 10$

51. $s: t \longmapsto |90 - t^2/30|$

52. $s: t \longmapsto 2 \sin \pi t/20$

---

53. The graph of the upper half of the unit circle can be obtained by graphing $y = \sqrt{1 - x^2}$ on the interval $[-1, 1]$. Using this and the illustration below, find the cubic spline approximation to the unit circle on the intervals $[-\frac{\sqrt{2}}{2}, 0]$ and $[0, \frac{\sqrt{2}}{2}]$. Graph your cubic spline over the interval $[-1, 1]$.

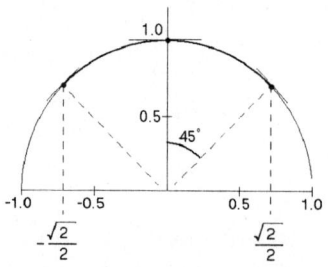

54. Graph both $y = e^{x/2}$ and the cubic spline of example 23 over the interval $[0, 3]$.

4.4 HIGHER ORDER DERIVATIVES

In exercises 55-61 you are given a function, f, and points x_0 and x_1. In each exercise, compute and graph both the cubic spline approximation to f with vertices x_0 and x_1, and the second order Taylor polynomial approximation to f at x_0. Compare these with f (i.e. graph the error) and find a point, $x_0 < c < x_1$, where the cubic spline is a better approximation and a point, $x_0 < c < x_1$, where the Taylor polynomial is a better approximation. If one or the other of these points doesn't exist, state that fact.

55. $f : x \longmapsto \sin(x)$ $\qquad x_0 = 0, x_1 = \pi/2$

56. $f : x \longmapsto \sqrt{x+1} - \sqrt{x} - 1$ $\qquad x_0 = 2, x_1 = 5.5$

57. $f : x \longmapsto \dfrac{1}{x^2 + 1}$ $\qquad x_0 = 0, x_1 = 2$

58. $f : x \longmapsto x^4$ $\qquad x_0 = 0, x_1 = 1$

59. $f : x \longmapsto 1 - \cos(x)$ $\qquad x_0 = 0, x_1 = 2\pi$

60. $f : x \longmapsto \tan(x)$ $\qquad x_0 = 0, x_1 = \pi/4$

61.
$$f : x \longmapsto \begin{cases} x & \text{for } x \leq 1 \\ 2 - x & \text{otherwise} \end{cases} \qquad x_0 = 0, x_1 = 2$$

CHAPTER

5

Applications of the Derivative

Calculus originated in an effort to solve particular problems. This chapter is devoted to examining some of the ways in which differential calculus can be used to solve various mathematical problems that arise in many fields. Perhaps the most important of these problems deal with *optimization*—that is, problems requiring us to find the "best" value of some quantity. This often translates to a problem of finding when a function achieves an extreme value, and we have already seen in the last chapter that differential calculus is a powerful tool for finding extrema.

In this chapter we'll extend the applicability of the derivative to measure rates of change of several related quantities. *Implicit differentiation* allows us to find the slopes of curves described in ways other than as function graphs. We can also use calculus to find the slopes of curves defined by *parametric* or *polar equations*.

5.1 PROBLEM SOLVING AND MATHEMATICAL MODELLING

Let's start out by considering mathematical problem solving in general. It is worthwhile to point out an important distinction between a mathematical *algorithm* and a *strategy*. By an algorithm we mean some specific formula or some explicit step-by-step procedure for obtaining a result or for making a decision. One example of a useful mathematical formula is the *quadratic formula*

$$x = \frac{-b \pm \sqrt{b^2 - 4ac}}{2a}$$

which can be used to find the solutions to a quadratic equation of the form

$$ax^2 + bx + c = 0.$$

5.1 PROBLEM SOLVING AND MATHEMATICAL MODELLING

Given any quadratic equation in x with real coefficients, this formula tells us how to find the solutions explicitly (and whether they are real or complex numbers). Unless we make a calculation error in applying the quadratic formula, we are guaranteed success in analyzing the solutions to the quadratic equation.

Another example is the use of **Pascal's triangle**:

$$
\begin{array}{c}
1 \\
1\ 1 \\
1\ 2\ 1 \\
1\ 3\ 3\ 1 \\
1\ 4\ 6\ 4\ 1 \\
1\ 5\ 10\ 10\ 5\ 1 \\
1\ 6\ 15\ 20\ 15\ 6\ 1 \\
\vdots
\end{array}
$$

to find the coefficients of the terms in the expansion of powers of the binomial

$$(x + h)^n.$$

As an illustration, the coefficients of

$$(x + h)^4 = x^4 + 4x^3 h + 6x^2 h^2 + 4xh^3 + h^4$$

are given in the fifth row of Pascal's triangle. We would not characterize Pascal's triangle as a *formula* in the same way as the quadratic formula, but it does give us a well-defined step-by-step procedure for determining the **binomial coefficients**.

On the other hand, a *strategy* is simply some *plan of action* devised for solving a problem. Strategies don't necessarily come with guarantees of success. A strategy may include the choice of one or more particular mathematical formulas or procedures, and those choices may or may not be appropriate for the problem at hand. (For example, it is not a good choice to use the quadratic formula to solve $ax^2 + bx + c = 0$ for a.) Also, a strategy might be appropriate, but difficult to carry out. Hence, the tools (for example, calculators and/or computers) and techniques we have at our disposal can greatly influence our strategy for solving a problem.

Polya's four steps in problem solving

George Polya (1887-1985) was considered by many as the greatest teacher of mathematical problem solving. In his work *How To Solve It*, Polya discusses in detail many aspects of the problem solving process and he provides several useful general strategies (*heuristics*) for mathematical problem solving.

Here are the four basic steps Polya outlined in the problem solving process.

POLYA'S FOUR STEPS IN PROBLEM SOLVING
1. **UNDERSTAND THE PROBLEM**
2. **DEVISE A PLAN**
3. **CARRY OUT THE PLAN**
4. **LOOK BACK**

Let's elaborate on these problem-solving steps.

1. UNDERSTAND THE PROBLEM

This means *understand what the problem is asking for.* While that may seem obvious, there are many times when we dive into a problem and waste a lot of time and effort that could have been saved by a few extra moments of reflection at the beginning. Ask yourself these questions: Do I understand all the terminology? What is given? What is the goal? Am I required to find something or to prove something? Is there enough information? Is there extraneous information? Have I seen a similar problem before? Rewriting the problem in your own words, drawing a figure, trying some examples are all ways to clarify a problem statement.

The full power of algebra and calculus can be unleashed if we can model a problem situation and its constraints as functional expressions or as equations or inequalities. We may be able to introduce a coordinate system for the purposes of graphing. The act of identifying and labelling variable quantities in and of itself may clarify aspects of the problem to us.

2. DEVISE A PLAN

Devise a plan of action for the problem. If you don't know where to begin, then try a general problem solving *heuristic* or strategy. Three very useful heuristics include:

Trial and Error. At worst, you may get a better feel for the constraints of the problem situation. At best, you may stumble on the answer directly. Trial and error doesn't necessarily mean blind guesswork—our early guesses can help guide us in making better guesses. Making a list of the results of our trials may reveal a pattern or relationship. Mathematics is sometimes called the science or art of finding patterns.

Try a Simpler Problem. If the original problem seems too complex or confusing, try simplifying it first and solving that version. The solution to the simpler problem may give insights on how to solve the original problem. Exactly how do you make a problem simpler? Some of the ways include:

substituting a smaller number in place of a larger one given in the problem;

substituting a specific numerical value for an unknown constant or parameter (0 or 1 are often good substitution choices);

making up a related problem that involves fewer dimensions or unknowns;

adding or dropping some of the problem constraints.

Try Extreme or Special Cases. We may get a special understanding from examining the problem situation in extreme or special cases. For example, if a problem involved the *elliptical* orbits of planets, we might benefit by considering the special case of a *circular* orbit. Substituting extreme values for an unknown constant or parameter can also give us useful information. For example, a question involving lines in the Cartesian plane can be examined for the special cases of horizontal (zero slope) and vertical (undefined slope) lines. Making a list of special cases may also reveal a pattern or relationship.

3. CARRY OUT THE PLAN

Carry out your plan of action. Implement the strategy you've chosen until the problem is solved or until a new course of action is suggested. Give yourself a reasonable period of time to solve the problem. Monitor yourself. If you feel that you've embarked on a dead-end road then consider a change of strategy. Don't be afraid of starting all over. Many times a fresh start and a new strategy lead to success. You can have a flash of insight when you least expect it!

4. LOOK BACK

Check your answer to see if it really satisfies the requirements of your problem. Looking back means more than just checking your answer, though. Also look at your method of solution. Can you see another way of coming up with the answer? Can you see how your method could be used on other problems? Look forward to how you might generalize or extend your solution.

Mathematical modelling—finding the function

The application of mathematics to solving "real world" problems may be thought of in terms of the process diagrammed in Figure 5.1.

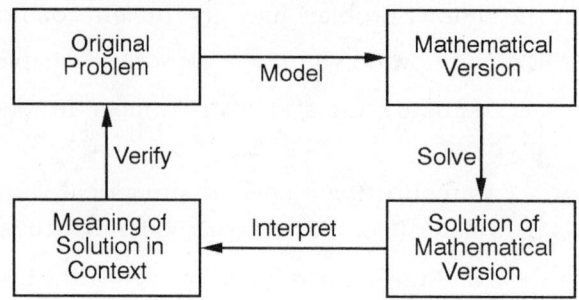

Figure 5.1 Applying mathematics to real world problems.

Mathematical training often deals only with the part of this process labelled "Solve." With the advent of computation devices capable of much of the rote work of solving, we can properly focus more attention on the mathematical modelling of problems, interpreting the mathematical solutions, and checking the reasonableness of the results.

In the modelling phase, a common sequence of steps is:

1. Gather paired data in a table (numerical representation).

2. Graph the pairs to obtain a picture (graphical representation).

3. Describe the functional relationship (or an approxiamtion of it) with a symbolic formula (symbolic representation).

In the interpretation phase, a common sequence of steps is:

1. Connect the various parts of the solution formula with their problem counterparts (symbolic representation).

2. Graph the solution to compare with the original data graph (graphical representation).

3. Make numeric predictions for comparison with the original problem (numeric representation).

In the verification phase, the symbolic model of the function can be tested by using it to predict new number pairs or graph points and then checking these predictions against actual gathered data.

The **mathematical modelling** of functional relationships is a cornerstone of scientific inquiry. Let's illustrate the mathematical modelling of a physical experiment.

An experiment—dropping a ball from a tower

In order to use mathematics to solve real world problems, the most important (and often the most difficult) step is to *identify* and *describe* a pattern or relationship in a way that allows us to take advantage of our mathematical tools. This activity is called *mathematical modelling*, and it is the bridge allowing us to use mathematics to solve applied problems in any discipline. In general, the procedure of gathering data, looking for patterns and relationships, devising mathematical models, using them to make predictions, and checking those predictions against new data is a hallmark of the scientific method. Let's illustrate this procedure with a discussion of an experiment involving a functional relationship.

Suppose we were in a tower and we dropped a ball repeatedly from several different heights, each time recording the time it takes for the ball to hit the ground (Figure 5.2).

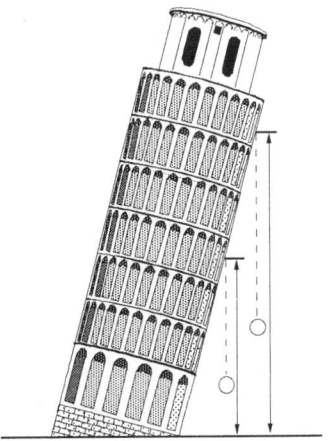

Figure 5.2 A ball-dropping experiment.

All other factors being equal (such as wind conditions), you would expect that the time it takes the ball to hit the ground would depend entirely on the height from which it was dropped. In other words, *the ball's drop time is a function of the height*.

How would this functional relationship between height and time be evident from the data we recorded? If the drop time depends entirely on the height, then *for repeated drops from the same height, we should get the same time reading*. That is, on every occasion that we drop the ball from, say, 20 feet, we should observe the same drop time. In actually performing this experiment, it's likely that we might observe slightly different times for repeated drops from the same height. The source of this variation could be imprecision in measuring the height and/or time. If the variation in readings is small enough that we could legitimately attribute it to these measurement errors, then our belief in the functional relationship between height and time would still be reasonable. However, if we could not adequately account for this variation, then we should reexamine our hypothesis regarding a functional relationship between height and time, or look for additional factors that could explain our observations.

Suppose we are satisfied that our experimental data supports our belief in a functional relationship between height and time. The next logical question would be, "what's the nature of the relationship," or, "exactly what kind of function is it?" A good first step in answering this question would be to arrange the data in a two-column table of numbers, with one column listing the different heights from which we dropped the ball, and the second column listing the corresponding drop times.

height in feet	time in seconds	height in feet	time in seconds
4	0.50	36	1.50
8	0.71	40	1.58
12	0.87	44	1.66
16	1.00	48	1.73
20	1.12	52	1.80
24	1.22	56	1.87
28	1.32	60	1.94
32	1.41	64	2.00

Table 5.1 Table of heights and drop times.

Table 5.1 shows some sample data from our ball-dropping activity. Each different height is recorded along with the corresponding time (or perhaps the average time if we dropped the ball repeatedly from the same height). Certainly, this table of values is not complete, for there are many other different heights from which we could have dropped the ball. We can look for patterns in the paired numbers, in the hope that we can discern some describable relationship that allows us to *predict* the time it would take the ball to drop from any given height. In particular, it would be very nice if we had a single formula that would allow us to simply compute the time t (in seconds) for any given height h in feet. That is, we are looking for a formula of the form

$$t = f(h),$$

where t represents the drop time in seconds, and $f(h)$ represents some expression whose value is determined by the value of h, when h is measured in feet. If we see what we think is such a pattern, we could check our formula by repeating the experiment at some new heights.

A powerful aid in our search for a pattern would be a visual display of the paired numbers representing the recorded heights and times. Figure 5.3 shows data from our table graphed as a set of points. The horizontal axis corresponds to the height in feet, and the vertical axis corresponds to the drop time in seconds. Each point corresponds to one of the pairs of numbers from our table.

5.1 PROBLEM SOLVING AND MATHEMATICAL MODELLING

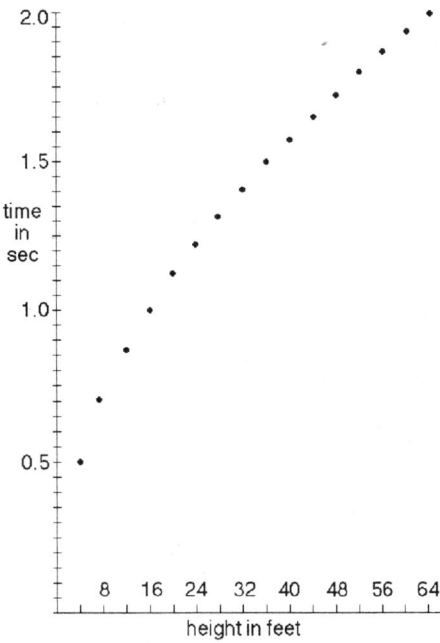

Figure 5.3 Graph of data points for the ball dropping experiment.

The **piece-wise linear** graph in Figure 5.4 has the same points connected with straight line segments.

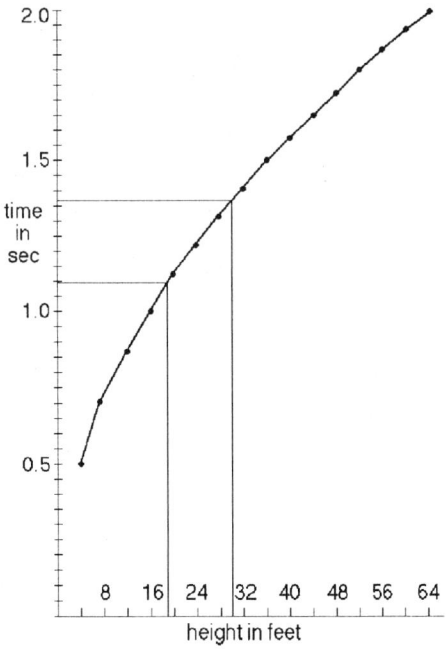

Figure 5.4 Piecewise linear graph of the ball dropping data.

The piece-wise linear graph can be used to make predictions of expected drop times for heights that fall between our recorded data points. For a new height that lies between two recorded heights, the strategy would be to predict a corresponding drop time so that the plotted point would lie on the piece-wise linear graph. This technique of prediction is called *linear interpolation*.

EXAMPLE 1   Using linear interpolation and the recorded data, predict the corresponding drop times from 30 feet and from 19 feet.

Solution   Since 30 feet is *halfway* between the recorded heights of 28 feet and 32 feet, the technique of linear interpolation would have us predict the expected drop time as approximately 1.365 seconds, which is *halfway* between the corresponding recorded drop times of 1.32 seconds and 1.41 seconds. Similarly, 19 feet is 3/4 of the way between the recorded heights of 16 feet and 20 feet. So, by linear interpolation, we would predict the corresponding drop time to be 1.09 seconds, which is 3/4 of the way between the corresponding drop times of 1.00 seconds and 1.12 seconds. Visually, these time predictions are made by locating the points on the piece-wise linear graph corresponding to 30 feet and 19 feet, respectively, and reading the time from the vertical axis as in Figure 5.4.   ■

We can generalize the procedure of linear interpolation for our example. Suppose h_1 and h_2 are two recorded heights with $h_1 < h_2$, and t_1 and t_2 are the corresponding drop times. If h_0 is an unrecorded height such that

$$h_1 < h_0 < h_2,$$

and we want to use linear interpolation to predict the corresponding drop time t_0, then

$$t_0 \approx t_1 + \frac{h_0 - h_1}{h_2 - h_1}(t_2 - t_1).$$

What assumption underlies the use of linear interpolation? We are simply assuming that between two inputs close together, the change in output is roughly proportional to the change in input. If we assumed that the drop time is a *differentiable* function of height, then this approximate local linearity is assured.

5.1 PROBLEM SOLVING AND MATHEMATICAL MODELLING

The piece-wise linear graph also gives us a visual approximation to a single smooth curve passing through all the points. If there is a simple formula $t = f(h)$ expressing drop time t (in seconds) as a function of the height h (in feet), then the shape of the piece-wise linear graph can provide a rough picture of the graph that formula would generate. In turn, that picture might help us make an educated *guess* about a formula describing the relationship between time and height.

For example, from the piece-wise linear graph, we would tend to rule out a *global* linear relationship between t and h, meaning that there is no *single* straight line will pass through all our data points. Symbolically, this means we cannot expect a functional relationship of the form

$$t = mh + b$$

to fit the data. Does the graph remind you of any other types of graphs you've seen before? For example, could we find part of a parabola that passes through the data points? This turns out to be a good guess (and somebody in history eventually made this guess). In fact, the upper half of the parabola with equation

$$16t^2 = h$$

does a very nice job of fitting our recorded data. (Try each recorded pair of values for t and h in the equation.) If we write this equation in the functional relationship form $t = f(h)$, we have

$$t = \sqrt{\frac{h}{16}}.$$

To test our formula, we should drop the ball from some new heights. Additional drop times are recorded in Table 5.2.

height in feet	time in seconds
19	1.09
30	1.37
70	2.09
120	2.75

Table 5.2 Additional drop times.

If we check our formula with this new data by substituting the heights and comparing the computed results with the experimental results, we find that the formula gives us excellent predictions. Until we are presented with experimental data that conflicts with our computed results, we appear to have a formula that effectively describes the functional relationship between height and drop time, at least for heights in the range of those of our experiment.

Modelling rates of change

Even when we have gathered and plotted data from a process, but do not have a symbolic formula that fits the data, we can still explore the rate of change of the process. Table 5.3 shows the population of a bacteria culture at one-hour intervals along with the increases in population computed by finding the successive hourly differences.

Time in Hours	Population	Growth Rate (Individuals per Hour)
1	210	
2	232	22
3	260	28
4	293	33
5	333	40
6	382	49
7	438	56
8	500	62
9	562	62
10	618	56
11	667	49
12	707	40
13	740	33
14	768	28
15	790	22
16	809	19
17	825	16
18	839	14
19	850	11
20	860	10

Table 5.3 Computing population growth rate for each hour.

Figure 5.5 illustrates a plot of the data. When we fit a curve to the data, we see a graph with a shape characteristic of many populations.

5.1 PROBLEM SOLVING AND MATHEMATICAL MODELLING

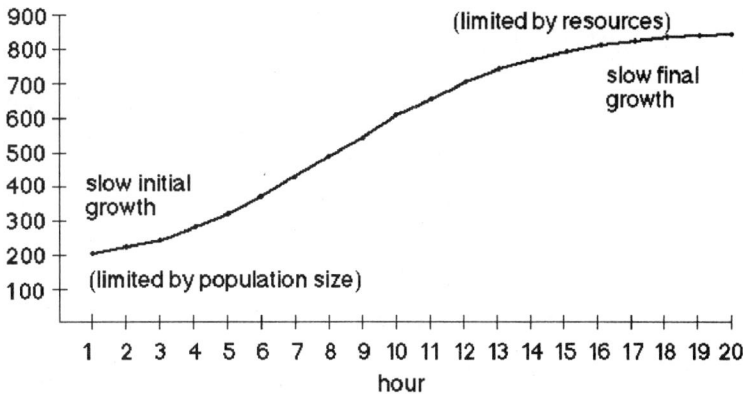

Figure 5.5 Population growth of a bacteria culture.

Population growth can be influenced by many factors, including the size of the existing population itself and the available resources (like food and light). When the population is small, the bacteria population's growth rate is slow. As the size increases, so does its growth rate, until the population approaches the limit that can be supported by the available resources.

We can visually approximate the growth rate at a particular time by estimating the slope of the graph at that point. We can also measure the *average* rate of change in population over each hour. If we plot the average growth rate each elapsed hour as a function of *time*, we obtain the plot in Figure 5.6.

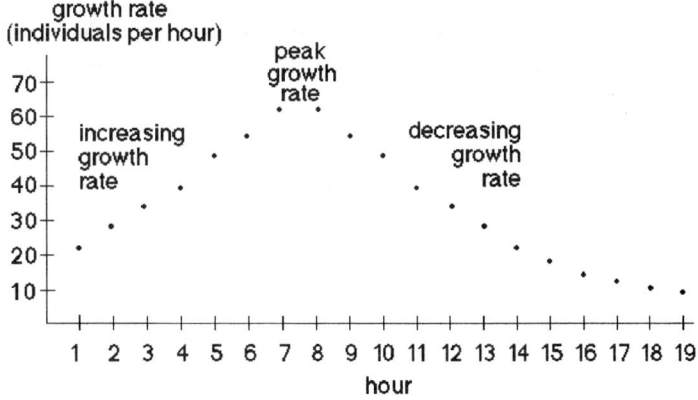

Figure 5.6 Population growth rate as a function of time.

Since the independent variable is again time, we can think of this plot as providing approximate values of the derivative for the population function graph in Figure 5.5. In other words, if we use P for population, and t for time in hours, then Figure 5.6 gives us approximate values for dP/dt. Note that the apparent inflection point in the original population plot (Figure 5.5) corresponds to the peak growth *rate* in this plot. Therein lies an observation worth emphasizing:

 For a differentiable function, the location of an inflection point is also the location of a local extremum of the *rate of change* **of the function.**

A function defined by a constraint

In many applications, the inputs and outputs of a function process may play the role of *causes* and *effects*. Keep in mind that any process that produces an output completely *determined* by the input can be called a function, even if it would be unreasonable to attribute a cause-and-effect connection. It might sound awkward to refer to the radius as "causing" the area of the circle, yet we would still say that the area is a function of the radius, since the area is completely determined by the radius through the formula $A = \pi r^2$. Sometimes a functional relationship arises between two variable quantities in the presence of special constraints. The following geometric example illustrates this.

EXAMPLE 2 Given a sphere of radius 5 cm, express the volume V of any nondegenerate cone inscribed in the sphere as a function of the cone's height h.

Solution By a nondegenerate cone, we mean one with positive volume, and not one that collapses to a single point or line segment. Figure 5.7 shows two typical cones inscribed in a sphere of radius 5 cm.

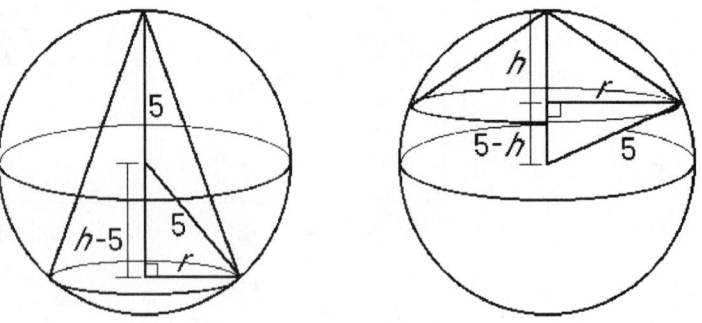

Figure 5.7 Cones inscribed in a sphere of radius $5\ cm$.

Let's agree to use the same units (cm) for the radius and height of the cone as for the radius of the sphere, so the volume of the cone will be in cubic cm. The volume of a cone is given by the formula

$$V = \frac{1}{3}\pi r^2 h$$

where V is the volume of the cone, r is the radius of the cone's base, and h is the height of the cone. In general, the volume of a cone is a function of both its height and radius, but in this situation our cone is constrained to be inscribed in a sphere of radius 5 cm. This constraint will force the values of r and h to be related. A very tall or very short cone will have a small radius while the medium cone (with the equator around its base) will have the largest radius of 5 cm.

We could treat this example much like our ball-dropping experiment of the previous example, by trying several different values h for the height of the cone, and computing the corresponding volume V for each value of h. For instance, a cone with a height h of exactly 5 cm will have its base at the equator of the sphere, so the radius r of the cone is also 5 cm. The volume of this particular cone is $V = \frac{1}{3}\pi 5^3 = \frac{125}{3}\pi$ cubic cm. If we repeat this process for several different heights h and compute the resulting volumes V, we could arrange the pairs of values (h, V) in a table and plot them in an attempt to find a pattern or relationship in the values.

Another way to find a functional relationship between V and h is to express r in terms of h, and then substitute this expression for r in the volume formula to get a functional expression solely in terms of h.

Since we are not going to consider degenerate cones, we know that the height h must be strictly greater than 0 cm and strictly less than 10 cm. Hence, the physical constraints on the problem also provide us with a natural *domain* for the possible heights h. In this case, we could also use the geometry of the situation to find this relationship.

Figure 5.7 illustrates the geometry of two typical cones, one with a height greater than the sphere's radius 5 cm and one with a smaller height extraction showing the relationship between the dimensions of the cone and the radius of the sphere. For a cone that is more than 5 cm tall, we can draw a right triangle having legs of r and $h - 5$ cm, respectively, and a hypotenuse of 5 cm (the radius of the sphere). In this case, the Pythagorean Theorem gives us

$$(h-5)^2 + r^2 = 5^2$$

or,

$$r^2 = 5^2 - (h-5)^2$$
$$= 25 - (h^2 - 10h + 25)$$
$$= 10h - h^2.$$

Since r^2 appears in the volume formula, we can simply substitute $(10h - h^2)$ directly for r^2 to obtain

$$V = \frac{1}{3}\pi(10h - h^2)h$$

$$= \frac{1}{3}\pi(10h^2 - h^3)$$

whenever 5 cm $< h <$ 10 cm.

For a cone that is less than 5 cm tall, we can draw a right triangle having legs of r and $5 - h$ cm, respectively, and a hypotenuse of 5 cm. Now we have

$$(5 - h)^2 + r^2 = 5^2$$

or,

$$r^2 = 5^2 - (5 - h)^2$$

$$= 25 - (25 - 10h + h^2)$$

$$= 10h - h^2.$$

This is exactly the same substitution value for r^2 that we obtained in the previous case, so we also have

$$V = \frac{1}{3}\pi(10h^2 - h^3)$$

whenever 0 cm $< h <$ 5 cm.

Since our formula $V = \frac{1}{3}\pi(10h^2 - h^3)$ worked for all other possible values of h, we might try it for $h = 5$ cm also:

$$V = \frac{1}{3}\pi(10(5^2) - 5^3)$$

$$= \frac{125}{3}\pi \; cubic \; cm.$$

So the formula works for all possible values of h. We can finally write

$$V : (0, 10) \longrightarrow \mathbb{R}$$

$$V : h \longmapsto \frac{1}{3}\pi(10h^2 - h^3)$$

where V is in cubic cm and h is in cm. So $V(h) = \frac{1}{3}\pi(10h^2 - h^3)$ for any h such that $0 < h < 10$. ∎

5.1 PROBLEM SOLVING AND MATHEMATICAL MODELLING

EXERCISES

Use linear interpolation to predict the population of the bacteria culture (Figures 5.5 and Table 5.3) at the times indicated in exercises 1-6.

1. 3 hours and 30 minutes
2. 10 hours and 45 minutes
3. 15 hours and 15 minutes
4. 6 hours and 20 minutes
5. 19 hours and 54 minutes
6. 7 hours and 30 minutes

Exercises 7-10 refer to the ball-dropping experiment data supplied in Table 5.1

7. Calculate the average time increase per foot for each interval determined by the measured heights. For example, for the first interval from $4\ ft$ to $8\ ft$, the drop time increased from 0.5 sec to 0.71 sec for an average time increase per foot of $(0.71 - 0.5)/(8 - 4) = .0525$ sec per foot.

8. Plot the computed values from exercise 7 as a function g of the initial measured height. For example, $g(4) = .0525$. How would you visually describe the resulting graph? Can you find a symbolic formula that generates a graph of g?

9. Calculate $f'(h) = df/dh$ where $f(h) = \sqrt{h/16}$ and compare the graph of f' to the graph of g from exercise 8.

10. Use the model $f(h) = \sqrt{h/16}$ to predict the time it would take the ball to drop from a height of 100 miles (remember, h is in feet). Calculate the ball's average velocity for this drop in miles per hour, and comment on whether or not the model still seems reasonable.

Below is a graph of hours of daylight as a function of time of year. Exercises 11-16 refer to this graph.

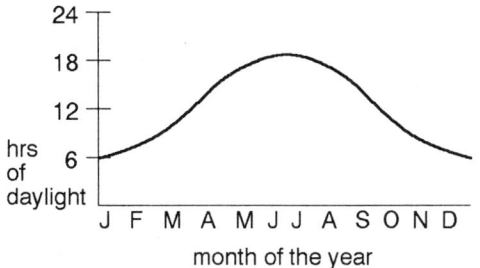

11. During what month are the hours of daylight increasing the fastest?

12. During what month are the hours of daylight decreasing the fastest?

13. Using an almanac, plot the *difference* in the time of sunset for the first of each month as compared to the first of the previous month. Relate the shape of this graph to the one above.

14. The vernal and autumnal equinoxes refer to the first day of spring and fall, respectively. If you plotted the *rate of change* of hours of daylight as a function of time, describe how the vernal and autumnal equinoxes could be spotted visually from this graph.

15. Elaborate stone formations used to find the summer and winter solstices are considered wonders of ancient civilizations. Why is it so much more difficult to pinpoint the first day of winter and summer than it is to pinpoint the vernal and autumnal equinoxes?

16. Let $H = A\sin(Bt+C)+D$, where H is the number of hours of daylight and t denotes the month of the year (January is $t = 1$ and December is $t = 12$). Find suitable constants A, B, C, and D so that the graph of $H(t)$ fits the graph above.

Exercises 17-25 describe geometric objects under various constraints. Find the domain and formula for the indicated function.

17. The surface area S of a cylinder inscribed in a sphere of radius 5 cm as a function of the height h of the cylinder.

18. The area A of a rectangle inscribed in a circle of radius 12 inches as a function of the width w of the rectangle.

19. The volume of a box with square base inscribed in a sphere of diameter 12 ft as a function of the box's height h.

20. The surface area S of a box with square base inscribed in a sphere of diameter 12 ft as a function of its height h.

21. The volume V of a cone circumscribed about a sphere of radius 6 m as a function of its own radius r.

22. The area A of an isosceles triangle inscribed in a circle of diameter 8 cm as a function of its vertex angle θ.

23. The area A of an isosceles triangle circumscribed about a circle of diameter 8 cm as a function of its vertex angle θ.

24. The perimeter P of an isosceles triangle inscribed in a circle of diameter 8 cm as a function of its base length b.

25. The diagonal ℓ of a rectangle inscribed in a circle of radius 12 inches as a function of the width w of the rectangle.

5.2 SOLVING OPTIMIZATION PROBLEMS—FINDING THE EXTREMA

We are always searching for the "best" in any field of endeavor, and to *optimize* means to find or achieve some best value. What exactly best means will depend on the situation, but in quantitative terms, to optimize most often means to *maximize* or *minimize* some variable quantity. For instance, a business strives to maximize profits and minimize costs. Manufacturers seek to maximize yields while minimizing resources used. In economics, engineering, and the physical, biological, and social sciences we can identify many problems of optimization.

If the variable quantity we want to maximize or minimize can be modelled as a function of other variables, then calculus can provide a powerful tool for solving the optimization problem. In the last chapter, we noted how the *absolute extrema* of a function of one variable must occur either at a critical point x_0 (a point in the domain of the function such that $f'(x_0) = 0$ or $f'(x_0)$ is undefined) or at an endpoint in the domain of the function. With that knowledge in hand, we can outline a step-by-step strategy for solving applied optimization problems with calculus.

Strategy for solving optimization problems

Step 1. *Identify the dependent variable.*

What variable quantity are we trying to maximize or minimize? Label it appropriately and express it in terms of the other variables in the process. A picture may be helpful.

Step 2. *Identify the constraints.*

The constraints are the boundaries and restrictions on the process. A constraint might be in the form of a particular interval of possible values for a variable or in the form of some relationship or connection that must hold between the variables.

Step 3. *Express the dependent variable as a function of a single independent variable, and identify its domain.*

Start with the expression (in terms of possibly several variables) from Step 1. Examine your constraints from Step 2 and use them to reformulate the expression in terms of one variable. The goal is to arrive at a relationship

$$y = f(x)$$

where y represents the quantity to be maximized or minimized and x is the single independent variable. Find the domain of acceptable values for x.

This may take the form of some interval of real numbers. Take special note whether or not the endpoints of such an interval are acceptable values.

Step 4. *If you can graph the dependent variable as a function of the independent variable,* **do it!**

You should insure that your graph contains all acceptable values of the independent variable.

One reason you might not be able to draw the graph is that the functional relationship you have derived involves unspecified constants (parameters). In this case, you should make a sample graph (or graphs) with some reasonable values chosen for these parameters. While you will not be able to use this graph to actually determine the maximum or minimum, it will still be useful as a "reality check" when you have found the maximum or minimum.

Step 5. *Identify the locations and values of the absolute maximum and/or minimum.*

If you have made a graph, it will give you a pretty good visual bearing on the location of the maximum and/or minimum. The maximum occurs where the graph reaches its highest point, and the minimum occurs where the graph reaches its lowest point.

To home in on the maximum or minimum, make a list containing all the candidates for maxima and minima. This will include all the end points and critical points in the acceptable region, together with their function values. In determining the acceptable region you have already determined the end points. Finding the critical points requires finding those x in the acceptable region where $f'(x) = 0$ or $f'(x)$ is *undefined*.

The first or second derivative test can be used to identify which critical points are local maxima, local minima, or neither. The function can be evaluated at the appropriate critical points and the endpoints to find the absolute extrema.

Step 6. *Interpret the results and use them to answer the particular questions posed by the problem.*

What exactly is being asked for in the problem? If we need to know the maximum or minimum value of the quantity, this means reporting the function value $f(x)$. If we need to know under what conditions this extremum occurs, then this means reporting one or more of the values of the other variables involved in the process.

Interpreting the result includes *checking its reasonableness*. Graphically, we can use a plot of the function obtained in Step 4 over the appropriate domain and check visually the reasonableness of either the values or locations of the extrema. We should also ask ourselves whether these

5.2 SOLVING OPTIMIZATION PROBLEMS—FINDING THE EXTREMA

values make sense in terms of the physical setting of the problem. Any conflict between our computations and this graphical or physical information needs to be resolved.

Optimization problem examples—a metalworks plant

The Platypus Metalworks Corporation makes and sells a large line of metal products. The company's physical plant is located on the banks of a river, and consists of three buildings—one for manufacturing, one for warehousing, and a business office. In one part of the manufacturing facility, sheets of metal of various sizes and rolls of wire of various thicknesses are produced. In the other part of the facility, these metal sheets and wire are trimmed, folded, and welded in various configurations for different products. The finished products are then warehoused until they are shipped out to the customer.

The examples in the remainder of this section and many of the exercises that follow provide a variety of optimization problems that might arise in the activities and operations of this metalworks plant. The intent is to provide a suitable "real-world" context for a wide range of practical optimization problems, while making clear the power of calculus to solve such problems.

EXAMPLE 3 Platypus is famous for its cookie sheets (they last a lifetime and longer, according to the ads). The cookie sheets come in a variety of sizes, and a special trim is applied to the perimeter of each. To standardize the trim length, all the rectangular cookie sheets have a perimeter of 60 inches. For what dimensions is the area of the sheet the greatest?

Solution *Step 1.* Identify the dependent variable.

We want to maximize the area of the sheet. If we label the area A, then we can write A in terms of the width w and the length ℓ of the sheet:

$$A = w\ell$$

Step 2. Identify the constraints.

The sheet must have a perimeter of 60 inches. In terms of the width and length of the sheet we must have

$$2w + 2\ell = 60.$$

Step 3. Express the dependent variable as a function of a single independent variable, and identify its domain.

We can use the constraint to solve for either w or ℓ in terms of the other variable. Solving for ℓ, we have $2\ell = 60 - 2w$, so that

$$\ell = 30 - w.$$

Now we can substitute this expression for ℓ in our functional expression

$$A = w(30 - w) = 30w - w^2.$$

We have now written A as a function of w. What is the domain of acceptable values for w? The algebraic expression $30w - w^2$ is certainly defined for any real number w, but the physical situation would require w to be a positive number ($w > 0$). How large can w be? If the perimeter is 60 inches, then the width must be strictly less than 30 inches. Hence, the acceptable values for w are

$$0 < w < 30.$$

Step 4. Graph the dependent variable as a function of the independent variable.

In this case, we graph A as a function of w. The acceptable values for w are between (but not including) 0 and 30.

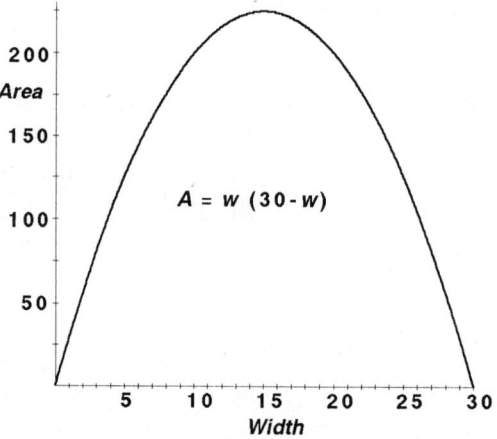

Figure 5.8 Graph of area A as a function of width w.

It is clear from the graph in Figure 5.8 that the maximum area occurs near the middle of the interval ($w \approx 15$.)

Step 5. Identify the locations and values of the absolute maximum and/or minimum.

The domain $(0, 30)$ is an *open* interval and does not include its endpoints. (Notice that if we had included the endpoints, $w = 0$ and $w = 30$ each result in $A = 0$).

5.2 SOLVING OPTIMIZATION PROBLEMS—FINDING THE EXTREMA

To find the critical points, we take the derivative with respect to w:

$$A'(w) = 30 - 2w.$$

This derivative is never undefined for any value of w. Setting $A'(w) = 30 - 2w = 0$ and solving for w, we have

$$w = 15.$$

This value $w = 15$ is in the domain of the function so $w = 15$ is the only critical point. This is then the only point on our list, and the area corresponding to $w = 15$ is $w(30 - w) = 225$. It is pretty clear from the graph that this is a maximum, but for insurance we'll apply the first derivative test to this critical point. For $w < 15$, the derivative $A'(w) = 30 - 2w > 0$ and for $w > 15$, the derivative $A'(w) = 30 - 2w < 0$. By the first derivative test, $A(15) = 30 \cdot 15 - 15^2 = 225$ is a *maximum* value for the function, and must be the absolute maximum. (If we applied the second derivative test, we find $A''(w) = -2$ for all values w. This also indicates that $w = 15$ must produce a maximum value.)

Step 6. Interpret the results and answer the question posed.

We have found the width $w = 15$ corresponded to the maximum area. We are asked to find the *dimensions* of the rectangle. To find the length ℓ, we can go back to Step 3 and substitute $w = 15$ to find

$$\ell = 30 - w = 30 - 15 = 15.$$

Therefore, the rectangle of perimeter 60 inches having maximum area is a *square* 15 inches by 15 inches. ∎

EXAMPLE 4 The Platypus cookie sheet custom shop is equally famous for its cookie sheets. They take a customer's special trim and make a custom cookie sheet with the trim applied to the perimeter. If the length of the customer's trim is P inches, for what dimensions (expressed in terms of P) is the area of the sheet going to be the greatest?

Solution *Step 1.* Identify the dependent variable.

We want to maximize the area of the sheet. If we label the area A, then we can write A in terms of the width w and the length ℓ of the sheet:

$$A = w\ell$$

Step 2. Identify the constraints.

The sheet must have a perimeter of P inches. In terms of the width and length of the sheet we must have

$$2w + 2\ell = P.$$

Step 3. Express the dependent variable as a function of a single independent variable, and identify its domain.

We can use the constraint to solve for either w or ℓ in terms of the other variable. Solving for ℓ, we have $2\ell = P - 2w$, so that

$$\ell = \frac{P}{2} - w.$$

Now we can substitute this expression for ℓ in our functional expression

$$A = w(\frac{P}{2} - w) = \frac{P}{2}w - w^2$$

We have now written A as a function of w. What is the domain of acceptable values for w? The algebraic expression $\frac{P}{2}w - w^2$ is certainly defined for any real numbers w and P, but the physical situation would require w and P to be positive numbers ($w > 0$ and $P > 0$). Of course, since the customer is in control of the length of the trim P, it is a parameter rather than an independent variable. How large can w be? If the perimeter is P inches, then the width must be strictly less than P inches. Hence, the acceptable values for w are

$$0 < w < \frac{P}{2}.$$

Step 4. Graph the dependent variable as a function of the independent variable.

In this case, we can't directly graph A as a function of w since we don't know in advance the value of P. However, we already have a sample graph for one reasonable value of P, namely $P = 60$. In that case, the maximum occurred in the middle of the graph ($w = \frac{P/2}{2}$.) Is this true for any positive P, or was our example a fluke?

Step 5. Identify the locations and values of the absolute maximum and/or minimum.

The domain $(0, \frac{P}{2})$ is an *open* interval and does not include its endpoints. (Notice that if we had included the endpoints, $w = 0$ and $w = \frac{P}{2}$ each result in $A = 0$).

To find the critical points, we take the derivative with respect to w:

$$A'(w) = \frac{P}{2} - 2w.$$

This derivative is never undefined for any value of w or P. Setting $A'(w) = \frac{P}{2} - 2w = 0$ and solving for w, we have

$$w = P/4.$$

This value $w = P/4$ is in the domain of the function so $w = P/4$ is the only critical point. This is then the only point on our list and in this case we'll

5.2 SOLVING OPTIMIZATION PROBLEMS—FINDING THE EXTREMA

need to apply the first derivative test to this critical point. For $w < P/4$, the derivative $A'(w) = P/2 - 2w > 0$ and for $w > P/4$, the derivative $A'(w) = P/2 - 2w < 0$. By the first derivative test, $A(P/4) = \frac{P}{4}(\frac{P}{2} - \frac{P}{4}) = \frac{P^2}{16}$ is a *maximum* value for the function, and must be the absolute maximum. (If we applied the second derivative test, we find $A''(w) = -2$ for all values w. This also indicates that $w = \frac{P}{4}$ must produce a maximum value.)

Step 6. Interpret the results and answer the question posed.

We have found the width $w = \frac{P}{4}$ corresponded to the maximum area. We are asked to find the *dimensions* of the rectangle in terms of P. To find the length ℓ, we can go back to Step 3 and substitute $w = \frac{P}{4}$ to find

$$\ell = \frac{P}{2} - w = \frac{P}{2} - \frac{P}{4} = \frac{P}{4}.$$

In other words, the maximum area cookie sheet is always a square. ∎

EXAMPLE 5 A rectangular tin box with closed bottom and open top is to be constructed out of 54 square inches of material. The length of the box is to be 2 times the width. If w, ℓ, and h are the width, length, and height (respectively) of the box, find the maximum possible volume of the box and its dimensions.

Solution *Step 1.* Identify the dependent variable.

We want to maximize the *volume* of the box. If we label the volume V, then we can write V in terms of the box's three dimensions:

$$V = w\ell h.$$

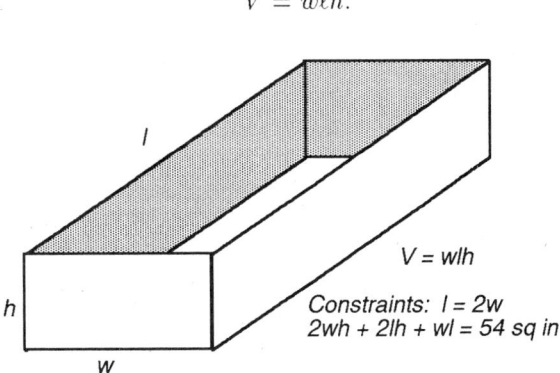

Figure 5.9 Maximizing the volume of an open box.

Step 2. Identify the constraints.

We have two constraints on our box. First, we know that the material of the box must have a total area of of 54 in². This means that the area of the four sides and the area of the bottom must total 54 in² ($2wh + 2\ell h + w\ell =$

54 in^2). Second, we know that the length must be twice the width ($\ell = 2w$). Figure 5.9 shows an illustration of the box along with these constraints.

Step 3. Express the dependent variable as a function of a single independent variable, and identify its domain.

We can substitute $2w$ for ℓ to rewrite the volume as

$$V = 2w^2 h$$

and we can also rewrite the other constraint as

$$2wh + 4wh + 2w^2 = 6wh + 2w^2 = 54.$$

If we solve this for h, we obtain

$$h = \frac{54 - 2w^2}{6w}.$$

Now we substitute $(54 - 2w^2)/6w$ for h in the volume formula:

$$V = 2w^2 \cdot \frac{54 - 2w^2}{6w} = \frac{108w^2 - 4w^4}{6w} = 18w - \frac{2w^3}{3}.$$

Now we have written V as a function of w alone. What is the domain of possible values for w? Certainly, $w > 0$. The larger w becomes, the shallower the box must be because of the constraint on our materials. The extreme of a flat sheet with no sides ($h = 0$) would leave only the bottom of the box taking up all the available material ($wl = 2w^2 = 54$). For this extreme $w = \sqrt{27} \approx 5.2$. Of course, this box has 0 volume and is not of interest for our optimization purposes, but we now know that our possible values for w are

$$0 < w < \sqrt{27} \approx 5.2.$$

Step 4. Graph the dependent variable as a function of the independent variable.

We are to graph $V = 18w - 2w^3/3$ over the w-interval $[0, 5.2]$. This will give us some idea of where the maximum value occurs. Figure 5.10 illustrates such a graph. On the basis of this picture, the maximum volume appears to be between 35 and 40 cubic inches, and corresponds to a width $w \approx 3$ inches. We can use this information to monitor the reasonableness of our results using calculus.

5.2 SOLVING OPTIMIZATION PROBLEMS—FINDING THE EXTREMA

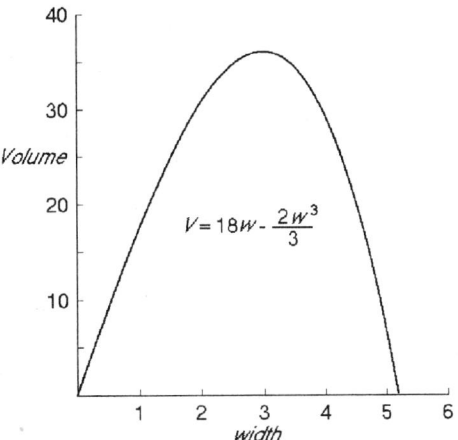

Figure 5.10 Graph of volume V as a function of width w.

Step 5. Identify the locations and values of the absolute maximum and/or minimum.

The endpoints of the domain $(0, \sqrt{27})$ are not included (even if they were, they would result in a minimum 0 volume instead of a maximum volume for the box).

To find the critical points, we take the derivative $V'(w)$

$$V'(w) = 18 - \frac{6w^2}{3} = 18 - 2w^2.$$

This derivative is never undefined for any value of w. If we set $V'(w) = 0$ we have

$$18 - 2w^2 = 0 \quad \text{so that} \quad w = \pm 3.$$

The value $w = -3$ is not acceptable, since we must have a positive width for the box. The only critical point is $w = 3$. We'll apply the first derivative test to this critical point. For $w < 3$, the derivative $V'(w) = 18 - 2w^2 > 0$ and for $w > 3$, the derivative $V'(w) = 18 - 2w^2 > 0$. By the first derivative test, $V(3) = 18 \cdot 3 - (2 \cdot 3^3)/3 = 54 - 18 = 36$ is a *maximum* value for the function, and must be the absolute maximum. (If we apply the second derivative test, we find $V''(w) = -4w$ and $V''(3) = -12 < 0$ also indicates that $w = 3$ must produce a maximum value.)

Step 6. Interpret the results and answer the question posed.

We have found the width $w = 3$ in corresponded to the maximum volume of 36 in^3. We are also asked to find the *dimensions* of the box. From the constraints, we have $\ell = 2w = 6$ in, and $h = (54 - 2w^2)/6w = 36/18 = 2$ in. Hence, the box of maximum volume has dimensions

$$3 \text{ in} \times 6 \text{ in} \times 2 \text{ in}$$

and a total volume of 36 in^3. ∎

EXAMPLE 6

A small rectangular sheet of metal 10 cm wide and 20 cm long is simply folded down the middle and triangles glued to the ends to produce a "V" shaped feed or water trough for a pet cage. What should the angle of the bend be to maximize the amount the trough can hold?

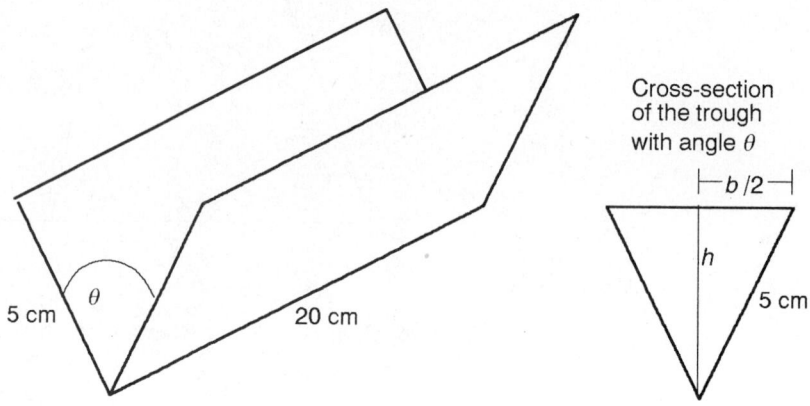

Figure 5.11 A trough made by folding a sheet of metal.

Solution *Step 1. Identify the dependent variable.* Figure 5.11 shows a picture of the piece used for the trough. The volume of the trough is determined by taking its cross-sectional area and multiplying by its length. Since the length is fixed at 20 cm, we need to maximize the cross-sectional area. From the picture, we can see that the cross-section will be an isosceles triangle with legs of 5 cm and a vertex angle of θ (the angle of the fold).

If we let C be this cross-sectional area, then our first step is to write C as a function of the single variable θ. From the formula for the area of a triangle, we can write

$$C = bh/2.$$

Step 2. Identify the constraints.

The constraints in this problem are defined by the side length of the triangle. Figure 5.11 shows how a half-base ($b/2$) and height h of the triangle can be viewed as legs of a right triangle with angle $\theta/2$ and hypotenuse 5 cm.

Step 3. Express the dependent variable as a function of a single independent variable, and identify its domain.

Using this information and some trigonometry, we can write

$$\sin(\theta/2) = \frac{opposite}{hypotenuse} = \frac{b/2}{5} \quad \text{and} \quad \cos(\theta/2) = \frac{adjacent}{hypotenuse} = \frac{h}{5}.$$

5.2 SOLVING OPTIMIZATION PROBLEMS—FINDING THE EXTREMA

If we solve for b and h and substitute these quantities into our cross-sectional area formula, we obtain C entirely in terms of θ:

$$C = 25 \sin(\theta/2) \cos(\theta/2)$$

The domain of possible values for θ are between $0°$ and $180°$ inclusive, or in terms of radians:

$$0 \leq \theta \leq \pi.$$

Step 4. Graph the dependent variable as a function of the independent variable.

In this case we want to graph $C = 25 \sin(\theta/2) \cos(\theta/2)$ for $0 \leq \theta \leq \pi$ radians.

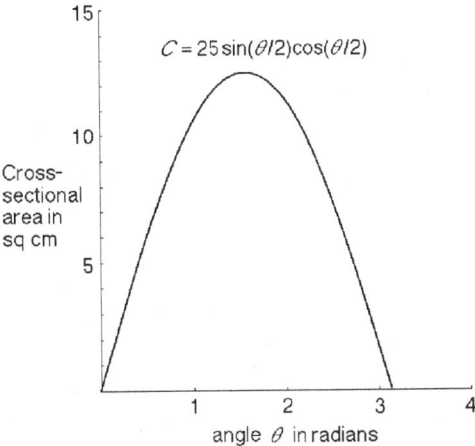

Figure 5.12 Graph of cross-section area C as a function of angle θ.

The graph in Figure 5.12 indicates that the maximum value occurs about midway between the endpoints or at $\theta \approx \dfrac{\pi}{2}$.

Step 5. Identify the locations and values of the absolute maximum and/or minimum.

Differentiating C with respect to θ gives us

$$dC/d\theta = 25(\cos(\theta/2) \cdot (1/2) \cdot \cos(\theta/2) + \sin(\theta/2) \cdot (-\sin(\theta/2)) \cdot (1/2)$$

$$= \frac{25}{2}(\cos^2(\theta/2) - \sin^2(\theta/2))$$

$$= \frac{25}{2}(\cos(\theta)).$$

If we set $dC/d\theta = 0$, we can see that the critical points will occur exactly when

$$\cos(\theta) = 0.$$

The only value θ in our domain satisfying this equation is

$$\theta = \pi/2 \quad \text{or} \quad 90°.$$

Since $\cos(\theta) > 0$ for $\theta < \pi/2$ and $\cos(\theta) < 0$ for $\theta > \pi/2$, the first derivative test tells us that $\theta = \pi/2$ will produce a local maximum for the cross-section. (You can check the second derivative test to see the same result.) The endpoints of our domain correspond to a flat sheet ($\theta = \pi$) or a completely folded sheet ($\theta = 0$), both of which result in 0 cross-sectional area.

Step 6. Interpret the results and answer the question posed.

Folding the sheet at a 90° angle is optimal to maximize the volume of the trough. ∎

EXAMPLE 7 A cylindrical can with closed bottom and closed top is to be constructed to have a volume of 1 gallon. The material used to make the bottom and top costs $0.06 per square inch, and the material used to make the curved surface costs $0.03 per square inch. Find the dimensions of the can (diameter and height) that minimize the total cost, and what that minimum cost is.

Solution *Step 1.* Identify the dependent variable.

We want to minimize the total cost C of the materials used to make the can. Figure 5.13 illustrates the three component surfaces of the can (top, bottom, and curved surface). The curved surface rolls out to a rectangle having the same height as the can and a length equal to the circumference of the can. The top and bottom are circles with the same diameter as the can.

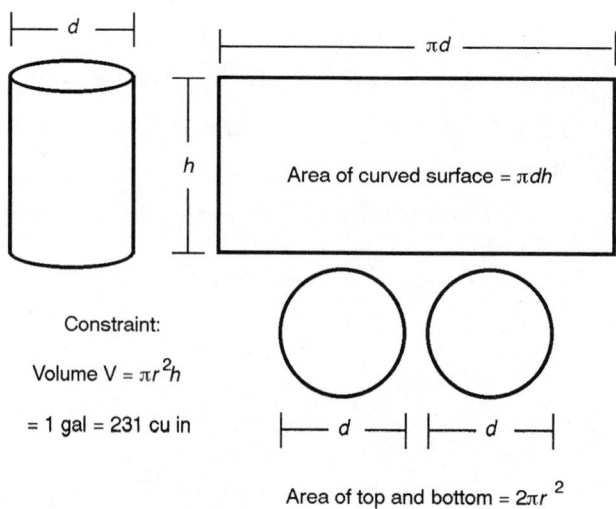

Figure 5.13 Minimizing the cost of a cylindrical can.

5.2 SOLVING OPTIMIZATION PROBLEMS—FINDING THE EXTREMA

If the diameter d and height h of the can are measured in inches, then we need to compute the area of each part of the can, multiply each by the corresponding cost per square inch, and total the results:

$$C = (.03)\pi dh + (.06)2\pi r^2,$$

where r is the radius ($d = 2r$), and C is the cost in dollars.

Step 2. Identify the constraints.

The can must hold exactly 1 gallon. Since we are measuring the dimensions of the can in inches, we need to convert the volume measurement to in^3. In this case, 1 gallon is equivalent to 231 cubic inches. Using the volume formula for a cylinder, we can now write the constraint as

$$V = \pi r^2 h = 231 \ in^3.$$

Step 3. Express the dependent variable as a function of a single independent variable, and identify its domain.

The constraint equation would allow us to solve for h in terms of r very easily,

$$h = \frac{231}{\pi r^2}$$

and since $d = 2r$, we can write the cost C entirely in terms of r:

$$C = (.03)\pi(2r)\frac{231}{\pi r^2} + (.06)2\pi r^2 = \frac{13.86}{r} + (.12)\pi r^2.$$

As for the domain of our independent variable r, we can see that $r > 0$. We could theoretically (though certainly not practically) make the radius of the can as large as we want by making the can extremely short (h close to zero). So, r could be any positive real number.

Step 4. Graph the dependent variable as a function of the independent variable.

In this case, we want to graph $C = \frac{13.86}{r} + (.12)\pi r^2$ for all positive values of r. Since this is not possible in any straight-forward way, we will graph it for some reasonable values of r, say over the interval $0 < r < 10$. This will at least give us some idea of the shape of the graph. The result is shown in Figure 5.14.

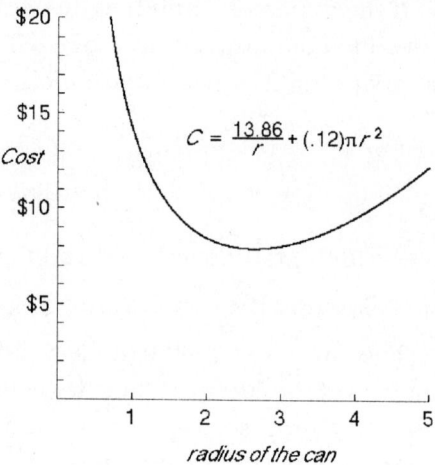

Figure 5.14 Graph of the cost function.

The graph appears to have a *local* minimum at $r \approx 2.5$, but we can't tell directly from the graph what happens for large values of r.

Step 5. Identify the locations and values of the absolute maximum and/or minimum.

Since our region of acceptable values has no endpoints, we need only consider the critical points.

Finding dC/dr, we have

$$dC/dr = -\frac{13.86}{r^2} + (.24)\pi r.$$

This expression is undefined when $r = 0$, but that value is not in the domain. The only critical points will occur when $dC/dr = 0$:

$$-\frac{13.86}{r^2} + (.24)\pi r = 0 \quad \text{or} \quad r^3 = \frac{13.86}{.24\pi}.$$

If we use $\pi \approx 3.1416$ as an approximation and take the cube root of both sides of this equation, we find

$$r \approx 2.64 \text{ in.}$$

This corroborates the conclusion we reached by looking at the graph. If we had chosen a region to graph which had not included this critical point, we would regraph now to insure that we included this point in the graph.

If we take the second derivative, we can see that

$$\frac{d^2C}{dr^2} = \frac{27.72}{r^3} + (.24)\pi$$

will be positive when $r \approx 2.64$, and that indicates that this represents a *minimum* for our cost function. Substituting $r = 2.64$ into our cost function, we find

$$C \approx \frac{13.86}{2.64} + (.12)(3.1416)(2.64)^2 \approx \$7.88.$$

Step 6. Interpret the results and answer the question posed.

We have found the minimum cost of the gallon can to be approximately $7.88. We were also asked to find the dimensions of this can. Since $r \approx 2.64$ in, we have

$$d = 2r \approx 5.28 \text{ in} \quad \text{and} \quad h = \frac{231}{\pi r^2} \approx 10.55 \text{ in}$$

as the diameter and height of the can, respectively. ■

We close by again summarizing the essential steps to solving an optimization problem with calculus.

Step 1. Identify the dependent variable.
Step 2. Identify the constraints.
Step 3. Express the dependent variable as a function of a single independent variable, and identify its domain.
Step 4. Graph the function over its domain if you can.
Step 5. Identify the locations and values of the absolute maximum and/or minimum by examining the end points and critical points.
Step 6. Interpret the results and answer the question posed.

EXERCISES

The following exercises continue to describe several optimization problems that come up in the operations of the Platypus metalworks plant. Suppose you have been hired by Platypus to help solve their optimization problems because of your knowledge of calculus.

1. To prevent animals from harming themselves on sharp pieces of scrap metal, Platypus wants to fence off a rectangular scrap yard. You are given 100 meters of fencing to enclose the yard. Since the company is near the river, you decide to fence a rectangular plot along the river so that you need only fence three sides (see picture below). What should be the dimensions of the yard to maximize the area enclosed?

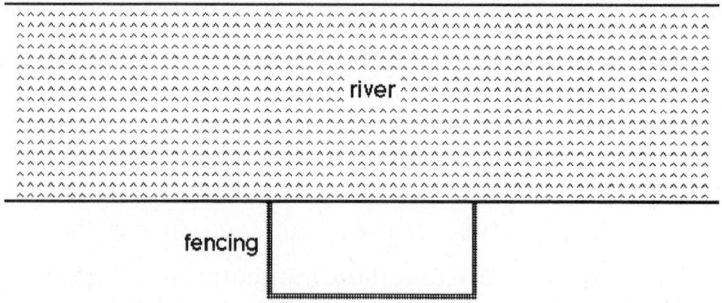

2. There is a concern that the fence you plan to put up will be unsightly to passers-by. You are now told that the yard must have at least 500 square meters area, but you must use decorative fencing along the front (parallel to the river). The decorative fencing costs $25 per meter, while the usual fencing costs $15 per meter. What dimensions will you use to enclose 500 square meters at a minimum cost to the company?

3. Upon contacting the decorative fence supplier, you find that the $25 figure is not firm. Determine in terms of D, the decorative fencing's eventual cost per meter, the dimensions you will use to enclose 500 square meters at a minimum cost to the company.

4. A rectangular box with square base and open top is constructed to have a volume of 270 cubic inches. The material used to make the bottom of the box costs $.06 per square inch. The material used to make the rest of the box costs $.03 per square inch. Find the dimensions of the box that minimize the total cost.

5. Your rectangular box project of the previous exercise was so successful that you are now to make a series of boxes with the same construction, but with varying volumes. Determine the dimensions which minimize the total cost in terms of V, the specified volume of the box in cubic inches.

6. Hundreds of thousands of copies of a page of advertising information are to be mailed worldwide. The page is to contain 24 square inches of printed material with margins of 1.5 inches at the top and bottom of the page, and margins of 1 inch at the sides. What should be the overall dimensions (length and width from edge to edge of the paper) to minimize the overall area of the page?

7. The ad agency tells you that they have decided that the margin sizes they provided you with are wrong and they haven't figured out the new ones yet, but that you should go ahead and find the overall dimensions in terms of T and S, the top/bottom and left/right margins, respectively.

8. A rectangular tin box with square base is constructed to have a volume of 96 cubic inches. The top of the box costs $.06 per square inch to paint. The rest of the box, including the bottom, costs only $.03 per square inch to paint. Find the dimensions of the box that minimize the total cost of painting.

5.2 SOLVING OPTIMIZATION PROBLEMS—FINDING THE EXTREMA

9. A cylindrical can with closed bottom and closed top is to be made from two kinds of material. The material used to make the bottom and top of the can costs $.06 per square inch, and the material used to make the curved outer surface of the can costs $.03 per square inch. The total cost of the can is $1.44. If r is the radius and h is the height of the can, find the value of r which maximizes the volume.

10. A rectangular box with square base and open top is constructed out of two types of material. The material used to make the bottom of the box costs $.10 per square inch. The material used to make the rest of the box costs $.06 per square inch. The total cost of the box is to be $3.00. If s is one side of the base of the box and h is the height, find the value of s that maximizes V.

11. Platypus makes lengths of rain gutter by taking a 12 inch wide strip of metal and folding it up at equal angles four inches from the edges into a trapezoidal shape. What should the angle be to maximize the capacity of the rain gutter?

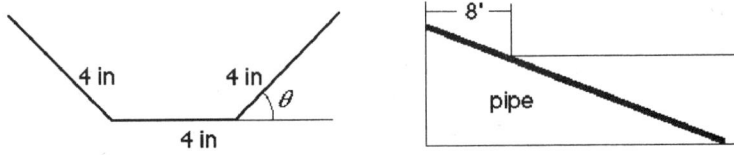

12. The boss tells you that she wants a sample of continuous pipe (6 inches in diameter) brought to the business office as an exhibit for an important shareholders' meeting. The instructions are clear: the bigger, the better. However, you are thinking ahead and realize that the business office has hallways that are only 8 feet across, and there is a corner you will need to negotiate along the way. What is the longest piece of pipe that you will be able to slide around the corner?

13. Platypus makes a variety of open-top boxes from a 20 cm by 30 cm rectangular flat sheet by simply cutting equal-sized squares off of each corner of the sheet and folding up the four sides (so the depth of the box is the same as the side length of the square). How big should the squares be to produce a box of maximum volume?

14. Platypus's sheet metal supplier makes it attractive for them to accept rectangular sheets with the same 20 cm width but varying lengths. Determine in terms of L, the length of the sheet, how big the squares should be to produce boxes of maximum volume.

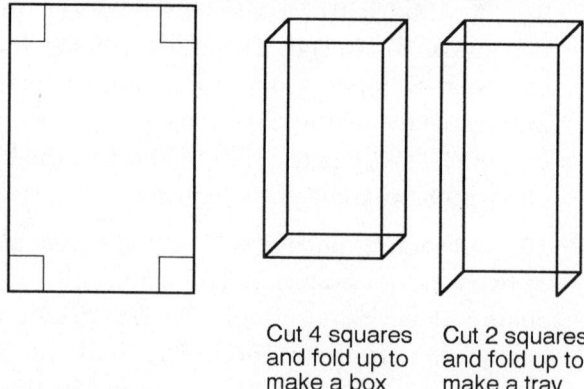

Cut 4 squares and fold up to make a box

Cut 2 squares and fold up to make a tray

15. Platypus also makes open-top and open-front trays from the same 20 cm by 30 cm rectangular flat sheet by cutting squares off the corners on only one end of the sheet and folding up three sides. How big should the squares be to produce a tray of maximum volume? (The illustration above shows one way to make the tray. You should also check the other way.)

16. A cylindrical can with closed bottom and open top is to be made from two kinds of material. The material used to make the bottom of the can costs $.05 per square inch, and the material used to make the curved surface of the can costs $.03 per square inch. The total cost of the can is to be no more than $7.35. Find the dimensions of the can with greatest volume.

17. The business office wants to put in Norman windows, which are windows in the shape of rectangles capped by a semicircle. These windows will have special decorative trim that is purchased by length, so it will pay to get the most light for the least perimeter of window. If 10 feet of trim is allotted for each Norman window, what dimensions maximize their area?

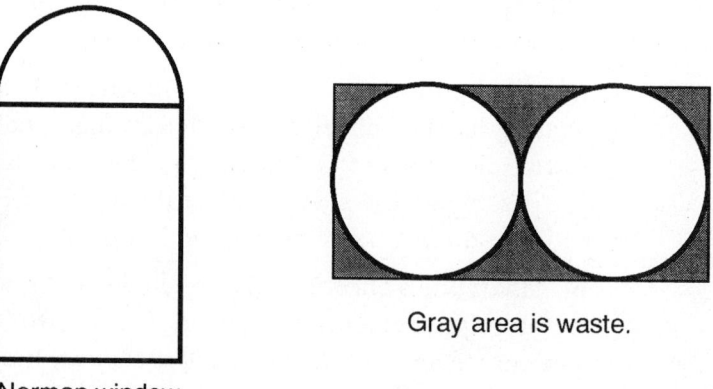

Norman window

Gray area is waste.

18. Suppose the top and bottom of the gallon can of Example 7 must be cut out of a rectangular piece of metal as shown above, and the leftover material must be discarded as waste. In other words, the cost of the bottom and top is now represented by the cost of this entire rectangle. What dimensions of the gallon can now minimize the cost?

5.2 SOLVING OPTIMIZATION PROBLEMS—FINDING THE EXTREMA

19. Platypus wants the power company to run a special line to the plant. The power company is located 3000 meters down the river on the opposite bank. Ground lines cost $20 per meter. The river is 500 meters wide and underwater line will cost four times as much per meter as the above ground line. Since Platypus owns the riverfront property on both sides of the river all the way to the power plant, you have the say as to where and how the line should cross the river. What should the path be to minimize the cost?

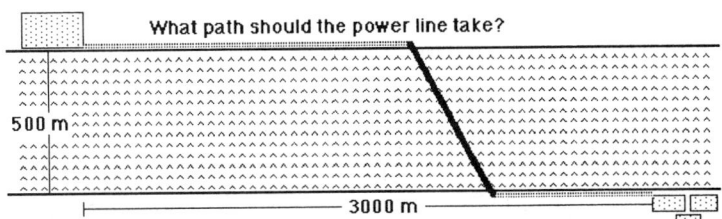

20. A wire 2 meters long is cut into two pieces. One piece is bent into a square for a stained glass frame while the other piece is bent into a circle for a TV antenna. To cut down on storage space, where should the wire be cut to minimize the total area of both figures? Where should the wire be cut to maximize the total area?

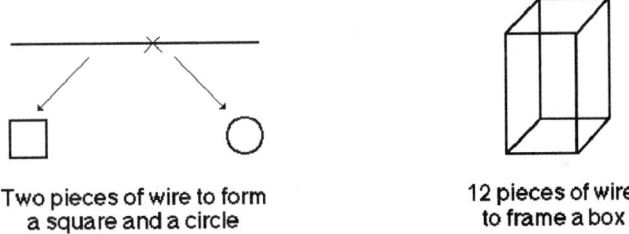

Two pieces of wire to form a square and a circle

12 pieces of wire to frame a box

21. A wire 6 meters long is cut into twelve pieces. These pieces are welded together at right angles to form the frame of a box with a square base. Where should the cuts be made to maximize the volume of the box? Where should the cuts be made to maximize the total surface area of the box?

22. The first steps in making a metal funnel are to take a circular piece of metal, cut out a sector (piece of pie), and connect the two radial edges to make an open cone. What should the angle of the sector be to maximize the volume of this cone?

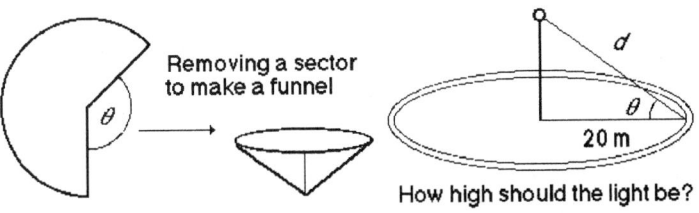

Removing a sector to make a funnel

How high should the light be?

23. There is a circular path around the open space between the warehouse and business office. Since Platypus has added a night shift, there is a need for better lighting of the path. You have decided to erect a light pole in the middle of the circular area that has a radius of 20 meters. You want to place the light at a height that will maximize the light intensity on the path, so you consult some reference books and find that for this light,

$$I = \frac{10\sin(\theta)}{d^2}$$

where I is the light intensity at a point on the ground, d is the straight line distance from the light to the point, and θ is the angle this line makes with the ground. How high should the light be to maximize I?

24. Platypus has just landed a big contract for steel wool. Because steel wool is relatively light, the shipping dock wants to pack as much as possible into each box they send out to cut down on handling charges. The trucking company allows boxes to have a combined length and girth (the distance measured around the box perpendicular to its length) of no more than 108 inches. If the shipping dock uses boxes with square ends, what box dimensions maximize the volume the box can hold? Now suppose the shipping dock uses cylindrical containers. What height and radius maximize the volume? Which, box or cylinder, is the best choice?

25. A chemical that Platypus uses in treating their metal is used at a steady rate of 1200 gallons per year. Any number of gallons can be ordered from the chemical company, but there is a set extra handling charge of $100 for any order, no matter what the size. On the other hand, it costs Platypus an average of $1 per gallon per year just to store the chemical in its warehouse. You are given instructions that the chemical should be reordered whenever the stock on hand gets down to 200 gallons. This means that if you order g gallons each time, you'll need to order $1200/g$ times a year and the company will be storing the equivalent of $g/2 + 200$ gallons throughout the year. How many gallons g should you order each time to minimize the extra handling and storage charges?

Exercises 26-34 give one geometric quantity as a function of another under a suitable constraint. Find the maximum and minimum values of each function.

26. The surface area S of a cylinder inscribed in a sphere of radius 5 cm as a function of the height h of the cylinder.

27. The area A of a rectangle inscribed in a circle of radius 12 inches as a function of the width w of the rectangle.

28. The volume of a box with square base inscribed in a sphere of diameter 12 ft as a function of the box's height h.

29. The surface area S of a box with square base inscribed in a sphere of diameter 12 ft as a function of its height h.

30. The volume V of a cone circumscribed about a sphere of radius 6 m as a function of its own radius r.

31. The area A of an isosceles triangle inscribed in a circle of diameter 8 cm as a function of its vertex angle θ.

32. The area A of an isosceles triangle circumscribed about a circle of diameter 8 cm as a function of its vertex angle θ.

33. The perimeter P of an isosceles triangle inscribed in a circle of diameter 8 cm as a function of its base length b.

34. The diagonal ℓ of a rectangle inscribed in a circle of radius 12 inches as a function of the width w of the rectangle.

5.3 IMPLICIT DIFFERENTIATION AND RELATED RATES

In this section we will extend the notion of derivative to allow us to measure the slope of more general curves than those that are the graphs of functions, and to functions of more than one variable. We'll apply some of those techniques to solving problems of finding rates of change of related quantities.

Implicit differentiation

A curve in the Cartesian plane can often be represented as the graph of a function f with domain D. In other words, the curve is the set of points

$$\{(x, y : x \in D, y = f(x)\}.$$

The slope of the curve at a single point (x_0, y_0) is given by the derivative $f'(x_0)$ or

$$\left. \frac{dy}{dx} \right|_x = x_0.$$

This slope also represents the instantaneous rate of change of the variable y with respect to x.

Not all curves in the plane can be described as the graph of a function $y = f(x)$. Nevertheless, it may still be possible to use calculus to calculate the slope at a point on such a curve.

An equation in two unknowns x and y has associated with it a set or **locus** of points satisfying the equation. When plotted, this locus takes on the visual appearance of a curve in the plane.

EXAMPLE 8   The locus of the equation

$$x^2 + y^2 = 9$$

is a circle with radius 3 and center at the origin (Figure 5.15). ∎

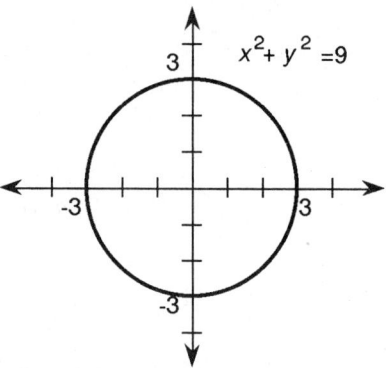

Figure 5.15 The circle $x^2 + y^2 = 9$.

Now, this relation $x^2 + y^2 = 9$ does not describe y as a function of x (the circle fails the vertical line test for function graphs). However, if we restrict our attention to only a part of the circle, we can find an explicit function whose graph *does* exactly match that part of the circle.

5.3 IMPLICIT DIFFERENTIATION AND RELATED RATES

EXAMPLE 9 Find explicit functions f and g whose graphs match the top and bottom halves of the circle in Figure 5.15.

Solution If we solve $x^2 + y^2 = 9$ for y, we find

$$y = \pm\sqrt{9 - x^2}.$$

Let $f : [-3, 3] \longrightarrow \mathbb{R}$ and $g : [-3, 3] \longrightarrow \mathbb{R}$ be defined by the formulas

$$f : x \longmapsto \sqrt{9 - x^2} \quad \text{and} \quad g : x \longmapsto -\sqrt{9 - x^2}.$$

Then the graphs of functions f and g are the top and bottom halves of the circle $x^2 + y^2 = 9$ as shown in Figure 5.16. ∎

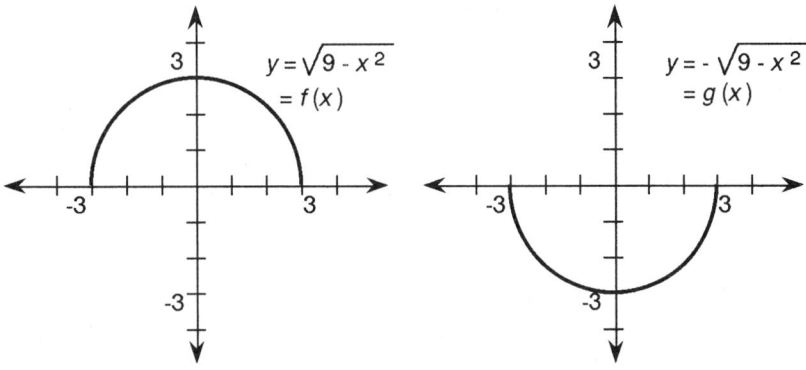

Figure 5.16 The graphs of f and g are semicircles.

This is only one of many ways we could have broken the circle up into function graphs. For instance, we could have defined *four* functions corresponding to the quarter circles bounded by the coordinate axes.

We can think of the original equation

$$x^2 + y^2 = 9$$

as *implicitly* defining y as a function of x. This simply means that by suitably restricting our attention to only part of the equation's locus, we can find an explicit functional relationship between y and x whose graph matches the locus in that region. The particular function may change as we move to different parts of the locus.

We could compute $y' = dy/dx$ (if it exists) at any point (x_0, y_0) by first finding an explicit function f that matches the graph in a region containing (x_0, y_0), and then computing $f'(x_0)$.

EXAMPLE 10 Find the slope of the tangent line to the circle $x^2 + y^2 = 9$ at the point $(1.8, 2.4)$.

Solution This point is on the graph of $f : x \longmapsto \sqrt{9 - x^2}$ since it lies on the top half of the circle. Hence, we can find the slope of the tangent line by computing $f'(1.8)$. Since

$$f(x) = (9 - x^2)^{1/2}$$

we have

$$f'(x) = \frac{1}{2}(9 - x^2)^{-1/2}(-2x) = \frac{-x}{\sqrt{9 - x^2}}$$

using the chain rule. Substituting $x = 1.8$ gives us

$$f'(1.8) = \frac{-1.8}{\sqrt{9 - (1.8)^2}} = -0.75.$$

So the slope of the tangent line at $(1.8, 2.4)$ is -0.75.

Since we are dealing with a circle, we have a nice way of checking this answer. From geometry, we know that the normal line to the circle at this point of tangency must pass through the center of the circle. Since the center of the circle is $(0, 0)$ and the point of tangency is $(1.8, 2.4)$, we can calculate directly the slope m of the normal line as $(2.4 - 0)/(1.8 - 0) = 2.4/1.8 = 4/3$. The slope of the tangent line should be the negative reciprocal $-1/m = -3/4 = -0.75$, just as we found using the derivative. ■

This approach is fine, but it depends on our ability to find a suitable explicit function whose graph includes the point in question. In general, there are infinitely many different functions that include a subset of the locus in their graphs—some are differentiable at the point in question, and some are not. For the example above, for values $-3 \leq x \leq 3$, we could choose $y = \sqrt{9 - x^2}$ when $x \leq 1.8$ and $y = -\sqrt{9 - x^2}$ when $x > 1.8$. This function includes the point $(1.8, 2.4)$ in its graph, but it has no derivative value at $x = 1.8$ because of the jump discontinuity there.

Fortunately, there is a way to find the derivative value dy/dx at a point (x_0, y_0) without having to explicitly solve for y as a suitable function of x.

Let's implicitly assume that y is a differentiable function of x in an appropriate region including the point (x_0, y_0). The original relationship between x and y still must hold, so we can differentiate both sides of the equation *with respect to* x while treating y as if it were a function of x. This process is called **implicit differentiation**, and it results in a relationship between x, y, and dy/dx. We can either attempt to solve this new equation for dy/dx in general, or we may substitute the specific values $x = x_0$ and $y = y_0$ and solve for the value of dy/dx at the particular point (x_0, y_0).

5.3 IMPLICIT DIFFERENTIATION AND RELATED RATES

EXAMPLE 11 Use implicit differentiation to find dy/dx for the relationship $x^2 + y^2 = 9$, and use it to calculate the slope of the tangent line at the point $(1.8, 2.4)$.

Solution If we assume y is a function of x, and differentiate both sides of

$$x^2 + y^2 = 9$$

with respect to x, we obtain

$$2x + 2y \frac{dy}{dx} = 0.$$

Note that on the left-hand side, we have $d(y^2)/dx = 2y(dy/dx)$, since the power rule applies to the *function* y raised to the second power.

We can solve for dy/dx to obtain

$$\frac{dy}{dx} = -\frac{x}{y}$$

and substituting the specific values $x = 1.8$ and $y = 2.4$ gives us

$$\left.\frac{dy}{dx}\right|_{(1.8, 2.4)} = -\frac{1.8}{2.4} = -0.75$$

as the slope of the tangent line to the circle at that point. This matches the value we obtained by using an explicit function. ∎

In fact, our derivative formula

$$\frac{dy}{dx} = -\frac{x}{y}$$

applies to every point on the circle for which $-x/y$ is defined. If $y = 0$ then the slope of the tangent line to the circle is undefined. Figure 5.17 shows that this reflects the fact that the tangent lines to the circle at $(3, 0)$ and $(-3, 0)$ are vertical.

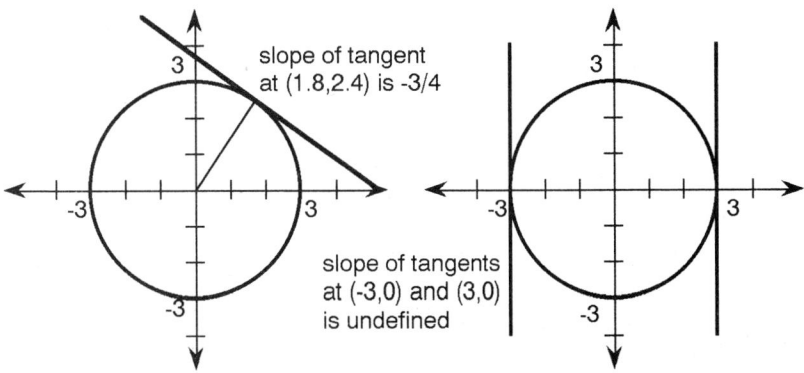

Figure 5.17 $dy/dx = -x/y$ is the slope of the tangent line to the circle.

☞ **When using implicit differentiation to find** dy/dx, **always remember to treat** y **as a function of** x.

EXAMPLE 12 Find $\dfrac{dy}{dx}\Big|_{(-4,2)}$ if $x^2 y^2 - \dfrac{y^3}{x} = 66$.

Solution We might start by noting that the point $(-4, 2)$ is indeed on the curve defined by this equation, since

$$(-4)^2 \cdot 2^2 - \frac{2^3}{-4} = 64 + 2 = 66.$$

Taking the derivative of both sides of the original equation with respect to x gives us

$$\left(2x \cdot y^2 + x^2 \cdot 2y\frac{dy}{dx}\right) - \frac{3y^2(dy/dx) \cdot x - y^3 \cdot 1}{x^2} = 0$$

where we have applied the product rule for the first term and the quotient rule for the second term. If substitute $x = -4$ and $y = 2$ into this equation, we obtain

$$-32 + 64\frac{dy}{dx} - \frac{-48(dy/dx) - 8}{16} = 0.$$

Solving for dy/dx gives us

$$\frac{dy}{dx} = \frac{63}{134}.$$

∎

EXAMPLE 13 Find a general formula for $y' = dy/dx$ when $\sin(xy) = 2x\cos^2(y)$.

Solution Differentiating both sides with respect to x (and writing y' for dy/dx):

$$\cos(xy) \cdot (1 \cdot y + x \cdot y') = 2\cos^2(y) + 2x \cdot (2\cos(y)(-\sin(y)) \cdot y').$$

After expanding:

$$y\cos(xy) + (x\cos(xy))y' = 2\cos^2(y) - (4x\cos(y)\sin(y))y'$$

and collecting terms involving y':

$$y'(x\cos(xy) + 4x\cos(y)\sin(y)) = 2\cos^2(y) - y\cos(xy)$$

we solve for y':

$$y' = \frac{2\cos^2(y) - y\cos(xy)}{x\cos(xy) + 4x\cos(y)\sin(y)}.$$

∎

5.3 IMPLICIT DIFFERENTIATION AND RELATED RATES

We can repeat the process of implicit differentiation to find higher order derivatives.

EXAMPLE 14 Find $y'' = \dfrac{d^2y}{dx^2}$ at $(1,-1)$ if $x^4 - y^3 = 2$.

Solution First, we find the value of $y' = dy/dx$ at $(1,-1)$, by implicitly differentiating $x^2 - y^3 = 2$ to obtain

$$4x^3 - 3y^2 \frac{dy}{dx} = 0.$$

Substituting $x = 1$ and $y = -1$ yields

$$\left.\frac{dy}{dx}\right|_{(1,-1)} = \frac{4}{3}$$

as the value of the first derivative y' at the point $(1,-1)$. Now, if we implicitly differentiate a second time, considering both y and dy/dx as functions of x, we get

$$12x^2 - 6y\frac{dy}{dx} \cdot \frac{dy}{dx} - 3y^2 \frac{d^2y}{dx^2} = 0.$$

To find the value of the second derivative y'' at $(1,-1)$, we can substitute $x = 1$, $y = -1$ and $dy/dx = 4/3$ to obtain

$$12 + \frac{32}{3} - 3\left.\frac{d^2y}{dx^2}\right|_{(1,-1)} = 0$$

and solve to find $\left.\dfrac{d^2y}{dx^2}\right|_{(1,-1)} = \dfrac{68}{9}.$ ∎

Related rates—finding the unknown rate of change

One of the most common uses of derivatives is in analyzing processes that change with time. In other words, time t is the independent variable. The interpretation of a derivative as a rate of change is particularly appropriate when the independent variable is time.

Often several quantities may be changing simultaneously with time. If these quantities are related to each other in some way, then clearly their rates of change will also be related.

As an illustration, consider a balloon being inflated over time. Simultaneously, its geometric attributes of volume, surface area, radius, diameter, circumference are all changing as well as other physical attributes such as its weight, internal pressure, and even its temperature. All of these variables are changing as an outcome of the single act of inflating the balloon. It stands to reason that since many of these variables are closely related to each other by geometric formulas or by physical laws, knowing the rates of change of some of the variables may allow us to determine the rates of other variables.

EXAMPLE 15 Suppose the balloon is considered to be a perfect sphere and its radius is changing at the rate of 3 cm/sec. How fast is the volume of the balloon changing when the radius is 20 cm?

Solution   The relationship between the balloon's radius r and its volume V is given by the formula for the volume of a sphere:

$$V = \frac{4}{3}\pi r^3.$$

In this situation, both the radius r and the volume V are functions of time. We might express this fact by writing

$$V(t) = \frac{4}{3}\pi r(t)^3.$$

If we differentiate both sides of the relationship with respect to t we arrive at

$$V'(t) = \frac{4}{3}\pi \cdot 3r(t)^2 r'(t) = 4\pi r(t)^2 r'(t).$$

(Note that we used the chain rule in differentiating the right-hand side.) We are interested in finding the rate of change of the volume (namely $V'(t)$) at the particular instant that the radius $r(t) = 20$ cm. Since $r'(t)$ represents the rate of change of the radius of the balloon we know that

$$r'(t) = 3 \text{ cm/sec}.$$

Substituting these known values above we have

$$V'(t) = 4\pi \cdot (20 \text{ cm})^2 \cdot 3 \text{ cm/sec} = 4800\pi \text{ cm}^3/\text{sec}.$$

Notice that carrying the units through our computation leads to the units for our final result. The units cm^3/sec make perfect sense, because $V'(t)$ represents change in volume cm^3 per (/) unit of time (sec). ∎

This example illustrates the solution of what is commonly called a *related rates problem*. A related rates problem generally refers to the task of determining the rate of change (with respect to time) of some variables based on their relationship to other variables whose rates of change are known.

5.3 IMPLICIT DIFFERENTIATION AND RELATED RATES

Strategy for solving related rates problems

Here's a step-by-step strategy for solving related rates problems.

Step 1. *Identify the variable and constant quantities involved in the process, along with any rates of change.*

What quantities change value with time? These will be variables and should be labelled appropriately. What quantities keep a fixed value? These are treated as constants (even though the particular fixed value may be unknown). What rates of change are given? The units of measurement can be extremely helpful in identifying rates of change. For example, units like *miles per hour* or *gallons per second* tell us the rate of change (d/dt) of other quantities measured in units of *miles* and *gallons*, respectively.

Step 2. *Find a relationship between the variables.*

How are the variable quantities related? Is there a formula or equation that involves the relevant variables? Do NOT substitute the numerical value of any variable at a specific time into the relationship.

Step 3. *Differentiate with respect to time to find a relationship between the variables and their rates of change.*

This means differentiating both sides of the equation from Step 2 with respect to time t. Make sure to treat those variables that change with time as functions of time, observing the chain rule and other rules of differentiation.

Step 4. *Substitute all known variable and rate values for the specific instant of time in question, and determine the unknown rates of change.*

This may merely be a matter of solving the resulting equation or may require using that information in some other way to determine the unknown rate(s) of change.

Step 5. *Interpret the results and use them to answer the particular questions posed by the problem.*

What exactly is being asked for in the problem? If we need to know a particular rate of change, then this may be given by Step 4. Analyzing both the *sign* (positive or negative) and the *units* ("something" per unit of time) can be helpful in judging the reasonableness of our answer.

EXAMPLE 16 An empty underground storage tank has the shape of a cone (vertex down) 20 feet deep and 40 feet in diameter. If we start pumping into the tank at the constant rate of 100 gallons per minute, then how fast is the depth of the fluid in the tank changing 10 minutes after we start pumping?

Solution *Step 1.* Identify the variable and constant quantities involved in the process, along with any given rates of change.

Figure 5.18 illustrates a picture of the tank. During the process of pumping fluid into the tank, the volume, the depth, and the radius of the circular surface of the fluid in the tank will all be changing. These are the variables in the process. Let's label them as follows:

V = volume at time t, h = depth at time t, r = radius at time t.

Given the dimensions of the tank, it appears that a suitable unit of measurement for r and h would be *feet*, in which case we should measure V in *cubic feet*. (For future reference, we note that the conversion for gallons to cubic feet to be 1 gal $\approx$ 0.13368 ft^3.)

The constant quantities in this problem are the dimensions of the tank—its total depth of 20 feet and its diameter of 40 feet.

The other information given in this problem tells us that $dV/dt = 100$ gal/min ≈ 13.368 ft^3/min at all times t. We wish to use this information to find dh/dt when $t = 10$ minutes.

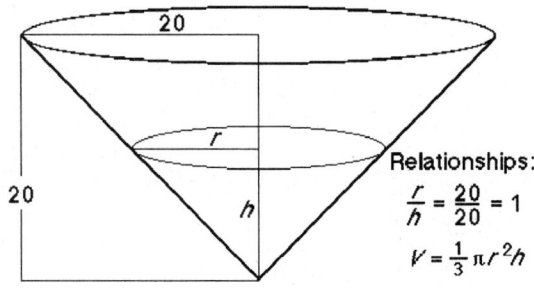

Figure 5.18 The conical underground storage tank.

Step 2. Find a relationship between the variables.

We can relate the volume V of the fluid to the radius and height by the formula for the volume of a cone:

$$V = \frac{1}{3}\pi r^2 h.$$

From Figure 5.18, we also note that r and h can be considered legs of a triangle that is similar to a larger triangle with equal legs of length 20 feet (the full radius and height of the tank). Thus, the ratios of these leg lengths must be the same:

$$\frac{r}{h} = \frac{20}{20} = 1 \quad \text{so that} \quad r = h.$$

With this information, we can relate V and h as functions of time:

$$V(t) = \frac{1}{3}\pi(h(t))^3.$$

Step 3. Differentiate with respect to time to find a relationship between the variables and their rates of change.

5.3 IMPLICIT DIFFERENTIATION AND RELATED RATES

$$\frac{dV}{dt} = \frac{1}{3}\pi(3h^2)\frac{dh}{dt} = \pi h^2 \frac{dh}{dt}.$$

Step 4. Substitute all known variable and rate values for the specific instant of time in question, and determine the unknown rate(s) of change.

Since fluid is being pumped in at a constant rate of 100 gallons per minute, we know that

$$\left.\frac{dV}{dt}\right|_{t=10} = 100 \text{ gal/min}.$$

After 10 minutes we will have pumped in 1000 gallons, so

$$V(10) = 1000 \text{ gal}.$$

From $V = \frac{1}{3}\pi h^3$, we can compute $h(10)$: First, since 1 gallon ≈ 0.13368 ft^3, we have

$$1000 \text{ gal} \approx 133.68 \text{ ft}^3 \approx \frac{3.14159}{3}(h(10))^3$$

from which we find that

$$h(10) \approx \sqrt[3]{\frac{401.4}{3.14159}} \approx 5.04 \text{ ft}.$$

We substitute

$$h(10) \approx 5.04 \text{ ft} \quad \text{and} \quad \left.\frac{dV}{dt}\right|_{t=10} = 100 \text{ gal/min} \approx 13.368 \text{ ft}^3/\text{min}$$

into the equation

$$\left.\frac{dV}{dt}\right|_{t=10} = \pi(h(10))^2 \left.\frac{dh}{dt}\right|_{t=10}$$

to obtain

$$13.368 \text{ ft}^3/\text{min} \approx (3.14159)(5.04 \text{ ft})^2 \cdot \left.\frac{dh}{dt}\right|_{t=10}$$

Solving for the unknown rate, we have

$$\left.\frac{dh}{dt}\right|_{t=10} \approx 0.1675 \text{ ft/min}.$$

Step 5. Interpret the results and use them to answer the particular questions posed by the problem.

After 10 minutes of pumping, the fluid is rising at a rate of approximately 0.1675 ft/min (a little over two inches each minute). ∎

EXAMPLE 17 Two roads intersect at right angles. Two cars leave simultaneously from the intersection. The first car travels at a speed of 30 mph traveling due North. The other car travels at a speed of 40 mph traveling due East. How fast is the distance between them changing 30 minutes later?

Solution *Step 1.* Identify the variable and constant quantities involved in the process, along with any given rates of change.

Figure 5.19 shows an illustration of the situation.

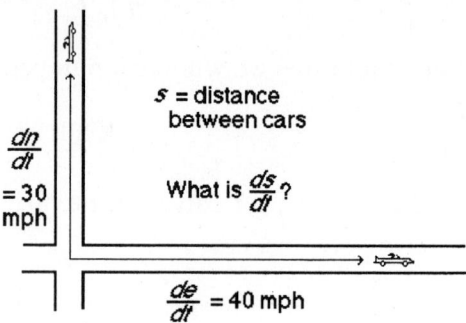

Figure 5.19 How fast is the distance between the two cars changing?

The distance each car travels from the starting point is changing with time as well as the distance between the two cars. If we label the distance traveled north by the first car as n, and the distance traveled east by the second car as e, then we know that

$$dn/dt = 30 \text{ mph} \quad \text{and} \quad de/dt = 40 \text{ mph}.$$

If we label the distance between the two cars as s, then we need to find ds/dt when $t = 0.5$ hours.

Step 2. Find a relationship between the variables. The three variables n, e, and s are related by the Pythagorean theorem:

$$n^2 + e^2 = s^2.$$

Step 3. Differentiate with respect to time to find a relationship between the variables and their rates of change.

$$2n\frac{dn}{dt} + 2e\frac{de}{dt} = 2s\frac{ds}{dt}.$$

Step 4. Substitute all known variable and rate values for the specific instant of time in question and determine the unknown rate(s) of change.

After 30 minutes (0.5 hours), the northbound car will have travelled 15 miles and the eastbound car will have travelled 20 miles. In other words,

5.3 IMPLICIT DIFFERENTIATION AND RELATED RATES

$$n(0.5) = 15 \text{ miles} \quad \text{and} \quad e(0.5) = 20 \text{ miles}.$$

Using this and the Pythagorean Theorem, we can compute

$$s(0.5) = \sqrt{15^2 + 20^2} = 25 \text{ miles}.$$

We substitute these and the known rates into the equation from Step 3 to obtain

$$2(15)(30) + 2(20)(40) = 2(25) \frac{ds}{dt}\bigg|_{t=0.5}$$

and solve to find

$$\frac{ds}{dt}\bigg|_{t=0.5} = \frac{900 + 1600}{50} = 50 \text{ miles per hour}.$$

Step 5. Interpret the results and use them to answer the particular questions posed by the problem.

Thirty minutes after the cars leave the intersection, the distance between the cars is increasing at a rate of 50 mph. ∎

WARNING! A common error in solving related rates problems is to substitute a particular value for t *before* **differentiating with respect to time** t**.**

This effectively changes the variable quantities to constants. (What do you think will happen when you differentiate?)

Differentiating the inverse trigonometric functions

The derivatives of the inverse trigonometric functions can be obtained by implicit differentiation.

EXAMPLE 18 Find the derivative of $x \longmapsto \arctan(x)$ by implicit differentiation.

Solution If $y = \arctan(x)$, then y satisfies
$$\tan(y) = x.$$
If we differentiate both sides with respect to x, we obtain
$$\sec^2(y)\frac{dy}{dx} = 1.$$
Solving for dy/dx gives us
$$\frac{dy}{dx} = \frac{1}{\sec^2(y)} = \cos^2(y).$$
Can we write $\cos^2(y)$ in terms of x? Since we know $\tan(y) = x$, a handy aid would be a right triangle reflecting this relationship. Figure 5.20 shows such a right triangle, with y as the relevant angle, x as the length of the opposite side, and 1 as the length of the adjacent side (so that $\tan(y) =$ opposite/adjacent $= x/1$). We can use the Pythagorean Theorem to compute the remaining side (the hypotenuse) to be $\sqrt{1 + x^2}$. Since $\cos(y) =$ adjacent/hypotenuse, we have
$$\frac{dy}{dx} = \cos^2(y) = \frac{1}{(\sqrt{1+x^2})^2} = \frac{1}{1+x^2}.$$

∎

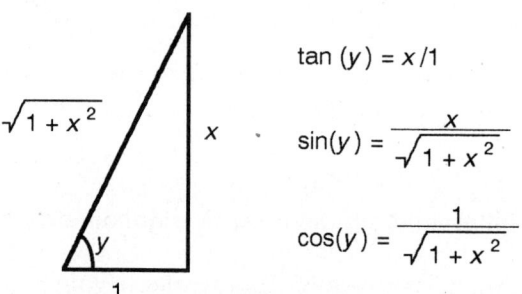

Figure 5.20 Right triangle with $\tan(y) = x$.

5.3 IMPLICIT DIFFERENTIATION AND RELATED RATES

EXAMPLE 19 Find the derivative of $x \longmapsto \arcsin(x)$ by implicit differentiation.

Solution If $y = \arcsin(x)$, then y satisfies

$$\sin(y) = x.$$

If we differentiate both sides with respect to x, we obtain

$$\cos(y)\frac{dy}{dx} = 1.$$

Solving for dy/dx gives us

$$\frac{dy}{dx} = \frac{1}{\cos(y)} = \sec(y).$$

Using a right triangle as an aid (Figure 5.21) as in the previous example, we label the relevant angle y, the opposite side x, and the hypotenuse 1 ($\sin(y)$ = opposite/hypotenuse). The Pythagorean Theorem gives us the remaining adjacent side as $\sqrt{1-x^2}$ and we can rewrite the derivative completely in terms of x:

$$\frac{dy}{dx} == \sec(y) = \frac{1}{\sqrt{1-x^2}}.$$

∎

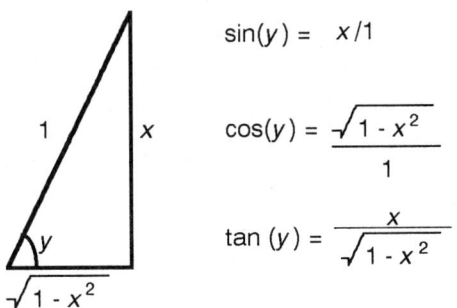

Figure 5.21 Right triangle with $\sin(y) = x$.

EXAMPLE 20 Find the derivative of $x \longmapsto \operatorname{arcsec}(x)$ by implicit differentiation.

Solution If $y = \operatorname{arcsec}(x)$, then y satisfies

$$\sec(y) = x.$$

If we differentiate both sides with respect to x, we obtain

$$\sec(y)\tan(y)\frac{dy}{dx} = 1.$$

Solving for dy/dx gives us

$$\frac{dy}{dx} = \frac{1}{\sec(y)\tan(y)} = \frac{1}{x\sqrt{x^2-1}}.$$

This time label the right triangle as follows: relevant angle y, hypotenuse x, and adjacent side 1. The Pythagorean Theorem gives us the remaining opposite side as $\sqrt{x^2-1}$, and we can rewrite the derivative as shown. ■

EXERCISES

In exercises 1-8, you are given an equation in two variables, x and y, and a specific point on the locus of the equation. Find dy/dx in terms of x and y, and evaluate dy/dx at the point indicated. Use that information to find the equations of the tangent line and the normal line at that point.

1. $4x^2 + 9y^2 = 36$; $(-3/2, \sqrt{3})$
2. $xy = 1$; $(-0.25, -4)$
3. $x^3 = y^2$; $(4, 8)$
4. $x^{2/3} + y^{2/3} = 13$; $(-27, -8)$
5. $(x+y)^2 = 9$; $(1, 2)$
6. $x = \sqrt{y} + \sqrt[3]{y}$; $(12, 8)$
7. $x^2y^2 = x^2 + y^2$; $(0, 0)$
8. $(x-y)/(x+y+1) = xy$; $(0, 0)$

In exercises 9-14, find the dy/dx in terms of x and y, using implicit differentiation.

9. $x = \sin(y)$
10. $x = \cos(y)$
11. $x = \tan(y)$
12. $x = \sec(y)$
13. $x = \csc(y)$
14. $x = \cot(y)$

5.3 IMPLICIT DIFFERENTIATION AND RELATED RATES

15. If f is a differentiable function with an inverse g, then we know that $x = g(y)$ whenever $y = f(x)$. Use implicit differentiation to show that $g'(y) = 1/f'(x)$.

16. Check the result of the previous exercise in the special case that $f(x) = \sin(x)$, $g(x) = \arcsin(x)$, and $(x, y) = (\pi/6, 1/2)$.

Find df/dt if x, y, and θ are functions of t (time) and f is the function indicated in exercises 17 - 22.

17. $f : t \longmapsto x^4 + 6\sin\theta$
18. $f : t \longmapsto 4x^3 + 10\cos\theta$
19. $f : t \longmapsto 5x^3 - 4\arcsin y$
20. $f : t \longmapsto 2x^3 - 5\arctan y$
21. $f : t \longmapsto \pi\sin(x^2\arcsin(y))$
22. $f : t \longmapsto \pi \cdot \sin^2(\theta) + \pi \cdot \cos^2(\theta)$

---

23. Find the derivative of $x \longmapsto \arccos(x)$ by implicit differentiation.

24. Find the derivative of $x \longmapsto \arccot(x)$ by implicit differentiation.

25. Find the derivative of $x \longmapsto \arccsc(x)$ by implicit differentiation.

Consider the inflating balloon of Example 15. Find the rates of change indicated in exercises 26-28.

26. How fast is the circumference of the balloon changing when the radius is 5 cm?

27. How fast is the surface area of the balloon changing when the radius is 10 cm?

28. Suppose the elasticity of the balloon at room temperature is given by the equation $E = 100 - \sqrt{V}$. The balloon explodes when the elasticity reaches 0. How fast is the volume of the balloon changing at the moment of explosion? How fast was the elasticity changing at that instant?

---

29. Suppose the two cars take off from the same intersection with the same speeds as given in Example 16. This time, the eastbound car leaves two hours earlier than the northbound car. How fast is the distance between them changing five hours after the eastbound car leaves?

30. Suppose now that the eastbound road is actually elevated 30 feet above the northbound road. How fast is the distance between the two cars changing four hours later, if they leave the intersection at the same moment?

31. At the end of one of the assembly lines at the Platypus corporation, metal shavings are emptied into a conical pile whose height is always the same as its diameter. If the shavings are spilled onto the pile at the rate of 10 cm^3/sec, how fast is the radius of the pile increasing when the height is 5 cm?

32. A weather balloon is tethered so that it stays at a constant height of 250 meters. The wind blows it horizontally at a rate of 5 meters/sec away from its original position. If the line is spooled out so that the altitude remains constant, how fast must the line be let out when the balloon lies over a point on the ground 300 meters from the spool?

33. A rectangular swimming pool has a depth ranging from 1 meter at the shallow end to 4 meters deep at the diving end. The pool is 50 meters long and 25 meters wide. The bottom slopes from the edge of the shallow area (10 meters long) to the diving area (also 10 meters long) as shown in the picture below. If water is pumped out of a full pool at the rate of 250 gallons per minute, how fast is the water level dropping when the depth at the diving end is 1 meters? 2 meters? 3 meters?

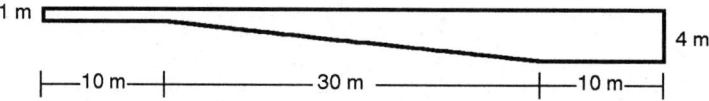

34. Platypus has accidentally spilled 30 ft^3 of chemicals into the river, causing a circular "slick" whose area is expanding while its thickness is decreasing. If the radius of the slick expands at the rate of 1 foot per hour, how fast is the thickness of the slick decreasing when the area is 100 ft^2?

35. A woman standing on the bank of a river is reeling in a fish. The tip of her fishing rod is 5 feet above the water's surface at the bank's edge. How fast is the fish approaching shore when there are 30 feet of line out from the tip of the rod and the woman is reeling in 3 inches per second?

36. Many manufacturing processes involve what is known as "economies of scale." That is, as you produce more items the cost per item decreases. This is due in part to fixed overhead costs that do not depend on the number of items produced. Suppose that the Platypus manufacturing plant has fixed overhead costs of $1000 per day plus a manufacturing cost of $0.10 per foot of wire. How fast will the overall cost per foot be decreasing as you begin increasing the number of feet produced upwards from 5000 feet per day? How fast will the overall cost per foot be decreasing when you cross the 10,000 feet per day mark?

5.4 PARAMETRIC AND POLAR EQUATIONS

Another way that a curve in the Cartesian plane may be described is through the use of **parametric equations**. Parametric equations describe both coordinates x and y as functions of a third variable t over some common domain D. A curve described in such a way is called a **parametrized curve** and the shared independent variable t is called the **parameter** for the curve. This is a different use for the word *parameter* than in earlier chapters. In this case, the parameter t really is the *independent variable*, while x and y are both *dependent variables*.

EXAMPLE 21 Graph the curve described by the parametric equations

$$x = t^2 \quad \text{and} \quad y = t^3 \quad \text{for} \quad 0 \leq t \leq 2.$$

Solution For each value t in the interval $[0, 2]$ we plot

$$(x(t), y(t)) = (t^2, t^3).$$

In Figure 5.22 we show a selected table of values for t, x, and y and the graph of the resulting parametrized curve. ∎

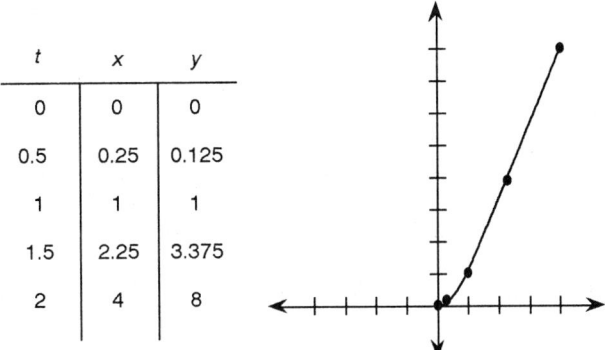

t	x	y
0	0	0
0.5	0.25	0.125
1	1	1
1.5	2.25	3.375
2	4	8

Figure 5.22 Table of values and graph of $(x(t), y(t)) = (t^2, t^3)$.

If we look only at the graph, notice that we cannot extract the value of t that corresponds to a point on the curve. All we see are the paired outputs $(x(t), y(t))$. The choice of the letter t is not accidental. It is meant to suggest an independent variable measured in units of *time*. A parametrized curve is often thought of as the path that a moving object traces out through time. The parametric equations giving us $x(t)$ and $y(t)$ tell us the position of the object by its coordinates. The path by itself cannot tell us *when* the object is at a certain point, since the curve is just a set of points. In fact, a completely different pair of parametric equations could give us exactly the same curve.

EXAMPLE 22 Graph the curve corresponding to the parametric equations

$$x = t^2 - 4t + 4 \quad \text{and} \quad y = t^3 - 6t^2 + 12t - 8 \quad \text{for} \quad 2 \leq t \leq 4.$$

Solution Even though the parametric equations have changed (as well as the interval of values t), the graph consists of exactly the same points as before, as shown in Figure 5.23. ∎

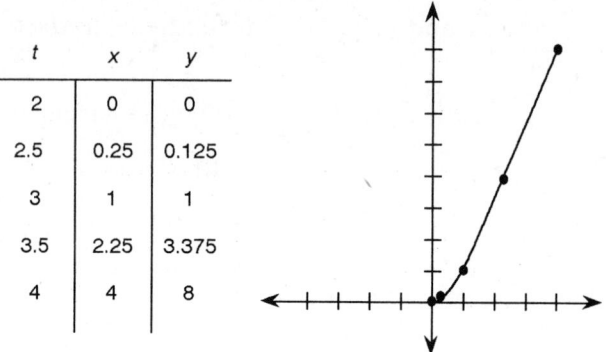

t	x	y
2	0	0
2.5	0.25	0.125
3	1	1
3.5	2.25	3.375
4	4	8

Figure 5.23 Table of values and graph of $(x(t), y(t)) = (t^2 - 4t + 4, t^3 - 6t^2 + 12t - 8)$.

EXAMPLE 23 Using the parametric equations in Example 20, express y as a function of x. Then use this to find

$$\left.\frac{dy}{dx}\right|_{t=0.5}$$

Solution Since $x = t^2$, and $t \geq 0$ on the interval $[0, 2]$, we can write $t = \sqrt{x}$ and substitute for t in the expression for y:

$$y = t^3 = (\sqrt{x})^3 = x^{3/2}.$$

From this, we can compute

$$\frac{dy}{dx} = \frac{3}{2}x^{1/2} = \frac{3\sqrt{x}}{2}.$$

When $t = 0.5$, we have $x = t^2 = (0.5)^2 = 0.25$. Substituting into dy/dx yields

$$\left.\frac{dy}{dx}\right|_{t=0.5} = \frac{3\sqrt{0.25}}{2} = \frac{3(0.5)}{2} = 0.75,$$

the slope of the curve at the point $(0.25, 0.125)$. ∎

5.4 PARAMETRIC AND POLAR EQUATIONS

This method is fine as long as we are successful in writing y as a function of x. (Or, if we could just find some relationship solely in terms of x and y, we might find dy/dx using implicit differentiation.) This may be quite difficult to do, even for fairly simple parametric equations. However, it is possible to find dy/dx directly in terms of t using the parametric equations for x and y.

The reasoning is based on the chain rule. If we could write

$$y = f(x),$$

then substituting x as a function of t gives us

$$y = f(x(t)).$$

If we take the derivative with respect to t we have (in Leibniz notation)

$$\frac{dy}{dt} = \frac{dy}{dx}\frac{dx}{dt}.$$

Algebraically, we could solve for dy/dx as

$$\frac{dy}{dx} = \frac{dy/dt}{dx/dt}$$

provided $dx/dt \neq 0$. This means we can compute dy/dx directly in terms of t, by taking the quotient of dy/dt and dx/dt. If we think of the curve as the path of a moving object, then the derivatives dx/dt and dy/dt tell us how fast the position of the object changes in the horizontal and vertical directions, respectively. The derivative dy/dx then gives us a rate of change of the y-coordinate relative to change in the x-coordinate.

EXAMPLE 24 Using the parametric equations from Example 22, calculate

$$\left.\frac{dy}{dx}\right|_{t=2.5}$$

Solution $\quad \dfrac{dy}{dx} = \dfrac{dy/dt}{dx/dt} = \dfrac{3t^2 - 12t + 12}{2t - 4}$, so we can compute

$$\left.\frac{dy}{dx}\right|_{t=2.5} = \frac{3(2.5)^2 - 12(2.5) + 12}{2(2.5) - 4} = 0.75.$$

This represents the slope of the parametrized curve at the point

$$(x(2.5), y(2.5)) = (0.25, 0.125).$$

∎

Note that even though this parametrization was different, we calculated the same slope for the same curve at the same point. The slope dy/dx at a point depends only on the curve itself, and not on a particular parametrization of that curve. Hence, if we use a graphing calculator or software to plot a curve given parametric equations, we will not be able to judge visually the values of dx/dt and dy/dt at a specific point, but we will be able to estimate the ratio $dy/dx = (dy/dt)/(dx/dt)$. In fact, any function graph of the form $y = f(x)$ can be considered a parametrized curve by simply substituting t for x:

$$x(t) = t \qquad y(t) = f(t)$$

Indeed, when we watch the screen as a graphing calculator or graphing software plots a function graph in "real time," we are seeing the dynamic trace of a curve parametrized in this way.

EXAMPLE 25 Suppose we parametrize the circle $x^2 + y^2 = 9$ using parametric equations

$$x = 3\cos(t) \quad \text{and} \quad y = 3\sin(t) \quad \text{for} \quad 0 \le t \le 2\pi.$$

Find the equations of the tangent line and of the normal line at $t = 2\pi/3$.

Solution $dy/dx = (dy/dt)/(dx/dt) = (3\cos t)/(-3\sin t) = -\cot(t)$. So,

$$\left.\frac{dy}{dx}\right|_{t=2\pi/3} = -\cot(2\pi/3) = \frac{1}{\sqrt{3}}.$$

The slope m of the tangent line is $1/\sqrt{3}$ and the slope $-1/m$ of the normal line is $-\sqrt{3}$. The point of tangency is

$$(x(t), y(t)) = (3\cos(2\pi/3), 3\sin(2\pi/3)) = (-3/2, 3\sqrt{3}/2).$$

From this information we can write down the equation of the tangent line:

$$y = \frac{1}{\sqrt{3}}\left(x + \frac{3}{2}\right) + \frac{3\sqrt{3}}{2}$$

and

$$y = -\sqrt{3}\left(x + \frac{3}{2}\right) + \frac{3\sqrt{3}}{2}$$

is the equation of the normal line. ■

5.4 PARAMETRIC AND POLAR EQUATIONS

Polar coordinates

A very useful alternative coordinate system for the plane is the **polar coordinate system**. To locate a point in the (two-dimensional) plane requires specifying two real numbers or coordinates. In the rectangular or Cartesian coordinate system, the two coordinates make reference to two perpendicular real axes intersecting at the origin.

We can imagine a point's coordinates (x, y) as specifying directions to a person at the origin. The x-coordinate gives the person the precise distance and direction to walk in the horizontal direction, and the y-coordinate gives the precise distance and direction to walk in the vertical direction. For example, we can locate the point $(3, -2)$ by walking 3 units to the right and 2 units down.

An alternative method of communicating a point's location would be to specify the point's distance from the origin and the direction to walk straight out to the point. For example, if we were given the information that a point located exactly 6 units from the origin along a line making a $40°$ angle measured counterclockwise from the positive x-axis, we can find the point precisely (see Figure 5.24).

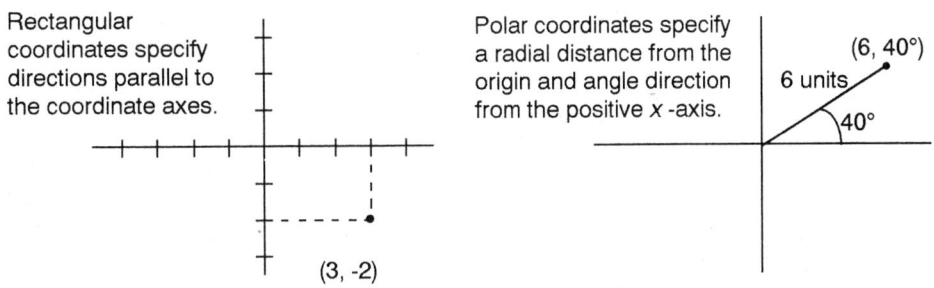

Figure 5.24 Rectangular and polar coordinate systems.

The convention for specifying polar coordinates is to list the distance r and angle θ as an ordered pair (r, θ). The angle θ could be measured in either degrees or radians, but we will usually use radian measure.

EXAMPLE 26 Locate the following points given their polar coordinates:

$$P = (2, \frac{\pi}{4}), \qquad Q = (5, \frac{2\pi}{3}), \qquad R = (3, -\frac{5\pi}{6}) \qquad S = (2, \frac{9\pi}{4}).$$

Solution The points are illustrated in Figure 5.25 below. Notice that the point $(2, \frac{9\pi}{4})$ represents the same point as $(2, \frac{\pi}{4})$. In other words, points P and S coincide. ∎

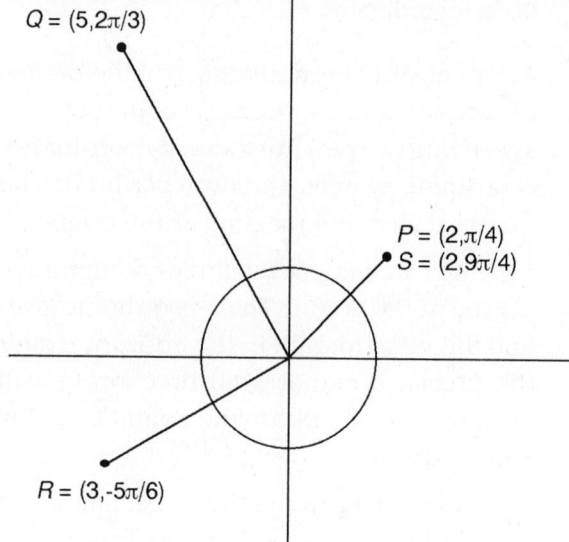

Figure 5.25 Locating points with polar coordinates

Both the Cartesian and polar coordinate systems require an ordered pair of real numbers to specify a point's location on the plane. An important difference between the systems has to do with *uniqueness of representations*. In the Cartesian system, every point has exactly one ordered pair (x, y) associated with it. However, in the polar coordinate system, a point has infinitely many representations (r, θ) because

$$(r, \theta + 2\pi n)$$

gives the same location as (r, θ) for any integer n.

EXAMPLE 27 $(2, \frac{\pi}{4})$, $(2, \frac{9\pi}{4})$, $(2, \frac{17\pi}{4})$, and $(2, -\frac{15\pi}{4})$ all represent the same point. ∎

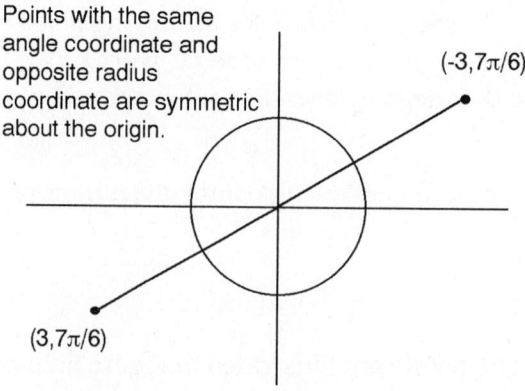

Figure 5.26 Comparing (r, θ) and $(r, -\theta)$.

5.4 PARAMETRIC AND POLAR EQUATIONS

A negative radius r can be specified as a polar coordinate by adopting the convention that $(-r, \theta)$ and $(r, \theta + \pi)$ represent the same point. Geometrically, the point $(-r, \theta)$ is the reflection of (r, θ) through the origin (see Figure 5.26).

A special case occurs when $r = 0$. Notice that $(0, \theta)$ represents the origin, regardless of the value of θ.

For points other than the origin, we could gain a unique polar coordinate representation by arbitrarily restricting the radius r to be positive and the angle measure θ to lie in some predetermined interval of length 2π, say

$$0 \leq \theta < 2\pi$$

or

$$-\pi < \theta \leq \pi.$$

Such restrictions may be useful in certain applications, but we will generally allow both r and θ to take on any real number values.

Conversions between coordinate systems

Given a point whose location is specified by polar coordinates (r, θ), we can find the corresponding rectangular coordinates (x, y) through some simple trigonometry. Figure 5.27 illustrates the situation for a point in the first quadrant.

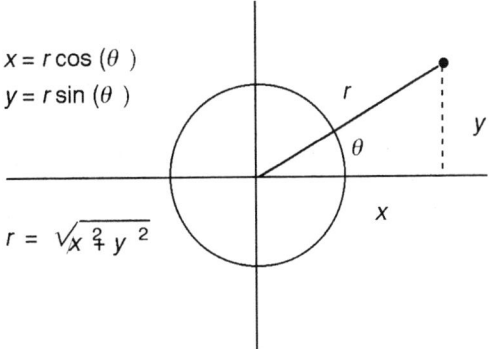

Figure 5.27 Conversions between rectangular and polar coordinates.

Specifically,

$$x = r\cos(\theta) \quad \text{and} \quad y = r\sin(\theta).$$

EXAMPLE 28 Convert the following from polar coordinates to rectangular coordinates:

$$(4, 3\pi/4) \quad \text{and} \quad (-2.5, -638°).$$

Solution For $(4, 3\pi/4)$, we have $r = 4$ and $\theta = 3\pi/4$, so

$$x = r\cos(\theta) = 4 \cdot \frac{-\sqrt{2}}{2} = -2\sqrt{2} \quad \text{and} \quad y = r\sin(\theta) = 4 \cdot \frac{\sqrt{2}}{2} = 2\sqrt{2}.$$

Hence, $(4, 3\pi/4)$ in polar coordinates corresponds to

$$(-2\sqrt{2}, 2\sqrt{2}) \approx (-2.8284, 2.8284)$$

in rectangular coordinates.

For $(-2.5, -638°)$, we have

$$x = (-2.5)\cos(-638°) \approx -0.3479 \quad \text{and} \quad y = (-2.5)\sin(-638°) \approx -2.4757,$$

so $(-2.5, -638°)$ in polar coordinates corresponds approximately to

$$(-0.3479, -2.4757)$$

in rectangular coordinates. ∎

Given a point whose location is specified with rectangular coordinates (x, y), we can find corresponding polar coordinates (r, θ) for the same point. First, we note that

$$r = \sqrt{x^2 + y^2}$$

by using the Pythagorean theorem.

If $r = 0$, then any value θ may be used. If $r \neq 0$, we can find a suitable value $0 \leq \theta < 2\pi$ by noting that

$$\sin(\theta) = y/r \quad \text{and} \quad \cos(\theta) = x/r.$$

More directly, if we compute $\arctan(y/x)$, then either

$$\theta = \arctan(y/x) \quad \text{or} \quad \theta = \arctan(y/x) + \pi,$$

depending on the quadrant (x, y) is in (the first value applies to Quadrants I and IV; the second value applies to Quadrants II and III).

5.4 PARAMETRIC AND POLAR EQUATIONS

EXAMPLE 29 Convert the following from rectangular coordinates to polar coordinates:

$$(3,4) \quad \text{and} \quad (-7.51,-6.28).$$

Solution For $(3,4)$ we have $x = 3$ and $y = 4$, so

$$r = \sqrt{3^2 + 4^2} = \sqrt{25} = 5.$$

Since $(3,4)$ is in the first quadrant, we can use

$$\theta = \arctan(y/x) = \arctan(4/3) \approx 0.9273 \text{ radians}$$

or, using degree measure, $\theta \approx 53.13°$. Thus, $(3,4)$ in rectangular coordinates corresponds approximately to

$$(5, 0.9273) \quad \text{or} \quad (5, 53.13°)$$

in polar coordinates.

For $(-7.51, -6.28)$, we have

$$r = \sqrt{(-7.51)^2 + (-6.28)^2} \approx 9.79.$$

Since $(-7.51, -6.28)$ is in the third quadrant, we can use

$$\theta = \arctan(y/x) + \pi = \arctan\left(\frac{-6.28}{-7.51}\right) + \pi \approx 3.838 \text{ radians}$$

or, using degree measure, $\theta \approx 219.9°$. Thus, $(-7.51, -6.28)$ in rectangular coordinates corresponds approximately to

$$(9.79, 3.838) \quad \text{or} \quad (9.79, 219.9°)$$

in polar coordinates. ∎

Polar graphs

Sometimes it is more convenient to describe a curve using polar coordinates rather than Cartesian coordinates. This is particularly true when the curve is defined in terms of distance to a reference point. The simplest example of such a curve is a circle. For instance, a circle of radius 3 centered at the origin has the equation

$$r = 3$$

in polar coordinates.

The grid markings on graph paper correspond to constant values of the two coordinates. For Cartesian coordinates, the grid lines are horizontal and vertical, corresponding to equations of the form

$$x = a \quad \text{and} \quad \theta = b.$$

The locus of $r = a$ is a circle, while the locus of $\theta = b$ is a line through the origin. Figure 5.28 shows a polar grid that could be used as an aid in plotting polar graphs by hand.

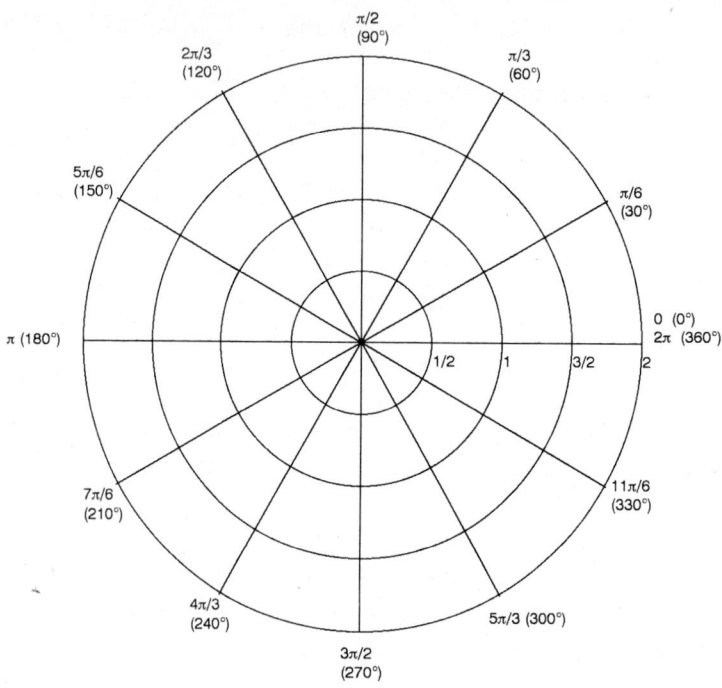

Figure 5.28 Polar grid lines are concentric circles and lines through the origin.

A functional equation of the form

$$r = r(\theta)$$

describes the distance r from the origin as a function of the angle θ. You might think of a radar screen with a beam emanating from the origin sweeping around in a circle, recording the distance r out to a curve for each value θ.

5.4 PARAMETRIC AND POLAR EQUATIONS

EXAMPLE 30 Plot $r = 2\sin(\theta)$.

Solution Figure 5.29 shows the plot of points for this polar equation. We can see that we have obtained what looks like a circle of radius 1 centered at the Cartesian point $(0, 1)$.

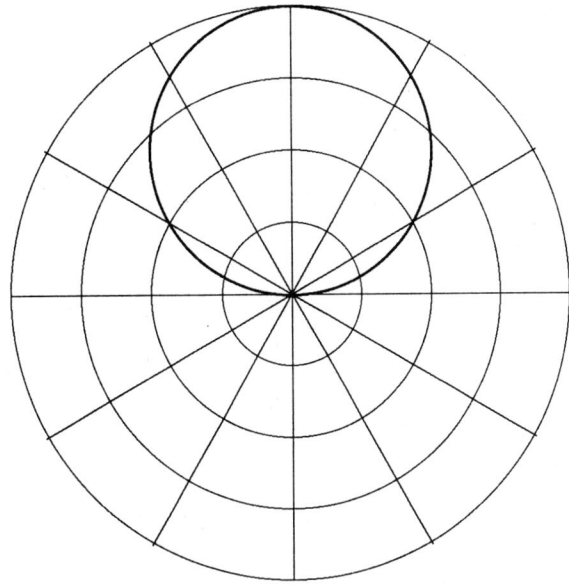

Figure 5.29 Graph of $r = 2\sin(\theta)$.

Can we be sure this is a circle? If we multiply both sides of the equation by r, we obtain

$$r^2 = 2r\sin(\theta).$$

Converting to Cartesian coordinates, we find that this is equivalent to

$$x^2 + y^2 = 2y.$$

On the other hand, a circle of radius 1 centered at $(0, 1)$ will have the equation

$$x^2 + (y-1)^2 = 1^2$$

or, after expanding,

$$x^2 + y^2 - 2y + 1 = 1.$$

We can see that this is equivalent to our polar equation. ∎

Some curves are much more easily described with polar coordinates than rectangular coordinates. Figure 5.30 shows the graph of a "hyperbolic spiral" $r = 2\pi/\theta$ for $0 < \theta \leq 4\pi$. The radius r is undefined for $\theta = 0$, and the spiral approaches the horizontal asymptote $y = 2\pi$ as $\theta \to 0^+$. In comparison, the rectangular description of this curve is much more complicated (try it!).

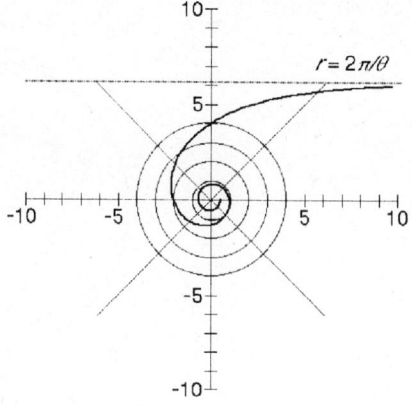

Figure 5.30 The hyperbolic spiral $r = 2\pi/\theta$ for $0 < \theta \leq 4\pi$.

When r is a periodic function of θ (like many functions made up from trigonometric functions through algebra or composition), then the graph of $r = r(\theta)$ will come back to its starting point over one period. In this case, the graph may determine one or more closed curves. The appearance of the resulting "loops" of such a curve provide the motivation for many colorful (and descriptive) names for particular polar graphs. For example, Figure 5.31 shows an example of a heart-shaped polar graph, aptly named a "cardioid."

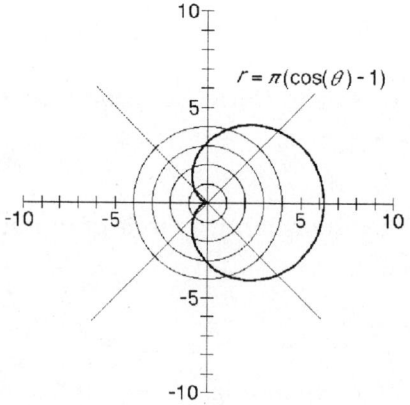

Figure 5.31 The cardioid $r = \pi(cos(\theta) - 1)$.

Polar graphs as parametric curves

We can think of polar plotting as a special case of parametric plotting. Recall that a curve can be described by expressing the coordinates as functions of a third variable (parameter) t:

$$(x(t), y(t)).$$

If we use t instead of θ as a parameter, then we can use the fact that $x = r\cos t$ and $y = r\sin t$ to express the curve

$$r = f(t)$$

parametrically as

$$x(t) = f(t)\cos(t)$$
$$y(t) = f(t)\sin(t).$$

EXAMPLE 31 Express the polar graph $r = 2\sin(t)$, $0 \le t \le 2\pi$ as a parametric curve $(x(t), y(t))$.

Solution Substituting into the usual conversion formulas for changing from polar to rectangular coordinates, we obtain

$$x(t) = 2\sin(t)\cos(t)$$
$$y(t) = 2\sin^2(t).$$

Try graphing this with a parametric plotter over the interval $0 \le t \le 2\pi$ to see that it does trace out the same circle of radius 1 with center at $(0, 1)$. ∎

Hence, any machine grapher capable of plotting parametric equations is automatically capable of plotting polar equations. Of course, some graphers have a special polar graphing feature.

The slope dy/dx of a polar graph $r = r(\theta)$ can be determined by using the chain rule and the two conversion formulas

$$x = r(\theta)\cos(\theta) \quad \text{and} \quad y = r(\theta)\sin(\theta)$$

to obtain

$$\frac{dy}{dx} = \frac{dy/d\theta}{dx/d\theta}.$$

Taking the indicated derivatives with respect to θ in the numerator and denominator yields

$$\frac{dy}{dx} = \frac{(dr/d\theta)\sin(\theta) + r(\theta)\cos(\theta)}{(dr/d\theta)\cos(\theta) - r(\theta)\sin(\theta)}.$$

EXAMPLE 32 Find the slope of the polar graph of $r = 2\sin(\theta)$ at $\theta = \pi/3$.

Solution Using the derivative formula for dy/dx we just developed, we have

$$\frac{dy}{dx} = \frac{2\cos(\theta)\sin(\theta) + 2\sin(\theta)\cos(\theta)}{2\cos(\theta)\cos(\theta) - 2\sin(\theta)\sin(\theta)} = \frac{2\cos(\theta)\sin(\theta)}{\cos^2(\theta) - \sin^2(\theta)} = \frac{\sin(2\theta)}{\cos(2\theta)} = \tan(2\theta).$$

For $\theta = \pi/3$, we have

$$\left.\frac{dy}{dx}\right|_{\theta = \pi/3} = \tan(2\pi/3) = -\sqrt{3}.$$

Examining the polar graph of $r = 2\sin(\theta)$ for $\theta = \pi/3$ shows that this slope value is reasonable (see Figure 5.29). ∎

EXERCISES

For each pair of parametric equations in exercises 1-8, graph the curve for the interval $0 \leq t \leq 10$ and find dy/dx in terms of t by first calculating dx/dt and dy/dt.

1. $x = 2\cos(t)$, $y = 3\sin(t)$
2. $x = -3\cos(t)$, $y = 2\sin(t)$
3. $x = 5 - t$, $y = 25 - 10t + t^2$
4. $x = 2t + 3$, $y = 3t + 2$
5. $x = t\cos(t)$, $y = t\sin(t)$
6. $x = \arctan(t)$, $y = t$
7. $x = |5 - t|$, $y = t$
8. $x = \sqrt{9 - t^2}$, $y = t$

9. Suppose $x(t)$ and $y(t)$ determine the coordinates of an object travelling around the unit circle. If dx/dt is positive when the object is in the third quadrant, is dy/dt positive, negative, or zero? Is the object travelling clockwise or counterclockwise?

10. If we have two parametrized curves (two pairs of parametric equations), then we should distinguish between an intersection point and a "collision" point. An intersection point is any point that lies on both curves. If this intersection point corresponds to the same value of t for both curves, we could call it a collision point. (An intersection of two roads may or may not be the site of a collision between two cars travelling on those roads.) Find all the intersection points for the pair of curves

$$x = t^3 - 2t^2 + t; y = t \qquad x = 5t; y = t^3$$

and indicate which intersection points are true collision points.

5.4 PARAMETRIC AND POLAR EQUATIONS

In exercises 11-18, convert the indicated polar coordinates to rectangular coordinates.

11. $(3, 60°)$ **12.** $(1.414, -765°)$
13. $(3, -60°)$ **14.** $(-3.5, -600°)$
15. (π, π) **16.** $(-5, -8\pi/3)$
17. $(0, 1000)$ **18.** $(1000, 0)$

In exercises 19-26, convert the indicated rectangular coordinates to polar coordinates, both with θ measured in degrees and radians.

19. $(3, 4)$ **20.** $(-5, -5)$
21. $(2.789, -3.254)$ **22.** $(-2.789, -3.254)$
23. (π, π) **24.** $(-5, -8\pi/3)$
25. $(0, 1000)$ **26.** $(1000, 0)$

In exercises 27-36, graph the given polar function over an appropriate interval (experiment with intervals having lengths that are multiples of π) for values of the parameter $a = 1/4, 1/2, 1, 2$ and 4. On the basis of these graphs, determine what characteristic of the graph a controls. Predict the appearance of the graph for $a = 1/3$ and for $a = 3$, and then check your predictions by graphing the polar function for these parameter values. (When the given graph has a special name, it appears in parentheses.)

27. $r = 2a\cos(\theta)$ (Circle)
28. $r = a\sin(\theta)\cos^2(\theta)$ (Bifolium)
29. $r = a(\cos(\theta) + 1)$ (Cardioid)
30. $r = a\sin(\theta)\tan(\theta)$ (Cissoid of Diocles)
31. $r = a\sec(\theta) + 1$ (Conchoid of Nicomedes)
32. $r = \sec(\theta) + a$ (Conchoid of Nicomedes)
33. $r = 1 + a\cos(\theta)$ (Limaçon of Pascal)
34. $r = a + \cos(\theta)$ (Limaçon of Pascal)
35. $r = a\cos(2\theta)\sec(\theta)$ (Strophoid)
36. $r = 2a\theta/\pi$ (Spiral of Archimedes)

Conic sections (parabolas, hyperbolas, and ellipses) can all be described using polar coordinates. In fact, a single polar equation can generate all three types of conic sections:

$$r = \frac{ae}{1 - e\cos(\theta)}$$

where a and e are parameters (e is *not* the special transcendental number used in exponential functions). The value of e is called the **eccentricity** of the conic section.

Let $a = 3$ and graph the polar function

$$r = \frac{3e}{1 - e\cos(\theta)}$$

over the interval $0 \leq \theta \leq 4\pi$ for values of the parameter $e = 1/4$, $1/2$, 1, 2 and 4. Exercises 37-40 refer to these graphs.

37. On the basis of these graphs, guess which type of conic is generated for $e = 1/3$ and for $e = 3$, and then check your predictions by graphing the polar function for these values e.

38. Graph the polar function for $e = .1$, $e = .01$, and $e = .001$. What kind of curve is the limiting case as $e \to 0$?

39. How will the answers to exercise 37 change if the value of a is changed from 3 to some other real value?

40. If $e = 1$, how does changing the value a change the appearance of the graph?

A "rose" is obtained by graphing either of the polar equations

$$r = a\cos(k\theta) \quad \text{or} \quad r = a\sin(k\theta).$$

Graph both of these polar functions over the interval $0 \leq \theta \leq 12\pi$ for values of the parameter $a = 1/2$, 1, and 2 and for $k = 1$, 2, 3, and 4 (twenty-four graphs in all). Exercises 41-44 refer to these graphs.

41. What does the value a control? (Look at the size of the leaves.)

42. What does the value k control? (Count the number of the leaves.)

43. What does the choice of $\sin(\theta)$ or $\cos(\theta)$ control? (Look at the location of the leaves.)

44. On the basis of your answers to exercises 41-43, predict the appearance of the graphs of $r = 3\cos(5\theta)$ and $r = 1.5\sin(6\theta)$, including size, number, and location of the leaves of the rose. Then check your prediction by graphing the polar equations.

Using as many examples from exercises 27-36 as you like, determine the graphical effect of each of the substitutions in exercises 45-51. In other words, how is the new graph related to the old graph?

45. Replace θ with $-\theta$.

46. Replace θ with $(\pi + \theta)$.

47. Replace θ with $(\pi - \theta)$.

5.4 PARAMETRIC AND POLAR EQUATIONS

48. Replace r by $-r$.
49. Replace r by $-r$ and θ with $-\theta$.
50. Replace r by $-r$ and θ with $(\pi - \theta)$.
51. Replace θ by $(\theta + \pi/2)$.

For each of the polar equations in exercises 52-60, convert from the polar description of the curve to a rectangular description in terms of x and y. Then, using the original polar form, replace θ by t and express in terms of parametric equations $x(t)$ and $y(t)$. Graph both the polar and the parametric forms to check your answers. Find $dr/d\theta$ and the slope dy/dx at $\theta = \pi/3$ and $\theta = 5\pi/4$. Judge the reasonableness of your slope values from the graph.

52. $r = 6\cos(\theta)$
53. $r = 3\sin(\theta)\cos^2(\theta)$
54. $r = -2(\cos(\theta) + 1)$
55. $r = 5\sin(\theta)\tan(\theta)$
56. $r = 4\sec(\theta) + 1$
57. $r = \sec(\theta) - 7$
58. $r = 1 + \frac{1}{3}\cos(\theta)$
59. $r = -3 + \cos(\theta)$
60. $r = \cos(2\theta)\sec(\theta)$

CHAPTER

6

The Integral

So far, our attention has been on the branch of calculus known as the *differential calculus.* In this chapter we turn to the other major branch, the **integral calculus**, including both types of integrals—*definite* integrals and *indefinite*.

The idea of a definite integral has its origins in problems of area measurement, and the *definite integral* of a function f is closely related geometrically to the area of the region between the graph of the function and the x-axis. We'll start this chapter by examining the problem of measuring the area of a circle. The method of solution used by the ancient Greeks captures the main idea behind definite integration. We will use this method to define definite integrals, and examine some of their key properties.

The process of finding the derivative f' of a function f is called differentiation. Naturally enough, the reverse process of finding a function F, whose derivative is the given function f (in other words, $F' = f$) is called **antidifferentiation**, and F is called an **antiderivative** of f. The problem of finding *indefinite integrals* is essentially the problem of finding antiderivatives. Just as we did with derivatives, we'll examine some ways of estimating antiderivatives numerically and graphically, and we'll develop some explicit formulas for antiderivatives as well.

At first glance, it seems unlikely that the problem of measuring area (definite integral) would be closely related to the problem of finding an antiderivative (indefinite integral). When Newton and Leibniz first realized the true nature of the relationship between definite and indefinite integrals, it marked not only the birth of modern calculus, but also the dawn of a new age of scientific and technological advancement. The theorems that establish the tie between area and antidifferentiation are rightfully called the *Fundamental Theorems of Calculus* and we'll provide a motivation and explanation of these cornerstones of calculus.

The chapter concludes with a look at some numerical techniques for approximating the value of a definite integral.

6. THE INTEGRAL

A problem of measurement

Many of our usual notions of measurement assume that we measure distances along straight lines. For example, a **polygon** is a figure made up of straight line segments. The measurement of the perimeter of the irregular polygon shown in Figure 6.1 can be accomplished by simply measuring the length of each side of the polygon and finding the sum of these lengths. Indeed, mathematically, this is the very definition of the perimeter of a polygon.

Similarly, the area enclosed by a polygon can be measured by first *triangulating* the region (i.e., subdividing it into nonoverlapping triangles) and using the familiar formula $A = \frac{1}{2}bh$ (Area = one-half the product of the base and height of the triangle) on each individual triangle and then summing those results. Figure 6.1 illustrates one possible triangulation of a polygon that could be used to find its area.

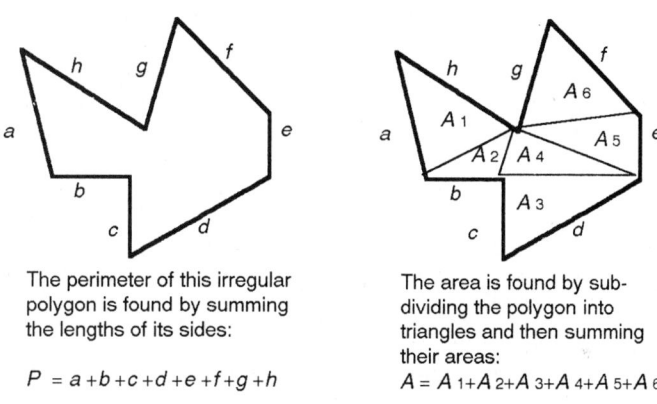

The perimeter of this irregular polygon is found by summing the lengths of its sides:

$P = a + b + c + d + e + f + g + h$

The area is found by subdividing the polygon into triangles and then summing their areas:

$A = A_1 + A_2 + A_3 + A_4 + A_5 + A_6$

Figure 6.1 Measuring the perimeter and area of a polygon.

Many of the area formulas from geometry are obtained using this simple principle of triangulation.

EXAMPLE 1 Derive the formula for the area of any trapezoid (a quadrilateral with two parallel sides), where a and b are the lengths of the two parallel sides and h is the height or distance between the parallel sides.

Solution Figure 6.2 shows how this formula can be derived by simply dividing the trapezoid into two nonoverlapping triangles and adding their areas. Using this simple technique of triangulation, we obtain the exact mathematical relationship

$$A = \frac{1}{2}h(a + b)$$

between a, b, h, and the area A of the trapezoid. ∎

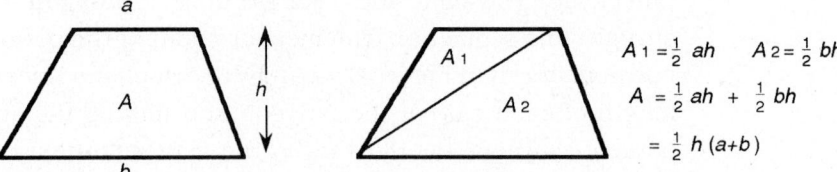

Figure 6.2 Deriving the area formula for a trapezoid.

How does one obtain an exact mathematical formula yielding the perimeter or area of a region whose boundary is not made of straight line segments? One of the most familiar examples of such a region is that of a circle (Figure 6.3). There are no straight line segments to measure and sum for its circumference (perimeter), and no number of line segments can be drawn to triangulate the circle (Figure 6.3).

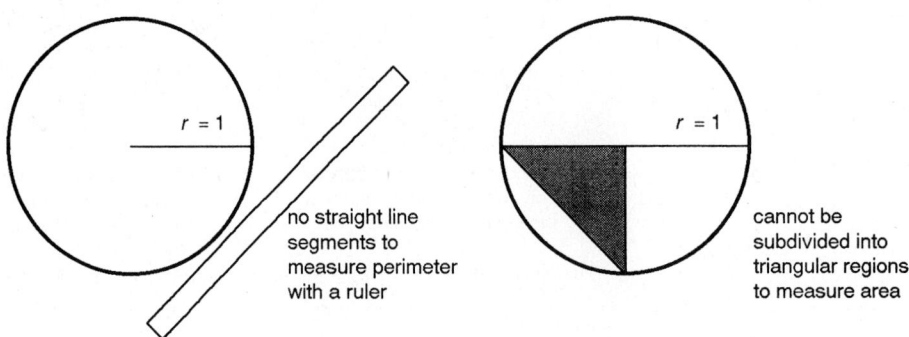

Figure 6.3 Measuring the perimeter and area of a circle.

Nevertheless, there *are* mathematical formulas that precisely relate the circumference C and area A of a circle to its radius r:

$$C = 2\pi r \qquad \text{and} \qquad A = \pi r^2.$$

The real number π is a *constant* appearing not only in these two formulas, but also in a multitude of other formulas involving geometric objects such as ellipses, cones, cylinders, and spheres.

What exactly is π?—The method of exhaustion

The number π is sometimes defined as

$$\pi = \frac{C}{d}$$

where, as before, C represents the circumference of a circle, and d ($= 2r$) represents the *diameter* of the same circle. If we compare this definition with the formula for the circumference of a circle, we can see that it is truly "circular" in the sense that we are no closer to knowing the value of π than before. We could also define π with the formula

$$\pi = \frac{A}{r^2},$$

but that is of no more help than the first definition.

We could *empirically* estimate π by drawing a circle, measuring its diameter, wrapping string snugly around the boundary of the circle, then measuring the string's length and dividing by the diameter. (Perhaps a similar activity was the source of the Biblical estimate $\pi \approx 3$.) This method of determining π is subject to unavoidable errors of measurement. What we seek is a mathematical method for determining π to any degree of precision we desire.

Archimedes (287-212 B.C.) used the **method of exhaustion** to calculate an approximation of π. The idea of the technique is fairly simple, and it captures the key idea behind definite integrals in calculus, as we'll soon see. We will illustrate the method by applying it to the problem of estimating π, much as Archimedes himself did.

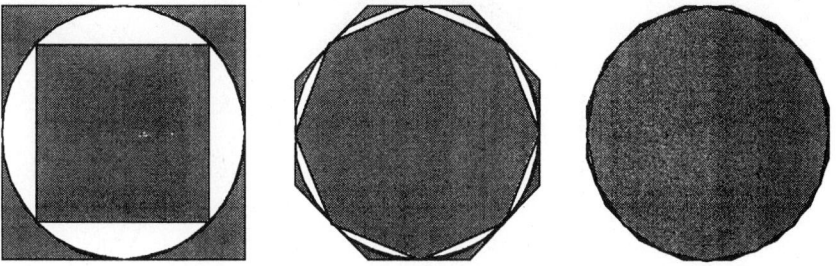

Figure 6.4 Approximating the area of a circle with regular polygons.

Suppose we have a *unit circle* (the radius $r = 1$ unit). If we can determine the area A of this circle, then we will have determined π, since

$$A = \pi \cdot (1)^2 = \pi.$$

Our strategy is to obtain a sequence of better and better approximations to π by measuring the areas of regular polygons inscribed in and circumscribed about this unit circle. Because the polygons can be triangulated, we can mathematically determine their exact areas. Figure 6.4 illustrates the situation for inscribed and circumscribed squares, regular octagons, and regular 16-sided polygons.

Let's use a_n to represent the area of an inscribed regular polygon having n sides, and A_n to represent the area of a circumscribed regular polygon with n sides. Since an inscribed polygon fits inside the circle, its area must be less than π. On the other hand, since the circle fits inside a circumscribed polygon, the polygon must have an area greater than π. What we have just noted is that the areas a_n and A_n provide a lower and an upper bound for the value of π:

$$a_n < \pi < A_n.$$

Now, let's compute a_n and A_n for the specific values $n = 4$, 8, and 16.

In the case of $n = 4$ (inscribed and circumscribed squares), we see that the diagonal of the inscribed square is a diameter of the circle and has length 2. A side of this square must have length $\sqrt{2}$ (use the Pythagorean Theorem) and so the area is $a_4 = (\sqrt{2})^2 = 2$. A square circumscribed about the circle will have *sides* of length 2, hence an area $A_4 = 2^2 = 4$. . The areas of the two squares trap the number π between them:

$$a_4 = 2 < \pi < 4 = A_4.$$

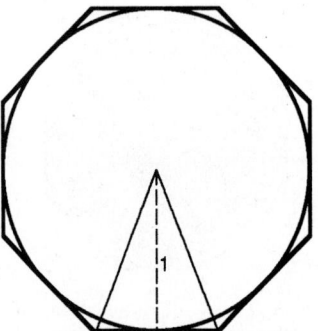 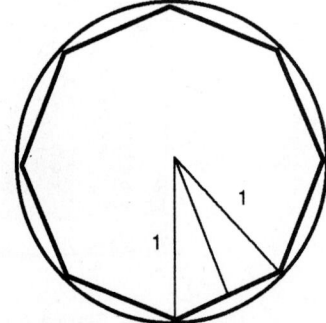

Area of the circumscribed octagon is 8 x the area of the isosceles triangle (height = radius = 1 unit).

Area of the inscribed octagon is 8 x the area of the isosceles triangle (leg length = radius = 1 unit).

Figure 6.5 Triangulating the octagon to find its area.

6. THE INTEGRAL

We can improve our approximations of π by using inscribed and circumscribed regular octagons ($n = 8$). We can calculate the area of a regular octagon by first subdividing it into eight congruent isosceles triangles (Figure 6.5) and summing their areas. When we calculate these areas, we obtain the following improved lower and upper bounds on the value of π:

$$2.8284 \approx a_8 = 2\sqrt{2} < \pi < 8(\sqrt{2} - 1) = A_8 \approx 3.3137.$$

Continuing, the fit is even better with the inscribed and circumscribed 16-sided regular polygons, and their areas provide tighter lower and upper bounds:

$$3.0614 \approx a_{16} < \pi < A_{16} \approx 3.1826.$$

If you are wondering why we have doubled the number of sides of the polygons we use at each step, it is because this allows a way of computing the areas of the new polygons using information from the polygons at the previous step (and without using trigonometry, which Archimedes did *not* have at his disposal). Archimedes himself started with regular inscribed and circumscribed *hexagons* and doubled the number of sides at each step until he obtained the approximations

$$3\frac{10}{71} \approx a_{96} < \pi < A_{96} \approx 3\frac{1}{7}.$$

One of these estimates, $\pi \approx 3\frac{1}{7}$, is still in common usage to this day. It is known, appropriately enough, as the Archimedean estimate of π, and is actually more accurate than the common decimal approximation to two places, $\pi \approx 3.14$.

The method of exhaustion applied to inscribed and circumscribed polygons can be considered an infinite process (even though such an idea was totally foreign to Archimedes). In the case of the area of the circle, we can view the method as producing an infinite sequence of lower bound approximations and an infinite sequence of upper bound approximations. Both sequences converge on a single value which we define as π. While we cannot write down the entire decimal expansion of π, we have a definite means of obtaining as close an approximation as we like. While it may seem computationally "exhausting," the number-crunching prowess of today's computers and calculators make the method of exhaustion a viable approximation strategy for attacking a wide array of measurement problems. In fact, the approximation of π to great precision is a common benchmark test of the speed and power of computational devices. Super computers have recently computed decimal approximations of π accurate to over a billion digits.

6.1 DEFINITE INTEGRALS—TERMINOLOGY AND EXAMPLES

Our interest in the method of exhaustion goes far beyond historical interest. The notion of computing the limiting value for an infinite sequence of approximations is at the heart of calculus. We have seen this idea before in the definition of a derivative as a limiting value of difference quotients.

What is a definite integral?—area under a function graph

The method of exhaustion captures the essence of what we call *definite integration* in calculus. In general, we want to apply the method to the problem of measuring the area between a function graph and the x-axis. This may seem to be a curious goal. Certainly, finding the area of regions like a circle has some practical use, but why would we even care to measure the area under a function graph?

A derivative has a geometric interpretation in terms of the slope of the original function graph, but its power as a problem solving tool comes from its various physical interpretations as a rate of change. Similarly, the area between a function graph and the x-axis may have many physical interpretations. For example, if our function f represents a variable *force* exerted on an object over some distance, then the area under the graph of f may represent the total *work* performed.

Computing the area under a function graph can be likened to finding the area of a "variable height" rectangle, meaning we have a region with a fixed length between the two boundaries $x = a$ and $x = b$, but the height $f(x)$ of the region varies as x varies over the interval $[a, b]$.

Suppose, for now, that f is a continuous function on the closed interval $[a, b]$ such that $f(x) \geq 0$ for every $x \in [a, b]$. Then we write

$$\int_a^b f(x)\,dx$$

to represent the area under the graph of $y = f(x)$ and above the x-axis between $x = a$ and $x = b$. The elongated "S" is the integral sign, and we call $f(x)$ the **integrand**, a and b the **limits of integration**, and dx denotes that the **variable of integration** is x. The $\int$ notation is due to Leibniz (who was also responsible for the d/dx notation for the derivative) and was meant to suggest a sum of *indivisibles*—the vertical line segments making up the region. The region in the Cartesian plane we've described is shown in Figure 6.6. We call this area the *definite integral of f with respect to x from a to b.*

6.1 DEFINITE INTEGRALS—TERMINOLOGY AND EXAMPLES

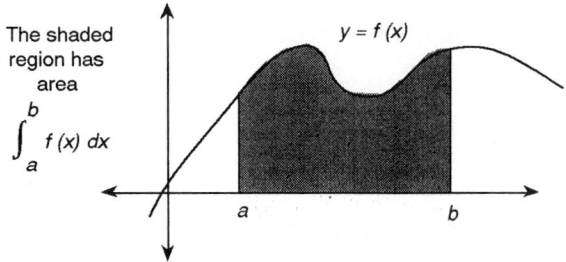

The shaded region has area
$$\int_a^b f(x)\, dx$$

Figure 6.6 Definite integral of a function

Some examples of definite integrals

For a positive constant function $f : x \longmapsto c$ where $c > 0$, the definite integral of f over an interval $[a, b]$ is simply

$$\int_a^b c\, dx = c(b - a),$$

the area of the rectangle with height c and length $b - a$ of the interval.

EXAMPLE 2 Evaluate $\int_{-1}^{4} 3\, dx$.

Solution The definite integral $\int_{-1}^{4} 3\, dx$ represents the area of the rectangle illustrated below in Figure 6.7.

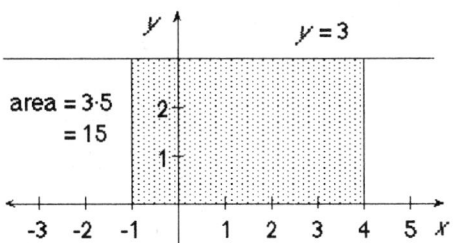

Figure 6.7 Area represented by $\int_{-1}^{4} 3\, dx$.

The area of this rectangle is the height 3 times its length $5 = 4 - (-1)$. Hence, $\int_{-1}^{4} 3\, dx = 15$. ∎

For a linear function f which is non-negative over the interval, the definite integral will simply be the area of either a triangle or a trapezoid.

EXAMPLE 3 Find the values of the following definite integrals.

$$\int_{-2}^{1} \frac{x+2}{3} \, dx \quad \text{and} \quad \int_{0}^{4} \frac{x+2}{3} \, dx.$$

Solution The graph of $y = (x+2)/3$ is shown below in Figure 6.8.

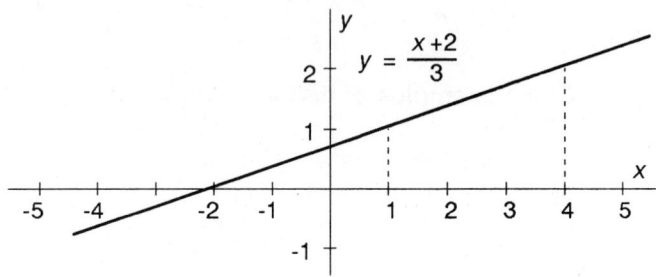

Figure 6.8 Graph of $y = (x+2)/3$.

The two definite integrals represent the area of a triangle and trapezoid, respectively.

area = (1/2) · 3 · 1 = 3/2

area = $\dfrac{(2/3) + 2}{2}$ · 4

= 16/3

Figure 6.9 Areas represented by $\int_{-2}^{1} \frac{x+2}{3} \, dx$ and $\int_{0}^{4} \frac{x+2}{3} \, dx$.

From Figure 6.9 we can see that

$$\int_{-2}^{1} \frac{x+2}{3} \, dx = \frac{3}{2} \quad \text{and} \quad \int_{0}^{4} \frac{x+2}{3} \, dx = \frac{16}{3}.$$

■

6.1 DEFINITE INTEGRALS—TERMINOLOGY AND EXAMPLES

The definite integral measures signed area

The definite integral

$$\int_a^b f(x)\,dx$$

can be interpreted as the area between the graph of $y = f(x)$ and the x-axis between $x = a$ and $x = b$ provided the following are true:

1) $f(x) \geq 0$ for x between a and b,

2) $a < b$,

3) f is continuous on the interval $[a, b]$.

How should we interpret the definite integral when one or more of these provisions do not hold?

1) In the case that $f(x) < 0$ (so that the graph of f dips below the x-axis), the definite integral has a value equal to the *negative* of the area between the graph and the x-axis (see Figure 6.10).

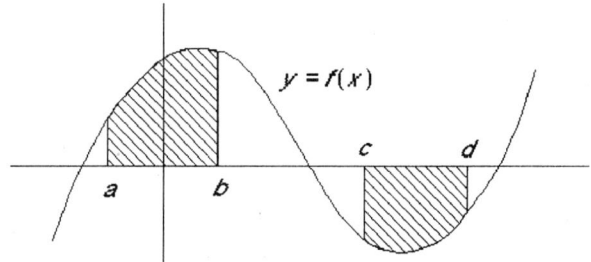

Figure 6.10 $\int_a^b f(x)\,dx$ is positive and $\int_c^d f(x)\,dx$ is negative.

EXAMPLE 4 Find $\int_{-5}^{-3} \frac{x+2}{3}\,dx$.

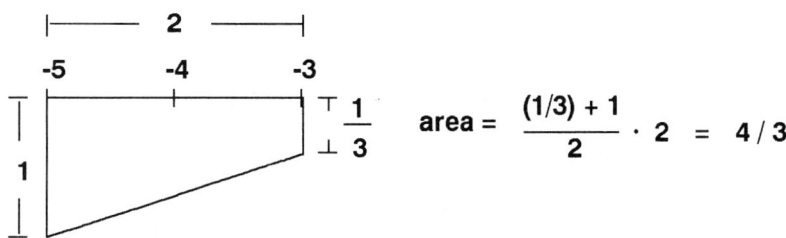

Figure 6.11 Trapezoidal region associated with $\int_{-5}^{-3} \frac{x+2}{3}\,dx$.

Solution Over the interval $[-5, -3]$, the graph is below the x-axis, so we need to calculate the area of the trapezoid in Figure 6.11, and take its *negative* for the purposes of integration

$$\int_{-5}^{-3} \frac{x+2}{3} \, dx = -\frac{4}{3}.$$

■

When a function takes on both negative and positive values over an interval, we can split its definite integral into its negative and positive components and sum the results.

EXAMPLE 5 Find $\int_{-5}^{4} \frac{x+2}{3} \, dx$.

Solution Using the graph in Figure 6.8 as a guide, we can write this definite integral as the sum of two definite integrals:

$$\int_{-5}^{4} \frac{x+2}{3} \, dx = \int_{-5}^{-2} \frac{x+2}{3} \, dx + \int_{-2}^{4} \frac{x+2}{3} \, dx.$$

The first definite integral represents the *negative* of the area of a triangle with base 3 and height 1. So

$$\int_{-5}^{-2} \frac{x+2}{3} \, dx = -\frac{3}{2}.$$

The second definite integral represents the *positive* area of a triangle with base 6 and height 2. So

$$\int_{-2}^{4} \frac{x+2}{3} \, dx = 6.$$

Our final result is

$$\int_{-5}^{4} \frac{x+2}{3} \, dx = -\frac{3}{2} + 6 = \frac{9}{2}.$$

■

2) If $a = b$, then we consider the definite integral as measuring the area of a single line segment, which is zero. In other words,

$$\int_{a}^{a} f(x) \, dx = 0$$

for any function f defined at $x = a$.

To handle the case where $b < a$, we agree to the convention

$$\int_a^b f(x)\,dx = -\int_b^a f(x)\,dx.$$

In other words, measuring area from right to left gives the opposite result as measuring from left to right.

EXAMPLE 6 Find $\int_3^3 \dfrac{x+3}{2}\,dx$ and $\int_4^{-5} \dfrac{x+3}{2}\,dx$.

Solution $\int_3^3 \dfrac{x+3}{2}\,dx = 0$ and $\int_4^{-5} \dfrac{x+3}{2}\,dx = -\int_{-5}^4 \dfrac{x+3}{2}\,dx = -\dfrac{3}{2}$. ∎

Can we find the definite integral of a function that is not continuous? It depends on how *many* and how *bad* the discontinuities are. If there are only finitely many discontinuities and they are removable (such as those representing holes and skips in the graph of the function), then we can safely ignore them. The reasoning is as follows: if we "fixed" the function by changing its values at a finite number of points to remove the discontinuities, then we would not affect the total area (since the area of each of the finitely many line segments is 0).

In fact, even if there were finitely many *jump* discontinuities, we could still find the definite integral by dividing the interval at the points where the jumps occur. For each of the individual subintervals produced, the former jumps have become removable one-sided discontinuities at the endpoints.

A consequence of this is that we can find the definite integral of any step function or other piece-wise linear function by subdividing the interval of integration at the vertices.

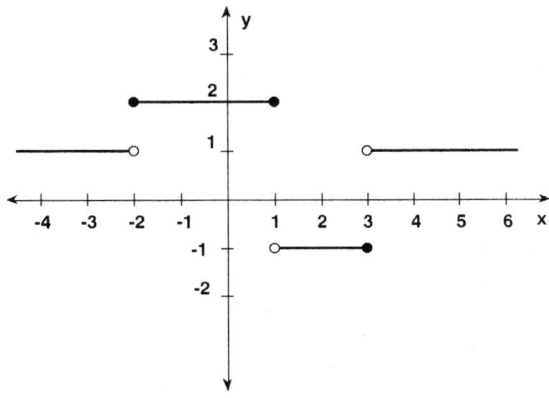

Figure 6.12 A step function.

EXAMPLE 7 Find $\int_{-1}^{4} f(x)\,dx$ where the function f is the step function whose graph is shown in Figure 6.12.

Solution If we divide the interval of integration $[-1, 4]$ at the points where the jumps occur, we have

$$\int_{-1}^{4} f(x)\,dx = \int_{-1}^{1} f(x)\,dx + \int_{1}^{3} f(x)\,dx + \int_{3}^{4} f(x)\,dx$$

$$= 2(1-(-1)) + (-1)(3-1) + 1(4-3)$$

$$= 4 - 2 + 1 = 3.$$

∎

EXAMPLE 8 Find $\int_{-2}^{5} \dfrac{|x-3|-4}{2}\,dx$.

Solution The function $x \longmapsto (|x-3|-4)/2$ is piece-wise linear (see Figure 6.13).

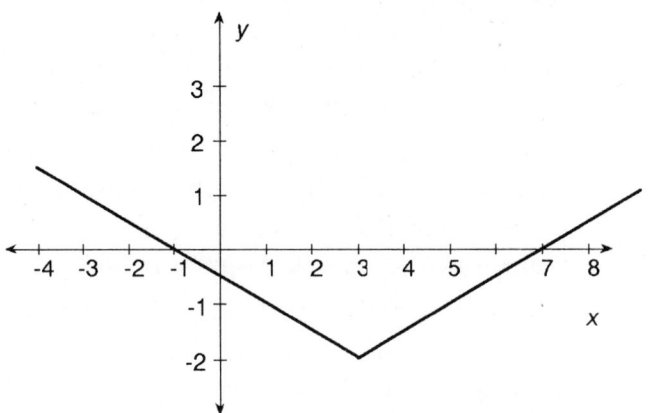

Figure 6.13 Graph of $y = \dfrac{|x-3|-4}{2}$.

We divide the interval up at $x = -1$ and $x = 3$ to aid our computation.

$$\int_{-2}^{-1} \frac{|x-3|-4}{2}\,dx = \frac{1}{2} \cdot 1 \cdot \frac{1}{2} = \frac{1}{4}$$

$$= \text{ area of triangle over } [-2, -1]$$

$$\int_{-1}^{3} \frac{|x-3|-4}{2}\,dx = -\left(\frac{1}{2} \cdot 4 \cdot 2\right) = -4$$

$$= -(\text{ area of triangle over } [-1, 3])$$

$$\int_{3}^{5} \frac{|x-3|-4}{2}\,dx = -\left(\frac{2+1}{2} \cdot 2\right) = -3$$

$$= -(\text{ area of trapezoid over } [3, 5]).$$

6.1 DEFINITE INTEGRALS—TERMINOLOGY AND EXAMPLES

The value of the definite integral is

$$\int_{-2}^{5} \frac{|x-3|-4}{2}\, dx = \frac{1}{4} - 4 - 3 = -\frac{27}{4}.$$

EXERCISES

Find the values of the definite integrals in exercises 1-10.

1. $\displaystyle\int_{1}^{4} (2x+3)\, dx$
2. $\displaystyle\int_{1}^{4} (3-2x)\, dx$
3. $\displaystyle\int_{-3}^{1} x\, dx$
4. $\displaystyle\int_{-3}^{2} |x|\, dx$
5. $\displaystyle\int_{-5}^{1} |3x+2|\, dx$
6. $\displaystyle\int_{-2}^{3} |2-3x|\, dx$
7. $\displaystyle\int_{-2}^{4} \lfloor x \rfloor\, dx$
8. $\displaystyle\int_{-4}^{2} \lceil x \rceil\, dx$
9. $\displaystyle\int_{-8}^{-3} (-2)\, dx$
10. $\displaystyle\int_{8}^{3} (-2)\, dx$

Using the graphs of functions f and g below, find the definite integrals in exercises 11-20.

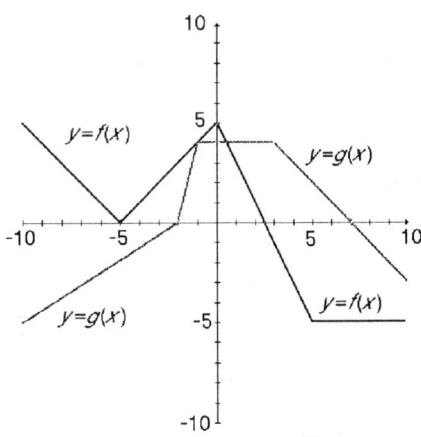

11. $\displaystyle\int_{-5}^{0} f(x)\, dx$
12. $\displaystyle\int_{5}^{10} f(x)\, dx$
13. $\displaystyle\int_{-10}^{2} g(x)\, dx$
14. $\displaystyle\int_{-2}^{7} g(x)\, dx$
15. $\displaystyle\int_{0}^{5} f(x)\, dx$
16. $\displaystyle\int_{3}^{3} f(x)\, dx$
17. $\displaystyle\int_{-5}^{10} g(x)\, dx$
18. $\displaystyle\int_{5}^{-2} f(x)\, dx$

19. $\int_{-5}^{5} f(x)\,dx$

20. $\int_{-5}^{5} g(x)\,dx$

Using the same graphs of functions f and g above, plot each of the functions h in exercises 21–30 over the interval $[-10, 10]$. Then use these graphs to find the definite integral

$$\int_{-5}^{5} h(x)\,dx.$$

21. $h(x) = f(x) + g(x)$
22. $h(x) = -g(x)$
23. $h(x) = f(x) + 2$
24. $h(x) = 3 - g(x)$
25. $h(x) = 2f(x)$
26. $h(x) = -3g(x)$
27. $h(x) = |f(x)|$
28. $h(x) = |g(x)|$
29. $h(x) = g(2x)$
30. $h(x) = g(x + 3)$

Exercises 31–36 ask you to compare some of your results from exercises 21–30 with those of exercises 19 and 20.

31. What does exercise 21 suggest to you regarding the definite integral of a sum of two functions?

32. What does exercise 22 suggest to you regarding the definite integral of the additive inverse of a function?

33. Calculate $\int_{-5}^{5} 2\,dx$ and $\int_{-5}^{5} 3\,dx$. Do the results of exercises 23 and 24 support your conclusion from exercises 21 and 22?

34. What do exercises 25 and 26 suggest to you regarding the definite integral of a constant multiple of a function?

35. Use exercises 27 and 28 to answer the question: Is the definite integral of the absolute value of a function the same as the absolute value of the definite integral?

36. Calculate $\int_{-10}^{10} g(x)\,dx$ and compare it to the result of exercise 29. Then compute $\int_{-10}^{10} f(x)\,dx$ and use it to predict the value of $\int_{-2}^{2} f(5x)\,dx$. Check your prediction by graphing $y = f(5x)$ and computing the definite integral.

37. Calculate $\int_{-2}^{8} g(x)\,dx$ and compare it to the result of exercise 30. For what values a and b will $\int_{a}^{b} f(x)\,dx$ have the same value as $\int_{-5}^{5} f(x-2)\,dx$? Check your result by graphing $y = f(x-2)$ and computing the definite integral from a to b.

6.2 COMPUTING AND ESTIMATING DEFINITE INTEGRALS

The idea behind the formal definition of a definite integral is based directly on an approximation process like the method of exhaustion Archimedes used to estimate π. Let's illustrate the process for a nonlinear function by approximating

$$\int_1^2 x^2\, dx$$

whose value corresponds to the area of the region illustrated in Figure 6.14.

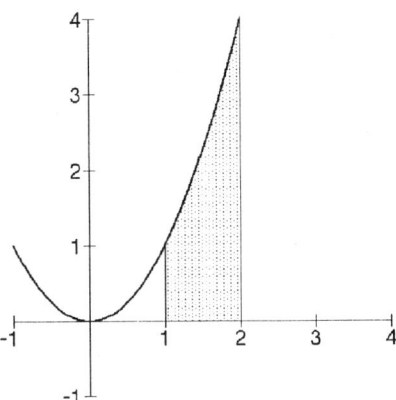

Figure 6.14 Area represented by $\int_1^2 x^2\, dx$.

Our strategy is to find lower and upper bounds for the area of the region by using inscribed and circumscribed rectangles. First, we can get very rough lower and upper bounds on the area by noting that because $y = x^2$ is strictly increasing over the interval $[1,2]$, its minimum value is $y = 1$ at the left endpoint ($x = 1$) and its maximum value is $y = 4$ at the right endpoint ($x = 2$). Geometrically, this means we can *inscribe* a rectangle of height 1 within the region, and *circumscribe* a rectangle of height 4 about the region, as shown in the first illustration of Figure 6.15.

Since the width of each of these two rectangles is 1 (the length of the interval $[1,2]$), the area of our region must be between $L_1 = 1 \cdot 1 = 1$ and $U_1 = 4 \cdot 1 = 4$:

$$1 < \int_1^2 x^2\, dx < 4.$$

We use the labels L_1 and U_1 to stand for the lower and upper approximations we obtain from one inscribed and one circumscribed rectangle. Since the area of the region is between L_1 and U_1, we could average them to find a single estimate:

$$\int_1^2 x^2\,dx \approx \frac{L_1 + U_1}{2} = \frac{1+4}{2} = 2.5.$$

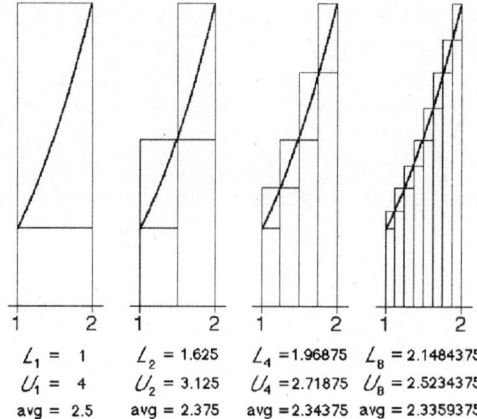

$L_1 = 1$ $L_2 = 1.625$ $L_4 = 1.96875$ $L_8 = 2.1484375$
$U_1 = 4$ $U_2 = 3.125$ $U_4 = 2.71875$ $U_8 = 2.5234375$
avg = 2.5   avg = 2.375   avg = 2.34375   avg = 2.3359375

Figure 6.15 Approximating $\int_1^2 x^2\,dx$ with inscribed and circumscribed rectangles.

To improve on our approximation, we can subdivide the interval $[1, 2]$ into two subintervals $[1, 1.5]$ and $[1.5, 2]$, each having length 0.5. Now we inscribe and circumscribe rectangles for each subinterval as shown in the second illustration of Figure 6.15.

We'll let L_2 represent the sum of the areas of the two inscribed rectangles and U_2 the sum of the areas of the two circumscribed rectangles.

The height of the first inscribed rectangle is $1 = 1^2$, and the height of the second inscribed rectangle is found by evaluating $y = (1.5)^2 = 2.25$. Since the width of each rectangle is 0.5, we have

$$L_2 = 1 \cdot (0.5) + (2.25)(0.5) = 0.5 + 1.125 = 1.625.$$

The height of the first circumscribed rectangle is also 2.25 and the height of the second circumscribed rectangle is 4, so we have

$$U_2 = (2.25)(0.5) + 4 \cdot (0.5) = 1.125 + 2 = 3.125.$$

Now we can write

$$1.625 < \int_1^2 x^2\,dx < 3.125,$$

and if we average the lower and upper approximations

$$\int_1^2 x^2\,dx \approx \frac{L_2 + U_2}{2} = \frac{1.625 + 3.125}{2} = 2.375.$$

We can try to improve the approximation even further by subdividing the interval into a greater number of equal-sized subintervals, and again calculating the areas of the inscribed and circumscribed rectangles. Figure 6.15 illustrates the tighter lower and upper bounds we obtain using 4 and 8 subintervals, respectively. If we continue to increase the number of inscribed and circumscribed rectangles, the lower and upper approximations (and their average) converge on a single limiting value. Since the true area of the region is always trapped between the lower and upper approximations, it must be that single value. In fact, later we'll be able to show that the limiting value is precisely

$$\int_1^2 x^2\, dx = 7/3 = 2.3333333\ldots,$$

so the average of L_8 and U_8 is already within .003 of the true area.

Formal definition of definite integral—Riemann sums

There are a variety of ways to define the definite integral of a function. We will present the **Riemann integral**, named after the mathematician Bernhard Riemann (1826-1866). It is general enough to allow a definite integral value to be defined for many functions, including every continuous function on a closed interval $[a, b]$.

A key ingredient in the definition is the notion of a **Riemann sum**. Here are the steps for producing a Riemann sum for a function f on an interval $[a, b]$.

Step 1. For some positive integer n, subdivide the interval $[a, b]$ into n subintervals of equal length. This forms what is called a **regular partition** of $[a, b]$ of size n. If we denote the subinterval length as Δx, then

$$\Delta x = \frac{b - a}{n}.$$

We can mark off the subdivision points by starting at a and stepping off lengths of Δx until we reach b. If we label our starting point $a_0 = a$, and label the subdivision points successively, we have

$$a_0 = a, \quad a_1 = a + \Delta x, \quad a_2 = a + 2\Delta x, \quad a_3 = a + 3\Delta x, \quad \ldots, \quad a_n = b = a + n\Delta x$$

as our partition points.

EXAMPLE 9 Find the subinterval length Δx and the partition points for a *regular* partition P of size 8 for the interval $[1,5]$.

Solution With $n = 8$, $a = 1$ and $b = 5$, we have subinterval length

$$\Delta x = \frac{b-a}{n} = \frac{5-1}{8} = 0.5.$$

The regular partition is the set of points

$$P = \{1.0,\ 1.5,\ 2.0,\ 2.5,\ 3.0,\ 3.5,\ 4.0,\ 4.5,\ 5.0\}.$$

For brevity we could denote a regular partition of size n for the interval $[a, b]$ as

$$P_n[a, b].$$

In this example, we found $P_8[1, 5]$. ■

Step 2. Choose one point x_i ($i = 1, 2, \ldots, n$) from each of the n subintervals. So

$$x_1 \in [a_0, a_1], \quad x_2 \in [a_1, a_2], \quad \ldots, \quad x_n \in [a_{n-1}, a_n].$$

Each point may be chosen from anywhere in its subinterval, including the endpoints.

Step 3. Calculate $f(x_i)$ for each x_i chosen.

Step 4. Multiply each function output $f(x_i)$ by the length of the subinterval Δx and sum up the results:

$$f(x_1) \cdot \Delta x \quad + \quad f(x_2) \cdot \Delta x \quad + \quad \ldots \quad + \quad f(x_n) \cdot \Delta x.$$

This is the value of the Riemann sum for this partition and choice of points.

Figure 6.16 illustrates the geometrical interpretation of a Riemann sum. For each subinterval, we can associate a rectangle with a base on the x-axis. The subinterval length corresponds to the common base length of each rectangle. The function outputs $f(x_i)$ correspond to the "heights" of these rectangles. Each product $f(x_i)\Delta x$ gives us a number corresponding to either the area of that rectangle or its negative (depending on whether $f(x_i)$ is *positive* or *negative*. The Riemann sum is the sum of these signed areas.

6.2 COMPUTING AND ESTIMATING DEFINITE INTEGRALS

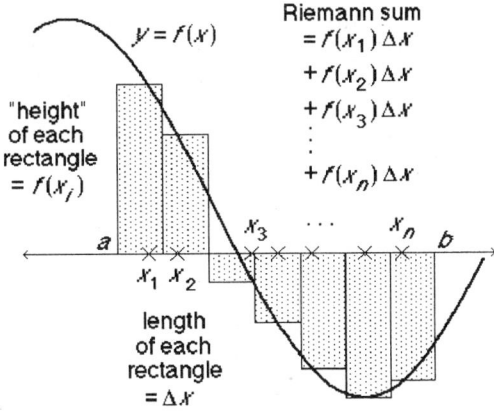

Figure 6.16 Geometrical meaning of a Riemann sum.

The sigma-notation is particularly convenient for denoting a Riemann sum:

$$\sum_{i=1}^{n} f(x_i) \Delta x$$

is a shorthand for the sum of the signed areas of the n rectangles generated by a partition of size n. To emphasize the interval $[a, b]$, we may also write

$$\sum_{a}^{b} f(x) \Delta x$$

for a Riemann sum as long as it is understood that an input x is chosen from each subinterval of length Δx to compute the function outputs $f(x)$.

EXAMPLE 10 Find the particular Riemann sum value for the function $f : x \longmapsto x^2$ over the interval $[0, 3]$ with a partition of $n = 4$ subintervals and with chosen points $x_1 = 0.5, x_2 = 1.0, x_3 = 2.0,$ and $x_4 = 2.7$.

Solution First, we calculate $\Delta x = (3 - 0)/4 = 0.75$. Using the given points x_i, we have

$$\sum_{i=1}^{4} f(x_i) \Delta x = (0.5)^2(0.75) + (1.0)^2(0.75) + (2.0)^2(0.75) + (2.7)^2(0.75) = 9.405.$$

For any given function defined on an interval $[a,b]$, there are *lots* of possible Riemann sums which can be formed and evaluated. First, there are infinitely many partition sizes (the number n of subintervals could be any positive integer). And, for each partition, there are infinitely many choices of points $x_1, x_2, \ldots, x_n$ to make.

Since a continuous function f always attains a minimum and maximum value over a closed interval (the Extreme Value Theorem), we could choose the points x_i so that $f(x_i)$ is the smallest output for each subinterval. The resulting Riemann sum is called the **lower Riemann sum** for that partition. For a positive function f, the lower Riemann sum corresponds to the sum of the areas of the inscribed rectangles. If we choose the points x_i so that $f(x_i)$ is the greatest output for each subinterval, then the resulting Riemann sum is called the **upper Riemann sum** for that partition. For a positive function f, the upper Riemann sum corresponds to the sum of the areas of the circumscribed rectangles. For our example $\int_1^2 x^2\,dx$, it was particularly easy to find the extreme values over each subinterval of the partition, because the function was strictly increasing and the minimum and maximum output values always occurred at the endpoints of each subinterval.

The lower and upper Riemann sums represent the two extreme values that all Riemann sums could take for a partition of given size n. If these lower and upper extreme approximations converge to a single value as we make the partition finer, then all the Riemann sums trapped between them will also converge to that same value. That becomes the criteria for the formal definition of a definite integral.

Definition 1

Let f be a function defined on a closed interval $[a,b]$. If the Riemann sums for f over $[a,b]$ converge to a single limiting value A as the size of the partitions $n \to \infty$ (in other words, as the number of subdivision points increases without bound and $\Delta x \to 0$), then we write

$$A = \lim_{\Delta x \to 0} \sum_a^b f(x)\Delta x = \int_a^b f(x)\,dx$$

and say f is **Riemann integrable** over the interval $[a,b]$.

6.2 COMPUTING AND ESTIMATING DEFINITE INTEGRALS

EXAMPLE 11 Compute $\int_1^2 x^2\,dx$ as a limit of lower and upper Riemann sums.

Solution If we partition $[1,2]$ into n subintervals of length $\Delta x = 1/n$, then the lower Riemann sum is

$$L_n = 1^2(\tfrac{1}{n}) + (1+\tfrac{1}{n})^2(\tfrac{1}{n}) + (1+\tfrac{2}{n})^2(\tfrac{1}{n}) + (1+\tfrac{3}{n})^2(\tfrac{1}{n}) + \cdots + (1+\tfrac{(n-1)}{n})^2(\tfrac{1}{n})$$

and the upper Riemann sum is

$$U_n = (1+\tfrac{1}{n})^2(\tfrac{1}{n}) + (1+\tfrac{2}{n})^2(\tfrac{1}{n}) + (1+\tfrac{3}{n})^2(\tfrac{1}{n}) + \cdots + (1+\tfrac{n}{n})^2(\tfrac{1}{n}).$$

With some algebraic effort, these expressions can be simplified to

$$L_n = \frac{14n^3 - 9n^2 + n}{6n^3} = \frac{7}{3} - \frac{3}{2n} + \frac{1}{6n^2}$$

and

$$U_n = \frac{14n^3 + 9n^2 + n}{6n^3} = \frac{7}{3} + \frac{3}{2n} + \frac{1}{6n^2}.$$

(You can check that these formulas give the lower and upper Riemann sums for $n = 1, 2, 4,$ and 8 by referring to Figure 6.15.) Now, as the partition becomes finer and finer (n grows larger and larger) the terms $3/2n$ and $1/6n^2$ grow smaller and smaller. In other words

$$\lim_{n\to\infty} L_n = \frac{7}{3} = \lim_{n\to\infty} U_n,$$

and we can conclude that

$$\int_1^2 x^2\,dx = \frac{7}{3}.$$

∎

Every continuous function f is Riemann integrable, but finding the limiting value A can be quite challenging. For many purposes, the approximations afforded by the Riemann sums may suit our needs.

EXAMPLE 12 Estimate $\int_{-2}^{1} \frac{1}{2^x} dx$ by averaging the upper and lower Riemann sum approximations using a regular partition of size $n = 5$.

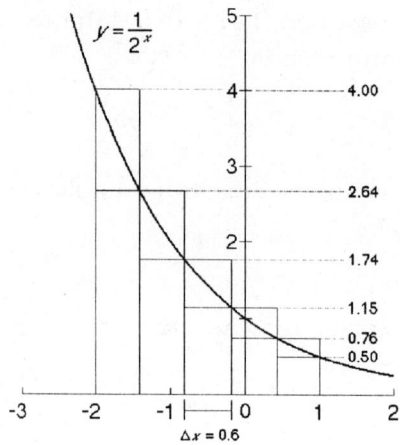

Figure 6.17 Graph of $y = 1/2^x$ over $[-2, 1]$.

Solution $\Delta x = (1 - (-2))/5 = 3/5 = 0.6$ and

$$P_5[-2, 1] = \{-2.0, \ -1.4, \ -0.8, \ -0.2, \ 0.4, \ 1.0\}.$$

Figure 6.17 shows the graph of $y = 1/2^x$ over the interval $[-2, 1]$. Because the function is strictly decreasing over the interval, we find our maximum values occurring at the left endpoints, and the minimum values occurring at the right endpoints of each subinterval. Figure 6.17 shows these endpoint values rounded to two decimal places.

We calculate

$$L_5 \approx 2.64(0.6) + 1.74(0.6) + 1.15(0.6) + 0.76(0.6) + (0.50)(0.6) = 4.074$$

and

$$U_5 \approx 4.00(0.6) + 2.64(0.6) + 1.74(0.6) + 1.15(0.6) + 0.76(0.6) = 6.174.$$

We conclude that

$$4.074 \leq \int_{-2}^{1} \frac{1}{2^x} dx \leq 6.174.$$

If we average our lower and upper approximations, we obtain an estimate of $\int_{-2}^{1} \frac{1}{2^x} dx \approx 5.124$. ■

Understand that the average of the lower and upper approximations is not automatically a better estimate. It may be the case that the true area of the region is very close to either the upper or the lower approximation.

How do we know how good our approximation is? Since we know the true area of the region is trapped between the lower and upper approximations, then the averaged estimate lying halfway between the lower and upper approximations cannot be off by more than half the distance between them.

EXAMPLE 13 What is the worst error possible if we used the approximation of the previous example as our estimate of the definite integral?

Solution In this case, our approximation of 5.124 is halfway between a lower bound of 4.074 and an upper bound of 6.174. Therefore,

$$\text{worst error} = \frac{6.174 - 4.074}{2} = 1.05.$$

Actually, our estimate is much closer than this to the true area. The point to be made is that we can quantify the largest possible error without advance knowledge of the true value. ■

The approximation process we have outlined is simple but increasingly tedious as n is taken larger and larger. Many calculators or computer software with definite integration capabilities approximate the values of a definite integral in a way very similar to this.

We can estimate the value of an integral or check the reasonableness of a definite integral computation by graphing the function f over the interval $[a, b]$ and using the graph to visually estimate the area. The last section of this chapter discusses some other numerical techniques for approximating the value of a definite integral.

Not all functions defined over an interval $[a, b]$ are Riemann integrable over that interval. For example, if a function has infinitely many skips over the interval $[a, b]$, then it may not be possible to assign a value to the definite integral. Also, even a single discontinuity, such as at a vertical asymptote, may make it impossible to assign a value to the definite integral of a function. Later in this book we'll examine the subject of *improper integrals* in more detail.

Properties of definite integrals

Many of the key properties of the definite integral follow directly from its interpretation as a measurement of signed area. As we discuss these properties, keep the example of the area of a rectangle in mind. Think of

$$\int_a^b f(x)\,dx$$

as representing the *area* of a region with $b - a$ as the *length* and $f(x)$ as the *height* at each point $x \in [a, b]$. Of course, the region in question won't be a rectangle unless f is a positive *constant* function, but the analogy will still serve us well.

Theorem 6.1 **Constant multiple property.** If c is a constant, then

$$\int_a^b cf(x)\,dx = c\int_a^b f(x)\,dx,$$

whenever the integrals on both sides of the equation exist.

In other words, we can factor a *constant* through the integral sign.

Reasoning If we multiply the height of a region at each point by a fixed value c while keeping its length the same, we end up multiplying its area by the same factor c. □

EXAMPLE 14 $\displaystyle\int_{-1}^{4} 2\sin(x)\,dx = 2\int_{-1}^{4} \sin(x)\,dx.$ ■

CAUTION! Don't abuse this property. Only constants can be safely factored through the integral sign. Do not factor variable quantities in or out of an integral expression.

6.2 COMPUTING AND ESTIMATING DEFINITE INTEGRALS

Theorem 6.2 | **Additive function property.**

$$\int_a^b (f(x) + g(x))\,dx = \int_a^b f(x)\,dx + \int_a^b g(x)\,dx.$$

In other words, the integral of a sum of two functions is the sum of their integrals.

Reasoning At each point $x \in [a,b]$, the height of the region represented by

$$\int_a^b (f(x) + g(x))\,dx$$

is the sum of the heights of the regions represented by

$$\int_a^b f(x)\,dx \quad \text{and} \quad \int_a^b g(x)\,dx.$$

Since all three regions have the same length, the area of the first region is the sum of the others. □

EXAMPLE 15 $\int_{-3}^{2} (\sin(x) + \cos(x))\,dx = \int_{-3}^{2} \sin(x)\,dx + \int_{-3}^{2} \cos(x)\,dx.$ ∎

The results of Theorems 6.1 and 6.2 are sometimes combined and simply referred to as the **linearity property** of the integral:

$$\int_a^b (c_1 f(x) + c_2 g(x))\,dx = c_1 \int_a^b f(x)\,dx + c_2 \int_a^b g(x)\,dx$$

for constants c_1 and c_2.

Theorem 6.3 | **Additivity over intervals.**

$$\int_a^b f(x)\,dx + \int_b^c f(x)\,dx = \int_a^c f(x)\,dx.$$

Reasoning If two regions are adjacent, but do not overlap, the total area can be found by summing the individual areas. □

EXAMPLE 16 $\int_0^2 \arctan(x)\, dx + \int_2^6 \arctan(x)\, dx = \int_0^6 \arctan(x)\, dx.$ ■

Theorem 6.4 **Comparison property.** If $f(x) \le g(x)$ for every $x \in [a,b]$, then

$$\int_a^b f(x)\, dx \le \int_a^b g(x)\, dx.$$

Reasoning   If two regions have the same length, and the height of one is always less than or equal to the height of the other at each point $x \in [a,b]$, then the area of the first region must also be less than or equal to the area of the second. □

The "heights" (positive and negative) of the rectangles associated with a Riemann sum represent the output values of a step function s (see Figure 6.18). Step functions are particularly easy to integrate. In this case, the definite integral of the step function is *exactly* the value of the Riemann sum.

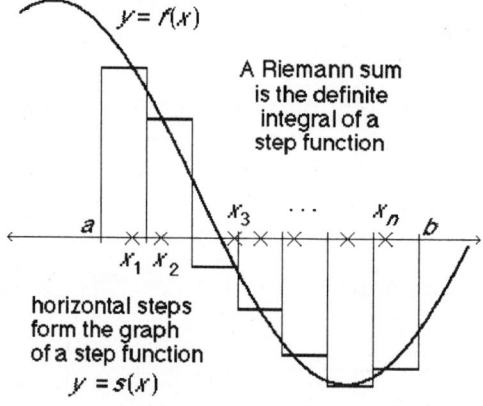

Figure 6.18 $\int_a^b s(x)\, dx$ approximates $\int_a^b f(x)\, dx.$

For a given regular partition of size n, the lower and upper Riemann sums of a function f over an interval $[a,b]$ can be thought of as the definite integrals of two such step functions. If we call the step function associated with the lower Riemann sum ℓ, and the step function associated with the upper Riemann sum u, then

$$\ell(x) \le f(x) \le u(x)$$

for all $x \in [a, b]$. The comparison property tells us that

$$\int_a^b \ell(x)\, dx \leq \int_a^b f(x)\, dx \leq \int_a^b u(x)\, dx.$$

The definite integrals of the two step functions

$$\int_a^b \ell(x)\, dx \quad \text{and} \quad \int_a^b u(x)\, dx$$

are precisely the values of the lower and upper Riemann sums, respectively.

You have already noticed that zooming in horizontally on the graph of a continuous function results in the graph looking like a horizontal line. In other words, over sufficiently small closed intervals, the outputs of a continuous function do not change drastically. We can think of this property as allowing us to approximate a continuous function f very well with a step function s whose graph is made up of horizontal line segments. The definite integral of the step function becomes a good approximation of the definite integral of the continuous function f.

EXERCISES

Find Δx and the partition points $a_0, a_1, \ldots, a_n$ for each of the partitions in exercises 1-4.

1. $P_6[-2, 10]$
2. $P_5[-3, 4]$
3. $P_8[7, 12]$
4. $P_{10}[-1, 3]$

Using a partition of size $n = 5$, find the lower and upper Riemann sums for the definite integrals in exercises 5-8. Then average your approximations and find the size of the worst possible error.

5. $\int_0^1 \sqrt{1 + x^2}\, dx$
6. $\int_{-1}^0 3^x\, dx$
7. $\int_{-1}^1 \arctan(x)\, dx$
8. $\int_1^4 \log_2(x)\, dx$

Use the information below to find the values of the integrals in exercises 9-16.

$$\int_a^b f(x)\,dx = 2.5 \qquad \int_b^c f(x)\,dx = -5.0 \qquad \int_c^d f(x)\,dx = 1.5$$

9. $\displaystyle\int_b^a f(x)\,dx$

10. $\displaystyle\int_a^c f(x)\,dx$

11. $\displaystyle\int_b^d f(x)\,dx$

12. $\displaystyle\int_a^d f(x)\,dx$

13. $\displaystyle\int_a^c 2f(x)\,dx$

14. $\displaystyle\int_c^a f(x)\,dx$

15. $\displaystyle\int_d^a f(x)\,dx$

16. $\left|\displaystyle\int_a^c f(x)\,dx\right|$

Exercises 17-20 ask you to estimate

$$\int_0^4 x(x-1)(x-2)(x-3)(x-4)\,dx$$

using various Riemann sums.

17. Find the Riemann sum corresponding to the partitions of size $n = 1$, 2, 4, and 8, using the left endpoint of each subinterval.

18. Find the Riemann sum corresponding to the partitions of size $n = 1$, 2, 4, and 8, using the right endpoint of each subinterval.

19. Find the Riemann sum corresponding to the partitions of size $n = 1$, 2, 4, and 8, using the midpoint of each subinterval.

20. Graph $y = x(x-1)(x-2)(x-3)(x-4)$ over the interval $[0, 4]$ and use the graph to explain the results of the previous three exercises.

In exercises 21-28, you are given the graph of a function f against a grid. Assuming that the grid lines are spaced 1 unit apart both horizontally and vertically, estimate the value of each of the following definite integrals by using $\Delta x = 1$ unit and counting "square units."

$$\int_1^5 f(x)\,dx, \qquad \int_{-3}^3 f(x)\,dx, \quad \text{and} \quad \int_6^{-2} f(x)\,dx.$$

If one or more of these definite integrals cannot be estimated for a particular graph, explain why.

6.2 COMPUTING AND ESTIMATING DEFINITE INTEGRALS

21.
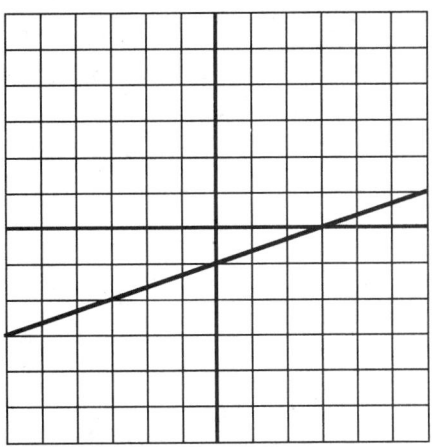

22.
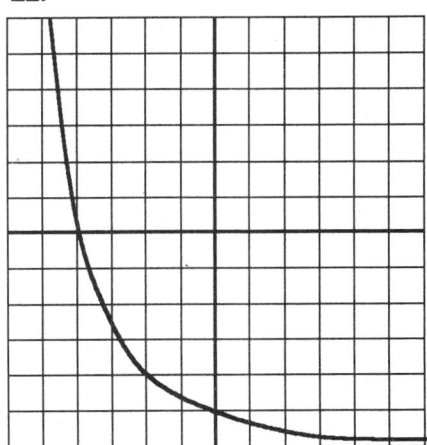

23.
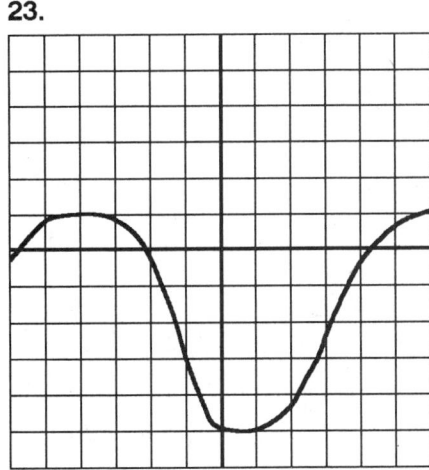

24.
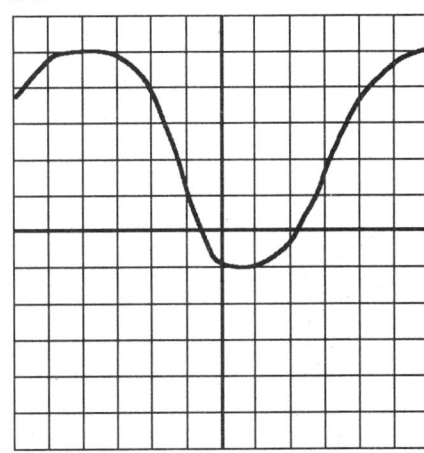

25.
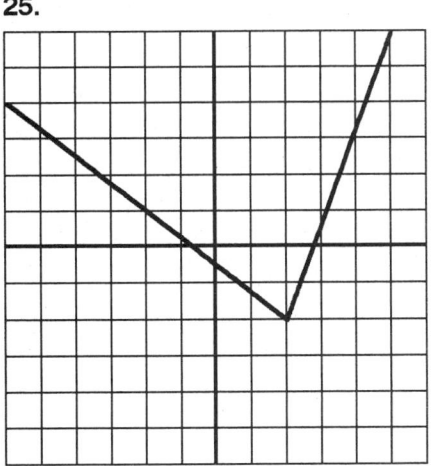

26.
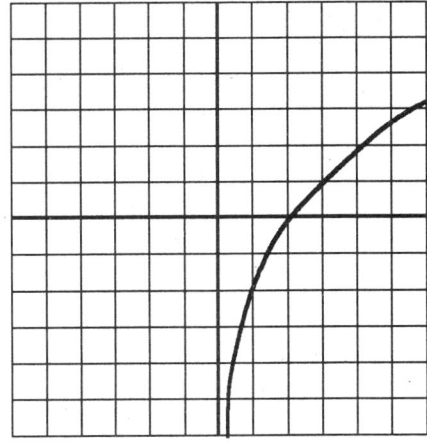

27.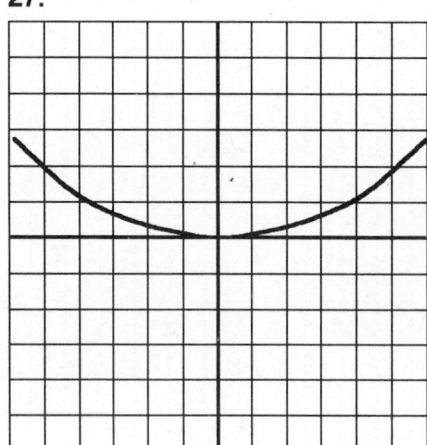

28.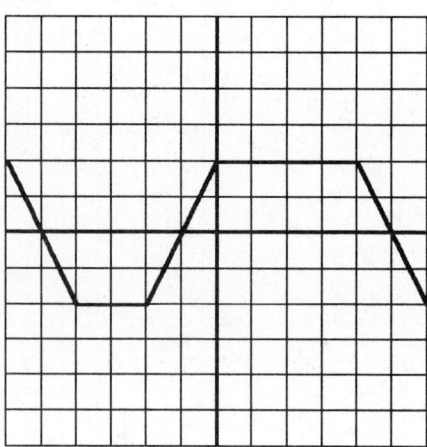

Suppose $f(x) = 1$ whenever x is an irrational number, and $f(x) = 0$ whenever x is a rational number. Given any interval of real numbers $[a, b]$ $(a < b)$, it is always possible to choose rational and irrational numbers belonging to the interval. Exercises 29-32 concern this function.

29. We could form a regular partition of $[2, 5]$ of size n, where we choose all the points $x_1, x_2, \ldots, x_n$ to be rational. What is the value of the resulting Riemann sum for f over this interval?

30. We could also form a regular partition of $[2, 5]$ where we choose all the points $x_1, x_2, \ldots, x_n$ to be irrational. What is the value of the resulting Riemann sum for f over this interval?

31. If we form a regular partition of $[2, 5]$ and always choose either an endpoint or midpoint of each subinterval, what is always the value of the Riemann sum?

32. Considering your results from exercises 29 and 30, does the definite integral

$$\int_2^5 f(x)\,dx$$

exist? (Will the lower and upper Riemann sums ever converge to the same value?)

6.3 INDEFINITE INTEGRALS—ANTIDERIVATIVES

The reverse process of differentiation is known, naturally enough, as **antidifferentiation**.

Definition 2 Given a function f, we say that F is an **antiderivative** of f provided that $F' = f$.

In other words, an antiderivative of f is a function F having f as *its* derivative.

We introduced the idea of derivative by using the example of a car's odometer, speedometer, and clock. The speedometer reading corresponds to a *derivative* reading—it gives us the instantaneous rate of change of distance (odometer reading) with respect to time (clock reading). One could approximate the speedometer reading from the odometer and clock, by computing the ratio of the distance covered to the length of a sufficiently small time interval (the difference quotient of change in distance over change in time). Graphically, the speedometer reading corresponds to the *local slope* on a graph of distance covered as a function of time. Let's go back to the illustration of a moving car to motivate the idea of an *antiderivative*.

Speedometers, odometers, and clocks—reversing the problem

Let's reverse the problem of determining speed from distance and time readings—our interest now is in determining the values of the distance function (odometer readings) from the car's speed over time. Instead of plotting distance versus time, we now plot *speed* versus time. For a car traveling at a *constant* rate of 40 mph over a two-hour period, the graph would be a horizontal line (Figure 6.19).

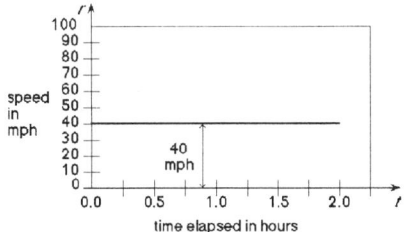

Figure 6.19 Constant rate vs. time for a car traveling 40 mph.

Now, the vertical height of this horizontal line represents the *slope* of the original graph of distance versus time. If we assume that the car's initial position (at $t = 0$) represents 0 distance covered, then we can reconstruct the distance graph by using 40 mph as the slope of the line (Figure 6.20).

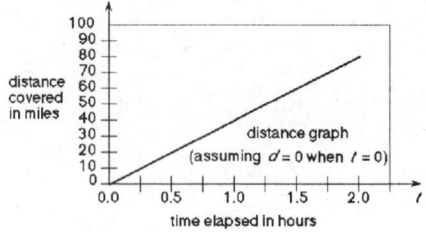

Figure 6.20 The distance graph for a car with initial position $d = 0$ miles.

Of course, if the car had already traveled a certain distance before the start of our time interval, our distance graph would still have the same slope of 40 mph, but would have a different d-intercept. For example, suppose the car had already traveled 15 miles at the time we started our stopwatch or clock at $t = 0$. If this car travels at a constant rate of 40 mph, then we could reconstruct its distance graph as shown in the second illustration in Figure 6.21.

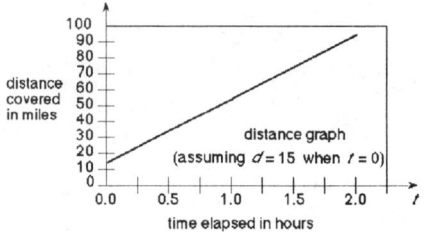

Figure 6.21 The distance graph for a car with initial position $d = 15$ miles.

This simply corresponds to the fact that the equation of a line

$$y = mx + b$$

cannot be determined by the slope m alone. We also need the y-intercept b (or some other point on the line which allows us to determine b). Knowing that our car travels at a constant rate of 40 mph is equivalent to knowing only the slope of the distance graph.

6.3 INDEFINITE INTEGRALS—ANTIDERIVATIVES

Now, let's suppose our car travels at varying speeds over the trip. Figure 6.22 shows the graph of speed readings for a second car traveling at varying speeds.

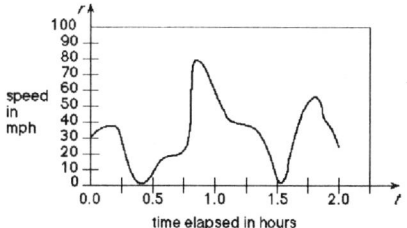

Figure 6.22 The speed graph for a car with varying speeds.

If we assume that the initial position of the car is $d = 0$, how can we use this speed graph to find the position of the car at any particular time t? One approximation strategy is outlined in the following example.

EXAMPLE 17 Assuming that the initial position of the car is $d = 0$, approximate the position of the car 15 minutes later, using the minute-by-minute speed readings from Table 6.1.

time t in minutes	speed V in mph	time t in minutes	speed V in mph
0	30	8	37
1	31	9	37.5
2	32	10	38
3	33	11	38
4	34	12	37
5	35	13	34
6	36	14	31
7	36.5	15	28

Table 6.1 Minute-by-minute speed readings.

Solution Our strategy is to use a linear approximation of the car's position function for each minute of the trip. In other words, we'll assume that the car's speed is *approximately constant* over each minute. Starting with the first minute of the trip, we read the initial velocity of the car from the graph to be 30 mph at $t = 0$. If the car does not change speed drastically during this minute, the car will travel approximately

$$30 \, \frac{\text{miles}}{\text{hour}} \cdot 1 \text{ minute} = 30 \, \frac{\text{miles}}{\text{hour}} \cdot \frac{1}{60} \text{ hours} = 0.5 \text{ miles}.$$

To calculate the distance covered during the second minute of the trip, we note the velocity at $t = 1$ minute (31 mph) and treat this as the car's approximate constant speed for the second minute of the trip. Thus, the car travels approximately

$$31\,\frac{\text{miles}}{\text{hour}} \cdot 1 \text{ minute} = 31\,\frac{\text{miles}}{\text{hour}} \cdot \frac{1}{60} \text{ hours} = 0.5017 \text{ miles}.$$

If we continue in this way for each minute of the trip, we obtain the approximate distance covered during each of the first 15 minutes. These results along with the approximate total distance covered by the car are shown in Table 6.2.

minute of trip	approximate distance (mi)	minute of trip	approximate distance (mi)
1	0.5000	9	0.6167
2	0.5167	10	0.6250
3	0.5333	11	0.6333
4	0.5500	12	0.6333
5	0.5667	13	0.6167
6	0.5833	14	0.5667
7	0.6000	15	0.5167
8	0.6083		
		total	8.6667

Table 6.2 Approximate distance covered minute-by-minute.

We can see that the car has travelled approximately $d = 8.667$ miles during the first 15 minutes of the trip. ∎

If we plot the running totals of the distance covered after each minute of the trip and connect the points with a smooth curve, we can obtain an approximate graph of the position function for this car. Of course, if the initial position ($t = 0$) of the car is not $d = 0$, then we must add the distance the car has already travelled to each of our measurements. Figure 6.23 shows position function plots for an initial position of $d = 0$ and $d = 15$ miles, respectively. You can see that the initial position information does not change the shape of the curve, but simply changes its vertical height.

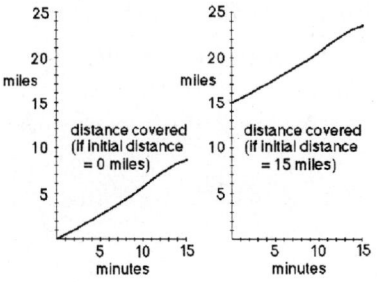

Figure 6.23 Position plots for initial positions of $d = 0$ and $d = 15$ miles.

6.3 INDEFINITE INTEGRALS—ANTIDERIVATIVES

Provided we know its initial position, we can estimate the values and sketch the graph of the original position function of a car from a knowledge of its velocity function (the derivative of the position function).

Estimating antiderivatives numerically and graphically

When we estimated the position of a car using a minute-by-minute table of speed readings, we essentially outlined a general strategy for estimating antiderivatives numerically. To review, our strategy consisted of using the speedometer readings (*instantaneous* rates of change) as estimates of the *constant* speed over each minute of the trip. Since the instantaneous rates of change varied, we needed to adjust these speed estimates to calculate the distance covered by the car each minute. Provided we knew the initial position of the car, we were able to estimate its position at the end of each minute. (Certainly, the accuracy of our estimates depended very much on the speed of the car not changing drastically during the minute-long time intervals.)

Let's use that same idea as a way of numerically approximating the output values of a function F, provided we know the outputs of its derivative $f = F'$ and at least one initial output $y_0 = F(x_0)$ for a specific input x_0. The steps are as follows:

Step 1. Look up the derivative value $F'(x_0) = f(x_0)$. We use this as the slope of the best linear approximation to F at our starting point (x_0, y_0) (where $y_0 = F(x_0)$).

Step 2. Pick a new input x_1 close to x_0, and use our best linear approximation to estimate $y_1 = F(x_1)$. Specifically,

$$y_1 = f(x_0)(x_1 - x_0) + y_0.$$

Step 3. Now repeat steps 1 and 2 at the new point (x_1, y_1). In other words, now use $f(x_1)$ as the slope of the best linear approximation to F at this point, and use it to estimate the output $y_2 = F(x_2)$ for another input x_2 close to x_1:

$$y_2 = f(x_1)(x_2 - x_1) + y_1.$$

By repeating this process successively, we obtain a sequence of points (x_0, y_0), (x_1, y_1), (x_2, y_2), (x_3, y_3), ..., that approximate points on the graph of F. In general, the closer together our inputs x_0, x_1, x_2, x_3, ..., are, the better our estimates of the function values.

EXAMPLE 18 Suppose a continuous function f has values as given in Table 6.3. Given that $F' = f$ and $F(1.0) = 4.5$, approximate the values of $F(x)$ to fill in the table.

x	1.0	1.1	1.2	1.3	1.4	1.5	1.6	1.7	1.8	1.9	2.0
f(x)	2.4	2.6	2.7	2.6	2.5	2.3	2.3	2.4	2.5	2.6	2.9
F(x)	4.5										

Table 6.3 What are the values of $F(x)$?

Solution Here $x_0 = 1.0$, $x_1 = 1.1$, $x_2 = 1.2$, ..., $x_{10} = 2.0$. To estimate $F(1.1)$, we use $f(1.0) = 2.4$ as the slope of the graph of F at 1.0. The *rise* of $F(x)$ from $x = 1.0$ to $x = 1.1$ is estimated by multiplying this slope by the *run* $(1.1 - 1.0 = 0.1)$. We add this to the *given initial value* $F(1.0) = 4.5$ to obtain

$$F(1.1) \approx 2.4(1.1 - 1.0) + 4.5 = 4.74.$$

Now, we repeat the process to estimate $F(1.2)$. We use $f(1.1) = 2.6$ as the slope of the graph of F at 1.1:

$$F(1.2) \approx 2.6(1.2 - 1.1) + 4.74 = 5.00.$$

Continuing, we have

$$F(1.3) \approx 2.7(1.3 - 1.2) + 5.00 = 5.27$$

$$F(1.4) \approx 2.6(1.4 - 1.3) + 5.27 = 5.53$$

$$F(1.5) \approx 2.5(1.5 - 1.4) + 5.53 = 5.78$$

$$F(1.6) \approx 2.3(1.6 - 1.5) + 5.78 = 6.01$$

$$F(1.7) \approx 2.3(1.7 - 1.6) + 6.01 = 6.24$$

$$F(1.8) \approx 2.4(1.8 - 1.7) + 6.24 = 6.48$$

$$F(1.9) \approx 2.5(1.9 - 1.8) + 6.48 = 6.73$$

$$F(2.0) \approx 2.6(2.0 - 1.9) + 6.73 = 6.99$$

■

We can imagine the function f as giving us directions to move on the graph of its antiderivative F. When we step off a small distance Δx horizontally from a point $(x_0, F(x_0))$, the value $f(x_0)\Delta x$ instructs approximately how far we need to step vertically to reach another point on the graph of F. We can then read off new directions.

6.3 INDEFINITE INTEGRALS—ANTIDERIVATIVES

More generally, the graphs of the set of antiderivatives for a given function form an infinite family of *parallel curves*. By parallel, we mean that the graphs all have the same shape, since their local slopes are identical for every input x (they all have the same derivative f). If we don't have an initial value from which to start, then we can still visualize the *shape* of the graph of an antiderivative, even though we cannot tell what its vertical height should be. This process can be accomplished by means of the **slope field** generated by the function f for its antiderivatives. The best way to explain this technique is through an example.

EXAMPLE 19 Suppose we have the following function values: $f(-4.5) = -5$,
$$f(-3) = -2, \quad f(-1.5) = -\frac{1}{3}, \quad f(0) = 2, \quad f(1.5) = 3, \quad f(3) = 3, \quad f(4.5) = 2.$$

Use these to sketch a slope field for the antiderivatives of f, and then sketch the graphs of some possible antiderivatives of f.

Solution The given function values $f(x)$ tell us the slope of any antiderivative's graph at the same input x. For example, since $f(0) = 2$, the graph of any antiderivative of f must have slope 2 at the point $x = 0$. We draw a family of small parallel tangent segments with common slope $m = 2$ above the input $x = 0$ to aid us. With each input x is associated such a family of parallel tangent segments whose common slope is determined by the value $f(x)$. This gives us the *slope field* for the antiderivatives of f. Figure 6.24 illustrates the part of the slope field we can determine from the given function values.

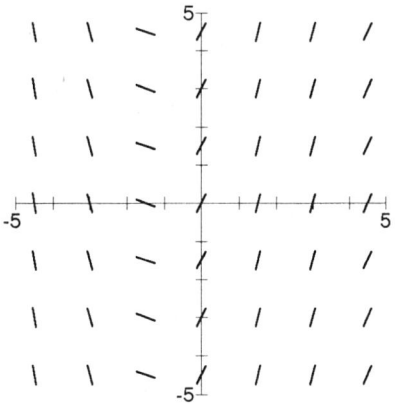

Figure 6.24 A slope field.

Now we can use the slope field to sketch the graphs of some of the antiderivatives of f. We simply pick a starting point and follow the directions indicated by the tangent segments. (A slope field is sometimes called a **direction field** for this reason.) Figure 6.25 illustrates several such curves, each obtained by using a different starting point. If you take any two of these curves, the *vertical distance* between them will be the same for any input value x.

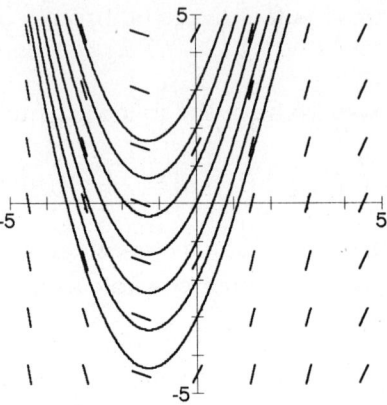

Figure 6.25 Family of antiderivative graphs.

If F is one of the antiderivatives of f, and an initial value $F(x_0)$ is specified, then this stipulates a starting point $(x_0, F(x_0))$ that pins down one of these curves. For example, if we were given the initial value $F(0) = 1$, then the third graph from the top pictured in Figure 6.25 would be our desired antiderivative. ■

Antiderivatives—Notation and examples

Symbolically, antidifferentiation can be thought of as filling in the unknown function expression in the statement

$$\frac{d}{dx}(?) = f(x).$$

EXAMPLE 20 Find an antiderivative of $f : x \longmapsto 2x$.

Solution Since $\frac{d}{dx}(x^2) = 2x$ we can see that $F : x \longmapsto x^2$ is an antiderivative of f. ■

6.3 INDEFINITE INTEGRALS—ANTIDERIVATIVES

EXAMPLE 21 Find an antiderivative of $g : x \longmapsto \cos(x)$.

Solution Since $\dfrac{d}{dx}(\sin(x)) = \cos(x)$ we can see that $G : x \longmapsto \sin(x)$ is an antiderivative of G. ∎

Notice that we have referred to "an antiderivative," and not "THE antiderivative" of a function, since a function can have infinitely many different antiderivatives. For example,

$$x \longmapsto x^2 \qquad x \longmapsto x^2 + 3 \qquad x \longmapsto x^2 - \pi \qquad x \longmapsto x^2 + \sqrt{17}$$

are all antiderivatives of the single function $x \longmapsto 2x$, since the derivative of each of these functions is $x \longmapsto 2x$.

These functions all have something in common, however. All are of the form

$$x \longmapsto x^2 + C$$

where C is some constant ($C = 0, 3, -\pi$, and $\sqrt{17}$ in the examples given). Certainly, *any* other arbitrary numerical value of the constant C will produce a perfectly good antiderivative of $x \longmapsto 2x$.

Is it possible that $x \longmapsto 2x$ also has other antiderivatives which are not of this particular form? The answer is NO. Recall that one of the consequences of the Mean Value Theorem was that any two functions which have the same derivative over an interval (a, b) can differ only by a constant function. If F is any function with a derivative $F'(x) = 2x$ for all real numbers x, then it *must* be of the form $F(x) = x^2 + C$ for some constant C.

The integral symbol $\int$ used without any limits of integration denotes the general *antiderivative* of a function. When used in this way, we call it an **indefinite integral** to distinguish from its other use as a notation for *definite integrals*. We read

$$\int f(x)\, dx$$

as the indefinite integral or general antiderivative of f with respect to x.

Given a continuous function f and any specific antiderivative F such that $F' = f$, we can write the general antiderivative of f as

$$\int f(x)\, dx = F(x) + C, \qquad C \text{ an arbitrary constant.}$$

This formula accounts for all the possible antiderivatives of the continuous function f. In other words, once we have found one antiderivative for f, we have essentially determined them *all* (up to the addition of the arbitrary constant).

EXAMPLE 22 $\displaystyle\int 2x\,dx = x^2 + C$, where C is an arbitrary constant. ∎

EXAMPLE 23 $\displaystyle\int \cos(x)\,dx = \sin(x) + C$, where C is an arbitrary constant. ∎

If we specify the output of the antiderivative at any particular input, this *initial condition* will determine a unique value of C.

EXAMPLE 24 Find the antiderivative F of $f : x \longmapsto 3x^2$ which satisfies the initial condition that $F(-2) = 5$.

Solution Since $\dfrac{d}{dx}(x^3) = 3x^2$, the antiderivative F must have the form

$$F : x \longmapsto x^3 + C$$

for some constant C. The initial condition specifies that

$$5 = F(-2) = (-2)^3 + C = (-8) + C$$

from which we can see that $C = 13$. Hence,

$$F : x \longmapsto x^3 + 13$$

is the desired antiderivative. ∎

☞ **The different uses of $\displaystyle\int$ produce very different results. With no limits of integration attached, the indefinite integral**

$$\int f(x)\,dx$$

is the general antiderivative of f, and represents an entire family of functions.

On the other hand, with limits of integration present, the definite integral

$$\int_a^b f(x)\,dx$$

is a specific real number A associated with the signed area under the graph of $y = f(x)$ over the interval $[a, b]$.

6.3 INDEFINITE INTEGRALS—ANTIDERIVATIVES

EXAMPLE 25 Find $\int (4x - 3)\, dx$ and $\int_1^2 (4x - 3)\, dx$.

Solution Since $\dfrac{d}{dx}(2x^2 - 3x) = 4x - 3$, the indefinite integral is

$$\int (4x - 3)\, dx = 2x^2 - 3x + C,$$

where C is an arbitrary constant. On the other hand, the definite integral

$$\int_1^2 (4x - 3)\, dx = 3,$$

the area of the trapezoidal region under the graph of $y = 4x - 3$ over the interval $[1, 2]$. ∎

Antiderivative formulas and properties

The derivative formula for any differentiable function automatically provides us with a corresponding *antiderivative* formula.

EXAMPLE 26 Find $\int x^r\, dx$, where r is a *constant* and $r \neq -1$.

Solution We know that $\dfrac{d}{dx}(x^{r+1}) = (r+1)x^r$ for any constant $r > 0$. Since derivatives have the property

$$\frac{d(cf)}{dx} = c\frac{df}{dx},$$

we can see that

$$\frac{d}{dx}\left(\frac{x^{r+1}}{r+1}\right) = x^r.$$

Here it is important that r is a constant, and that $r \neq -1$ to avoid division by zero. We conclude that

$$\int x^r\, dx = \frac{x^{r+1}}{r+1} + C$$

where C is an arbitrary constant. ∎

WARNING! The use of the arbitrary constant C in an indefinite integral formula

$$\int f(x)\,dx = F(x) + C$$

is under the assumption that the original function f is continuous. If f has any discontinuities, then its general antiderivative can involve a *different* arbitrary constant for each interval over which f is continuous.

EXAMPLE 27   Find the general antiderivative F of $f : x \longmapsto 1/x^2$, where $F(x)$ is defined for all $x \neq 0$.

Solution   If we write $f(x) = 1/x^2 = x^{-2}$, we can see that the formula from the previous example can be applied:

$$\int x^{-2}\,dx = \frac{x^{-1}}{-1} + C = -\frac{1}{x} + C,$$

where C is an arbitrary constant.

However, since f is undefined at $x = 0$, the graph of its antiderivative could have a break at $x = 0$. Therefore, the most general antiderivative is

$$F(x) = \begin{cases} -1/x + C_1 & \text{if } x < 0 \\ -1/x + C_2 & \text{if } x > 0 \end{cases}$$

Here C_1 and C_2 are two possibly different arbitrary constants.   ■

For example, if $C_1 = 7$ and $C_2 = -3$, then

$$F(x) = \begin{cases} -1/x + 7 & \text{if } x < 0 \\ -1/x - 3 & \text{if } x > 0 \end{cases}$$

is a function certainly satisfying $F'(x) = 1/x^2$ for all $x \neq 0$.

A way to visualize the necessity for different arbitrary constants is with a slope field. Figure 6.26 shows the slope field for the antiderivatives of $f : x \longmapsto 1/x^2$ along with several potential graphs of antiderivatives.

6.3 INDEFINITE INTEGRALS—ANTIDERIVATIVES

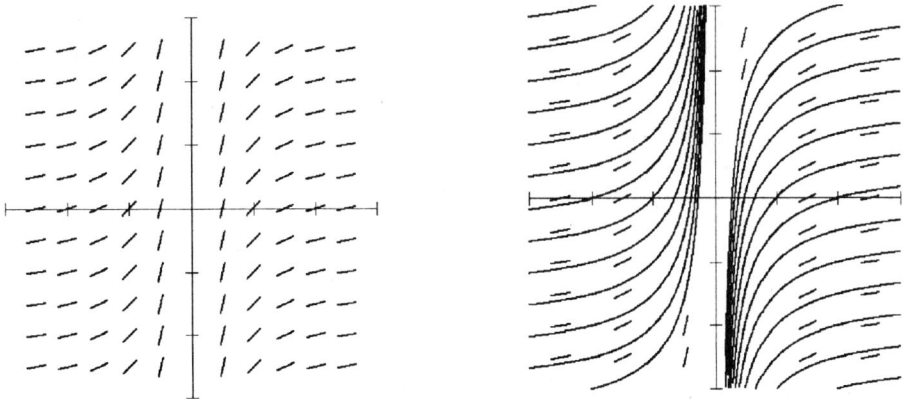

Figure 6.26 Slope field for antiderivatives of $1/x^2$.

Note that $x = 0$ is a boundary that the curves do not cross. The graph of a particular antiderivative of $1/x^2$ can be obtained by choosing any one curve to the left of $x = 0$ and any other curve to the right of $x = 0$.

The linearity properties of the derivative naturally imply linearity properties for antiderivatives. In other words,

$$\int cf(x)\, dx = c\int f(x)\, dx$$

$$\int (f(x) + g(x))\, dx = \int f(x)\, dx + \int g(x)\, dx$$

The linearity properties, combined with the result of the previous example, allow us to find the antiderivative of any polynomial function easily.

EXAMPLE 28 Find $\int (5x^4 - 3x^3 + 7x^2 + x - 8)\, dx$.

Solution We simply *antidifferentiate* term by term factoring out the coefficients and using the formula $\int x^r\, dx = \dfrac{x^{r+1}}{r+1} + C$:

$$\int (5x^4 - 3x^3 + 7x^2 + x - 8)\, dx$$

$$= 5\int x^4\, dx - 3\int x^3\, dx + 7\int x^2\, dx + \int x\, dx - 8\int dx$$

$$= 5 \cdot \frac{x^5}{5} - 3 \cdot \frac{x^4}{4} + 7 \cdot \frac{x^3}{3} + \frac{x^2}{2} - 8x + C$$

$$= x^5 - \frac{3x^4}{4} + \frac{7x^3}{3} + \frac{x^2}{2} - 8x + C.$$

Note that the term $\int dx = \int 1\,dx = x$ is an antiderivative of the constant function $x \longmapsto 1$. Also note that we need only one arbitrary constant C in the final expression, since the arbitrary constants arising from each term will sum together as a single constant.

We can check our answer by taking the derivative and comparing it to the original integrand:

$$\frac{d}{dx}\left(x^5 - \frac{3x^4}{4} + \frac{7x^3}{3} + \frac{x^2}{2} - 8x + C\right) = 5x^4 - 3x^3 + 7x^2 + x - 8.$$

∎

 This is a certain way of checking any antiderivative: differentiate and compare to the original function.

EXAMPLE 29 Find $\int \left(\dfrac{4x^{2/3}}{\sqrt{7}} - 5\sqrt{x} + \dfrac{\pi}{3x^2}\right) dx.$

Solution This becomes simpler if we write each term using rational exponents:

$$\int \left(\frac{4x^{2/3}}{\sqrt{7}} - 5\sqrt{x} + \frac{\pi}{3x^2}\right) dx$$

$$= \int \left(\frac{4}{\sqrt{7}} \cdot x^{2/3} - 5x^{1/2} + \frac{\pi}{3}x^{-2}\right) dx$$

$$= \frac{4}{\sqrt{7}} \cdot \frac{x^{5/3}}{5/3} - 5 \cdot \frac{x^{3/2}}{3/2} + \frac{\pi}{3} \cdot \frac{x^{-1}}{-1} + C$$

$$= \frac{12x^{5/3}}{5\sqrt{7}} - \frac{10x^{3/2}}{3} - \frac{\pi}{3x} + C.$$

∎

Other antiderivative formulas can be obtained directly by reversing the derivative formulas for the other basic functions. The antiderivative formulas for trigonometric functions are summarized below:

$$\int \cos(x)\,dx = \sin(x) + C \qquad \int -\sin(x)\,dx = \cos(x) + C$$

$$\int \sec^2(x)\,dx = \tan(x) + C \qquad \int -\csc(x)\cot(x)\,dx = \csc(x) + C$$

$$\int \sec(x)\tan(x)\,dx = \sec(x) + C \qquad \int -\csc^2(x)\,dx = \cot(x) + C$$

We can reverse the differentiation formulas for the inverse trigonometric formulas to obtain

6.3 INDEFINITE INTEGRALS—ANTIDERIVATIVES

$$\int \frac{1}{\sqrt{1-x^2}}\, dx = \arcsin(x) + C \qquad \int \frac{-1}{\sqrt{1-x^2}}\, dx = \arccos(x) + C$$

$$\int \frac{1}{1+x^2}\, dx = \arctan(x) + C \qquad \int \frac{-1}{1+x^2}\, dx = \operatorname{arccot}(x) + C$$

Provided $x \geq 1$,

$$\int \frac{1}{x\sqrt{x^2-1}}\, dx = \operatorname{arcsec}(x) + C \qquad \int \frac{-1}{x\sqrt{x^2-1}}\, dx = \operatorname{arccsc}(x) + C$$

As for the exponential and logarithmic differentiation formulas, reversing them gives us

$$\int \ln(a) \cdot a^x\, dx = a^x + C \qquad \int \frac{1}{x \ln(a)}\, dx = \log_a(x) + C$$

for $a > 0$, $a \neq 1$.

Antidifferentiation as pattern matching

For differentiation, the linearity properties and the product, quotient, and chain rules allow us (or a machine) to differentiate virtually any function made up of the basic functions through algebra or composition. Unfortunately, while antiderivatives also enjoy the linearity property, there are no counterpart rules for indefinite integrals to allow us to mechanically antidifferentiate products, quotients, and compositions of basic functions.

Finding symbolic formulas for antiderivatives is very much an art of *pattern matching*. Given a function f, we attempt to recognize what function F would satisfy the requirement that $F' = f$. This can be a very challenging task. While there are only a handful of basic derivative formulas needed along with the differentiation rules, there are entire books filled with indefinite integral formulas. In a later chapter, we will discuss some general hints for recognizing antiderivatives and for using tables of integral formulas.

Some computer algebra systems have many indefinite integral formulas stored in their "libraries." When faced with an indefinite integral, a computer algebra system must search for the right pattern. In many cases, there simply are *no* closed-form formulas to be found for the antiderivatives of some functions. For example, the function $f : x \longmapsto e^{-x^2}$ has no simple formula for its antiderivative. However, numerical and graphical strategies, like the use of a slope field, can give us very valuable information about the behavior of the antiderivative.

EXERCISES

The table of values below are the minute-by-minute speed readings of a car over the second 15 minutes of its trip. Use this table to answer exercises 1-3.

time t in minutes	speed V in mph	time t in minutes	speed V in mph
15	28	23	0
16	24	24	1
17	20	25	1
18	16	26	2
19	12	27	3
20	9	28	4
21	6	29	6
22	3	30	8

1. Estimate the distance covered over the interval $[15\ min, 30\ min]$.

2. Plot the running totals of distance covered each over the interval $[15, 30]$, assuming the car has already traveled 8.667 miles at $t = 15$ minutes.

3. Are there any inflection points in your graph from exercise 2? If so, at what time or times t?

---

4. Using the table of values below for the continuous function f, estimate the corresponding values for its antiderivative F, given that $F(1.27320) = -2.76548)$.

x	f(x)
1.27319	3.81456
1.27320	3.86714
1.27321	3.90551
1.27322	3.92017
1.27323	4.34405
1.27324	4.66292
1.27325	4.69141
1.27326	4.65674
1.27327	4.61993
1.27328	4.59550
1.27329	4.58799
1.27330	4.52556

Find a formula of the form $F(x) + C$ for each of the indefinite integrals indicated in exercises 5-20.

5. $\displaystyle\int (x^2 + x + 1)\,dx$

6. $\displaystyle\int (3x^2 + 2x + 1)\,dx$

6.3 INDEFINITE INTEGRALS—ANTIDERIVATIVES

7. $\int \dfrac{x^4}{2}\,dx$

8. $\int \sqrt{17}\,dx$

9. $\int \sqrt{x}\,dx$

10. $\int \dfrac{2}{x^4}\,dx$

11. $\int \dfrac{3}{\sqrt{x}}\,dx$

12. $\int \pi x^{100}\,dx$

13. $\int \dfrac{x^3 - x^{-3}}{3}\,dx$

14. $\int \sqrt[5]{x}\,dx$

15. $\int x^{7/3}\,dx$

16. $\int (5\sin(x) - 3\cos(x))\,dx$

17. $\int (\sin^2(x) + \cos^2(x))\,dx$

18. $\int \dfrac{4}{\sqrt{1-x^2}}\,dx$

19. $\int \dfrac{1}{4+4x^2}\,dx$

20. $\int \dfrac{1}{x \ln(3)}\,dx$

In exercises 21-28, you are given the graph of a function f against a grid. Assuming that the grid lines are spaced 1 unit apart both horizontally and vertically, sketch the slope field generated by each derivative. Then use this slope field to sketch the graph of an antiderivative F of f over the interval $[0, 5]$, assuming $F(0) = -2$.

21.

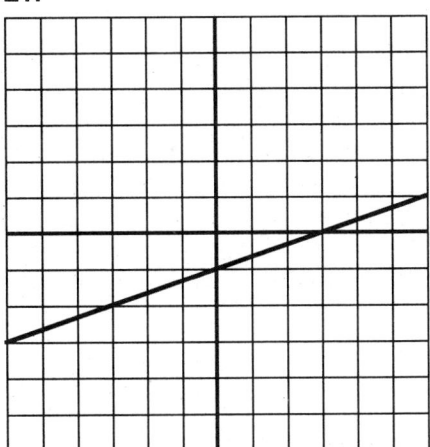

22.

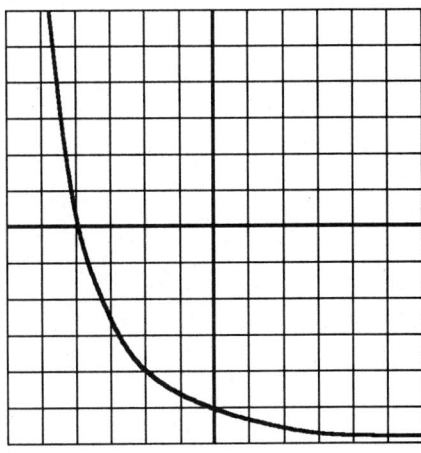

23.

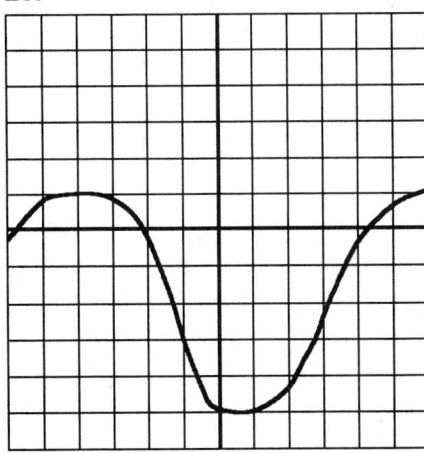

24.

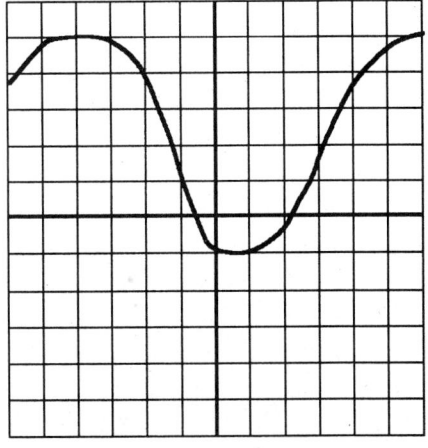

25.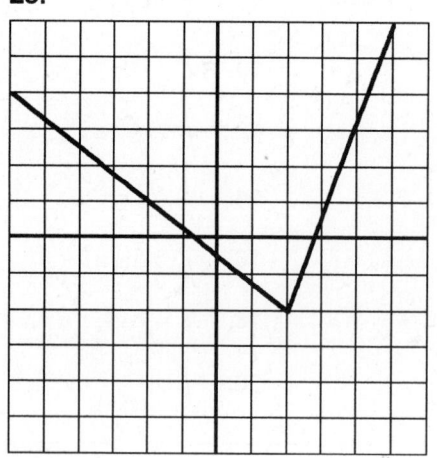

26.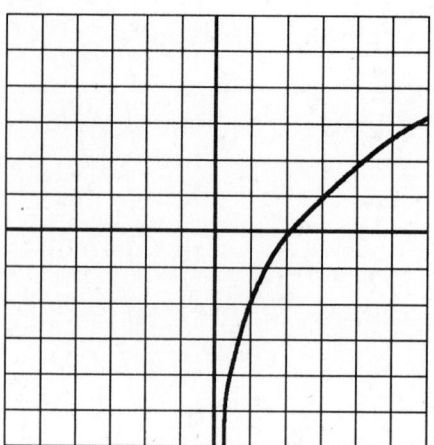

27.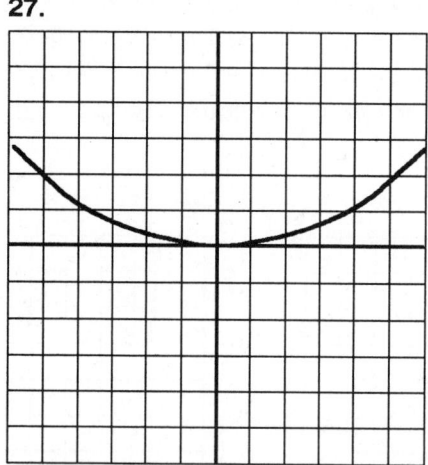

28.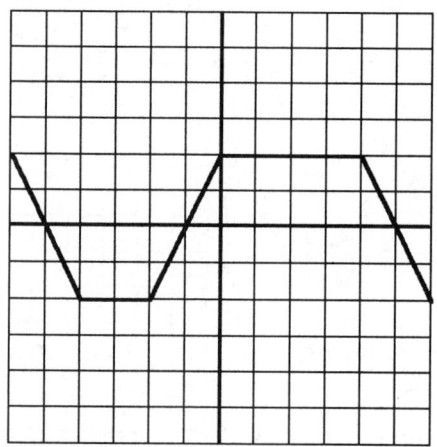

In each of exercises 29-40, generate a slope field for each function over the indicated interval and sketch the graphs of some antiderivatives of the function over this interval.

29. $f : x \longmapsto 1/x, \quad [1, 5]$
30. $f : x \longmapsto 1/x, \quad [-5, -1]$
31. $f : x \longmapsto \tan(x), \quad (-\pi/2, \pi/2)$
32. $f : x \longmapsto \cot(x), \quad (0, \pi)$
33. $f : x \longmapsto \sec(x), \quad (-\pi/2, \pi/2)$
34. $f : x \longmapsto \csc(x), \quad (0, \pi)$
35. $f : x \longmapsto \arctan(x), \quad [0, 4]$
36. $f : x \longmapsto \text{arccot}(x), \quad [0, 4]$
37. $f : x \longmapsto \arcsin(x), \quad [0, 1]$
38. $f : x \longmapsto \text{arcsec}(x), \quad [1, 2]$
39. $f : x \longmapsto \log_4(x), \quad [0.5, 2]$
40. $f : x \longmapsto e^{-x^2}, \quad [-2, 2]$

6.4 THE FUNDAMENTAL THEOREMS OF CALCULUS

In the first two sections of this chapter, we discussed *definite integrals*, which are closely related to *area measurement*. In the last section, we discussed *indefinite integrals*, which are best described in terms of *antiderivatives*. Except for a similarity in the notation, definite and indefinite integrals seem to have little to do with each other. In this section, we'll see that area measurement and antiderivatives are intimately related. The tie between the two types of integration are the subject of the *Fundamental Theorems of Calculus*.

Area under the speed graph

Let's return once again to the illustration of a car's speed over time. If a car travels at a constant speed, then the graph of its speed over time is simply a horizontal line. The distance covered over any elapsed time interval is

$$d = rt,$$

where r is the rate (the height of the constant speed graph) and t is the elapsed time (the length of the time interval). We can see that the d, the distance covered by the car, is graphically represented by the rectangular *area* under this line over that time interval.

For example, from $t = 1$ hour to $t = 1.5$ hours we have an elapsed time interval of 0.5 hours, so the car would travel

$$d = rt = (40 \text{ mph})(0.5 \text{ hours}) = 20 \text{ miles}.$$

Figure 6.27 illustrates this distance covered as an area under the speed graph between $t = 1$ hour and $t = 1.5$ hours.

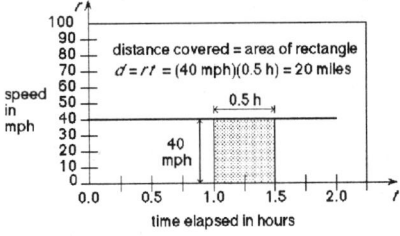

Figure 6.27 The area under a constant speed graph represents distance covered.

The total area under the graph over the two-hour time interval represents the total distance covered of 80 *miles*.

Of course, the formula $d = rt$ is not valid if the rate r changes or varies over the time interval. In Figure 6.28 we have plotted speedometer readings over time for a second car. The *height* of this graph over any particular value of t represents the speed of the car at that *instant* in time t. In the graph of constant rate over time, the distance covered corresponds to the *area under the graph*. Could we generalize and say that the distance covered by this car traveling at varying speeds between $t = 1$ hour to $t = 1.5$ hours *still* corresponds to the area under the graph for that time interval (the shaded region)?

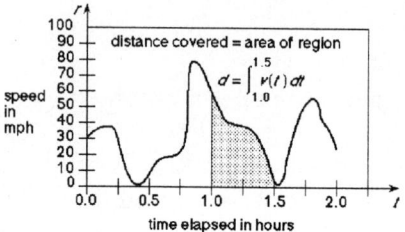

Figure 6.28 Rate vs. time for a car traveling at variable speeds.

If so, then the problem of determining the position of the car boils down to finding the definite integral of the velocity function over the time interval. In other words, if we use $v(t)$ to represent the velocity of the car at time t, then the distance covered by the car over the interval $1 \leq t \leq 1.5$ hours corresponds to the definite integral

$$\int_1^{1.5} v(t)\,dt.$$

We know that the velocity function can be found by taking the derivative of the position function. It appears that we might be able to use the definite integral to antidifferentiate the velocity function and find the original position function!

The insight of Newton and Leibniz that this inverse relationship does indeed exist between the derivative and the definite integral is considered one of the greatest intellectual achievements in the history of mathematics and science, and is the central idea behind the *Fundamental Theorems of Calculus*.

When a theorem is called "fundamental," it means that its statement is considered a cornerstone for the mathematical subject at hand. In number theory, the *Fundamental Theorem of Arithmetic* states that every positive integer greater than 1 is either prime or can be factored as a product of primes in essentially only one way. (For example, $18 = 2 \cdot 3 \cdot 3$ is the only factorization of 18 into primes). This property of positive integers is a key result in the foundations of number theory.

6.4 THE FUNDAMENTAL THEOREMS OF CALCULUS

The Fundamental Theorems of Calculus provide the tie between differential calculus and integral calculus. In some sense, these two branches of calculus developed quite separately from each other, but now we can see that area measurement problems like those that Archimedes pondered and the multitude of physical problems involving rates of change turn out to be intimately related. When the connection between the two was discovered, it unleashed an avalanche of new results and applications for calculus. We now turn to understanding the precise statements of, and reasoning behind these two key theorems.

The First Fundamental Theorem of Calculus

The derivative of a function f is the *slope function* for the graph of f. Now we'll use the definite integral to make an *area function* for the graph of f.

Suppose $[a, b]$ is an interval and f is a continuous function for which the definite integral

$$\int_a^x f(t)\, dt$$

exists for every real number $a \leq x \leq b$. Notice that we're using x as the upper limit of integration and t for the variable of integration. Figure 6.29 illustrates the graph of $y = f(t)$ and the region corresponding to a particular value x.

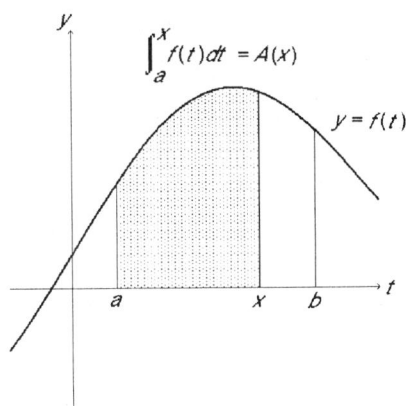

Figure 6.29 The value of the area function A depends on x.

This definite integral measures the (signed) area under the graph of $y = f(t)$ from a to x. If we keep a as a fixed reference point and vary the upper limit of integration x, then the value of this definite integral will also vary. In other words, we can think of this definite integral expression as describing a function of x. We'll call this function A for *area*:

$$A(x) = \int_a^x f(t)\, dt \qquad \text{for} \qquad a \leq x \leq b.$$

EXAMPLE 30 Set up and graph the area function A for $f : t \longmapsto 1/t$ over the interval $[1, 3]$.

Solution The area function A in this case has the evaluation formula

$$A(x) = \int_1^x 1/t \, dt \quad \text{for} \quad 1 \leq x \leq 3.$$

Using the definite integration capabilities of a calculator, we can compute $A(x)$ for several values of x (Table 6.4) and plot them (Figure 6.30).

x	A(x)	x	A(x)
1.0	0.000	2.1	0.742
1.1	0.095	2.2	0.788
1.2	0.182	2.3	0.833
1.3	0.262	2.4	0.875
1.4	0.336	2.5	0.916
1.5	0.405	2.6	0.955
1.6	0.470	2.7	0.993
1.7	0.531	2.8	1.030
1.8	0.588	2.9	1.065
1.9	0.642	3.0	1.099
2.0	0.693		

Table 6.4 Area under the graph of $y = 1/t$ from 1 to x.

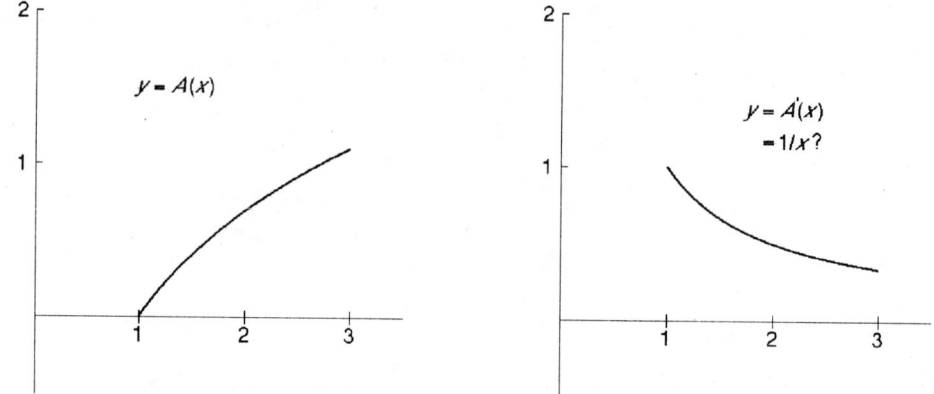

Figure 6.30 Plots of $y = A(x)$ and $y = A'(x)$ from Table 6.4.

What's the rate of change of the area function? In other words, what is $A'(x)$? If we sketch the graph of the derivative A' using the graph of $y = A(x)$ as a guide, then we get a picture remarkably like that of the graph of $y = 1/x$ over the interval $[1, 3]$ (see second illustration in Figure 6.30).

Is it a coincidence that we have come back to our original function f with x in place of t? In fact, the First Fundamental Theorem of Calculus states that this will always be the case.

6.4 THE FUNDAMENTAL THEOREMS OF CALCULUS

Theorem 6.5

The First Fundamental Theorem of Calculus.
Hypothesis: Suppose f is a continuous function for which

$$A(x) = \int_a^x f(t)\,dt$$

exists for every real number $a \leq x \leq b$.
Conclusion: If $a < x < b$, then $A'(x) = f(x)$.

Reasoning To estimate the derivative $A'(x)$ with a difference quotient, we need to examine

$$\frac{\Delta A}{\Delta x} = \frac{A(x + \Delta x) - A(x)}{\Delta x}$$

for a small change Δx. The numerator of this difference quotient corresponds to the area of the dark strip in Figure 6.31, and is given by the definite integral

$$\int_x^{x+\Delta x} f(t)\,dt.$$

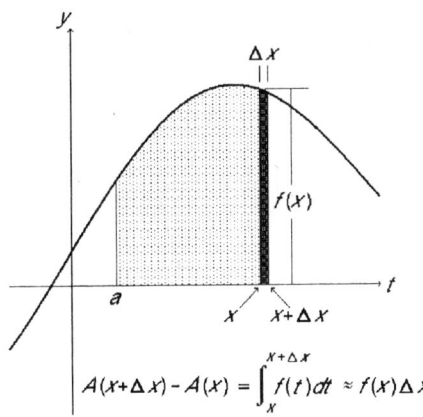

Figure 6.31 Finding the difference quotient of the area function.

The area of this region is approximately the area of a tall thin rectangle with width Δx and height $f(x)$. Hence,

$$A'(x) \approx \frac{A(x + \Delta x) - A(x)}{\Delta x} \approx \frac{f(x)\Delta x}{\Delta x} = f(x).$$

(This is assuming that the function value $f(x)$ and Δx are both positive as suggested by the illustration. However, you can check that the product $f(x)\Delta x$ still gives us the approximate change in the value of the definite integral in case either or both are negative.)

This approximation improves as Δx gets smaller. The true change ΔA lies between the areas of the inscribed and circumscribed rectangles over this small interval. If L is the height of the inscribed rectangle, and U is the height of the circumscribed rectangle, then we can say that ΔA lies between $L\Delta x$ and $U\Delta x$. Now, dividing through by Δx shows that

$$L \leq \frac{\Delta A}{\Delta x} \leq U.$$

As $\Delta x \to 0$, our interval $[x, x+\Delta x]$ shrinks to a single point x. By hypothesis, the function f is continuous, so both the maximum function value U and the minimum function value L over this interval must approach $f(x)$. Since $\Delta A/\Delta x$ is trapped between L and U, we must have

$$\frac{\Delta A}{\Delta x} \to f(x) \quad \text{as} \quad \Delta x \to 0.$$

Thus, $A'(x) = f(x)$ as stated in the conclusion of the theorem. □

The First Fundamental Theorem says that we can produce an antiderivative for *any* continuous function f by simply building an area function A for it. The choice of where to begin measuring the area (the reference point a) is entirely arbitrary. If we make a new choice of a, then we may be adding or subtracting some fixed amount from all our area function measurements. In other words, changing a may change the antiderivative by a constant.

EXAMPLE 31 Find an antiderivative for $f : x \longmapsto \frac{\cos(x)}{2^x}$.

Solution The area function defined by the formula

$$A(x) = \int_0^x \frac{\cos(t)}{2^t} dt$$

is an antiderivative of f by the First Fundamental Theorem of Calculus. Hence, $A'(x) = \frac{\cos(x)}{2^x}$. We could have selected a different lower limit than 0:

$$F(x) = \int_{-\pi}^x \frac{\cos(t)}{2^t} dt$$

is another antiderivative of f. The functions A and F differ by a constant. Specifically,

$$F(x) = A(x) + C$$

where $C = \int_{-\pi}^0 \frac{\cos(t)}{2^t} dt$. ■

6.4 THE FUNDAMENTAL THEOREMS OF CALCULUS

EXAMPLE 32 Find an antiderivative G for $g : x \longmapsto e^{-x^2}$ such that $G(3) = 0$.

Solution The First Fundamental Theorem of Calculus tells us that any function G having a formula

$$G(x) = \int_a^x e^{-t^2} \, dt$$

will have a derivative

$$G'(x) = e^{-x^2} = g(x).$$

The initial condition will force us to make a particular choice of a. If we substitute $x = 3$, we have

$$0 = G(3) = \int_a^3 e^{-t^2} \, dt.$$

What value a makes this definite integral have value 0? Certainly, $a = 3$ does the job, so we have

$$G(x) = \int_3^x e^{-t^2} \, dt$$

as an antiderivative meeting the condition $G(3) = 0$. ■

Area functions are treated like any other differentiable functions when it comes to the rules of differentiation.

EXAMPLE 33 Suppose $F(x) = \int_1^{x^3 - 7x} \cos^4(t) \, dt$.

Solution The function F can be thought of as a composition

$$F(x) = A(x^3 - 7x)$$

where A is the area function $A(x) = \int_1^x \cos^4(t) \, dt$. Using the chain rule, we have

$$F'(x) = A'(x^3 - 7x) \cdot (3x^2 - 7) = \cos^4(x^3 - 7x) \cdot (3x^2 - 7). \qquad ■$$

The Second Fundamental Theorem of Calculus

The Second Fundamental Theorem of Calculus is closely related to the First Fundamental Theorem. It gives us a powerful way of computing definite integrals, provided we can find an antiderivative of the integrand.

Theorem 6.6 | **The Second Fundamental Theorem of Calculus.**
Hypothesis: F is any antiderivative of a continuous function f.
Conclusion: $\int_a^b f(x)\,dx = F(b) - F(a)$.

This theorem says that one way we can compute the definite integral of a function f over an interval $[a, b]$ is to first find an antiderivative F for the function, and then simply take the difference of its value at the endpoints: $F(b) - F(a)$.

EXAMPLE 34 Evaluate $\int_1^2 x^2\,dx$.

Solution One antiderivative of $f : x \longmapsto x^2$ is $F : x \longmapsto x^3/3$. The Second Fundamental Theorem simply says that

$$\int_1^2 x^2\,dx = F(2) - F(1) = \frac{2^3}{3} - \frac{1^3}{3} = \frac{8}{3} - \frac{1}{3} = \frac{7}{3}.$$

Notice that any other antiderivative of f would have produced the same result. For instance, $G : x \longmapsto x^3/3 + 47$ is also an antiderivative of $f : x \longmapsto x^2$, but

$$G(2) - G(1) = (\frac{2^3}{3} + 47) - (\frac{1^3}{3} + 47) = \frac{8}{3} + 47 - \frac{1}{3} - 47 = \frac{7}{3}.$$

Since the constant 47 appears as a term in the evaluation of both $G(2)$ and $G(1)$, it simply cancels out when we take the difference. The same cancellation occurs if we replace 47 by any other arbitrary constant C. ∎

A notation often used to denote the difference $F(b) - F(a)$ is

$$F(x)\Big]_{x=a}^{x=b} \quad \text{or} \quad F(x)\Big]_a^b$$

6.4 THE FUNDAMENTAL THEOREMS OF CALCULUS

We could write down the Second Fundamental Theorem as

$$\int_a^b f(x)\,dx = \left[\int f(x)\,dx\right]_a^b.$$

In other words, the definite integral of f is the difference in the value of the indefinite integral at the limits of integration.

EXAMPLE 35 Find the area under the graph of $y = 1/(1+x^2)$ from $x = 0$ to $x = 1$.

Solution Since $y = 1/(1+x^2)$ is positive, the desired area under its graph corresponds to the definite integral

$$\int_0^1 \frac{1}{1+x^2}\,dx = \left[\int \frac{1}{1+x^2}\,dx\right]_0^1 = \arctan(x)\Big]_0^1 = \arctan(1) - \arctan(0) = \pi/4.$$

We used an antiderivative formula from the previous section to complete the solution. ∎

Let's see how the Second Fundamental Theorem follows from the First Fundamental Theorem.

Reasoning Suppose we have a continuous function f and any antiderivative F. We can set up an area function A such that

$$A(x) = \int_a^x f(t)\,dt.$$

From this we can see that $A(a) = 0$ and

$$A(b) = \int_a^b f(t)\,dt.$$

The First Fundamental Theorem of Calculus tells us that A is also an antiderivative of f. Since two antiderivatives of the same continuous function can differ only by a constant, there must be some value C such that

$$A(x) = F(x) + C.$$

Let's try to find the value C. If we substitute $x = a$, we see that

$$A(a) = 0 = F(a) + C.$$

From this, we can conclude that $C = -F(a)$. Now we substitute $x = b$ to find

$$A(b) = \int_a^b f(t)\,dt = F(b) + C = F(b) - F(a)$$

as stated in the conclusion of the theorem. □

The First Fundamental Theorem gives a recipe for constructing an antiderivative of any continuous function f: we simply pick a point a and form the area function

$$A(x) = \int_a^x f(t)\,dt.$$

This function satisfies the requirement

$$A'(x) = f(x).$$

The Second Fundamental Theorem gives a technique for computing definite integrals $\int_a^b f(x)\,dx$, provided we can find an antiderivative F of the function f:

$$\int_a^b f(x)\,dx = F(b) - F(a).$$

Historical notes

The English mathematician Isaac Newton (1642-1727) is considered by many to be the greatest genius who ever lived. His three-volume work, *Principia Mathematica*, is lauded as the most influential scientific treatise in history. Legend would have us believe that Newton discovered the law of gravity on the occasion of an apple falling on his head. Whatever the truth of this particular anecdote, Newton, along with the German mathematician Gottfried Leibniz (1646-1716), are credited with being the first people to fully comprehend the importance of the fundamental relationship between the slope of one graph (like that of distance versus time) and the area under the graph of those slopes (instantaneous speed versus time). There is fairly strong evidence that Newton and Leibniz had essentially the same insight, but independently of each other. Their ideas were quickly recognized as ushering in a mathematical revolution. (It is a shame that these two men were also drawn into a bitter plagiarism controversy regarding primacy of the invention of calculus. For one account of this "calculus controversy," see David Burton's *The History of Mathematics*.) Although Newton and Leibniz made their breakthrough observations in the seventeenth century, we can recognize that many of the key ideas of calculus can be traced back much further to Archimedes and the ancient Greeks. The insights of Newton and Leibniz were the culmination of centuries of development and refinement of those ideas. In recognition of his debt to Archimedes and other scientists who preceded him, Newton once said, "If I have seen farther than others, it is because I have stood on the shoulders of giants."

6.4 THE FUNDAMENTAL THEOREMS OF CALCULUS

EXERCISES

In exercises 1-6, use the First Fundamental Theorem of Calculus to find an antiderivative F of the given function f, having the specified initial value.

1. $f : x \longmapsto 2^{-x^2}$ $F(1) = 0$
2. $f : x \longmapsto \log_2(x)$ $F(2) = 0$
3. $f : x \longmapsto \cos^2(x)$ $F(-\pi) = 0$
4. $f : x \longmapsto 1/(x^3 + 8)$ $F(-1) = 0$
5. $f : x \longmapsto 1/x$ $F(\sqrt{17}) = 0$
6. $f : x \longmapsto \arctan(x)$ $F(-\sqrt{3}) = 0$

Suppose G is the area function with formula

$$G(x) = \int_2^x \frac{\sin(2t)}{1+t^2} \, dt.$$

In exercises 7-12, find dy/dx.

7. $y = G(x)$
8. $y = G(x^2)$
9. $y = x^2 G(x)$
10. $y = G(x)/x^2$
11. $y = (G(x))^2$
12. $y = G'(x)$

Use the Second Fundamental Theorem of Calculus to calculate each of the definite integrals in exercises 13-20.

13. $\displaystyle\int_{-1}^{3} (x^2 + x + 1) \, dx$
14. $\displaystyle\int_{1}^{2.5} (3x^2 + 2x + 1) \, dx$
15. $\displaystyle\int_{4}^{9} \sqrt{x} \, dx$
16. $\displaystyle\int_{\pi/2}^{\pi} (5\sin(x) - 3\cos(x)) \, dx$
17. $\displaystyle\int_{-2\pi}^{2\pi} 2\pi(\sin^2(x) + \cos^2(x)) \, dx$
18. $\displaystyle\int_{0}^{0.5} \frac{4}{\sqrt{1-x^2}} \, dx$
19. $\displaystyle\int_{0}^{1} \frac{1}{4+4x^2} \, dx$
20. $\displaystyle\int_{-4}^{-1} \frac{1}{x^2} \, dx$

In exercises 21-28, you are given the graph of a function f against a grid. Assuming that the grid lines are spaced 1 unit apart both horizontally and vertically, sketch the graph of the area function

$$A(x) = \int_1^x f(t) \, dt$$

over the interval $[1,5]$ for each function. Then sketch the graphs of the derivative of each area function and compare it to the original function's graph.

21.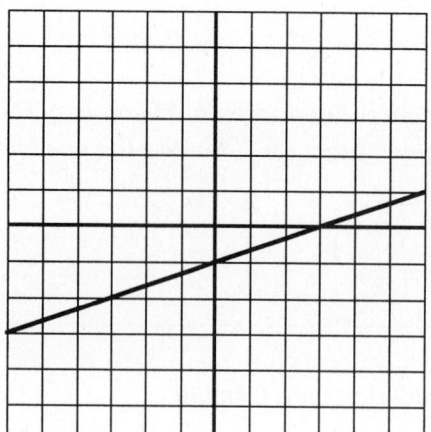

22.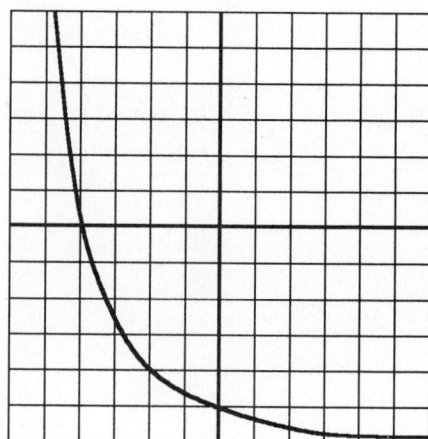

23.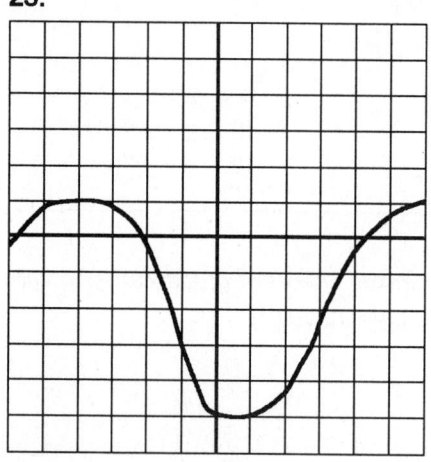

24.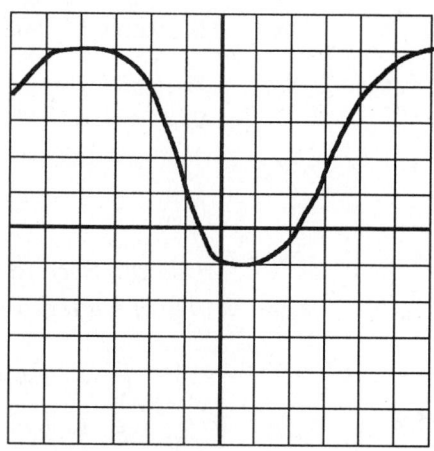

25.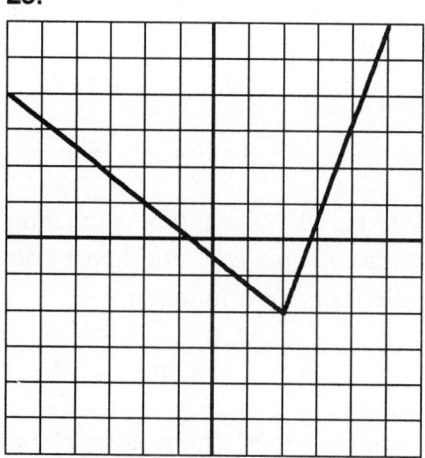

26.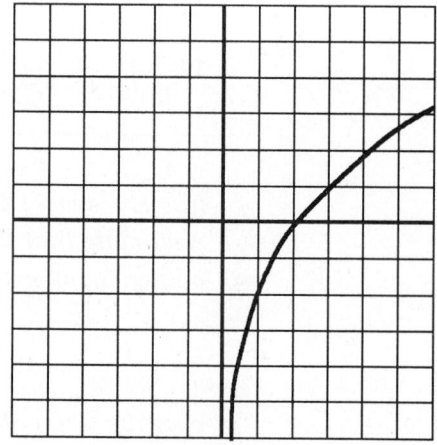

6.4 THE FUNDAMENTAL THEOREMS OF CALCULUS

27.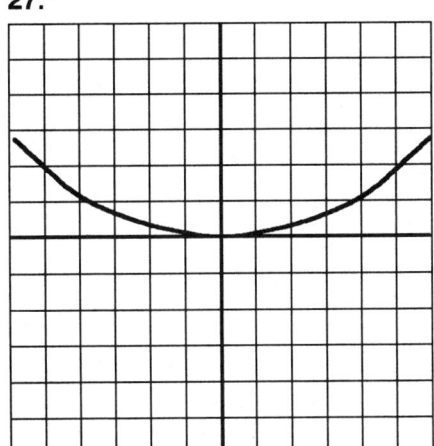

28.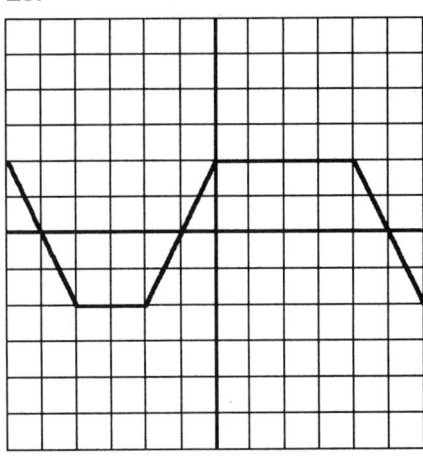

29. A student uses the Second Fundamental Theorem of Calculus to calculate

$$\int_{-1}^{1} \frac{1}{x^2}\, dx = -\frac{1}{x}\Big]_{-1}^{1} = -1 - 1 = -2.$$

Graph $y = 1/x^2$ and explain why this answer is unreasonable. Why isn't this a contradiction of the Second Fundamental Theorem of Calculus?

30. The function

$$A(x) = \int_{1}^{x} \frac{1}{t}\, dt$$

is quite special, for it fills in the one gap in our indefinite integral formula

$$\int x^r\, dx = \frac{x^{r+1}}{r+1} + C \quad (\text{for } r \neq -1)$$

with

$$\int x^{-1}\, dx = A(x) + C.$$

Using the definite integral capabilities of a calculator or computer, find the value x_0 such that $A(x_0) = 1$ to within 5 decimal places. (Note: from Table 6.4, we can see that $2.7 < x_0 < 2.8$.)

6.5 NUMERICAL INTEGRATION TECHNIQUES

The Second Fundamental Theorem of Calculus allows us to evaluate a definite integral provided we can find an antiderivative for the integrand:

$$\int_a^b f(x)\,dx = F(b) - F(a)$$

where $F' = f$.

If a machine such as a calculator or computer is capable of finding an antiderivative F for the function f, then the machine can simply use the Second Fundamental Theorem to evaluate the definite integral.

Sometimes it is difficult or even impossible to find a nice closed form formula for the antiderivative F. Of course, the First Fundamental Theorem guarantees that we can always find an antiderivative for a continuous function, but that puts us back where we started—trying to calculate the signed area under the graph of f.

In this section, we examine some of the numerical approximation techniques that can be used in evaluating definite integrals. The key idea to keep in mind is that a definite integral

$$\int_a^b f(x)\,dx$$

is *defined* to be the limiting value of Riemann sums

$$\lim_{\Delta x \to 0} \sum_a^b f(x)\Delta x,$$

where the interval $[a, b]$ is partitioned into subintervals of length Δx, one input x is chosen from each subinterval, and we sum up the products $f(x)\Delta x$ for all the subintervals.

One way to approximate the value of a definite integral is to simply evaluate such a Riemann sum for a particular subinterval size Δx and a particular choice of inputs from those subintervals. This is indeed the strategy in the first three methods of numerical integration we discuss. Other approximation techniques can be thought of as improvements achieved by averaging these results.

6.5 NUMERICAL INTEGRATION TECHNIQUES

Left endpoint, right endpoint and midpoint rules

The left endpoint, right endpoint, and midpoint rules for approximating definite integrals are named for the choice of input we make from each subinterval. The advantage of making these particular selections is that they are easy to "automate." That is, it is not difficult to program a machine to make these selections and then carry out the computation. Let's make the procedures for carrying out these techniques explicit.

Step 1. Choose a number n of subintervals in the partition of $[a, b]$.

Step 2. Calculate $\Delta x = \dfrac{b-a}{n}$.

Step 3. Locate the n inputs $x_1, x_2, ..., x_n$.

Step 4. Evaluate f at each input x_i and find the Riemann sum:

$$\sum_{i=1}^{n} f(x_i)\Delta x = f(x_1)\Delta x + f(x_2)\Delta x + f(x_3)\Delta x + \cdots + f(x_n)\Delta x.$$

For the left endpoint rule, our inputs will be

$$x_1 = a, \quad x_2 = a + \Delta x, \quad x_3 = a + 2\Delta x, \quad \ldots, \quad x_n = a + (n-1)\Delta x.$$

For the right endpoint rule, our inputs will be

$$x_1 = a + \Delta x, \quad x_2 = a + 2\Delta x, \quad x_3 = a + 3\Delta x, \quad \ldots, \quad x_n = a + n\Delta x = b.$$

For the midpoint rule, our inputs will be

$$x_1 = a + \frac{\Delta x}{2}, \quad x_2 = a + \frac{3\Delta x}{2}, \quad x_3 = a + \frac{5\Delta x}{2}, \quad \ldots, \quad x_n = a + \frac{(2n-1)\Delta x}{2}.$$

Note that our inputs are equally spaced apart under all three rules, so that once we know the first input x_1, the rest are determined:

$$x_2 = x_1 + \Delta x, \quad x_3 = x_2 + \Delta x, \quad x_4 = x_3 + \Delta x, \quad \ldots, \quad x_n = x_{n-1} + \Delta x.$$

EXAMPLE 36 Approximate $\displaystyle\int_{.5}^{3.5} \sin^3(x)\, dx$ using each of the three rules for a partition of size $n = 6$.

Solution Here $a = 0.5, b = 3.5, n = 6$, and $f(x) = \sin^3(x)$. The subinterval size is

$$\Delta x = \frac{b-a}{n} = \frac{3.5 - 0.5}{6} = 0.5.$$

If we calculate the output values of the function at each point of the partition, we find

$f(0.5) \approx 0.1102$ $\qquad f(1.0) \approx 0.5958$
$f(1.5) \approx 0.9925$ $\qquad f(2.0) \approx 0.7518$
$f(2.5) \approx 0.2144$ $\qquad f(3.0) \approx 0.0028$
$f(3.5) \approx -0.0432$

The left endpoint rule leads to the approximation

$$(0.5)(f(0.5) + f(1.0) + f(1.5) + f(2.0) + f(2.5) + f(3.0)) \approx 1.334$$

The right endpoint rule leads to the approximation

$$(0.5)(f(1.0) + f(1.5) + f(2.0) + f(2.5) + f(3.0) + f(3.5)) \approx 1.257.$$

The midpoint rule requires us to calculate the output values of the function at the midpoint of each subinterval:

$f(0.75) \approx 0.3167$ $\qquad f(1.25) \approx 0.8546$
$f(1.75) \approx 0.9527$ $\qquad f(2.25) \approx 0.4710$
$f(2.75) \approx 0.0556$ $\qquad f(3.25) \approx -0.0013$

This leads to the approximation

$$(0.5)(f(0.75) + f(1.25) + f(1.75) + f(2.25) + f(2.75) + f(3.25)) \approx 1.325.$$

For comparison, a machine computation yields $\int_{.5}^{3.5} \sin^3(x)\,dx \approx 1.315$, so in this case the midpoint rule gave the best approximation. ∎

The rectangles whose (signed) areas are represented by the terms in each Riemann sum are illustrated in Figures 6.32 through 6.34. The right and left endpoint rules are sometimes called **right** and **left rectangle rules**.

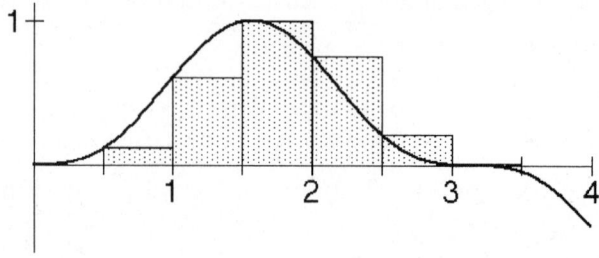

Figure 6.32 Left rectangle approximation of $\int_{.5}^{3.5} \sin^3(x)\,dx$ for a partition of size $n = 6$.

6.5 NUMERICAL INTEGRATION TECHNIQUES

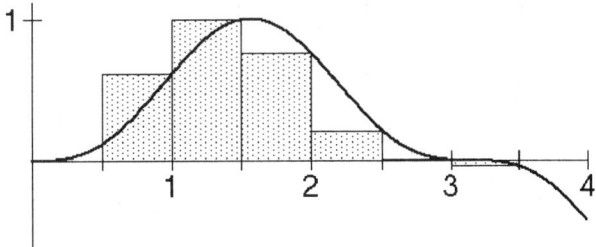

Figure 6.33 Right rectangle approximation of $\int_{.5}^{3.5} \sin^3(x)\,dx$ for a partition of size $n = 6$.

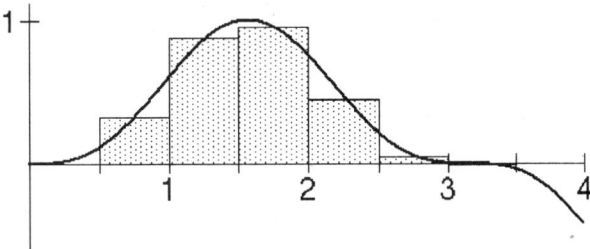

Figure 6.34 Midpoint approximation of $\int_{.5}^{3.5} \sin^3(x)\,dx$ for a partition of size $n = 6$.

Notice that if a function f is *increasing* over a given subinterval, then the area of the left rectangle is an underestimate of the area under the graph, while the area of the right rectangle is an overestimate. In this case, the area of the midpoint rectangle will be between this underestimate and overestimate (but not necessarily closer to the true area than each of the endpoint approximations). The situation is reversed when the function f is *decreasing* over the subinterval. If the function f is not monotonic over a subinterval, then the relationship between the true area and the areas of the left, right, and midpoint rectangles cannot be determined without closer examination.

As the partition becomes finer and finer (in other words n is chosen larger and larger) one expects that these Riemann sum approximations should get closer and closer to the actual value of the definite integral. In general, this must be true as long as the definite integral exists. However, for certain choices of n, it is possible that the particular sampling of inputs obtained by using the left, right, or midpoint rules may give a *terrible* approximation. Since all three of these methods sample at regular spaced intervals, a function that is periodic or has some graphical symmetry will be approximated badly for certain partition sizes n. Indeed, it is possible for a smaller value of n to produce a better result than a larger value in some cases.

Trapezoidal Rule and Simpson's Rule

The next two techniques we discuss approximate the value of a definite integral by using *averages* of Riemann sums. These techniques generally produce better results than the right, left, and midpoint techniques.

The **trapezoidal rule** for a partition of size n is simply the average of the results obtained by using the right and left endpoint rules for partitions of size n.

EXAMPLE 37 Approximate $\int_{0.5}^{3.5} \sin^3(x)\,dx$ using the trapezoidal rule for a partition of size $n = 6$.

Solution We have already determined the results of the right and left endpoint rules for this definite integral for a partition of size $n = 6$. Therefore the trapezoidal rule approximation is $(1.334 + 1.257)/2 = 1.296$. ∎

The idea behind the trapezoidal rule is that over intervals for which f is monotonic (either increasing or decreasing,) then one endpoint rule will give an underestimate and the other endpoint rule will give an overestimate. Hence, averaging the results from these two endpoint rules should give a better approximation to the value of the definite integral. The trapezoidal rule derives its name from the fact that the average of the areas of the right and left rectangles can be thought of as the area of a trapezoid with the same base (see Figure 6.35).

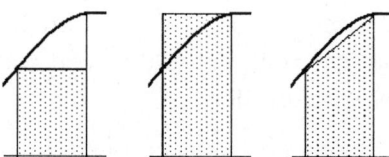

Figure 6.35 Trapezoidal rule is the average of the left and right endpoint rules.

The easiest way to remember the trapezoidal rule is as the average of the left and right rectangle rules. Figure 6.36 illustrates the trapezoidal approximation for the definite integral of the previous example.

6.5 NUMERICAL INTEGRATION TECHNIQUES

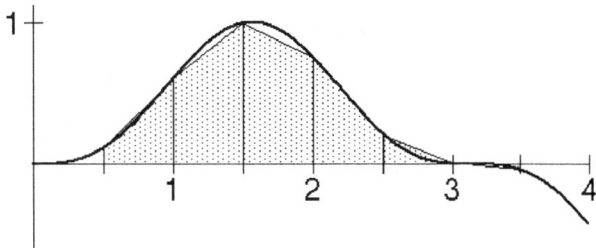

Figure 6.36 Trapezoidal approximation of $\int_{.5}^{3.5} \sin^3(x)\, dx$ for a partition of size $n = 6$.

An even better approximation to the value of a definite integral can be obtained by taking a *weighted* average of the midpoint and trapezoidal rules. The motivation behind this strategy is to take into account the concavity of the function's graph over each subinterval.

Figure 6.37 illustrates the midpoint and trapezoidal approximations over subintervals where the function graph is concave down and concave up.

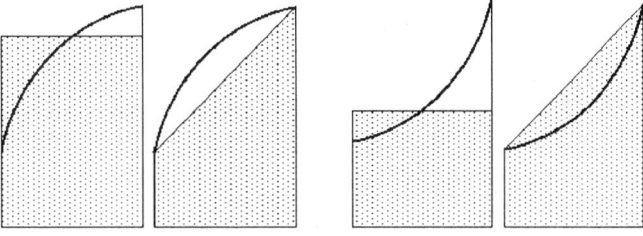

Figure 6.37 Comparing midpoint and trapezoidal approximations.

Look closely at Figure 6.37. Note that for the concave down graph (the two pictures on the left), the midpoint rule overestimates the area while the trapezoidal rule underestimates the area. Of the two estimates, the trapezoidal estimate appears to be approximately twice as far off as the midpoint estimate. For the concave up graph (the two pictures on the right), the midpoint rule now underestimates the area while the trapezoidal rule overestimates the area. Again, the trapezoidal estimate appears to be approximately twice as far off as the midpoint estimate.

This suggests that the value whose distance to the midpoint estimate is half as far as the distance to the trapezoidal estimate is a much better approximation of the actual area under the graph. This is precisely the motivation behind the approximation technique known as **Simpson's rule**, named after the English mathematician Thomas Simpson (1710-61).

The Simpson's rule approximation value can be computed by using a weighted average of the trapezoidal and midpoint estimates, namely

$$\text{Simpson's estimate} = \frac{1}{3}(\text{Trapezoidal estimate}) + \frac{2}{3}(\text{Midpoint estimate}).$$

EXAMPLE 38 Approximate $\int_{0.5}^{3.5} \sin^3(x)\, dx$ using Simpson's Rule for a partition of size $n = 6$.

Solution We have already determined the results of the midpoint rule and trapezoidal rules for this definite integral for a partition of size $n = 6$. The Simpson's rule approximation is

$$\frac{1.296}{3} + \frac{2(1.325)}{3} \approx 1.315.$$

Note that this approximation is accurate to three decimal places of the actual value of the definite integral. ∎

Here's another way of thinking about the trapezoidal rule and Simpson's Rule. When we subdivide the interval $[a, b]$ into n subintervals, we obtain the graph of a piecewise linear function by connecting the points $(x_i, f(x_i))$ corresponding to the endpoints of these subintervals. The trapezoidal rule is simply the approximation we obtain when we integrate this piecewise linear function instead of our original function f.

Now, instead of connecting the graph points at the ends of each subinterval with a straight line segment, suppose we find a *parabola* which passes through both of these points *and* the middle graph point. (Three noncollinear points determine a parabola just as two points determine a line; if the two endpoints and the midpoint line up, we can just connect them with a straight line instead of a parabola.) Figure 6.38 illustrates this idea for a given subinterval.

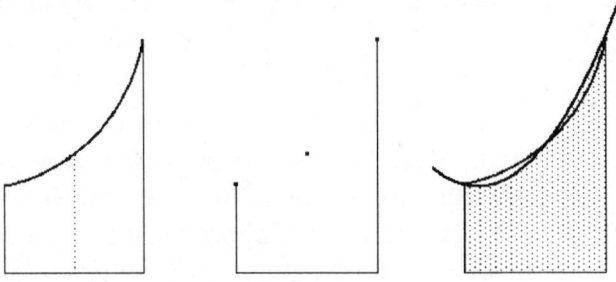

Figure 6.38 Approximating the graph with a parabola through three points.

6.5 NUMERICAL INTEGRATION TECHNIQUES

Notice how closely the parabola approximates the graph. If we connect these parabolic pieces end-to-end for each subinterval, we obtain the graph of a **piece-wise quadratic function**. Simpson's Rule is simply the approximation we obtain when we integrate this piece-wise quadratic function instead of our original function. Figure 6.39 shows the piece-wise quadratic fit for the example we computed earlier. The graph appears virtually identical to the original graph, and consequently, the values of the definite integrals agreed to three decimal places.

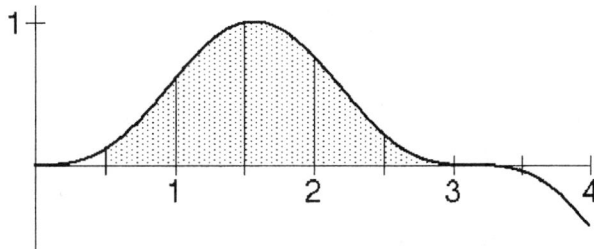

Figure 6.39 Simpson's approximation of $\int_{.5}^{3.5} \sin^3(x)\,dx$ for a partition of size $n = 6$.

On the other hand, Simpson's rule requires us to sample almost twice as many inputs as the trapezoidal rule for the same partition. The approximation could be improved even more by sampling more inputs and using a piece-wise cubic function, or by using the first derivative values at the endpoints to create a cubic spline approximation.

Summary of the numerical integration techniques

Let's summarize by introducing some convenient notation. Suppose we partition a closed interval $[a, b]$ into n equal-sized subintervals of length $\Delta x = (b-a)/n$. Let's label the function outputs at the n midpoints with odd subscripts:

$$y_1 = f(a + \frac{\Delta x}{2}), \quad y_3 = f(a + \frac{3\Delta x}{2}), \quad y_5 = f(a + \frac{5\Delta x}{2}), \quad \ldots$$

up to

$$y_{2n-1} = f(a + \frac{(2n-1)\Delta x}{2}) = f(b - \frac{\Delta x}{2}).$$

Now let's label the function outputs at the endpoints of the subintervals with even subscripts, starting with 0:

$$y_0 = f(a), \quad y_2 = f(a + \Delta x), \quad y_4 = f(a + 2\Delta x), \quad \ldots$$

up to

$$y_{2n} = f(a + n\Delta x) = f(b).$$

With this labeling scheme, we can write down an explicit formula for each of the rules discussed in this section.

Numerical approximations of $\int_a^b f(x)\,dx$ **for a partition of size** n

Left rectangle rule:

$$\int_a^b f(x)\,dx \approx L_n = \Delta x(y_0 + y_2 + \cdots + y_{2n-2}).$$

Right rectangle rule: $\int_a^b f(x)\,dx \approx R_n$, where

$$\int_a^b f(x)\,dx \approx R_n = \Delta x(y_2 + y_4 + \cdots + y_{2n}).$$

Midpoint rule:

$$\int_a^b f(x)\,dx \approx M_n = \Delta x(y_1 + y_3 + \cdots + y_{2n-1}).$$

Trapezoidal rule:

$$\int_a^b f(x)\,dx \approx T_n = \frac{L_n + R_n}{2} = \frac{\Delta x}{2}(y_0 + 2y_2 + 2y_4 + \cdots + 2y_{2n-2} + y_{2n}).$$

Simpson's rule:

$$\int_a^b f(x)\,dx \approx S_{2n} = \frac{2M_n + T_n}{3}.$$

The subscript $2n$ in the notation S_{2n} for Simpson's rule is to indicate that we have effectively subdivided our interval $[a,b]$ into $2n$ subintervals, since we make use of both the endpoints and midpoints of each of our original n subintervals. The explicit formula for Simpson's rule is

$$S_{2n} = \frac{\Delta x}{6}(y_0 + 4y_1 + 2y_2 + 4y_3 + 2y_4 + \cdots + 2y_{2n-2} + 4y_{2n-1} + y_{2n}).$$

Uses of numerical integration

Calculators and computers may use numerical integration techniques to calculate the values of definite integrals. When an antiderivative cannot be found so that the Second Fundamental Theorem can be used, there is no recourse but to use numerical techniques. Often, Simpson's rule or a closely related method is used, but there are other methods (you might encounter these in a more advanced course in numerical analysis).

As we mentioned earlier, when equally spaced inputs are sampled for a function having symmetry in its graph, the definite integral may not be approximated well by the techniques described here. One way to alleviate this problem is to sample irregularly spaced inputs. Another idea is to transform the function in some way that is likely to remove the symmetry, yet preserve the area under the graph.

We used the integral $\int_{0.5}^{3.5} \sin^3(x)\,dx$ as an example to illustrate the various techniques of numerical integration. However, one of the primary uses of numerical integration techniques is when the function has no known formula. Perhaps only a partial table of inputs and outputs is available (this is often the case when data is gathered from some experimental process) or perhaps we have a graphical representation of the function. Since the techniques described in this section only depend on a sampling of outputs from the function, we can just as easily use them when a function has only a numerical or graphical representation.

If we desire a certain level of accuracy in the value of a definite integral, how fine a partition should we use? In other words, how can we (or a machine) decide how large n should be ahead of time without knowing the actual value of the definite integral? (If we know the actual value of the definite integral, there is little point in approximating it.) In practice, a machine may simply follow these steps:

1. Compute the approximation of $\int_a^b f(x)\,dx$ for a starting value n.

2. Double the size of n, and compute the new approximation.

3. Repeat until three consecutive approximations are within the desired accuracy *of each other*.

Note that the criteria is how close together the approximations are, and not their proximity to an unknown target value. Of course, even if the approximations are close together, it is possible they all may be far away from the true value.

EXERCISES

Using a partition of size $n = 5, 10, 20,$ and 40, find the left, right, and midpoint estimates for the definite integrals in exercises 1-6.

1. $\int_0^1 \sqrt{1+x^2}\,dx$
2. $\int_{-1}^0 3^x\,dx$
3. $\int_{-1}^1 \arctan(x)\,dx$
4. $\int_1^4 \log_2(x)\,dx$
5. $\int_0^1 e^{-x^2}\,dx$
6. $\int_{0.5}^{3.5} \sin(x^3)\,dx$

Using a partition of size $n = 5, 10, 20,$ and 40, and the results from exercises 1-6, find the trapezoidal and Simpson's rule estimates for exercises 7-12. Calculate each integral with a machine to five decimal place accuracy and compare these values with the five estimates you have obtained.

7. $\int_0^1 \sqrt{1+x^2}\,dx$
8. $\int_{-1}^0 3^x\,dx$
9. $\int_{-1}^1 \arctan(x)\,dx$
10. $\int_1^4 \log_2(x)\,dx$
11. $\int_0^1 e^{-x^2}\,dx$
12. $\int_{0.5}^{3.5} \sin(x^3)\,dx$

The table of values below are the minute-by-minute speed readings of a car over the second 15 minutes of its trip. Use this table to answer exercises 13-14.

time t in minutes	speed V in mph	time t in minutes	speed V in mph
15	28	23	0
16	24	24	1
17	20	25	1
18	16	26	2
19	12	27	3
20	9	28	4
21	6	29	6
22	3	30	8

13. Estimate the distance covered over the interval $[15, 30]$ using $\Delta t = 1$ minutes and the trapezoidal rule.

14. Estimate the distance covered over the interval $[20, 30]$ using $\Delta t = 2$ minutes and Simpson's Rule.

6.5 NUMERICAL INTEGRATION TECHNIQUES

15. Using the table of values below for the continuous function f, estimate the definite integral of f over the entire interval using the trapezoidal rule.

x	f(x)
1.27319	3.81456
1.27320	3.86714
1.27321	3.90551
1.27322	3.92017
1.27323	4.34405
1.27324	4.66292
1.27325	4.69141
1.27326	4.65674
1.27327	4.61993
1.27328	4.59550
1.27329	4.58799
1.27330	4.52556

16. Use Simpson's rule to find 5 terms of a sequence which approximates

$$\int_0^1 \frac{4}{1+x^2}\, dx.$$

Let the number of subdivisions of $[0, 1]$ be $1, 2, 3, 4, 5$. What does the limit of this sequence appear to be? How many decimal places of accuracy does a_5 give? Using the trapezoidal approximation and 50 subdivisions of $[0, 1]$, approximate the same integral. How many decimal places of accuracy do you get?

Exercises 17-20 ask you to estimate

$$\int_0^4 x^2(x-1)(x-2)(x-3)(x-4)\, dx$$

using various approximation techniques.

17. Find the trapezoidal estimates for partitions of size $n = 16$ and $n = 32$.

18. Find the Simpson's rule estimates for partitions of size $n = 4$ and $n = 8$.

19. Expand the polynomial $x^2(x-1)(x-2)(x-3)(x-4)$ and integrate over $[0, 4]$ using the Second Fundamental Theorem of Calculus.

20. Which estimate is better, the trapezoidal estimate for $n = 32$ or Simpson's rule for $n = 8$?

In exercises 21-28, you are given the graph of a function f against a grid. Assuming that the grid lines are spaced 1 unit apart both horizontally and vertically, estimate the value of

$$\int_1^5 f(x)\,dx$$

by using $\Delta x = 1$ unit and the trapezoidal rule.

21.

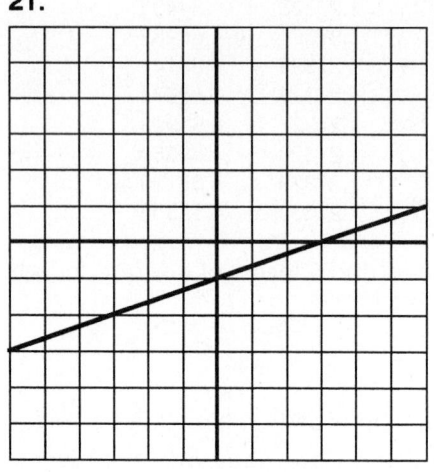

22.

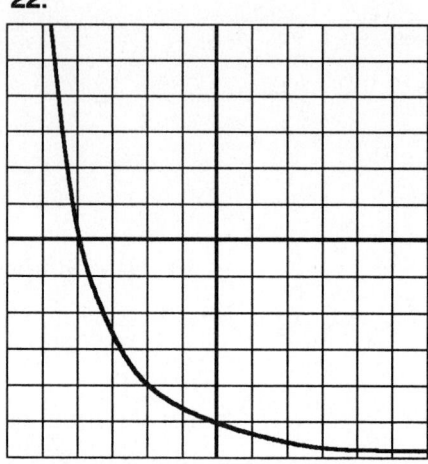

23.

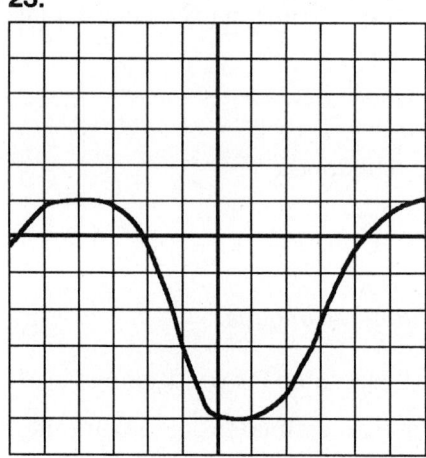

24.

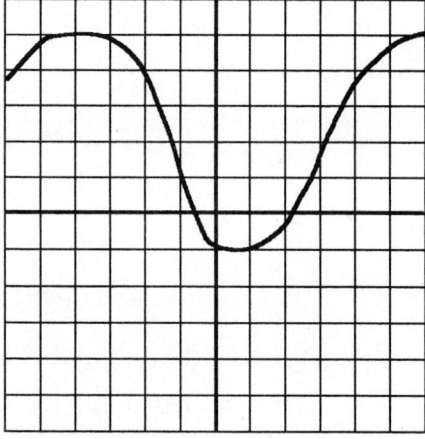

6.5 NUMERICAL INTEGRATION TECHNIQUES

25.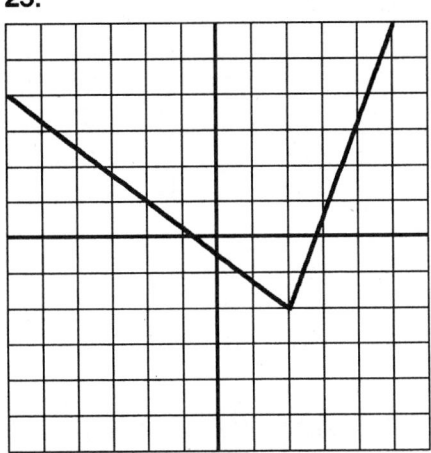

26.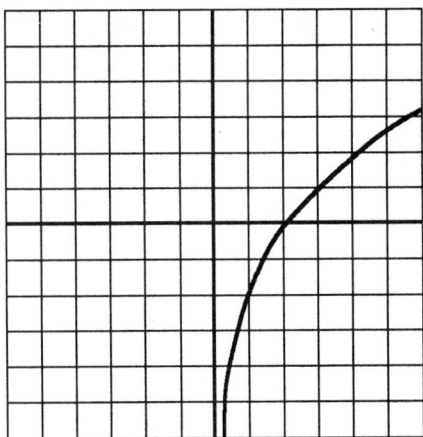

27.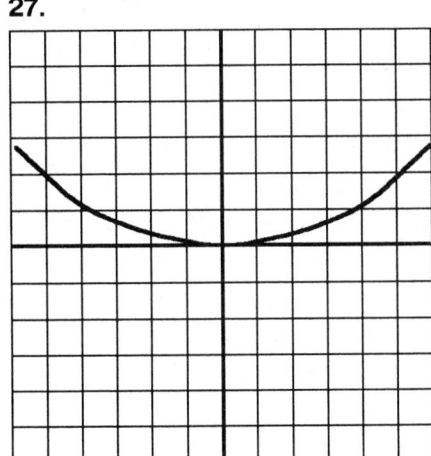

28.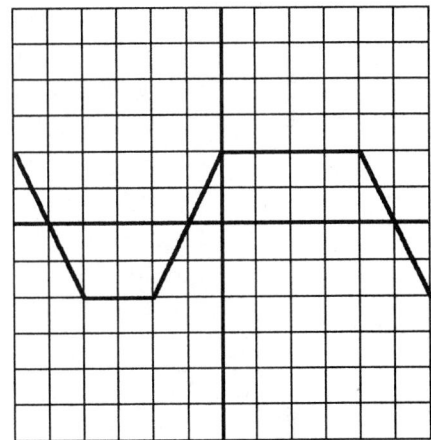

29. For which of the graphs in exercises 21-28 is the trapezoidal estimate exactly correct? Why?

---

30. Suppose three points (x_i, y_i), $(x_i + \Delta x/2, y_{i+1})$ and $(x_i + \Delta x, y_{i+2})$ are given. Find a quadratic $g(x) = ax^2 + bx + c$ that contains all three points and find $\int_{x_i}^{x_i + \Delta x} g(x)\,dx$ in terms of Δx, y_i, y_{i+1}, and y_{i+2}.

31. Using the results of the previous exercise, derive the Simpson's rule estimate for S_{2n} where $y_0, \ldots, y_{2n}$ are the outputs sampled from inputs evenly spaced at intervals of length $\Delta x/2$.

CHAPTER 7

Differential Equations

Functions built from addition, subtraction, multiplication, division, roots, and powers by means of composition are called algebraic functions. Although almost all of the functions we have used up to this point have been algebraic, many important functions are not algebraic and cannot be expressed as a composition of these simple functions. These are called **transcendental** functions. All the transcendental functions we will consider are only one step removed from algebraic functions. Instead of being determined by an algebraic relationship between the input and the output, these functions are determined by an algebraic relationship between the function's input, its output, *and its slope*. Examples of such a relationship include

$$y'(x) = x^2$$
$$y'(x) = (y(x))^2$$
$$y'(x) = y(x) + x.$$

These are called **differential equations**.

In the first section we will investigate a few kinds of differential equations and symbolic, numeric, and graphical techniques for solving them. Then we will turn to some of the most important differential equations that define the exponential and logarithmic functions. These differential equations arise quite naturally, and they govern such diverse phenomena as population growth and radioactive decay. The chapter concludes by examining some of the symbolic techniques available for finding antiderivatives.

7.1 WHAT IS A DIFFERENTIAL EQUATION?

The word *differential* in differential equation is used to indicate a relationship between a function's input x, output y, and slope y'. A relationship between inputs and outputs, such as $y + x^2 = 0$, can sometimes determine a function, in this case $y : x \longmapsto -x^2$. Similarly, a relationship between the input, the output, and the slope may also determine a function.

Determining a function from the differential equation is called *solving* the differential equation. Let's review through a simple example some of the tools we have for solving differential equations.

EXAMPLE 1 Find a function $y : \ x \longmapsto y(x)$ whose derivative $y' : \ x \longmapsto y'(x)$ satisfies the differential equation

$$y'(x) = x^2$$

and also satisfies the **initial condition**

$$y(0) = 2.$$

Solution If we can simply recognize a solution to the differential equation we are nearly done. In this simple case, the most general solution to the differential equation is the antiderivative of x^2, namely

$$y(x) = \frac{x^3}{3} + C$$

where C is an arbitrary constant. The constant is not arbitrary if we wish to satisfy the initial condition

$$y(0) = \frac{0^3}{3} + C = 2$$

so that $C = 2$ and the function sought is

$$y : x \longmapsto \frac{x^3}{3} + 2.$$

Suppose, on the other hand, that we didn't immediately recognize the general solution to the differential equation, or perhaps that the solution could not be written as a "closed form" formula. Would we be stuck? Absolutely not. There are still graphical and numeric methods, and even additional symbolic methods for obtaining more information about the solution of the differential equation.

Using slope fields

It is always a good idea to graph the slope field corresponding to the differential equation in order to get some feeling for what the solution's graph should look like. In the case of our differential equation $y'(x) = x^2$, the slope y' depends only on the input x. In Figure 7.1, we've made up a table of slope vs. input for selected inputs and then plotted line segments with the given slope over the given input. Again, since the slope, y', depends only on x and not explicitly on y, we can plot parallel copies of the line segment along vertical lines.

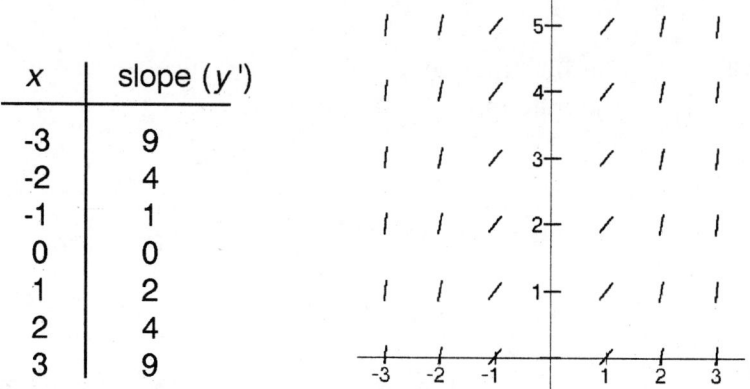

x	slope (y')
-3	9
-2	4
-1	1
0	0
1	2
2	4
3	9

Figure 7.1 Table of slope vs. input and graph of the slope field for $y'(x) = x^2$.

The graph of any solution to the differential equation must be tangent to these line segments and we can use them as a guide for graphing solutions to the differential equation.

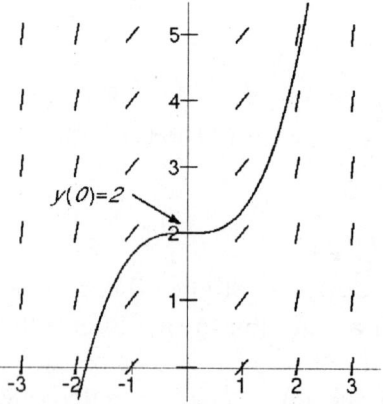

Figure 7.2 Graphing the solution of $y'(x) = x^2$ with initial condition $y(0) = 2$.

7.1 WHAT IS A DIFFERENTIAL EQUATION?

The initial condition $y(0) = 2$ forces our graph to pass through the point $(0, 2)$ as shown in Figure 7.2. Starting from this point we can sketch the graph to the right and to the left by following the "flow" of the slope field. Notice that the resulting graph is exactly what you would expect the graph of our explicit solution $y(x) = \dfrac{x^3}{3} + 2$ to look like.

Using the First Fundamental Theorem of Calculus

The First Fundamental Theorem of Calculus tells us that if we know the derivative $y'(x)$ of a function in some region (and y' is continuous) then the difference between the function values, $y(x_1) - y(x_0)$, at any two points x_0 and x_1 in the region can be computed as the (signed) area under the graph of y' between x_0 and x_1. In other words

$$y(x_1) - y(x_0) = \int_{x_0}^{x_1} y'(t)\,dt$$

or

$$y(x_1) = y(x_0) + \int_{x_0}^{x_1} y'(t)\,dt.$$

By the differential equation, $y'(x) = x^2$, and by the initial condition $y(0) = 2$ so that by the fundamental theorem

$$y: \quad x \mapsto y(0) + \int_0^x y'(t)\,dt = 2 + \int_0^x t^2\,dt = 2 + \left.\frac{x^3}{3}\right]_{t=0}^{t=x} = 2 + \frac{x^3}{3}.$$

Even if we got no further symbolically than setting up the integral

$$y(x) = 2 + \int_0^x t^2\,dt,$$

we could use Riemann sums or other numerical integration techniques to get approximate values for $y(x)$ for any desired x. ∎

In this example, we were able to find a closed form formula for the solution of the differential equation. While all continuous functions have antiderivatives (by the First Fundamental Theorem of Calculus), many quite simple functions have antiderivatives that cannot be written in algebraic form.

EXAMPLE 2 Use the First Fundamental Theorem of Calculus to find the solution of the differential equation

$$y' = \sqrt{\sqrt{x^2+1}+1}$$

with initial condition

$$y(1) = 5$$

and estimate $y(2)$ to two significant digits.

Solution In this case, we don't recognize $x \mapsto \sqrt{\sqrt{x^2+1}+1}$ as the derivative of anything we know (in fact, it is not the derivative of anything that can be written in algebraic form). Even so, since we are given the derivative $y'(x)$ as a function of x and this derivative is continuous, the Fundamental Theorem of Calculus tells us that

$$y(x) = y(1) + \int_1^x \sqrt{\sqrt{t^2+1}+1}\,dt = 5 + \int_1^x \sqrt{\sqrt{t^2+1}+1}\,dt.$$

Using five-point upper and lower Riemann sums to estimate the integral, we find that

$$y(2) = 5 + \int_1^2 \sqrt{\sqrt{x^2+1}+1}\,dx$$

is between 6.6503 and 6.6994, which certainly gives an estimate to two significant digits. ∎

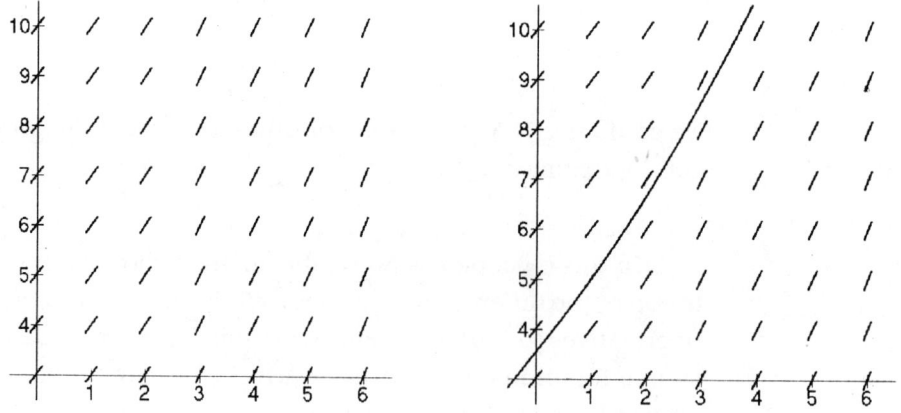

Figure 7.3 Slope field for $y' = \sqrt{\sqrt{x^2+1}+1}$ and solution graph

7.1 WHAT IS A DIFFERENTIAL EQUATION?

We can check the reasonableness of our answer by graphing the slope field of this differential equation as shown in Figure 7.3 and sketching the graph of the solution starting at the given initial condition. This equation defines $y(x)$ and and we can use machine numeric integration to graph it as shown in Figure 7.3. All of these reinforce the reasonableness of our computed answer.

The function

$$y : x \longmapsto 5 + \int_1^x \sqrt{\sqrt{t^2+1}+1}\, dt$$

is an example of a transcendental function. As can be seen from its graph, there is nothing very special about this function, and it looks quite ordinary. We know its derivative explicitly, we can use it as a building block in constructing other functions even though we don't have its definition in simple algebraic form.

We've seen other examples of differential equations arise from implicit differentiation. If we start with the equation of a circle of radius 3 centered at the origin, $x^2 + y^2 = 9$, implicit differentiation leads to the differential equation

$$2x + 2yy' = 0$$

or, solving algebraically for y' assuming $y \neq 0$

$$y' = -\frac{x}{y}.$$

Suppose now that we started with this differential equation and a single point on the circle (our initial condition). How could we work backward to recover the circle?

EXAMPLE 3 Sketch the slope field corresponding to the differential equation

$$y' = -\frac{x}{y}$$

in the region $[-3, 3] \times [-3, 3]$. Sketch both the solution with initial condition $y(0) = 3$ and the solution with initial condition $y(0) = -3$ in this same region.

Solution   In this case the slope depends not only on x but also on y. We will need to compute the slope y' at a grid of points in the x-y plane. Choosing points with integer coordinates, and computing the slope at each (for example $y' = -(-2/3) = 2/3$ at the point $(-2, 3)$ gives the slope field pictured in Figure 7.4.

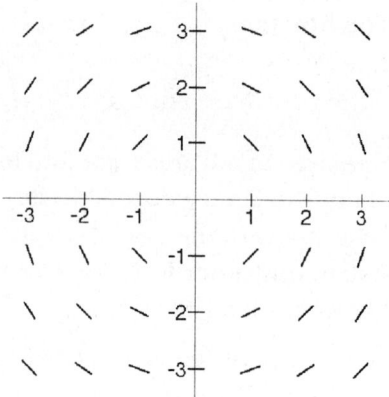

Figure 7.4   Slope field corresponding to $y' = -x/y$.

Sketching the solution of the differential equation with initial condition $y(0) = 3$ and then with initial condition $y(0) = -3$ results in the curves shown in Figure 7.5.

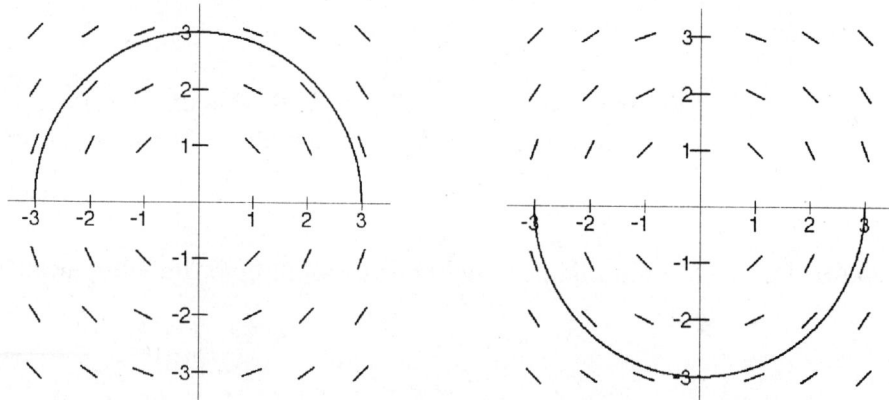

Figure 7.5   Solutions of $y' = -x/y$ with $y(0) = 3$ and with $y(0) = -3$.

These are two halves of the circle. The reason we didn't get the whole circle with either initial condition is that in forming the differential equation we divided by y, assuming $y \neq 0$. The slope field is therefore undefined along the x-axis. If we joined the two solutions together we have a graph that doesn't pass the vertical line test for function graphs. ∎

7.1 WHAT IS A DIFFERENTIAL EQUATION?

In this case we started out with the the equation of a circle and produced a differential equation. This can give us some clues for solving other differential equations. Thinking back to the process of implicit differentiation, if we see (or can arrange to see) a pattern of the form

$$y^n y' = f(x)$$

this should make us think of implicitly differentiating y^{n+1} to get $(n+1)y^n y'$.

EXAMPLE 4 Solve the differential equation

$$y' = \frac{x}{y^2}$$

with initial condition

$$y(0) = 2.$$

Solution By multiplying both sides by y^2 (we will need to assume that $y \neq 0$) we obtain

$$y^2 y' = x.$$

The pattern $y^2 y'$ suggests that we think of this as $\frac{1}{3}(y^3)'$ so that the differential equation can be written as

$$\frac{1}{3}(y^3)' = x$$

or

$$(y^3)' = 3x.$$

Now we can recognize that

$$y^3 = \frac{3}{2}x^2 + C$$

for some constant C, and use the initial condition to find C. Since $y(0) = 2$, and

$$(y(0))^3 = \frac{3}{2}0^2 + C,$$

we have $C = 8$ and

$$y(x) = \sqrt[3]{\frac{3}{2}x^2 + 8}.$$

Note that y is never zero, which accords with our assumption. ■

Again, we don't actually need to be able to find the integral in closed form to "find" the solution of the differential equation. Using slope fields we can sketch the graph of the solution and we can use Riemann sums or other numerical integration techniques to approximate the value for any given input.

EXAMPLE 5 Sketch the slope field of the differential equation

$$y' = \frac{\sqrt{\sqrt{x^2+1}+1}}{y^2}$$

in the region $[0,5] \times [0,3]$ using integer grid points and sketch the solution with initial condition $y(1) = 2$ in the same region. Also, use the Fundamental Theorem of Calculus to express the solution as an integral.

Solution The slope field and solution for the initial condition $y(1) = 2$ are shown in Figure 7.6. Note that the slopes are undefined when $y = 0$, so that we can omit these points, and the solution graph doesn't hit the x-axis in this region.

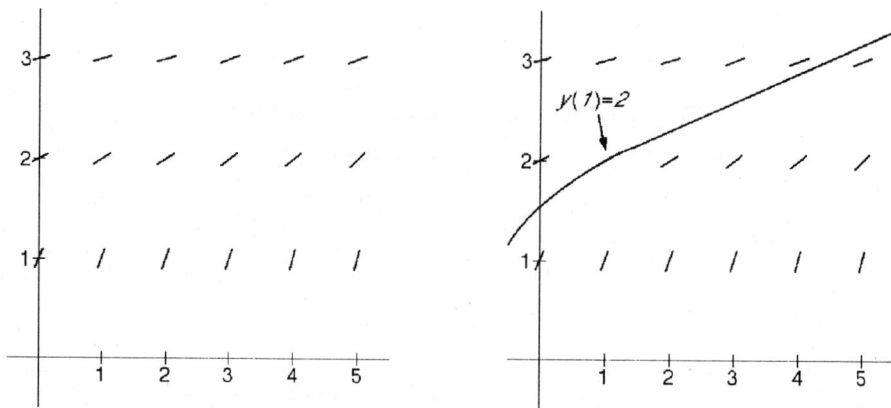

Figure 7.6 Slope field for $y' = \dfrac{\sqrt{\sqrt{x^2+1}+1}}{y^2}$ and solution graph with initial condition $y(1) = 2$.

Bringing y^2 to the other side of the equation (assuming $y \neq 0$), we obtain

$$y^2 y' = \sqrt{\sqrt{x^2+1}+1} \qquad \text{so that} \qquad \frac{1}{3}(y^3)' = \sqrt{\sqrt{x^2+1}+1}.$$

7.1 WHAT IS A DIFFERENTIAL EQUATION?

Using the First Fundamental Theorem of Calculus, we obtain

$$y^3 = (y(1))^3 + \int_1^x \sqrt{\sqrt{t^2+1}+1}\, dt \quad \text{or} \quad y = \sqrt[3]{8 + \int_1^x \sqrt{\sqrt{t^2+1}+1}\, dt}.$$

Since the integrand is always positive, $y \neq 0$ for $x \geq 1$. For $x < 1$, y will stay positive until the integral value reaches -8. At this point, our formula will no longer be valid. (It appears from the sketch that y will reach 0 somewhere between $x = -1$ and $x = 0$.) ∎

EXERCISES

Use the First Fundamental Theorem of Calculus to solve the differential equations with the given initial conditions in exercises 1-5.

1. $y' = 2\sqrt{x}$; $\quad y(2) = 7$

2. $y' = 1 + x$; $\quad y(0) = 0$

3. $y' = ((x+1)^2 + 1)^2$; $\quad y(-1) = 5$

4. $y' = \sin(\sin(x))$; $\quad y(0) = 0$

5. $y' = \dfrac{1}{x^2+1}$; $\quad y(1) = 4$

For exercises 6-11, sketch the slope field of each of the differential equations in the region $[-3, 3] \times [0, 3]$ using integer grid points. Sketch the graph of the differential equation with the given initial condition in the same region.

6. $y' = 0.2\sqrt{x+5}$; $\quad y(2) = 2.5$

7. $y' = 1 + x$; $\quad y(0) = 1$

8. $y' = 0.1((x+1)^2 + 1)^2$; $\quad y(-1) = 1$

9. $y' = \sin(\sin(x))$; $\quad y(0) = 0$

10. $y' = \dfrac{1}{x^2+1}$; $\quad y(1) = 1$

11. $y' = x - y$; $\quad y(-1) = 2$

In exercises 12-17, use the fact that $(y^{n+1})' = (n+1)y^n y'$ to solve the given differential equation with the given initial condition.

12. $y^2 y' = 1;$ $\quad y(0) = 4$

13. $y' = -x^2 y^2;$ $\quad y(4) = 4$

14. $y' = \dfrac{1+x}{\sqrt{y}};$ $\quad y(2) = 9$

15. $y' = \dfrac{y^3}{x^2};$ $\quad y(3) = 2$

16. $y' + yy' = x;$ $\quad y(0) = 1$

17. $y' = \dfrac{1}{y^3 + y};$ $\quad y(1) = 1$

In each of exercises 18-22, sketch the slope field on an integer grid of the the indicated differential equation. Use a viewing window of $[-4, 4] \times [-2, 2]$. Then sketch the solution of the differential equation with the indicated initial condition.

18. $y' = \dfrac{1}{1+x^2};\ y(0) = 0.$

19. $y' = \dfrac{1}{\sqrt{1-x^2}};\ y(0) = 0.52360.$

20. $y' = \dfrac{-1}{1+x^2};\ y(0) = 1.5708.$

21. $y' = \dfrac{-1}{x\sqrt{x^2-1}};\ y(2) = 1.04720.$

22. $y' = \dfrac{-1}{x\sqrt{x^2-1}};\ y(-2) = 2.09440.$

7.2 EXPONENTIAL AND LOGARITHMIC FUNCTIONS

The most important transcendental functions can be thought of as solutions to particularly simple differential equations. In the last section, we saw that slope fields can often provide us with a graphical means of estimating the solution of a differential equation. Also, the First Fundamental Theorem of Calculus can provide us with a solution to many differential equations by simply setting up the appropriate area function. We'll use

7.2 EXPONENTIAL AND LOGARITHMIC FUNCTIONS

these tools in this section as we take a closer look at exponential and logarithmic functions. Even though we assume that you have some familiarity with these functions, some of their most important properties can only be fully appreciated in the context of calculus and differential equations.

Exponential functions

By an exponential function, we mean a function f of the form

$$f : x \longmapsto a^x$$

where a is a *fixed, positive real number*. The number a is called the **base** for the exponential function f, and it will be convenient to stipulate $a \neq 1$ to exclude the constant function $x \longmapsto 1$ from being called an exponential function. The value $a = 0$ is excluded for a similar reason, and *negative* values a are excluded so that we avoid some particularly sticky difficulties in evaluation.

Some examples of exponential functions are

$$x \longmapsto 2^x \qquad x \longmapsto 3^x \qquad x \longmapsto (2/3)^x \qquad x \longmapsto (1/4)^x \qquad x \longmapsto \pi^x.$$

When the input x is a *rational number* (in other words, $x = p/q$ where p and q are integers and $q \neq 0$), then we can evaluate a^x as

$$a^{p/q} = \sqrt[q]{a^p} = (\sqrt[q]{a})^p$$

since the base a is positive. But what if x is *not* a rational number? For instance, exactly how do we interpret or evaluate something like

$$3^{\sqrt{2}} \qquad \text{or} \qquad (1/2)^\pi.$$

One approach is to simply define these numbers in terms of limits. We might write

$$3^{\sqrt{2}} = \lim_{p/q \to \sqrt{2}} 3^{p/q}$$

to mean that $3^{\sqrt{2}}$ is the limit of the rational powers $3^{p/q}$ as the rational numbers p/q approach $\sqrt{2}$, as long as such a limiting value exists. In fact, these limits do exist for any positive base a and any real number x. Indeed, with this interpretation of evaluation, exponential functions are continuous over the entire set of real numbers. Four examples of exponential functions and their graphs are shown in Figure 7.7 and 7.8.

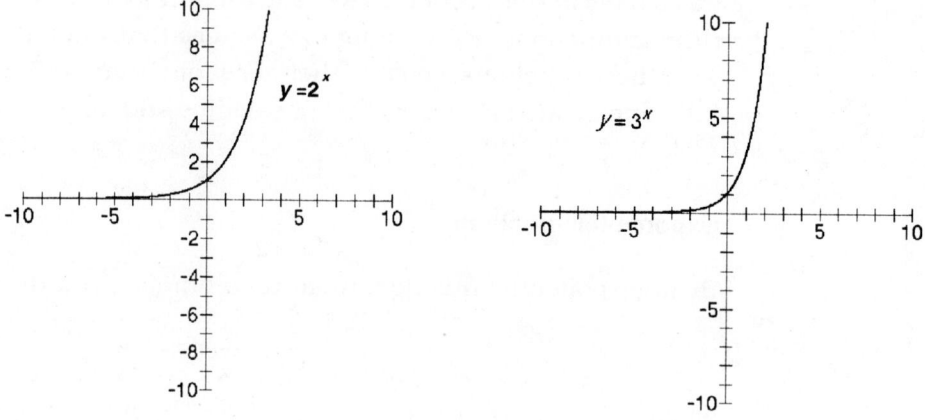

Figure 7.7 Graphs of exponential functions $y = 2^x$ and $y = (3)^x$

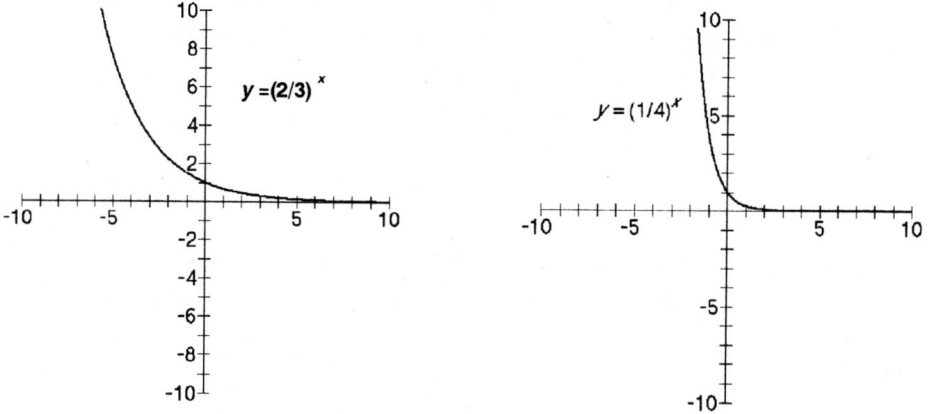

Figure 7.8 Graphs of exponential functions $y = (2/3)^x$ and $y = (1/4)^x$

Note that the function $x \longmapsto a^x$ is *strictly increasing* when $a > 1$, and is *strictly decreasing* when $0 < a < 1$.

Some familiar properties of exponents become interesting function properties for exponential functions $f : x \longmapsto a^x$. If c is any constant, then

$$f(x + c) = a^{x+c} = a^x a^c = f(x)f(c)$$

and

$$f(cx) = a^{cx} = (a^x)^c = (f(x))^c.$$

We also note that $f(0) = 1$ for any $a > 0$.

7.2 EXPONENTIAL AND LOGARITHMIC FUNCTIONS

The differential equations $y' = ky$

If we are faced with a differential equation along with an initial value condition

$$y' = f(x), \quad y(a) = b$$

where f is continuous over an interval containing $x = a$, then our problem essentially requires us to antidifferentiate the function f and use the initial condition to pin down a specific value of the arbitrary constant. Even if we do not recognize a "nice" formula for $y = F(x)$ satisfying these requirements, we can always use the First Fundamental Theorem of Calculus to write down a solution of the form

$$y = F(x) = \int_a^x f(t)\,dt + b$$

satisfying our requirements over the same interval.

On the other hand, if we have a more general differential equation of the form

$$y' = \text{expression in terms of } x \text{ and } y, \quad y(a) = b$$

then we may not be able to solve it even with the help of the First Fundamental Theorem of Calculus. However, we can plot a slope field using this expression and use it to visualize the graph of the solution function y.

Perhaps the simplest such differential equations are of the form

$$y' = ky, \quad y(0) = A,$$

where k and A are given constants. Remarkably, such differential equations arise very naturally in applications as diverse as population growth, heat transfer, and radioactive decay.

If we interpret the independent variable to represent time t, then we can translate this differential equation to say that y represents some quantity whose rate of growth $y'(t)$ at any time t is directly proportional to its value $y(t)$ at that same time. The initial condition $y(0) = A$ under this interpretation simply tells us the amount of the quantity at time $t = 0$ is A. To solve this differential equation, we seek a function $y = F(x)$ whose derivative is a constant multiple k of the original function.

Let's use a slope field to visualize the solution to a special case of this equation. Suppose we start out with one unit of material ($A = 1$), and the constant of proportionality is $k = 1$, so that we have

$$y' = y, \quad y(0) = 1.$$

The slope field generated by this differential equation consists of tangent line segments arranged in *horizontal* rows, since the slopes at each point should be the same as the y-value of that point. A sketch of the slope field is shown in Figure 7.9.

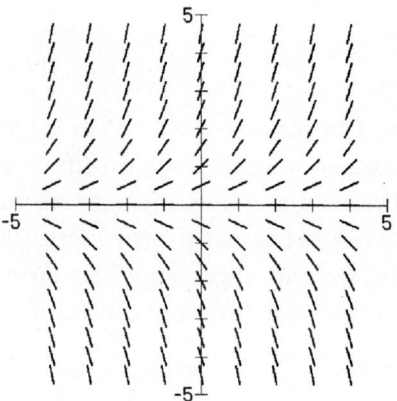

Figure 7.9 The slope field generated by $y' = y$.

For example, the row of parallel tangent line segments along $y = 1$ all have slope 1 and the row of parallel tangent line segments along $y = -2$ all have slope -2. The initial condition $y(0) = 1$ specifies that the point $(0, 1)$ is on the graph of the solution. Using this and following the directions indicated by the slope field, we obtain a graph like that in Figure 7.10.

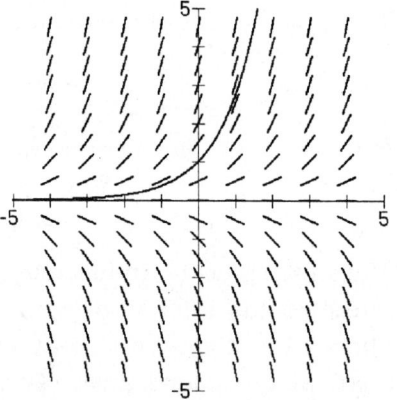

Figure 7.10 The solution graph for $y' = y$ with initial condition $y(0) = 1$.

This looks remarkably like the graph of an exponential function $y = a^x$. In fact, if we superimpose the graphs of $y = 2^x$ and $y = 3^x$ with the graph obtained from this slope field and initial condition, it suggests that we might find some real number $2 < a < 3$ that gives a very good fit of $y = a^x$ to the slope field (see Figure 7.11).

7.2 EXPONENTIAL AND LOGARITHMIC FUNCTIONS

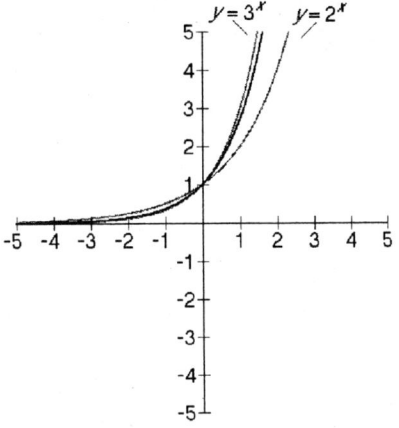

Figure 7.11 Comparing the solution curve to $y = 2^x$ and $y = 3^x$.

Let's investigate this further by taking a closer look at the derivative of exponential functions in general. Writing the derivative of $x \longmapsto a^x$ as a limit of difference quotients, and using the property $a^{x+h} = a^x \cdot a^h$, we have

$$\frac{d}{dx}(a^x) = \lim_{h \to 0} \frac{a^{x+h} - a^x}{h} = \lim_{h \to 0} \frac{a^x a^h - a^x}{h} = a^x \lim_{h \to 0} \frac{a^h - 1}{h} = ka^x,$$

provided $k = \lim_{h \to 0} \frac{a^h - 1}{h}$ exists.

EXAMPLE 6 Numerically investigate the growth rates of $x \longmapsto 2^x$ and $x \longmapsto 3^x$.

Solution Table 7.1 suggests that $k \approx 0.693$ for 2^x and $k \approx 1.099$ for 3^x.

h	$\frac{2^h - 1}{h}$	$\frac{3^h - 1}{h}$
0.1	0.71773	1.16123
0.01	0.69556	1.10467
0.001	0.69339	1.09922
0.0001	0.69317	1.09867
-0.1	0.66967	1.04042
-0.01	0.69075	1.09260
-0.001	0.69290	1.09801
-0.0001	0.69312	1.09855

Table 7.1 Table of values of $\frac{2^h - 1}{h}$ and $\frac{3^h - 1}{h}$ for small h.

Thus, $\frac{d}{dx}(2^x) \approx (0.693)2^x$ and $\frac{d}{dx}(3^x) \approx (1.099)3^x$. ∎

Since the growth rate of $x \longmapsto 2^x$ is less than 1 and that of $x \longmapsto 3^x$ is greater than 1, we are lead to believe that there is a number $2 < e < 3$ such that $x \longmapsto e^x$ has growth rate exactly $k = 1$. The function $y : x \longmapsto e^x$ will satisfy our differential equation and initial condition, since

$$y'(x) = \frac{d}{dx}(e^x) = 1 \cdot e^x = y(x) \quad \text{and} \quad y(0) = e^0 = 1$$

This number, e, does exist and has value

$$e \approx 2.718281828459045.$$

 The function $x \longmapsto e^x$ has a remarkable property—it is equal to its own derivative!

This function is called the **natural exponential function**. The notation $\exp$ is sometimes use for the natural exponential function. That is,

$$\exp(x) = e^x.$$

One advantage of the $\exp(x)$ notation is that it reminds us that the exponent x in $x \longmapsto e^x$ represents the input to the function. (The base e is a *constant*.).

Using the chain rule, we note that

$$\frac{d}{dx}e^{f(x)} = e^{f(x)} \cdot f'(x)$$

$$\frac{d}{dx}\exp(f(x)) = \exp(f(x)) \cdot f'(x).$$

EXAMPLE 7 Calculate $\frac{d}{dx}(e^{2x})$, $\frac{d}{dx}(\exp(x^2 + 1))$, and $\frac{d}{dx}((\exp(x))^2)$.

Solution

$$\frac{d}{dx}(e^{2x}) = e^{2x}\frac{d}{dx}(2x) = 2e^{2x}$$

$$\frac{d}{dx}(\exp(x^2 + 1)) = \exp(x^2 + 1)\frac{d}{dx}(x^2 + 1) = 2x\exp(x^2 + 1)$$

$$\frac{d}{dx}((\exp(x))^2) = 2\exp(x)\frac{d}{dx}(\exp(x)) = 2(\exp(x))^2.$$

■

From this example we can see that $\frac{d}{dx}(e^{kx}) = ke^{kx}$ for any k, so that $x \longmapsto e^{kx}$ is a solution to the differential equation

$$y' = ky.$$

7.2 EXPONENTIAL AND LOGARITHMIC FUNCTIONS

If we multiply this function by any constant A the resulting function will also be a solution:

$$\frac{d}{dx}(Ae^{kx}) = k(Ae^{kx}).$$

Since $e^0 = 1$, we can choose A to satisfy the initial condition $y(0) = A$. We can collect these observations as a theorem.

Theorem 7.1 The function $y : x \longmapsto Ae^{kx}$ is a solution to the differential equation $y' = ky$ with initial condition $y(0) = A$.

□

The natural logarithm

We have just seen that the natural exponential function $x \longmapsto e^x$ is the solution to a very special differential equation $y' = y$ and satisfies the initial condition $y(0) = 1$. Likewise, the **natural logarithmic function** can be thought of as the solution to a special differential equation

$$y'(x) = \frac{1}{x}$$

satisfying the initial condition

$$y(1) = 0.$$

The First Fundamental Theorem of Calculus provides us with a solution directly, namely

$$y(x) = \int_1^x \frac{1}{t}\, dt.$$

This solution holds at least for $x > 0$, since $x \longmapsto 1/x$ is not continuous at $x = 0$. While the differential equation $y'(x) = 1/x = x^{-1}$ looks simple enough, it is the one exception to the general antiderivative formula for powers:

$$\int x^r\, dx = \frac{x^{r+1}}{r+1} + C$$

for any power r except $r = -1$. There simply is no algebraic closed form formula for the antiderivative of $1/x$. Hence, the solution y represents a transcendental function that we'll call the **natural logarithm** and denote by ln. In other words, by definition,

$$\ln(x) = \int_1^x \frac{1}{t}\, dt.$$

Geometrically, the value $\ln(x)$ corresponds to the area under the graph of $y = 1/t$ between $t = 1$ and $t = x$ (see Figure 7.12). We can make several observations regarding the values of $\ln(x)$ directly from this picture.

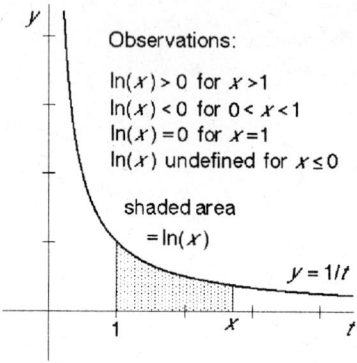

Figure 7.12 Interpreting $\ln(x)$ as an area function.

To actually graph $y = \ln(x)$, we can tabulate several of these area values for selected positive values x and then plot the resulting ordered pairs, or we can appeal to the slope field generated by the original differential equation $y'(x) = 1/x$ shown in Figure 7.13 and use the initial condition $y(0) = 1$.

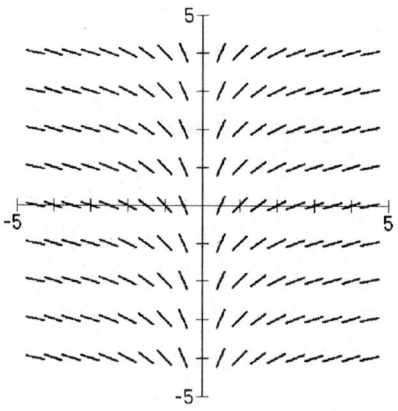

Figure 7.13 Slope field for $y'(x) = 1/x$.

Either way, we will obtain a graph like that shown in Figure 7.14.

7.2 EXPONENTIAL AND LOGARITHMIC FUNCTIONS

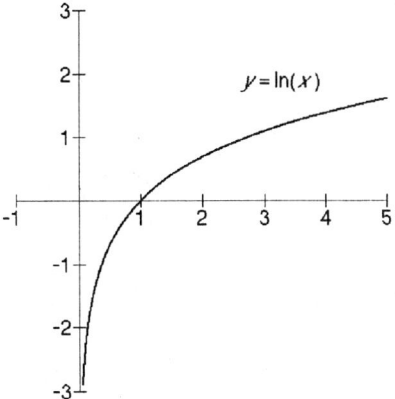

Figure 7.14 Graph of the natural logarithm function.

By its definition,

$$\frac{d}{dx}(\ln(x)) = \frac{1}{x}.$$

Therefore, the chain rule tells us that

$$\frac{d}{dx}(\ln(f(x))) = \frac{f'(x)}{f(x)}$$

for any differentiable function f.

EXAMPLE 8 $\quad \dfrac{d}{dx}(\ln(\arctan(x))) = \dfrac{1}{(1+x^2)\arctan(x)}.$ ∎

Logarithmic functions $x \longmapsto \log_a(x)$

We are used to thinking of a **logarithmic function** $x \longmapsto \log_a(x)$ as the inverse of the corresponding exponential function $x \longmapsto a^x$. That is, the output value

$$y = \log_a(x)$$

is determined to be the unique value y satisfying the exponential equation

$$a^y = x.$$

By this definition, logarithmic functions reverse the exponentiation process and vice-versa, but we should note that

$$\log_a a^x = x$$

for all real numbers x, but

$$a^{\log_a x} = x$$

only for positive values x.

EXAMPLE 9 Find $\log_2 2^7$ and $\log_2 4^3$.

Solution $\log_2 2^7 = 7$. Since $4^3 = (2^2)^3 = 2^6$, we have $\log_2 4^3 = \log_2 2^6 = 6$. ∎

The graph of an exponential function $y = a^x$ and the corresponding logarithmic function $y = \log_a(x)$ look like mirror images of each other through the line $y = x$. Figure 7.15 illustrates this symmetry property between the graphs of logarithmic functions and exponential functions.

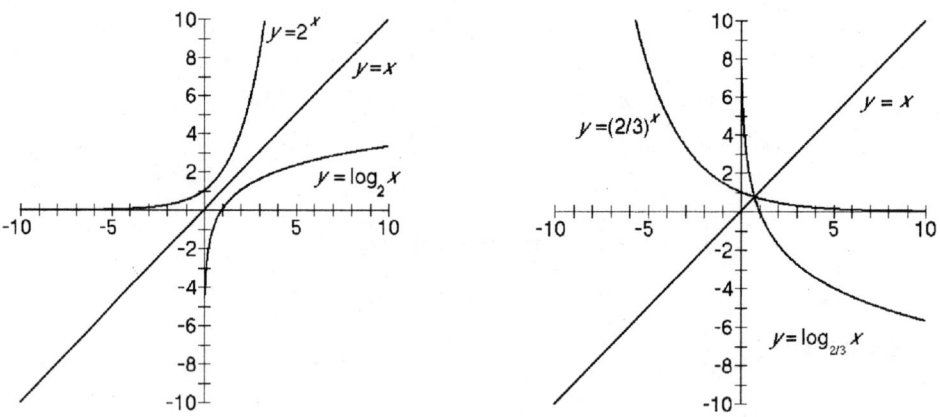

Figure 7.15 Graphs of exponential and logarithmic functions.

From the multiplicative property of exponentials we have

$$\log_a(b \cdot c) = \log_a b + \log_a c$$

for positive b and c, and from the power property

$$\log_a(b^c) = c \log_a b$$

for positive b.

Now, the natural logarithmic function $x \longmapsto \ln(x)$ seems to bear little resemblance to these other logarithmic functions until we examine its graph (Figure 7.14). Let's examine closely the slope fields generated by the differential equations for the natural exponential and the natural logarithm functions (Figures 7.9 and 7.13).

7.2 EXPONENTIAL AND LOGARITHMIC FUNCTIONS

Take any *row* of parallel line segments from the slope field for $y'(x) = y(x)$ and reflect it in the line $y = x$. The resulting *column* of parallel line segments belongs to the slope field for $y'(x) = 1/x$. Moreover, the initial conditions are reversed: $y = 1$ when $x = 0$ for the natural exponential, and $x = 1$ when $y = 0$ for the natural logarithm. Taken together, this means the graphs of the natural exponential and natural logarithm functions are reflections of each other in the line $y = x$, and these functions are inverses of each other!

In other words,

$$\ln(x) = \log_e(x).$$

Thus, we have

$$\ln(\exp(x)) = \ln(e^x) = x$$

for all x, and

$$\exp(\ln(x)) = e^{\ln(x)} = x$$

for positive x.

In fact, we can express any exponential or logarithmic function in terms of the natural exponential and logarithmic functions. First, we can see that

$$a^x = (e^{\ln a})^x = e^{x \ln a}.$$

Using this relationship between exponential functions, we can write

$$\ln x = \ln(a^{\log_a x}) = \ln(e^{\ln a \log_a x}) = \ln a \log_a x$$

so that

$$\log_a(x) = \frac{\ln(x)}{\ln a}.$$

Derivatives and the exponential and logarithm functions

These relationships give a simple method of determining the derivative of any exponential or logarithmic function.

EXAMPLE 10 Find $\dfrac{d}{dx}a^x$.

Solution

$$\frac{d}{dx}a^x = \frac{d}{dx}e^{x\ln a} = \ln(a)\cdot e^{x\ln a} = \ln(a)a^x.$$

$$\frac{d}{dx}\log_a x = \frac{d}{dx}\Big(\frac{1}{\ln a}\ln x\Big) = \Big(\frac{1}{\ln a}\Big)\Big(\frac{1}{x}\Big).$$

EXAMPLE 11 Express $\displaystyle\lim_{h\to 0}\frac{5^h-1}{h}$ in terms of the natural logarithm.

Solution Since

$$\frac{d}{dx}5^x = 5^x \lim_{h\to 0}\frac{5^h-1}{h}$$

and

$$\frac{d}{dx}5^x = \ln(5)\cdot 5^x,$$

we have

$$\lim_{h\to 0}\frac{5^h-1}{h} = \ln(5).$$

EXAMPLE 12 Compute $\dfrac{d}{dx}\log_3(x^2+1)$.

Solution

$$\frac{d}{dx}\log_3(x^2+1) = \Big(\frac{1}{\ln 3}\Big)\Big(\frac{1}{x^2+1}\Big)\frac{d}{dx}(x^2+1)$$

$$= \Big(\frac{1}{\ln 3}\Big)\Big(\frac{1}{x^2+1}\Big)2x.$$

7.2 EXPONENTIAL AND LOGARITHMIC FUNCTIONS

When the input x appears in the exponent of a functional expression, a technique known as **logarithmic differentiation** may be useful. We'll illustrate the technique with an example.

EXAMPLE 13 If $y = x^x$, find y'.

Solution First, we take the natural logarithm of both sides of the formula and use a property of logarithms:

$$\ln(y) = \ln(x^x) = x \ln(x).$$

Now, we *implicitly differentiate*:

$$\frac{y'}{y} = 1 \cdot \ln(x) + x \cdot \frac{1}{x} = \ln(x) + 1.$$

Finally, we can solve for y', and substitute for y to get a formula entirely in terms of x:

$$y' = y \cdot (\ln(x) + 1) = x^x (\ln(x) + 1).$$

■

Integration and the exponential and logarithm functions.

The fact that $\frac{d}{dx}(e^x) = e^x$ immediately supplies the antidifferentiation formula for the exponential function:

$$\int e^x \, dx = \int \exp(x) \, dx = e^x + C.$$

Similarly, since $\frac{d}{dx}(e^{kx}) = k \cdot e^{kx}$,

$$\int k \cdot e^{kx} \, dx = \int k \exp(kx) \, dx = e^{kx} + C$$

or

$$\int e^{kx} \, dx = \int \exp(kx) \, dx = \frac{1}{k} e^{kx} + C.$$

EXAMPLE 14 $\displaystyle\int_0^5 e^{-3x} \, dx = -\frac{1}{3} e^{-3x} \Big]_{x=0}^{x=5} = -\frac{1}{3} e^{-15} + 1.$ ■

EXAMPLE 15

$$\int (7.5)^x \, dx = \int \exp(x \ln(7.5)) \, dx$$

$$= \frac{1}{\ln(7.5)} \exp(x \ln(7.5)) + C$$

$$= \frac{1}{\ln(7.5)} (7.5)^x + C.$$

∎

The natural logarithm fills a gap in our table of antiderivatives of powers of x. Recall that for $r \neq -1$,

$$\int x^r \, dx = \frac{x^{r+1}}{r+1} + C$$

on any region where x^r is continuous. Since, on its domain of positive numbers, $\frac{d}{dx} \ln(x) = \frac{1}{x}$, it follows that

$$\int x^{-1} \, dx \equiv \int \frac{1}{x} \, dx = \ln(x) + C$$

again on the domain of positive numbers.

What happens when $x < 0$? We can examine this by looking at the slope field for the differential equation $y' = 1/x$ (see Figure 7.15). One solution of the differential equation (the one satisfying $y(1) = 0$) is $x \longmapsto \ln(x)$. All other solutions on the right half-plane are $x \longmapsto \ln(x) + C$ for various constants C. Note that the slope field on the left half-plane is simply the mirror image of that in the right half-plane. The solutions in the left half-plane will also be mirror images of those in the right half-plane. In other words, any function

$$f : (-\infty, 0) \longrightarrow \mathbb{R}$$

$$f : x \longmapsto \ln(-x) + C$$

will be a solution in the left half-plane and

$$\int x^{-1} \, dx \equiv \int \frac{1}{x} \, dx = \ln(-x) + C$$

on the domain of negative numbers. We can combine this with the formula valid on the domain of positive numbers by using the absolute value notation:

$$\int x^{-1} \, dx = \int \frac{1}{x} \, dx = \ln(|x|) + C$$

on any region where x^{-1} is continuous.

7.2 EXPONENTIAL AND LOGARITHMIC FUNCTIONS

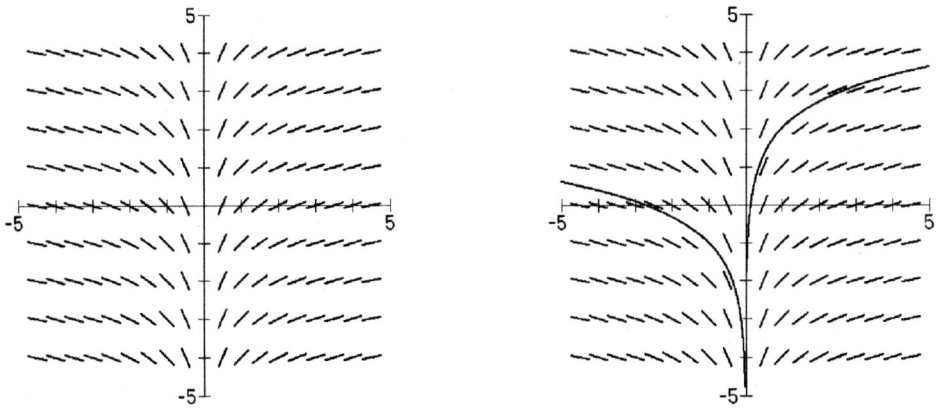

Figure 7.16 Slope field of $y' = 1/x$ and solutions in the left and right half-plane.

EXAMPLE 16 Solve the differential equation $y' = \dfrac{3}{x}$ with initial condition $y(-2) = 5$.

Solution From the fact that $\int x^{-1}\, dx = \int \dfrac{1}{x}\, dx = \ln(x) + C$ we have that

$$y = 3\ln(|x|) + C$$

and from the initial condition we have that

$$5 = y(-2) = 3\ln(|-2|) + C = 3\ln(2) + C$$

so that

$$C = 5 - 3\ln(2)$$

and

$$y: \quad x \longmapsto 3\ln|x| + 5 - 3\ln(2).$$

The domain is the set of all negative real numbers (see Figure 7.16). ∎

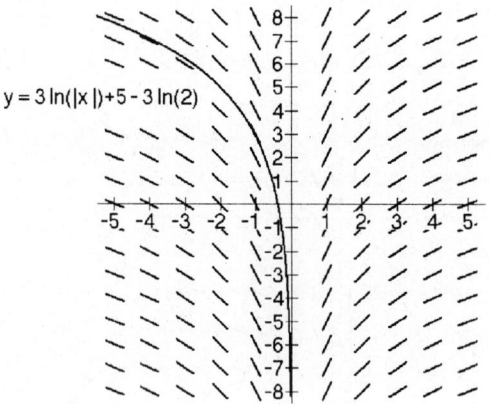

Figure 7.17 Slope field of $y' = 3/x$ and solution with $y(-2) = 5$.

We can also add the pattern $\dfrac{y'}{y}$ to the set of patterns to look for in attempting to solve differential equations. That is, if we see or can arrange to see $\dfrac{y'}{y}$ in the differential equation, we should think of this as $(\ln(|y|))'$.

EXAMPLE 17 Solve $\dfrac{y'}{y} = x$ with initial condition $y(2) = 7$.

Solution Since $\dfrac{y'}{y} = (\ln(|y|))'$ we can think of the differential equation as

$$(\ln(|y|))' = x$$

so that

$$\ln(|y|) = \frac{x^2}{2} + C$$

and exponentiating both sides yields

$$|y| = \exp(\frac{x^2}{2} + C)$$

$$= \exp(\frac{x^2}{2}) \cdot \exp(C).$$

Using the initial condition $7 = y(2) = \exp(2) \cdot \exp(C)$, we see that $\exp(C) = 7/e^2$ and we have

$$y : x \longmapsto \frac{7}{e^2} \exp(\frac{x^2}{2}).$$

∎

7.2 EXPONENTIAL AND LOGARITHMIC FUNCTIONS

EXAMPLE 18 Find $\int \dfrac{2x}{x^2+1}\,dx$.

Solution Since $2x = \dfrac{d}{dx}(x^2+1)$, this is another example of the pattern $\dfrac{y'}{y} = (\ln(|y|))'$.

$$\int \frac{2x}{x^2+1}\,dx = \int \left(\frac{d}{dx}\ln(|x^2+1|)\right)dx$$
$$= \ln(|x^2+1|) + C.$$

■

EXERCISES

In exercises 1-8, express the indicated quantity in terms of numbers and the other quantities given. (Assume that $a > 0$.)

1. $\dfrac{1}{\exp(5a)}$ in terms of e^a.

2. $\ln(a^4 + 2a^2 + 1)$ in terms of $\log_2 a^2 + 1$.

3. a^2 in terms of $\ln a$ and e^2.

4. $\log_3 a$ in terms of $\log_5 a$.

5. 2.75^a in terms of 0.9^a.

6. $\exp(2a+1)$ in terms of 3^a.

7. $\lim\limits_{h \to 0} \dfrac{a^{2h} - 1}{h}$ in terms of $\ln(a)$.

8. $\lim\limits_{h \to 0} \dfrac{\left(\frac{1}{a^h}\right) - 1}{h}$ in terms of $\ln(a)$.

In exercises 9-16, solve the indicated differential equations with initial conditions. Include the domain of definition of the solution.

9. $y' = 4y$; $y(1) = 42$

10. $\dfrac{3y'}{y} = -1$; $y(0) = 1$

11. $\dfrac{y'}{y} = \dfrac{1}{x}$; $y(0.1) = 0.1$

12. $y' = \dfrac{y}{x}$; $y(-1) = 0.75$

13. $yy' = \dfrac{1}{x}; \quad y(2) = 1$

14. $y' = y + 2; \quad y(0) = 0$

15. $y' = \dfrac{x}{x^2 + 1}; \quad y(0) = 0$

16. $y' = \dfrac{-2x}{x^2 + 1}; \quad y(0) = 0$

In each of exercises 17-22 a pair of functions is given. Both of the functions are solutions to a single differential equation with different initial conditions. For each exercise, find the (not necessarily unique) differential equation and initial conditions.

17. $y : x \longmapsto 3\exp(2x); \quad y : x \longmapsto -2\exp(2x).$

18. $y : x \longmapsto 3\exp(2x); \quad y : x \longmapsto 3\exp(2x) + 5.$

19. $y : x \longmapsto 5^{(3x+2)}; \quad y : x \longmapsto 125^x.$

20. $y : x \longmapsto 5^{(x-1)}; \quad y : x \longmapsto 0.2\exp(x\ln(5)) - 0.7.$

21. $y : x \longmapsto \log_2(x); \quad y : x \longmapsto \log_2(x) + 5.$

22. $y : x \longmapsto \log_3(x); \quad y : x \longmapsto \dfrac{1}{\ln(3)}(\ln(-x)).$

7.3 APPLICATIONS OF THE EXPONENTIAL MODEL

When we gather output data from a process over time, we can attempt to mathematically model the process as a function of time $y = f(t)$.

Linear models. One very important type of mathematical model is the **linear model**. If the outputs from a process appear to increase or decrease at a constant rate, then a linear model is indicated. Numerically, we have evidence of a linear model if the *difference* in outputs is observed to be proportional to the difference in time readings. We can think of this constant rate of change property in terms of the very simple differential equation

$$y' = m.$$

7.3 APPLICATIONS OF THE EXPONENTIAL MODEL

The solution of this differential equation has the familiar form

$$y(t) = mt + b.$$

Graphically, an easy way to recognize a linear process is to plot the graph of data outputs versus time and see if the points line up.

Once we have determined that a process behaves linearly (or nearly so), we want to determine its *parameters*. In the case of a linear model, the parameters are the slope m, and the y-intercept b. The slope tells us about the growth rate ($m > 0$) or decay rate ($m < 0$) of the outputs over time, while the y-intercept communicates an initial condition $y(0) = b$. These can be determined from two data pairs $(t, y(t))$ (two points on the line). After we determine the parameters, we can check the linear model against other observed data or use it to make predictions.

Exponential models

Another very important type of mathematical model is the **exponential model**. If the outputs from a process appear to increase or decrease at a rate proportional to the outputs themselves, then an exponential model is indicated. Numerically, we have evidence of an exponential model if the *ratio* of outputs is proportional to the difference in time readings. We can think of this proportional rate of change property in terms of the differential equation

$$y' = ky.$$

is indicated turns out to be an appropriate model for the observed growth and decay over time in a variety of physical, biological, and social phenomena. Using the results of the previous section, we know that this differential equation has solution

$$y(t) = Ae^{kt},$$

or, if we choose to use $c = e^k$ as the base,

$$y(t) = Ac^t.$$

Graphically, it may be more difficult to recognize an exponential model, but if the plot of outputs over time has the shape of one of the two exponential graphs in Figure 7.15, then we might consider the exponential model.

Once we have determined that an exponential model describes the behavior of a process, we want to determine its parameters. In the case of the exponential model, the parameter k tells us about the growth rate ($k > 0$) or decay rate ($k < 0$) of the outputs over time, while the parameter A communicates an initial condition $y(0) = A$. (If we choose to write $c = e^k$, then $c > 1$ indicates growth and $c < 1$ indicates decay. See Figure 7.15.) These parameters can also be determined from two data pairs (two points on the exponential curve). Again, after we determine the parameters, we can check the exponential model against other observed data or use it to make predictions.

Over the short term (in other words, a short time interval), many processes appear to be approximately linear. For example, if you charted the world's daily population over ten days, the graph would be almost straight. Indeed, the slope of this line gives an approximation of the derivative of the world's population function. However, if you charted the world's population every 100 years over the last millenium, you would obtain a very different shape of graph, indicating some other model. (Is it exponential?)

In the rest of this section, we'll discuss several phenomena that the exponential model seems to fit quite well.

Population growth

In a colony of micro-organisms supplied with abundant food, each *individual* reproduces at a predictable, and constant, rate. This being the case, the number of new individuals produced in a time period will be proportional to the total number of individuals. Thus, an exponential model works well for this type of population growth.

EXAMPLE 19 Using the data of population vs. time given in Table 7.2, predict the time at which the population will reach 1000000 and the time at which the population was 30.

time	individuals
9:00	372
10:00	4295
11:00	49531

Table 7.2 Population vs. clock time

Solution   Let t be the time in hours past $9:00$ and $y(t)$ be the number of individuals. First, note that the exponential model does fit the data, since

$$\frac{y(1)}{y(0)} \approx 11.55 \quad \text{and} \quad \frac{y(2)}{y(1)} \approx 11.53.$$

7.3 APPLICATIONS OF THE EXPONENTIAL MODEL

The solution is of the form

$$y(t) = Ae^{kt} = Ac^t \quad \text{where } c = e^k$$

for some A and k. Since $A = y(0) = 372$, we know that

$$y(t) = 372c^t.$$

To find c, we need to use another reading, say at $t = 1$ hour.

$$y(1) = 372c^1 = 372c = 4295 \quad \text{so} \quad c \approx 11.55.$$

Note that c is the factor by which the number of individuals multiply over each hour. The solution is

$$y(t) \approx 372(11.55)^t.$$

As a consistency check, we note that the model predicts that

$$y(2) \approx 372(11.55)^2 \approx 49626,$$

which is very close to the observed number of individuals at time $t = 2$ (11 : 00). To solve $y(t) = 1000000$ for t, we set up the equation

$$372(11.55)^t = 1000000.$$

Dividing by 372 and taking the natural logarithm of both sides:

$$t \ln(11.55) = \ln(1000000) - \ln(372)$$

and dividing by $\ln(11.55)$ we obtain

$$t = \frac{\ln(1,000,000) - \ln(372)}{\ln(11.55)} \approx 3.23.$$

The clock time when the population reaches 1000000 is therefore about 12 : 14.

To find the time when the population was 30, we only need replace 1000000 by 30 in our calculation:

$$t = \frac{\ln(30) - \ln(372)}{\ln(11.55)} \approx -1.03.$$

So the clock time was about 7 : 57. ∎

Continuously compounded interest

Suppose a savings account in a bank pays 5.25% yearly compounded quarterly. That is, starting with $100, three months later your accrued interest is credited to your account and you will have

$$\$100 + (\frac{0.0525}{4})\$100 = \$101.3125.$$

Three more months later and you will have

$$\$101.3125 + (\frac{0.0525}{4})\$101.3125 \approx \$102.6422$$

and so on. In general, the bank account balance A can be expressed in terms of the original balance P, the interest rate r, the number of times n it is compounded yearly, and the time t in years. In the case just discussed, $P = \$100$, $r = .0525$, and $n = 4$. If P, r, and n are known, we can consider A as a function of t:

$$A(t) = P(1 + \frac{r}{n})^{nt}.$$

Many savings accounts offer *daily* compounding of interest. However, rather than using $n = 365$ in the formula above, most banks use an *effective* interest rate that approximates *continuous* compounding. That is, they use a flat annual rate that gives approximately the same interest payment as obtained with the stated rate r and by letting the number of compounding times $n \to \infty$. Equivalently, you can think of this as letting r be the *instantaneous* interest rate, so that the instantaneous rate of change of our account balance is simply

$$A'(t) = rA(t).$$

This means that the account balance under continuous compounding follows an exponential model, with solution

$$A(t) = Pe^{rt}.$$

The effective annual interest rate for continuous compounding is $e^r - 1$.

7.3 APPLICATIONS OF THE EXPONENTIAL MODEL

EXAMPLE 20 Suppose $100 is invested for two years at 5.25% compounded quarterly and another $100 is invested at 5.25% compounded continuously. Compare the ending account balances. What is the effective annual interest rate for continuous compounding at 5.25%?

Solution For quarterly compounding, the ending account balance in dollars is

$$(100)(1 + (0.0525/4))^{4 \cdot 2} \approx 111.00,$$

while for continuous compounding, the ending account balance is

$$100 e^{0.0525 \cdot 2} \approx 111.07.$$

The effective annual interest rate for continuous compounding at 5.25%? is $e^{0.0525} - 1 \approx .0539$ or 5.39%. ∎

Radioactive decay

An atom of radioactive material has a fixed probability of undergoing radioactive decay in any given time period. It is as if in each second, all of the atoms rolled their own multi-sided die, and all those whose number came up transmute to a different element or isotope. In every second a fixed fraction of the remaining material decays, so that the rate of radioactive decay is proportional to the amount present. It has become customary to express the decay rate in terms of the amount of time it takes half of the material to decay, that is, the **half-life** of the material. If some radioactive material has a half-life of N years, the amount $y(t)$ present as a function of time t, also in years, is given by

$$y(t) = A(1/2)^{t/N}.$$

Every N years, the amount is decreased by half. This is why the storage of radioactive waste is such a difficult problem—its rate of decay gets slower as it decays, and hence storage facilities must be able to protect it from the environment for extended periods of time.

EXAMPLE 21   The isotope carbon-14, ^{14}C, whose nucleus consists of 6 protons and 8 neutrons, has a half-life of approximately 5580 years. If a sample of material is found to have 0.027 g of ^{14}C now, how many years ago did it have 1.26 g?

Solution   If we let $y(t)$ be the amount (in grams) of ^{14}C t years from the present, then

$$y(t) = A \cdot (1/2)^{(t/5580)}.$$

Since $0.027 = y(0) = A \cdot (\frac{1}{2})^{(0/5580)} = A$, we know that

$$y(t) = 0.027 \cdot (1/2)^{(t/5580)}.$$

To find the time that 1.26 g were present, we need to solve

$$1.26 = 0.027 \cdot (1/2)^{(t/5580)}$$

for t. Dividing both sides by 0.027, we obtain

$$\frac{1.26}{0.027} = (1/2)^{(t/5580)},$$

and taking the logarithm (base 1/2) of both sides yields

$$\log_{1/2}(1.26/0.027) = t/5580.$$

Hence, $t = 5580 \cdot \log_{1/2}(1.26/0.027) \approx -30937$, and we conclude that the sample had 1.26 g of ^{14}C about 30937 years ago. ∎

Drug concentrations

Drug levels in the body follow the same kind of pattern as radioactive decay. On each pass through the liver, kidneys, and other organs a certain fraction of any particular foreign substance in the blood is removed. This is another area where the usual unit of measurement is the half-life.

EXAMPLE 22   Blood samples taken 1 hour apart show concentrations 7.57 μg cm^3 and 5.80 μg cm^3 of a particular non-steroidal anti-inflammatory drug (of which aspirin is an example.) What is the half-life of this drug in the blood?

Solution   The concentration C as a function of time t in hours past the first sample is of the form

$$C(t) = A(1/2)^{t/k},$$

where k is the half-life in hours.

7.3 APPLICATIONS OF THE EXPONENTIAL MODEL

Since $7.57 = C(0) = A(1/2)^{0/k} = A$, we see that we must solve

$$5.80 = C(1) = 7.57(1/2)^{1/k}$$

for k. Dividing by 7.57 and then taking logarithms (base 1/2) gives us

$$\log_{1/2}\left(\frac{5.80}{7.57}\right) = \frac{1}{k},$$

so $1/k approx 0.3842$ and $k \approx 2.60$.

The half-life is about 2.6 hours. ∎

Heat transfer

The rate of heat (thermal energy) exchange between two objects in contact is proportional to the temperature *difference* between the two (heat flows from the hotter to the colder object). The constant of proportionality depends on the thermal contact (or insulation) between the two. Newton's law of cooling states that

> *the rate of change of the temperature difference between an object and its surroundings is proportional to the temperature difference itself.*

EXAMPLE 23 The temperatures $68.00\,°F$, $66.50\,°F$, and $65.07\,°F$ are recorded indoors 5 minutes apart. During this time, all sources of heat (furnace, etc.) are shut off and the outdoor temperature is constant. What is the outdoor temperature?

Solution Let $i(t)$ be the indoor temperature, t the the time in minutes past the first temperature reading, T the outdoor temperature, and $y(t) = i(t) - T$ be the difference between the indoor and outdoor temperature. Newton's law of cooling states that $y(t) = Ae^{kt}$ for some A and k. If we write $c = e^k$, then

$$y(t) = Ac^t$$

and

$$i(t) = Ac^t + T.$$

Our temperature readings tell us

$$i(0) = Ac^0 + T = A + T = 68, \quad i(5) = Ac^5 + T = 66.50, \quad i(10) = Ac^{10} + T = 65.07.$$

Note that

$$i(5) - i(0) = Ac^5 - A = -1.5 \quad \text{and} \quad i(10) - i(5) = Ac^{10} - Ac^5 = -1.43,$$

so that
$$\frac{Ac^{10} - Ac^5}{Ac^5 - A} = \frac{c^5(Ac^5 - A)}{Ac^5 - A} = c^5 = \frac{-1.43}{-1.5}$$

from which we can solve for c. Now, $Ac^5 - A = -1.5$, so
$$A = \frac{-1.5}{c^5 - 1} = \frac{-1.5}{\frac{-1.43}{-1.5} - 1} \approx 32.14.$$

Since $A + T = 68$, we have $T = 68 - A \approx 68 - 32.14 = 35.86$ degrees. ∎

EXERCISES

In each of exercises 1-6 experimental bacterial population samples taken 1 hour apart are presented. In some cases, the population was growing exponentially. In other cases, where an antibiotic was added to the culture, the population was decreasing exponentially. In other cases, where the bacteria were simply migrating, the populations were linear functions of time. Decide which are which and for each predict the populations 3 hours after the last sample.

1. 4129, 4501, 4906, and 5347.

2. 4129, 4831, 5652, and 6613.

3. 4129, 4501, 4872, and 5244.

4. 4129, 4061, 3992, and 3924.

5. 4129, 4061, 3994, and 3928,

6. 4129, 2890, 2023, and 1416.

A hot object is plunged into a beaker of ice water maintained at $32\,°F$. One and two minutes later the temperature of the object is measured to be $87\,°F$ and $76.2\,°F$, respectively. Use this information to answer exercises 7-10.

7. What will the temperature be one minute after the last measurement?

8. What was the temperature when the object was plunged in the ice water?

9. If the water had been maintained at $40\,°F$ instead, what would the measurements have been after one and two minutes?

10. If the initial temperature of the object had been $120\,°F$, and the water had been maintained at $32\,°F$, what would the measurements have been after one and two minutes?

7.4 METHODS OF INTEGRATION

11. After 10 years, it is found that only 30% of the initial amount of a radioactive substance remains. What is the half-life of this substance?

12. An artifact is found containing only 4% of the amount of ^{14}C it had when it was created. Using a half-life of 5580 years for ^{14}C, determine the age of the artifact.

A certain therapeutic drug has a half-life of 3.3 hours in the bloodstream. The minimal therapeutic dose requires 1.5 mg in the bloodstream. Use this information to answer exercises 13-15.

13. How often should a dose of 3 mg be injected to guarantee that the minimum dose is maintained?

14. If a dose is to be injected every hour, how big should it be?

15. If, starting with no drug in the bloodstream, 3 mg is injected every 3 hours, how much is in the bloodstream just before the fifth injection? How much just before the 100th?

Exercises 16-20 assume that $100 is invested at 5.25% compounded quarterly.

16. What interest rate r would yield the same account balance at the end of one year if the interest is compounded continuously?

17. How many years (rounded to the next quarter) will it take to double your initial balance?

18. What would the yearly interest rate need to be to double your money in 10 years?

19. Another competing bank pays 5.28% but only compounds twice a year. Which account offers a better deal? What interest rate would make the two banks comparable?

20. A third bank offers a 7% yearly interest rate compounded quarterly but also has a $.10 per quarter service charge, which is deducted from your account at the same time the interest is credited. If your starting balance is $100, which bank offers the best deal? If your starting balance is $1000, which bank offers the best deal?

7.4 METHODS OF INTEGRATION

The general problem of indefinite integration, or antidifferentiation, is to find a function F satisfying

$$\int f(x)\,dx = F(x) + C.$$

This can be thought of as a problem of "derivative recognition." That is, if we can recognize f as the derivative of another function F, that is

$$f(x) = F'(x),$$

then F is the function we're looking for.

Certainly, if f is continuous over an interval $[a, b]$, then the First Fundamental Theorem of Calculus gives us a guaranteed solution, namely

$$F(x) = \int_a^x f(t)\,dt,$$

which will satisfy $F'(x) = f(x)$ for $a < x < b$. However, evaluation of these area functions can require a great deal of computing power just to achieve approximate outputs. Finding a nice "closed form" formula for F has its advantages. This section is devoted to some techniques and methods for finding such formulas.

Integral formulas

We can develop several integral formulas by simply reversing basic derivative formulas. Let's review these:

$$\int x^r\,dx = \frac{x^{r+1}}{r+1} + C \text{ for } r \neq -1 \qquad \int \frac{1}{x}\,dx = \ln|x| + C$$

$$\int \sin x\,dx = -\cos x + C \qquad \int \cos x\,dx = \sin x + C$$

$$\int \sec^2 x\,dx = \tan x + C \qquad \int \sec x \tan x\,dx = \sec x + C$$

$$\int \csc^2 x\,dx = -\cot x + C \qquad \int \csc x \cot x\,dx = -\csc x + C$$

$$\int e^x\,dx = e^x + C \qquad \int a^x\,dx = \frac{a^x}{\ln(a)} + C$$

$$\int \frac{1}{\sqrt{1-x^2}}\,dx = \arcsin(x) + C \qquad \int \frac{-1}{\sqrt{1-x^2}}\,dx = \arccos(x) + C$$

$$\int \frac{1}{1+x^2}\,dx = \arctan(x) + C \qquad \int \frac{-1}{1+x^2}\,dx = \arccot(x) + C$$

$$\int \frac{1}{x\sqrt{x^2-1}}\,dx = \arcsec|x| + C \qquad \int \frac{-1}{x\sqrt{x^2-1}}\,dx = \arccsc|x| + C$$

7.4 METHODS OF INTEGRATION

 CAUTION! These and any other integral formulas are valid only for those intervals over which the integrand is continuous.

For example, $\int \frac{1}{x^3} dx = \int x^{-3} dx = \frac{x^{-2}}{-2} + C = -\frac{1}{2x^2} + C$ only for intervals that do not include $x = 0$, since $x \longmapsto \frac{1}{x^3}$ is not continuous at $x = 0$. Similarly,

$$\int \sec^2 x \, dx = \tan x + C$$

is valid only for intervals that do not include any value $x = \frac{\pi}{2} + n\pi$ (n any integer) since $\sec^2 x$ is undefined at each of these values.

Notice that the integral formula

$$\int \frac{1}{x} = \ln |x| + C$$

involves the absolute value. Since $1/x$ is undefined at $x = 0$, there are really two integral formulas involved: one for $x < 0$ and one for $x > 0$. We know that for $x > 0$,

$$\frac{d}{dx}(\ln(x)) = \frac{1}{x}.$$

If we examine the slope field generated by $1/x$ (see Figure 7.18), we can see that the antiderivative graphs for $x < 0$ are mirror images of those for $x > 0$ (reflected through the y-axis).

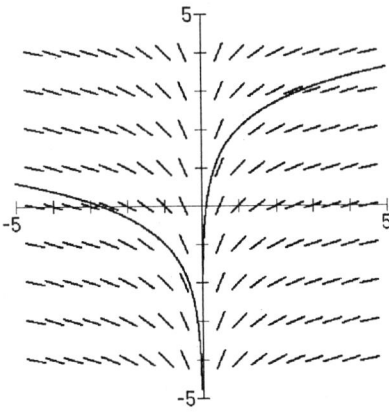

Figure 7.18 Antiderivatives of $1/x$ depend on the sign of x.

Hence, for $x < 0$, we also have

$$\frac{d}{dx}(\ln(-x)) = \frac{1}{x}.$$

Since $|x| = x$ when $x > 0$ and $|x| = -x$ when $x < 0$, our integral formula handles both cases at once. Similarly, the integral formulas

$$\int \frac{1}{x\sqrt{x^2-1}}\,dx = \operatorname{arcsec}|x| + C \quad \text{and} \quad \int \frac{-1}{x\sqrt{x^2-1}}\,dx = \operatorname{arccsc}|x| + C$$

handle both of the separate cases where $x > 1$ and $x < -1$.

If an interval includes points of discontinuity for a function f, then an antiderivative may or may not be defined at those points. In any case, a different arbitrary constant must be designated for each subinterval determined by the points of discontinuity of the original function f.

EXAMPLE 24 Find the most general antiderivative of $x \longmapsto 1/x$ over the set of inputs

$$\{x : -1 < x < 3,\ x \neq 0\}.$$

Solution We write

$$\int \frac{1}{x}\,dx = \ln(-x) + C_1 \text{ for } -1 < x < 0 \quad \text{and} \quad \int \frac{1}{x}\,dx = \ln(x) + C_2 \text{ for } 0 < x < 3$$

to express the fact that two different arbitrary constants are involved for the two subintervals $-1 < x < 0$ and $0 < x < 3$. ∎

For differentiation of many functions, one need only know the derivatives of a few basic functions and how to apply the basic differentiation rules like the product and chain rules. Unfortunately, there are no product and chain rules for integration, but there are some techniques we can use to find many more antiderivatives than those listed above.

The method of substitution

The **method of substitution** is essentially a technique used to help us recognize a derivative produced by the *chain rule*.

If we take the derivative of the composition of two functions $f \circ g$, then the chain rule gives us

$$\frac{d}{dx}f(g(x)) = f'(g(x))g'(x).$$

This means that if we encounter $f'(g(x))g'(x)$ as an integrand, we can reverse the process:

$$\int f'(g(x))g'(x)\,dx = f(g(x)) + C.$$

7.4 METHODS OF INTEGRATION

EXAMPLE 25 Use the fact that $\frac{d}{dx}\sin(x^2) = \cos(x^2) \cdot 2x$ to integrate $\int x\cos(x^2)\,dx$.

Solution Since $x\cos(x^2) = \frac{1}{2}(\cos(x^2) \cdot 2x)$, and we know that $\frac{d}{dx}(\sin(x^2)) = \cos(x^2) \cdot 2x$, we must have

$$\int x\cos(x^2)\,dx = \frac{1}{2}\int \cos(x^2) \cdot 2x\,dx = \frac{1}{2}\sin(x^2) + C.$$

■

EXAMPLE 26 Use the fact that $\frac{d}{dx}[\tan^3(x)] = 3\tan^2(x)\sec^2(x)$ to integrate $\int \tan^2(x)\sec^2(x)\,dx$.

Solution Since $\tan^2(x)\sec^2(x) = \frac{1}{3}(3\tan^2(x)\sec^2(x))$, we must have

$$\int \tan^2(x)\sec^2(x)\,dx = \frac{1}{3}\int (3\tan^2(x)\sec^2(x))\,dx = \frac{1}{3}\tan^3(x) + C.$$

■

In both of these examples, finding an anitiderivative boiled down to recognizing the integrand as being a constant multiple of a derivative we had just seen.

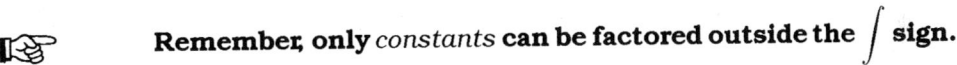 **Remember, only** *constants* **can be factored outside the** $\int$ **sign.**

Sometimes, a change in the variable of integration can put the integrand in a more recognizable form. This is the main idea behind the *method of substitution*. If we write $u = g(x)$, then the chain rule can be written as

$$\frac{d}{dx}f(u) = f'(u)\frac{du}{dx},$$

and the corresponding integral formula becomes

$$\int f'(u)\frac{du}{dx}\,dx = f(u) + C.$$

If we agree to use the shorthand

$$du = \frac{du}{dx}dx$$

(we refer to du as the **differential** of u with respect to x), then we can write this as

$$\int f'(u)\,du = f(u) + C$$

which is a valid integral formula entirely in terms of the new variable u. If f' happens to be a derivative we more easily recognize, then this change of variables has been useful.

Let's carefully outline the method of substitution, and then apply it to several examples.

METHOD OF SUBSTITUTION

Step 1 Let $u = g(x)$.

Good choices of substitutions to try are suggested when $g(x)$ appears raised to a power, in the denominator, or in general, as an "inside" function of a composition.

Step 2 Compute $du = g'(x)\,dx$.

Step 3 Substitute u for $g(x)$ and $\dfrac{du}{g'(x)}$ for dx in the integrand. Try to express the new integrand entirely in terms of u and du (no appearances of x should be left). If not, a different substitution must be made (go back to Step 1 and try a different choice for $g(x)$).

Step 4 Integrate and substitute $g(x)$ back for u. (Don't forget the arbitrary constant C.)

Step 5 Check your solution by differentiating and comparing to the original integrand.

You should note that the method of substitution carries with it no guarantee of success. Ultimately you (or a machine) must be able to recognize the integrand as the derivative of some function.

EXAMPLE 27 Find $\int \sin^2(x) \cos(x)\,dx$.

Solution Let $u = \sin(x)$. Then $du = \cos(x)\,dx$. Substituting u for $\sin(x)$ and $\dfrac{du}{\cos(x)}$ for dx yields

$$\int \sin^2(x) \cos(x)\,dx = \int u^2 \cos(x) \frac{du}{\cos(x)}.$$

Simplifying, we have

$$\int u^2\,du = \frac{u^3}{3} + C = \frac{\sin^3(x)}{3} + C.$$

7.4 METHODS OF INTEGRATION

We can check our answer by differentiating:

$$\frac{d}{dx}\left(\frac{\sin^3(x)}{3} + C\right) = \sin^2(x)\cos(x).$$

∎

EXAMPLE 28 Find $\int x^4 e^{x^5}\, dx.$

Solution Let $u = x^5$. Then $du = 5x^4 dx$. Substituting u for x^5 and $\dfrac{du}{5x^4}$ for dx yields

$$\int x^4 e^{x^5}\, dx = \frac{1}{5}\int e^u\, du = \frac{e^u}{5} + C = \frac{e^{x^5}}{5} + C.$$

We can check our answer by differentiating:

$$\frac{d}{dx}\left(\frac{e^{x^5}}{5} + C\right) = x^4 e^{x^5}.$$

∎

EXAMPLE 29 Find $\int \dfrac{x^2}{5 + x^3}\, dx.$

Solution Let $u = 5 + x^3$. Then $du = 3x^2 dx$. Substituting u for $5 + x^3$ and $\dfrac{du}{3x^2}$ for dx yields

$$\int \frac{x^2}{5 + x^3}\, dx = \int \frac{x^2}{u}\, \frac{du}{3x^2}.$$

Simplifying, we have

$$\frac{1}{3}\int \frac{1}{u}\, du = \frac{\ln|u|}{3} + C = \frac{\ln|5 + x^3|}{3} + C.$$

We can check our answer by differentiating:

$$\frac{d}{dx}\left(\frac{\ln|5 + x^3|}{3} + C\right) = \frac{x^2}{5 + x^3}.$$

∎

EXAMPLE 30 Find $\int \csc^3(x) \cot(x) \, dx$.

Solution Let $u = \csc(x)$. Then $du = -\csc(x)\cot(x)dx$. Substituting u for $\csc(x)$ and $\dfrac{du}{-\csc(x)\cot(x)}$ for dx yields

$$\int \csc^3(x) \cot(x) \, dx = \int u^3 \cot(x) \frac{du}{-\csc(x)\cot(x)}.$$

Simplifying, we have $\int u^3 \dfrac{du}{-\csc(x)}$. It appears that we may not have been successful (x still appears in the integrand). However, $u = \csc(x)$, so we can substitute to obtain

$$\int u^3 \frac{du}{-\csc(x)} = -\int u^2 \, du = -\frac{u^3}{3} + C = -\frac{\csc^3(x)}{3} + C.$$

(Check the answer by differentiating.) ∎

EXAMPLE 31 Find $\int \dfrac{\cos\sqrt{x}}{\sqrt{x}} \, dx$.

Solution Let $u = \sqrt{x} = x^{1/2}$. Then $du = \dfrac{1}{2}x^{-1/2}dx = \dfrac{dx}{2\sqrt{x}}$. Substituting u for $\sqrt{x}$ and $2\sqrt{x}\,du$ for dx yields

$$\int \frac{\cos\sqrt{x}}{\sqrt{x}} \, dx = \int \frac{\cos(u)}{\sqrt{x}} 2\sqrt{x} \, du.$$

Simplifying, we have

$$2\int \cos(u) \, du = 2\sin(u) + C = 2\sin(\sqrt{x}) + C.$$

(Check the answer by differentiating.) ∎

EXAMPLE 32 Find $\int \dfrac{x}{x+1} \, dx$.

Solution Let $u = x + 1$. Then $du = dx$. Substituting u for $x + 1$ and du for dx yields

$$\int \frac{x}{x+1} \, dx = \int \frac{x}{u} \, du.$$

If we solve $u = x + 1$ for x, we can remove the remaining appearance of x in the integrand:

$$\int \frac{x}{u} \, du = \int \frac{u-1}{u} \, du = \int (1 - 1/u) \, du = u - \ln|u| + C = x + 1 - \ln|x+1| + C.$$

7.4 METHODS OF INTEGRATION

Since C is an arbitrary constant, there is no real need to include the additional constant term 1 ($C+1$ is as arbitrary as C), so we can write the final form as

$$x - \ln|x+1| + C.$$

We can check our answer by differentiating:

$$\frac{d}{dx}(x - \ln|x+1| + C) = 1 - \frac{1}{x+1} = \frac{x+1}{x+1} - \frac{1}{x+1} = \frac{x}{x+1}.$$

■

To apply the method of substitution to a *definite integral*

$$\int_a^b f(x)\,dx$$

we have a couple of options.

<u>Option 1</u> Carry out the method of substitution all the way on the corresponding indefinite integral, then compute the difference of the values at $x = b$ and $x = a$ as usual.

EXAMPLE 33 Compute $\displaystyle\int_1^2 \frac{x}{(x^2+3)^3}\,dx$.

Solution We first find the antiderivative $\displaystyle\int \frac{x}{(x^2+3)^3}\,dx$. Let $u = x^2 + 3$ and $du = 2x\,dx$. Substitute u for (x^2+3) and $\dfrac{du}{2x}$ for dx:

$$\int \frac{x}{(x^2+3)^3}\,dx = \int \frac{x}{u^3}\frac{du}{2x} = \int \frac{du}{2u^3} = \frac{1}{2}\int u^{-3}\,du = \frac{1}{2}\frac{u^{-2}}{-2} = \frac{u^{-2}}{-4} = \frac{-1}{4(x^2+3)^2}.$$

We've suppressed the arbitrary constant C in the antiderivative because it will cancel anyway when we evaluate the definite integral.

$$\int_1^2 \frac{x}{(x^2+3)^3}\,dx = \frac{-1}{4(x^2+3)^2}\bigg]_{x=1}^{x=2} = \frac{-1}{196} - \frac{-1}{64} = \frac{33}{3136} \approx .010523.$$

■

<u>Option 2</u> After making the substitutions for $u = g(x)$ and $dx = \dfrac{du}{g'(x)}$, also substitute for the limits of integration: $u = g(a)$ when $x = a$ and $u = g(b)$ when $x = b$.

EXAMPLE 34 Compute $\int_1^2 \frac{x}{(x^2+3)^3}\,dx$.

Solution Let $u = x^2 + 3$ and $du = 2x\,dx$. When $x = 1, u = 1^2 + 3 = 4$. When $x = 2, u = 2^2 + 3 = 7$.

$$\int_1^2 \frac{x}{(x^2+3)^3}\,dx = \int_4^7 \frac{du}{2u^3} = \left.\frac{u^{-2}}{-4}\right]_{u=4}^{u=7} = \frac{-1}{196} + \frac{1}{64} = \frac{33}{3136} \approx 0.10523.$$

■

Integration by parts

The method of substitution helps us recognize derivatives produced by the chain rule. The method known as **integration by parts** helps us recognize derivatives produced by the *product rule*.

The product rule for the derivative of fg gives us

$$\frac{d}{dx}(f(x)g(x)) = f'(x)g(x) + f(x)g'(x).$$

If we write u for $f(x)$ and v for $g(x)$ we have

$$\frac{d}{dx}(uv) = \frac{du}{dx}v + u\frac{dv}{dx}.$$

Following our convention of writing $du = \frac{du}{dx}dx$ and $dv = \frac{dv}{dx}dx$, we can reverse the product rule to be a statement about antiderivatives:

$$uv = \int v\,du + \int u\,dv.$$

When we subtract $\int v\,du$ from both sides, the resulting statement is commonly known as the **integration by parts formula:**

$$\int u\,dv = uv - \int v\,du.$$

Given an integral of the form $\int u\,dv$, the practical use of the formula is realized if the integral $\int v\,du$ that appears on the right-hand side is much easier to recognize than the original integral $\int u\,dv$.

Let's outline the method of integration by parts and then illustrate it with several examples.

7.4 METHODS OF INTEGRATION

INTEGRATION BY PARTS

<u>Step 1</u> Let $u = f(x)$ and $dv = g(x)dx$ where $f(x)g(x)dx$ is the original integrand. Good choices to make are suggested when $\int dv = \int g(x)dx$ is easy to integrate.

<u>Step 2</u> Compute $du = f'(x)dx$ and $v = \int g(x)dx$.

<u>Step 3</u> Substitute u, v, du and dv into the formula

$$\int u\,dv = uv - \int v\,du.$$

<u>Step 4</u> Calculate $uv - \int v\,du$.

If $\int v\,du$ is difficult or impossible to integrate, then go back to Step 1 and consider other choices for u and dv.

<u>Step 5</u> Check your solution by differentiating and comparing to the original integrand.

As with the method of substitution, integration by parts comes with no guarantee of success. However, it is a technique that works well on many integrals involving products of functions, particularly if transcendental functions appear as factors.

EXAMPLE 35 Find $\int xe^x\,dx$.

Solution Let $u = x$ and $dv = e^x\,dx$. We find that $du = dx$ and

$$v = \int dv = \int e^x\,dx = e^x.$$

Substituting into the integration by parts formula, we have

$$\int xe^x\,dx = \int u\,dv = uv - \int v\,du = xe^x - \int e^x\,dx.$$

The last integral that appears is simple to integrate, and we have

$$\int xe^x\,dx = xe^x - e^x + C.$$

We can check our answer by differentiating:

$$\frac{d}{dx}(xe^x - e^x + C) = (1 \cdot e^x + x \cdot e^x) - e^x = xe^x.$$

∎

EXAMPLE 36 Find $\int x^2 e^x \, dx$.

Solution Let $u = x^2$ and $dv = e^x \, dx$. We find that $du = 2x \, dx$ and

$$v = \int dv = \int e^x \, dx = e^x.$$

Substituting into the integration by parts formula, we have

$$\int x^2 e^x \, dx = \int u \, dv = uv - \int v \, du = x^2 e^x - \int 2x e^x \, dx.$$

The last integral can be written as $2 \int x e^x \, dx$ and we successfully integrated this by parts in the previous example. Substituting that result, we obtain

$$\int x^2 e^x \, dx = x^2 e^x - 2x e^x + 2 e^x + C.$$

(Check the answer by differentiating.) ■

EXAMPLE 37 Find $\int \ln(x) \, dx$.

Solution Let $u = \ln(x)$ and $dv = 1 \, dx$. We find that $du = \dfrac{1}{x} \, dx$ and

$$v = \int dv = \int 1 \, dx = x.$$

Substituting into the integration by parts formula, we have

$$\int \ln(x) \, dx = \int u \, dv = uv - \int v \, du = x \cdot \ln(x) - \int x \cdot \frac{1}{x} \, dx.$$

After simplifying the last integrand and integrating, we have

$$\int \ln(x) \, dx = x \ln(x) - x + C.$$

We can check our answer by differentiating:

$$\frac{d}{dx}(x \ln(x) - x + C) = (1 \cdot \ln(x) + x \cdot \frac{1}{x}) - 1 = \ln(x).$$

■

7.4 METHODS OF INTEGRATION

EXAMPLE 38 Find $\int \arctan(x)\,dx$.

Solution Let $u = \arctan(x)$ and $dv = 1\,dx$. We find that $du = \dfrac{1}{1+x^2}\,dx$ and

$$v = \int dv = \int 1\,dx = x.$$

Substituting into the integration by parts formula, we have

$$\int \arctan(x)\,dx = \int u\,dv = uv - \int v\,du = x \cdot \arctan(x) - \int x \cdot \frac{1}{1+x^2}\,dx.$$

Now, this last integral can be integrated using the method of substitution: Let $w = 1 + x^2$, so that $dw = 2x\,dx$. Then

$$\int \frac{x}{1+x^2}\,dx = \int \frac{x}{w}\frac{dw}{2x} = \frac{1}{2}\int \frac{1}{w}\,dw = \frac{1}{2}\ln|w| + C = \frac{1}{2}\ln|1+x^2| + C.$$

Using this result, we can now write

$$\int \arctan(x)\,dx = x\arctan(x) - \frac{1}{2}\ln|1+x^2| - C.$$

Because $1 + x^2$ is positive for all real values x, we can drop the absolute value signs from this expression. Also, subtracting an arbitrary constant C is the same as adding an arbitrary constant (C could be any real number), so we might as well write this as

$$\int \arctan(x)\,dx = x\arctan(x) - \frac{\ln(1+x^2)}{2} + C.$$

(Check the answer by differentiating.) ∎

Other methods of integration—using tables of integrals

The method of substitution and integration by parts are the most widely used general integration techniques. We have seen that they can be helpful in recognizing the results of the chain and product rules of differentiation. Many calculus books include a number of other techniques of integration. However, in many cases, these other techniques turn out to be simply special cases of the method of substitution combined with some "clever algebraic and/or trigonometric manipulations. At one time, these "tricks of the trade" were covered at length in beginning calculus courses, for they were the only tools available. Even then, the tricks cover a relatively small special set of functions. Now, technology provides us with additional tools to study both indefinite integrals (antiderivatives can be investigated graphically with slope fields) and definite integrals (using a machine's numerical integration routines).

Nevertheless, closed-form expressions for antiderivatives can be quite useful when we can find them. Many books (and software) contain long tables of integral forms that have been compiled to catalog many of the types of integrands encountered in applications. Quite often these tables of integrals are written in terms of u and du with the expectation that the user of the table will first make a substitution to bring about a "match" with one of the entries in the table. Even when a computer algebra system is available, a change of variables may aid the system in making the match.

For example, here is a formula from an integral table:

$$\int e^{au} \sin(bu) \, du = \frac{e^{au}}{a^2 + b^2}(a \sin(bu) - b \cos(bu)) + C.$$

In this formula, it is understood that a and b are constants, and u represents some differentiable function. The next example shows how we can utilize this formula using the method of substitution.

EXAMPLE 39 Find $\int 3x e^{7x^2} \sin(5x^2) \, dx$.

Solution First, we'll make the substitution $u = x^2$ (so $du = 2x \, dx$ and $dx = du/2x$):

$$\int 3x e^{7x^2} \sin(5x^2) \, dx = \int 3x e^{7u} \sin(5u) \frac{du}{2x} = \frac{3}{2} \int e^{7u} \sin(5u) \, du.$$

Now we can use the integral formula with $a = 7$ and $b = 5$:

$$\frac{3}{2} \int e^{7u} \sin(5u) \, du = \frac{3}{2} \frac{e^{7u}}{7^2 + 5^2}(7 \sin(5u) - 5 \cos(5u)) + C$$

and then substitute $u = x^2$ and simplify:

$$\int 3x e^{7x^2} \sin(5x^2) \, dx = \frac{3 e^{7x^2}}{148}(7 \sin(5x^2) - 5 \cos(5x^2)) + C.$$

∎

Some integral formulas involving powers of a function are called *reduction formulas*. They generally allow an integral to be expressed in terms of integrals involving smaller powers of the same function. Repeated application of the formula can therefore reduce the problem to one whose solution is known.

Here is an example of a reduction formula from an integral table:

$$\int \sin^n(u) \, du = -\frac{1}{n} \sin^{n-1}(u) \cos(u) + \frac{n-1}{n} \int \sin^{n-2}(u) \, du.$$

7.4 METHODS OF INTEGRATION

We illustrate how this formula is applied with the following example:

EXAMPLE 40 Find $\int \sin^5(x)\, dx$.

Solution Here $n = 5$, so a direct application of the formula gives us

$$\int \sin^5(x)\, dx = -\frac{1}{5}\sin^4(x)\cos(x) + \frac{4}{5}\int \sin^3(x)\, dx.$$

Notice that we can apply the formula again to the last integral with $n = 3$:

$$\int \sin^3(x)\, dx = -\frac{1}{3}\sin^2(x)\cos(x) + \frac{2}{3}\int \sin(x)\, dx.$$

If we substitute this right-hand expression for $\int \sin^3(x)\, dx$ above, we have

$$\int \sin^5(x)\, dx = -\frac{1}{5}\sin^4(x)\cos(x) - \frac{4}{15}\sin^2(x)\cos(x) + \frac{8}{15}\int \sin(x)\, dx.$$

Finally, we have reduced the problem to one involving an integral we recognize immediately, so the final answer is

$$\int \sin^5(x)\, dx = -\frac{1}{5}\sin^4(x)\cos(x) - \frac{4}{15}\sin^2(x)\cos(x) - \frac{8}{15}\cos(x) + C.$$

∎

EXERCISES

Find the indefinite integrals in exercises 1-10 using the method of substitution.

1. $\int x \cos(x^2)\, dx.$
2. $\int \sin(x) \cos^2(x)\, dx.$
3. $\int \dfrac{x^4}{x^5 - 17}\, dx.$
4. $\int (.05x - \pi)^{132}\, dx.$
5. $\int \dfrac{x^5}{\sqrt[3]{2x^3 + 7}}\, dx.$
6. $\int \sin(x) \sqrt{2\cos(x) + 3}\, dx.$
7. $\int x e^{-x^2}\, dx.$
8. $\int \dfrac{e^x}{1 + e^{2x}}\, dx.$
9. $\int \dfrac{\ln(x)}{x}\, dx.$
10. $\int \tan^3(x) \sec^2(x)\, dx.$

Find the indefinite integrals in exercises 11-20 using integration by parts.

11. $\int \arcsin(x)\, dx.$
12. $\int \text{arccot}(x)\, dx.$
13. $\int x^3 e^x\, dx.$
14. $\int x e^{-x}\, dx.$

15. $\int \arccos(x)\, dx$.

16. $\int x^3 e^{-x}\, dx$.

17. $\int x \ln(x)\, dx$.

18. $\int x \cos(5x)\, dx$.

19. $\int x^2 \sin(x)\, dx$.

20. $\int x(\ln(x))^2\, dx$.

Verify the trigonometric integration formulas in exercises 21-24 using the indicated substitution.

21. $\int \tan(x)\, dx = -\ln|\cos(x)| + C$. (Write $\tan(x) = \dfrac{\sin(x)}{\cos(x)}$ and use the substitution $u = \cos(x)$.)

22. $\int \cot(x)\, dx = \ln|\sin(x)| + C$. (Write $\cot(x) = \dfrac{\cos(x)}{\sin(x)}$ and use the substitution $u = \sin(x)$.)

23. $\int \sec(x)\, dx = \ln|\sec(x) + \tan(x)| + C$. (First, multiply the integrand by $\dfrac{\sec(x) + \tan(x)}{\sec(x) + \tan(x)}$ and then use the substitution $u = \sec(x) + \tan(x)$.)

24. $\int \csc(x)\, dx = \ln|\csc(x) - \cot(x)| + C$. (First, multiply the integrand by $\dfrac{\csc(x) - \cot(x)}{\csc(x) - \cot(x)}$ and then use the substitution $u = \csc(x) - \cot(x)$.)

The hyperbolic sine and cosine functions are defined as follows:

$$\sinh(x) = \frac{e^x - e^{-x}}{2} \quad \text{and} \quad \cosh(x) = \frac{e^x + e^{-x}}{2}.$$

The rest of the hyperbolic functions are defined in a way analagous to the corresponding trigonometric functions:

$\tanh(x) = \dfrac{\sinh(x)}{\cosh(x)}$ $\qquad$ $\coth(x) = \dfrac{\cosh(x)}{\sinh(x)}$

$\text{sech}(x) = \dfrac{1}{\cosh(x)}$ $\qquad$ $\text{csch}(x) = \dfrac{1}{\sinh(x)}$.

In exercises 25-28, graph the indicated hyperbolic function over the interval $[-10, 10]$. Then calculate both its derivative and antiderivative and express (if possible) in terms of other hyperbolic functions.

25. $\sinh(x)$

26. $\cosh(x)$

27. $\tanh(x)$

28. $\coth(x)$

7.4 METHODS OF INTEGRATION

29. Look up $\int \operatorname{arcsec}(x)\, dx$ and $\int \operatorname{arccsc}(x)\, dx$ in an integral table or on a computer algebra system. Graph the slope fields of generated by arcsec and arccsc and graph the antiderivatives you find. Do the results make sense?

30. Sometimes integration by parts leads us back to an expression involving the original integral. Use integration by parts twice on

$$\int e^x \cos(x)\, dx$$

and then *solve* for the original integral.

In exercises 31-40 you are given definite integrals whose integrands correspond to the indefinite integrals in exercises 1-10. For each one, graph the integrand over the interval of values indicated by the limits of integration and estimate visually the value of the definite integral. Compute the new limits of integration under the substitution used in the corresponding exercise (1-10). Calculate each definite integral using both options discussed in the text and use a computer or calculator to compute the definite integral and compare results.

31. $\int_0^\pi x \cos(x^2)\, dx.$

32. $\int_0^{2\pi} \sin(x) \cos^2(x)\, dx.$

33. $\int_{-1}^1 \dfrac{x^4}{x^5 - 17}\, dx.$

34. $\int_0^\pi (.05x - \pi)^{132}\, dx.$

35. $\int_{-1}^1 2 \dfrac{x^5}{\sqrt[3]{2x^3 + 7}}\, dx.$

36. $\int_0^{\pi/2} \sin(x)\sqrt{2\cos(x) + 3}\, dx.$

37. $\int_1^{\ln(2)} xe^{-x^2}\, dx.$

38. $\int_0^{\ln(\pi/4)} \dfrac{e^x}{1 + e^{2x}}\, dx.$

39. $\int_1^e \dfrac{\ln(x)}{x}\, dx.$

40. $\int_{\pi/3}^{2\pi/3} \tan^3(x) \sec^2(x)\, dx.$

---

41. Use the integral formula

$$\int e^{au} \cos(bu)\, du = \dfrac{e^{au}}{a^2 + b^2}(a\cos(bu) + b\sin(bu)) + C$$

to find

$$\int \dfrac{e^{-\arctan(x)} \cos(4\arctan(x))}{1 + x^2}\, dx.$$

42. Use the reduction formula

$$\int \sec^n(x)\, dx = \dfrac{\sec^{n-2}(x)\tan(x)}{n-1} + \dfrac{n-2}{n-1}\int \sec^{n-2}(x)\, dx$$

to find $\int \sec^3(x)\, dx.$

CHAPTER 8

Integral as Measurement Tool

Integrals provide powerful tools for measurement in a wide variety of applications. By its very definition, a definite integral can be used to measure certain areas. However, the utility of definite integration goes far beyond this particular geometric measurement problem.

In general, you can think of definite integrals as providing a means of measuring the *net effect* of a functional quantity over an interval. For example, we might think of the area of a rectangle as being the net effect of its height over an interval representing its length. Similarly, the definite integral

$$\int_a^b f(x)\,dx$$

can be thought of as the net effect of the function values $f(x)$ (the height of the graph) over the interval $[a, b]$ (see Figure 8.1.)

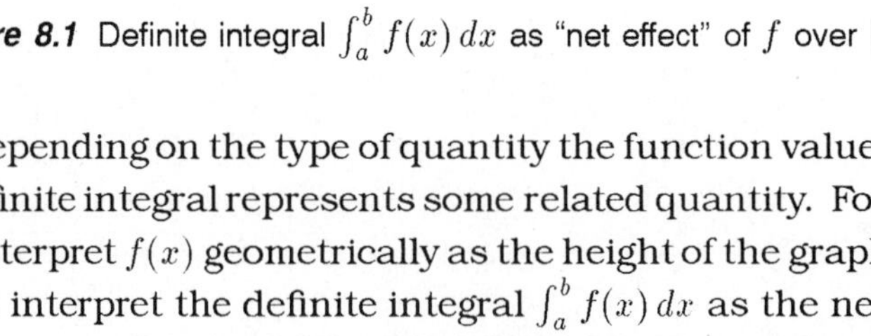

Area of a rectangle is the net effect of its height over an interval.

Definite integral is the net effect of $f(x)$ over an interval $[a, b]$.

Figure 8.1 Definite integral $\int_a^b f(x)\,dx$ as "net effect" of f over $[a, b]$.

Now, depending on the type of quantity the function values $f(x)$ represent, the definite integral represents some related quantity. For example, if we simply interpret $f(x)$ geometrically as the height of the graph of $y = f(x)$, then we can interpret the definite integral $\int_a^b f(x)\,dx$ as the net area *above* the x-axis (as usual, we count area lying below the x-axis as negative). However, if we interpret $f(x)$ as representing something other than height (like velocity or force or area or density), then we can, in turn, interpret the definite integral as something other than net area (like distance or work or volume or mass).

When is a definite integral the right measurement tool?

There are many situations where a quantity can be expressed as a *constant* times the length of an interval. As we just mentioned, a rectangle's area is its *constant height* times its length. In physics, the notion of *work* can be thought of as a *constant force* times a distance traversed. The distance travelled by a moving object with *constant velocity* during a given time interval is simply that velocity times the length of the time interval. In general, the total change in any quantity increasing or decreasing at a *constant rate* over a time interval is the product of that rate times the length of the time interval.

In any of these cases, if we replace the constant with a *function* whose outputs vary continuously over the interval in question, then a definite integral provides the appropriate measurement tool. The reasoning makes use of the fact that the outputs of a continuous function are "relatively constant." You have probably experienced this visually at times while scaling the viewing window for a function graph on a calculator or computer screen. If one zooms in horizontally (in other words, x-axis scaling only) on the graph of a continuous function, then eventually the graph resembles a flat line (see Figure 8.2).

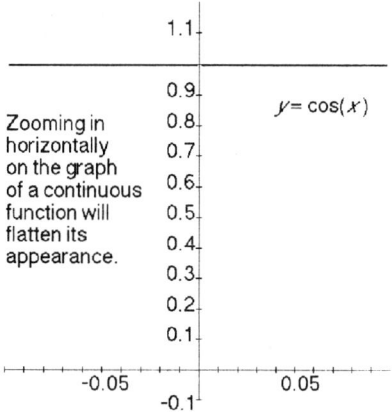

Figure 8.2 Graph of $y = \cos(x)$ under horizontal zoom.

 Do not confuse this property of continuous functions with the idea of local linearity. When we zoom in (using *equal* **scale factors for both the** x**-axis and** y**-axis) on the graph of a** *differentiable function* **we see the graph straighten out and we can estimate its local slope at a point. Differentiability is a** *stronger* **property than continuity.**

EXAMPLE 1   To see the distinction, try zooming in on the graph of $y = |x-1| + 1$ at the point $(1,1)$. If you use equal scale factors both horizontally and vertically, then the corner persists (the function *is not differentiable* at $x = 1$). However, if you zoom in only horizontally, the graph flattens out (the function *is continuous* at $x = 1$). ∎

If we partition an interval $[a, b]$ into sufficiently small subintervals of length Δx, then for many purposes we can treat a continuous function's outputs $f(x)$ as being relatively constant over each subinterval. This means that we can choose a *single* input x from each subinterval and use the corresponding function output $f(x)$ as the (approximately) constant value of the function over that subinterval. If the quantity we want to measure is approximated by

$$f(x)\Delta x$$

for each subinterval, then we can add up all the products to approximate the total quantity. What we have formed is a *Riemann sum approximation* for the definite integral of the continuous function f over the interval $[a, b]$. We might denote this Riemann sum approximation as

$$\sum_{a}^{b} f(x)\, \Delta x,$$

as long as we understand that the value $f(x)$ can change for each of the tiny subintervals, and that we are summing the products $f(x)\Delta x$ for all the subintervals of $[a, b]$.

This Riemann sum approximation becomes more accurate as the partition becomes finer (as Δx is chosen smaller and smaller), and we define

$$\int_{a}^{b} f(x)\, dx = \lim_{\Delta x \to 0} \sum_{a}^{b} f(x)\, \Delta x.$$

We can see the motivation behind the Leibniz notation of the definite integral. The $\int$ symbol denotes the limit of a *sum*, just as the Leibniz notation

$$\frac{dy}{dx} = \lim_{\Delta x \to 0} \frac{\Delta y}{\Delta x}$$

denotes the limit of a ratio.

8.1 USING DEFINITE INTEGRALS TO MEASURE AREA

In applying definite integration as a measurement tool in any situation, we need to identify both an interval of values $[a, b]$ and a functional quantity $f(x)$ defined over that interval. This may require us to set up a coordinate system for reference in a physical situation. When trying to identify the appropriate function over an interval, a good strategy to follow is outlined below:

Step 1. Pick some arbitrary value x in the interval $[a, b]$ and focus your attention on a tiny subinterval $[x, x + \Delta x]$.

Step 2. Calculate the desired quantity just for this subinterval. Any values depending on x should be written in terms of x. The goal is to express the result for this subinterval in the form

$$f(x)\,\Delta x$$

for some functional expression f.

Step 3. Calculate $\int_a^b f(x)\,dx$. If the function f has no simple antiderivative (so that you can use the Fundamental Theorem of Calculus), you will need to use a computer or calculator equipped with the means to numerically approximate the value of the definite integral.

Step 4. Check the reasonableness of the result. If physical quantities are involved, check the units of the result. This can be found from Step 2: the units of the result are found by taking the product of the units of $f(x)$ and the units of x (which are the same as the units of Δx). Do these units make sense? For example, if the value of the integral represents distance travelled, then units of length should be expected. Check the magnitude of the result against some rough upper and lower estimates. A graph of f over the interval $[a, b]$ can be helpful in this regard.

 Of course, there is nothing sacred about the use of the letter x for the independent variable in a physical application. You may find it more convenient to use a different symbol for the independent variable in many cases.

In this chapter, we will examine the use of definite integration to the measurement of area, volume, and curve length. In the next chapter, we'll discuss the use of definite integrals to compute various "averages," and a number of other applications for integration in such fields as physics and probablity.

8.1 USING DEFINITE INTEGRALS TO MEASURE AREA

Cavalieri's principle (named after the Italian mathematician Bonaventur Cavalieri) simply states that the area of a region in a plane is completely determined by the heights of its cross sections (slices made perpendicular to the axis along its length). Roughly speaking, if we know the height of a region at every point along its length, then we should be able to determine the area of the region. We can illustrate the idea of Cavalieri's principle as shown in Figure 8.3.

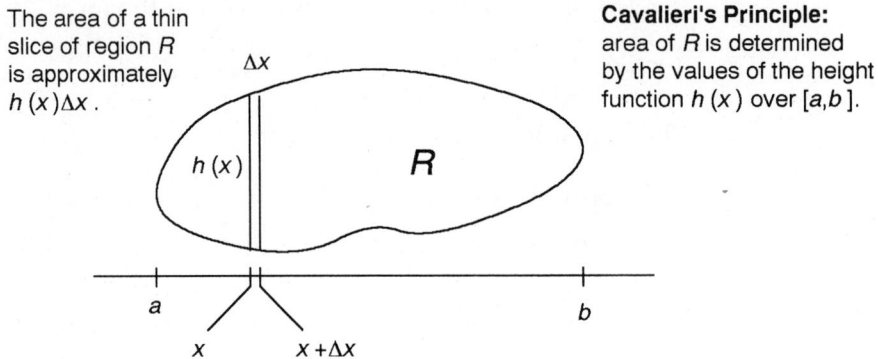

Figure 8.3 Cavalieri's principle: area of $R = \int_a^b h(x)\,dx$.

If we set up a coordinate axis along the length of the region as shown in Figure 8.3 so that the region lies over the interval $[a,b]$, then we can mathematically describe the cross-sectional height $h(x)$ as a function of the points x in $[a,b]$.

In terms of Riemann sums, we could imagine cutting a thin slice of the region between two points x and $x + \Delta x$ as shown in Figure 8.3. If the height h is a continuous function of x, then the area of the slice is approximately $h(x)\Delta x$. Adding up the areas of the slices from a to b and taking the limit as $\Delta x \to 0$ we have

$$\lim_{\Delta x \to 0} \sum_a^b h(x)\Delta x = \int_a^b h(x)\,dx.$$

This description allows us to write down the area of the region R as a definite integral:

$$\text{area of } R = \int_a^b h(x)\,dx.$$

Strictly speaking, there are regions with such strange shapes that either the cross-sectional height function h does not exist or it cannot be integrated. However, for any region whose height varies continuously over

8.1 USING DEFINITE INTEGRALS TO MEASURE AREA

its length (or can be broken down into a finite number of subregions satisfying this property), we could *define* the area of the region as $\int_a^b h(x)\,dx$. In the rest of this section we apply this principle to several examples.

In the simplest case, a single definite integral may give us the desired area.

EXAMPLE 2 Find the area of the region bounded by the x-axis, the vertical lines $x = 1$ and $x = 3$, and the graph of $y = x^2$.

Solution If we graph $y = x^2$ between the vertical lines $x = 1$ and $x = 3$, we can recognize the area of the described region as being given by the definite integral

$$\int_1^3 x^2\,dx = \frac{26}{3} \text{ square units.}$$

(See Figure 8.4.)

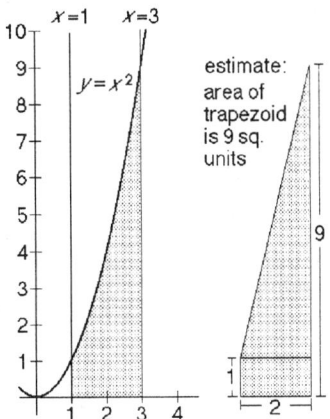

Figure 8.4 $\int_1^3 x^2\,dx$ gives the area of the region.

It is an excellent idea to judge the reasonableness of your area calculation by estimating the area visually. In this example, we can roughly estimate the area to be somewhat less than 9 square units by simply approximating the region with a trapezoid (or, if you prefer, the sum of the areas of a rectangle and triangle) as shown in Figure 8.4. Our exact calculation of $8\frac{2}{3}$ square units looks very reasonable in light of this estimate.

Powerful numerical integration capabilities exist in many computer software packages and on some hand-held calculators. These machines allow us to quickly calculate many definite integrals very easily to high precision without the need for finding an antiderivative. However, a machine cannot tell us whether the definite integral we ask it to compute is the *right* one. Visually estimating the area as a check is not foolproof, but it can often help you catch errors like a wrong button push or an incorrect set-up. For example, if we had accidentally integrated $\int_1^3 x\,dx$ instead of $\int_1^3 x^2\,dx$, we would have obtained an area of 4. This is much smaller than our estimate of 9, and so this should prompt us to check our work very carefully. With luck, we might spot the missing exponent in our integrand.

It's important to keep in mind that when a function's graph lies below the x-axis for an interval $[a, b]$ (in other words, $f(x) < 0$) then the definite integral $\int_a^b f(x)$ is *negative*.

EXAMPLE 3 Find the total area enclosed between the graph of $y = \sin(x)$ and the x-axis for $0 \le x \le 2\pi$.

Solution If we simply integrate over $[0, 2\pi]$ we find

$$\int_0^{2\pi} \sin(x)\,dx = -\cos(x)\Big]_0^{2\pi} = -\cos(2\pi) - (-\cos(0)) = -1 - (-1) = 0.$$

What's going on here? If we look at the graph of $y = \sin(x)$ over this interval, we see the problem (Figure 8.5).

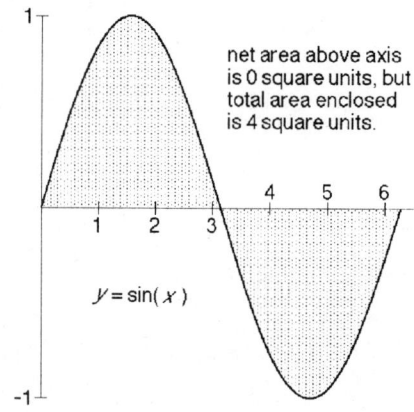

net area above axis is 0 square units, but total area enclosed is 4 square units.

Figure 8.5 Net area versus total area.

The definite integral of $\sin(x)$ over $[0, \pi]$ is exactly cancelled out by its definite integral over $[\pi, 2\pi]$. If we want the actual area enclosed by the graph and the x-axis over $[0, 2\pi]$, then we'll need to take the absolute value of the definite integral over $[\pi, 2\pi]$ in our calculation:

$$\int_0^{\pi} \sin(x)\,dx = -\cos(x)\Big]_0^{\pi} = -\cos(\pi) - (-\cos(0)) = -(-1) - (-9) = 2,$$

8.1 USING DEFINITE INTEGRALS TO MEASURE AREA

while $\int_{\pi}^{2\pi} \sin(x)\,dx = -2$. Thus, the total area is $2 + |-2| = 4$. ∎

In general, calculating the area of a region bounded by two function graphs $y = f(x)$ and $y = g(x)$ may first require finding the intersection points of the two graphs (solve $f(x) = g(x)$ for x) to determine the appropriate limits of integration. The value of the height function $h(x)$ will be either $f(x) - g(x)$ (in the case that $f(x) \geq g(x)$) or $g(x) - f(x)$ (in the case that $f(x) < g(x)$). In either case, we could simply write down the area of the region as

$$\int_a^b |f(x) - g(x)|\,dx$$

where a and b are the first and last intersection points, and the absolute value signs in the integrand take care of both possibilites at once.

Our next area example exploits the graphing, root-finding, and integration capabilities of a calculator or computer.

EXAMPLE 4 Find the area bounded by the graphs of $y = e^{-x^2}$ and $y = x^3 - x + 1$.

Solution First, we graph both functions simultaneously to obtain a plot like that pictured in Figure 8.6.

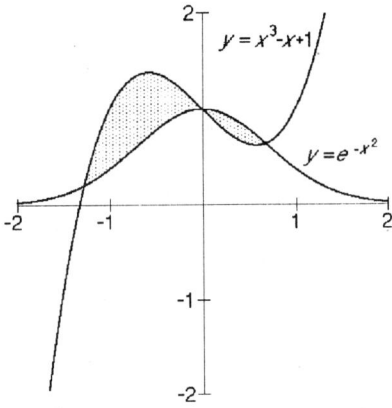

Figure 8.6 Graphs of $y = x^3 - x + 1$ and $y = e^{-x^2}$.

By using a machine's root-finder on the difference of the two functions $(x^3 - x - 1) - e^{-x^2}$, we find three intersection points, at $a \approx -1.276638$, at 0, and at $b \approx 0.676228$. The intersection at $x = 0$ is easy to check, and the other two look quite reasonable from the graphs. (Can you tell why these are the only three intersection points?)

To calculate the area bounded by these two curves, we used a machine's numerical integrator to calculate

$$\int_{-1.276638}^{0.676228} |(x^3 - x + 1) - e^{-x^2}| \, dx \approx 0.6902.$$

Without using absolute values, we need to express the area as the sum of two definite integrals:

$$\int_{-1.276638}^{0} ((x^3 - x + 1) - e^{-x^2}) \, dx + \int_{0}^{0.676228} (e^{-x^2} - (x^3 - x + 1)) \, dx.$$

∎

Sometimes curves are more easily or naturally described as graphs of the form $x = f(y)$. To calculate the area enclosed by two curves of this type, we can integrate the "width function" of the region over an interval of y-values.

EXAMPLE 5 Find the area of the region bounded by the graphs of $x = y^2$ and $x = 4 - y^2$.

Solution The graphs of $x = y^2$ and $x = 4 - y^2$ are shown in Figure 8.7.

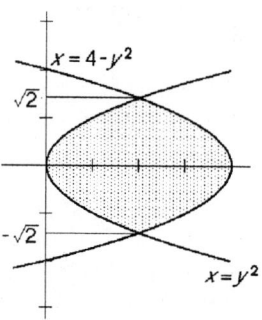

Figure 8.7 Region bounded by the graphs of $x = y^2$ and $x = 4 - y^2$.

The two curves intersect when $y = \pm\sqrt{2}$. For any value y such that $-\sqrt{2} \leq y \leq \sqrt{2}$, the horizontal width of the region is $(4 - y^2) - y^2 = 4 - 2y^2$. We can calculate the area of the region using the definite integral

$$\int_{-\sqrt{2}}^{\sqrt{2}} (4 - 2y^2) \, dy = (4y - 2y^3/3) \Big]_{-\sqrt{2}}^{\sqrt{2}} = 16\sqrt{2}/3 \approx 7.5425.$$

∎

8.1 USING DEFINITE INTEGRALS TO MEASURE AREA

EXERCISES

In exercises 1-15, find the area of the indicated region using definite integration. Graph each region and judge the reasonableness of your answer. Some of the exercises may require the help of a machine.

1. Region enclosed between the graph of $f : x \mapsto x$ and the x-axis between $x = 0$ and $x = 5$.

2. Region enclosed between the graph of $f : x \mapsto x^2 - 9$ and the x-axis.

3. Region enclosed between the graph of $f : x \mapsto \frac{1}{2}\sin x$ and the x-axis between $x = 0$ and $x = \pi$.

4. Region enclosed between the graph of $f : x \mapsto e^{x-1}$ and the x-axis between $x = 0$ and $x = 3$.

5. Region enclosed between the graph of $f : x \mapsto \frac{|x|}{4} - 6$ and the x-axis.

6. Region enclosed between the graph of $f : x \mapsto e^{3x} - 3x - 1$ and the x-axis.

7. Region enclosed between the graph of $f : x \mapsto |2 - x|$ and the line $y = 1$.

8. Region enclosed between $y = \ln(x)$, the vertical line $x = 2$, and the horizontal line $y = -2$.

9. Region bounded by $x = 8 - y^2$ and $x = -y + 2$.

10. Region bounded by $x = y^3 - 3$ and $x = 1$.

11. Region bounded by $y = \arcsin(x)$ and $y = \arccos(x)$ between $y = \pi/6$ and $y = \pi/3$. (Integrate in terms of y.)

12. Region bounded by $y = \sin \frac{1}{2}x$ and $y = \frac{1}{2}\sin x$ between 0 and π.

13. Region bounded by $y = \arctan(x)$, $y = \operatorname{arccot}(x)$ and the y-axis.

14. Region bounded by $y = \dfrac{3}{1 + x^2}$ and $y = e^{x^2}$.

15. Region bounded by $y = \cos(x)$ and $y = (x + 1)/3$.

---

16. Find the constant a so that the area bounded by $y = x^2 - a^2$ and $y = a^2 - x^2$ is exactly 72.

17. The area bounded by the curve $y = x^2$ and the line $y = 4$ is divided into two equal portions by the line $y = c$. Find c.

18. Calculate $\int_{-1}^{1} \sqrt{1 - x^2}\, dx$. The number π is sometimes *defined* as a certain integer multiple of this integral. What integer multiple should be used? Graph $y = \sqrt{1 - x^2}$ and describe the region whose area the definite integral represents.

19. Find the area enclosed by the ellipse $1 = \dfrac{x^2}{25} + \dfrac{y^2}{49}$.

20. Find the area enclosed by the ellipse $1 = \dfrac{x^2}{a^2} + \dfrac{y^2}{b^2}$ where a and b are positive constants.

In exercises 21-28, you are given the graph of a function f against a grid. Assuming that the grid lines are spaced 1 unit apart both horizontally and vertically, estimate the value of $\int_{-6}^{6} |f(x)|\, dx$ for each function. (If the graph does not appear over certain parts of the interval, assume that $f(x) = 0$ for these inputs.)

21.

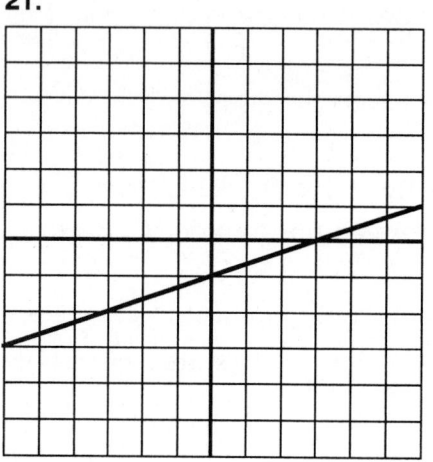

22.

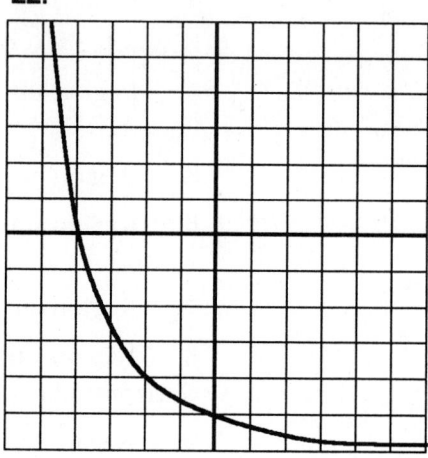

23.

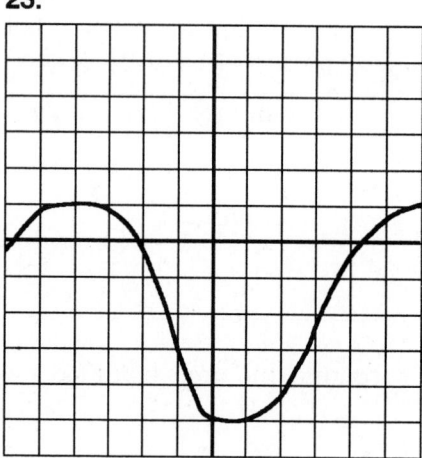

24.

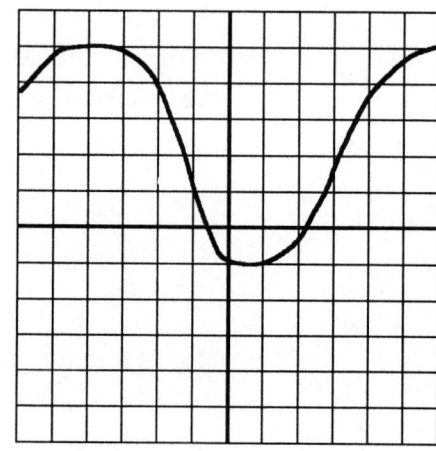

25.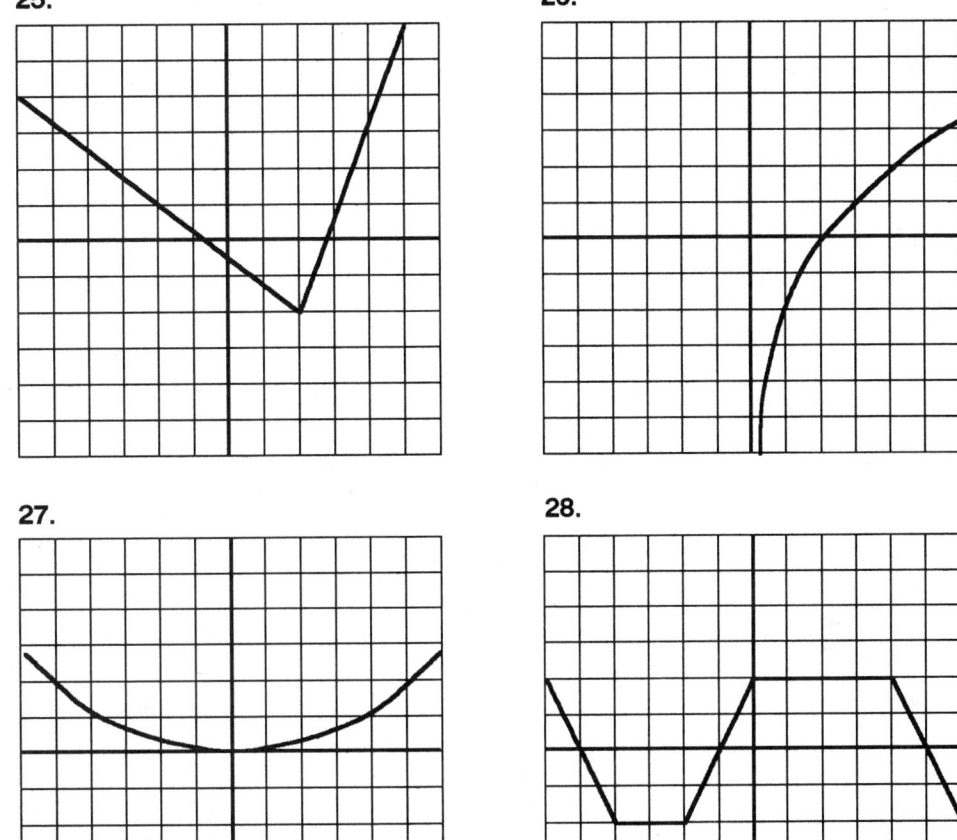

26.

27.

28.

29. One can of paint will cover 50 ft². How many cans of paint should you buy to cover the region bounded by $y = -x^2 + 49$, the x-axis, and y-axis, assuming that units along the axes represent feet?

8.2 USING DEFINITE INTEGRALS TO MEASURE VOLUME

Cavalieri's principle can also be stated for objects or regions in space: the volume of a region in space is completely determined by the areas of its cross-sections (sliced perpendicular to an axis along its length).

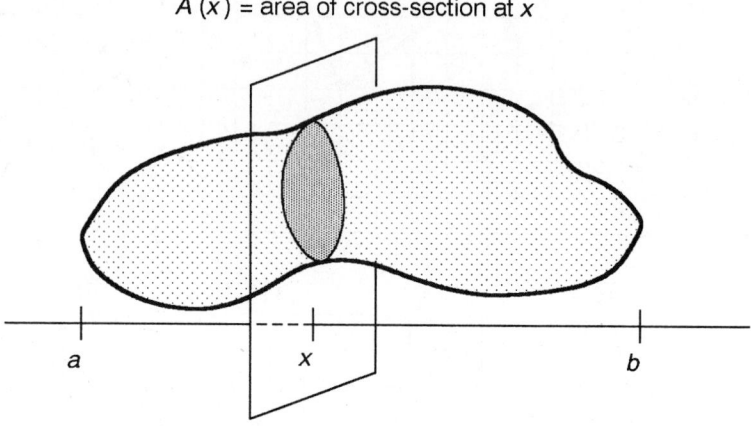

Figure 8.8 The volume of the region is $\int_a^b A(x)\,dx$.

Figure 8.8 illustrates a solid object over the interval $[a, b]$. The cross-sectional area $A(x)$ is the area of the plane region determined by slicing the object with a plane perpendicular to the axis at the point x. In terms of a definite integral the volume of the region is given by Volume $= \int_a^b A(x)\,dx$.

The reasoning is similar to that given for finding areas. If we slice a very thin "slab" of the object with two planes at x and $x + \Delta x$, then the volume of the slab is its base area times its thickness (see Figure 8.9).

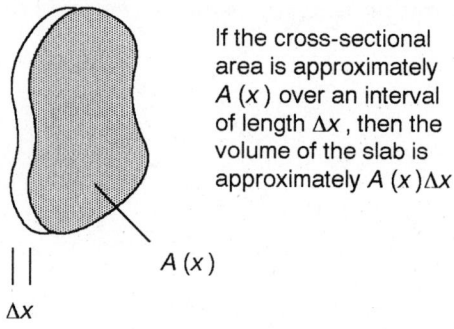

Figure 8.9 Approximating the volume of a thin slice of the region.

8.2 USING DEFINITE INTEGRALS TO MEASURE VOLUME

Summing up the volumes of all the slabs from a to b and taking the limit as $\Delta x \to 0$ we have

$$\lim_{\Delta x \to 0} \sum_a^b A(x)\Delta x = \int_a^b A(x)\,dx.$$

EXAMPLE 6 An object lies over the interval $[2, 5]$ and its cross-section at each point is a square of length x. Find its volume.

Solution Although there is not enough information to really tell exactly what the object looks like (there are infinitely many objects fitting the description), there is enough information to determine the volume of the object (all such objects have exactly the same volume). You might think of a deck of square cards as an illustration: the volume of the deck doesn't depend on how neatly the deck is stacked up.

The cross-sectional area function is given as $A(x) = x^2$ over the interval $[2, 5]$, so we can calculate the volume as

$$\int_2^5 x^2\,dx = \left.\frac{x^3}{3}\right]_2^5 = \frac{125}{3} - \frac{8}{3} = \frac{117}{3} = 39 \text{ cubic units.}$$

■

EXAMPLE 7 Find the volume of a sphere of radius 5.

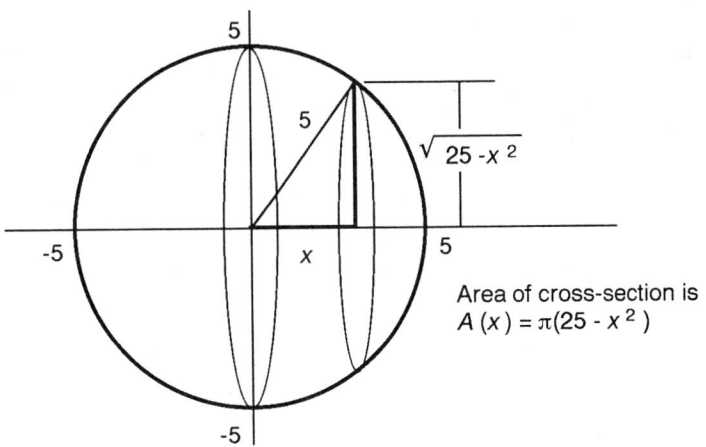

Figure 8.10 Cross-sections of a sphere are circles.

Solution Let's imagine an axis running along a diameter of the sphere with its origin at the center, so that the sphere lies along the interval $[-5, 5]$. The cross-sections of the sphere perpendicular to this axis are circular disks, and for each $x \in [-5, 5]$, the radius of the corresponding cross-section is seen to be $r(x) = \sqrt{25 - x^2}$ using the Pythagorean Theorem (see Figure 8.10). Note that the radical symbol indicates the positive square root, so this formula is valid for both positive and negative values $x \in [-5, 5]$.

Hence, the area of each cross-section is $A(x) = \pi r(x)^2 = \pi(25 - x^2)$. A good check of the formula is to try some extreme or special values x: $A(-5) = A(5) = 0$, indicating the cross-sectional area is 0 at the two ends of the sphere; $A(0) = 25\pi$, indicating that the cross-sectional area at the center of the interval is that of a circle of radius 5.

We can find the volume (in cubic units) of the sphere by integrating $A(x)$ over the interval $[-5, 5]$:

$$V = \int_{-5}^{5} \pi(25 - x^2)\, dx = \pi\left(25x - \frac{x^3}{3}\right)\Bigg]_{-5}^{5} = \frac{250\pi}{3} - \left(-\frac{250\pi}{3}\right) = \frac{500\pi}{3}.$$

We can see that this matches exactly the value we obtain using the volume formula for a sphere $V = 4\pi R^3/3$ with $R = 5$. ∎

In fact, we could use definite integration to *establish* this volume formula by replacing the specific radius 5 in this example with an unspecified spherical radius R. Now the sphere lies along the interval $[-R, R]$, and for each x, the radius of the circular cross-section is $\sqrt{R^2 - x^2}$. The area function is given by

$$A(x) = \pi(R^2 - x^2).$$

If we integrate A over the interval $[-R, R]$, we obtain

$$V = \int_{-R}^{R} \pi(R^2 - x^2)\, dx = \pi\left(R^2 x - \frac{x^3}{3}\right)\Bigg]_{-R}^{R} = \frac{2\pi R^3}{3} - \left(-\frac{2\pi R^3}{3}\right) = \frac{4\pi R^3}{3}.$$

Many volume formulas can be established similarly with definite integration.

EXAMPLE 8 Find the volume of a spike six inches long whose cross-sections are equilateral triangles with sides of length 1/6 the distance from the point of the spike (so the side length at the blunt end is exactly 1 inch).

Solution If we set up a coordinate axis with the point of the spike at $a = 0$ and the other end at $b = 6$, then the cross-section at any point $0 \leq x \leq 6$ is an equilateral triangle with side length $x/6$ inches. We can calculate the height of this triangle using the Pythagorean Theorem to be $h(x) = \sqrt{3}x/12$ (see Figure 8.11).

8.2 USING DEFINITE INTEGRALS TO MEASURE VOLUME

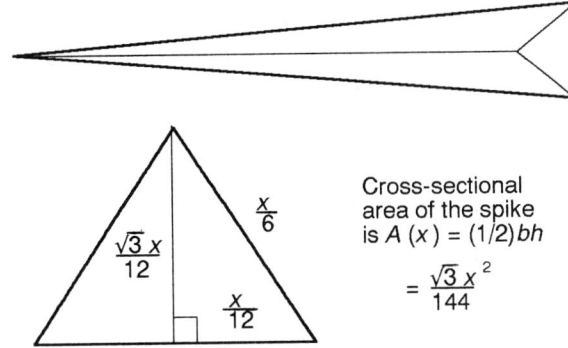

Figure 8.11 Finding the cross-sectional area function for the spike.

Thus, the area of the triangle is

$$A(x) = (1/2)(x/6)(\sqrt{3}x/12) = \sqrt{3}x^2/144.$$

We integrate $A(x)$ over the interval $[0, 6]$ to obtain the volume V of the spike:

$$V = \int_0^6 \sqrt{3}x^2/144\, dx = \sqrt{3}x^3/432 \Big]_0^6 = \sqrt{3}/2 \text{ cubic inches.}$$

Solids of revolution

The sphere is an example of a solid of revolution. A solid of revolution is any object obtained by rotating a plane region through space about some axis. For example, if we rotate the region enclosed by a circle about the axis of its diameter, we obtain a sphere. If we rotate a rectangle about an axis along one of its sides, we get a cylinder (see Figure 8.12).

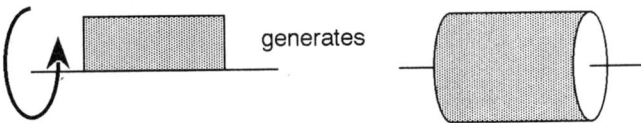

Figure 8.12 A cylinder is a solid of revolution generated by a rectangle.

Objects produced on machine or wood lathes or on pottery wheels are examples of more general solids of revolution. If we slice a solid of revolution perpendicular to the axis we always get one of two shapes.

1) If the axis of rotation is in contact with the region, then the cross-section will be a "disk." The radius r of the disk is determined by the furthest distance from the axis to the outer edge of the region, and the area of the disk is πr^2.

2) If there is any space between the region and the axis of rotation then the cross-section will be an annulus or "washer." In the case of an annulus, the area is the difference $\pi R^2 - \pi r^2$, where r is the inner radius (of the hole) and R is the outer radius.

In Figure 8.13 we can see the possible resulting cross-sections that can be obtained by rotating a line segment about a perpendicular axis.

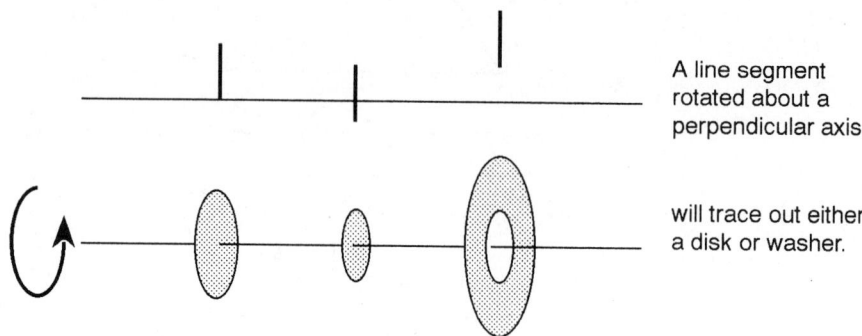

Figure 8.13 Possible cross-sections for a solid of revolution.

To calculate the value of the cross-sectional area function $A(x)$ at each point x along the axis of a solid of revolution, we need to determine first whether the cross-section is a disk or washer, and then we must express the radius ($r(x)$ in the case of a disk) or radii ($R(x)$ and $r(x)$ in the case of a washer) in terms of x in the appropriate area formula. Once we have determined the area function A, we can integrate it from one end of the solid to the other to find its volume.

No single simple formula handles all possible regions that might be rotated about an axis. Given a region and an axis of rotation, you may find it difficult to visualize all at once what the resulting solid of revolution looks like. However, if you keep in mind that only two kinds of cross-sections are possible, then you should be able to visualize what a single cross-section looks like at any specific point. For the purposes of determining the volume of the solid, this is exactly what we need to determine the cross-sectional area value $A(x)$ at the point x.

EXAMPLE 9 Find the volume of the solid of revolution obtained by rotating the region bounded by $y = x^2$, the x-axis, the vertical lines $x = 1$ and $x = 2$ about the x-axis.

Solution The region is pictured in Figure 8.14. Since the region sets upon the axis of revolution, each cross-section is a disk having radius $r(x) = x^2$ for $1 \leq x \leq 2$. The area of each disk is

$$A(x) = \pi(r(x))^2 = \pi(x^2)^2 = \pi x^4.$$

8.2 USING DEFINITE INTEGRALS TO MEASURE VOLUME

The volume of the solid of revolution can be computed by integrating:

$$\int_1^2 \pi x^4 \, dx = \pi \frac{x^5}{5} \Big]_1^2 = \frac{32\pi}{5} - \frac{\pi}{5} = \frac{31\pi}{5} \text{ cubic units.}$$

∎

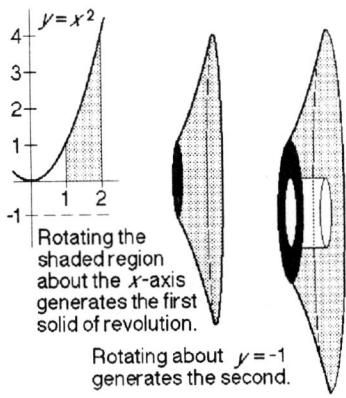

Figure 8.14 Two solids of revolution generated by the same region.

EXAMPLE 10 Find the volume of the solid of revolution obtained by rotating the region bounded by $y = x^2$ the x-axis, the vertical lines $x = 1$ and $x = 2$ about the line $y = -1$.

Solution The region is the same as in the previous example (see Figure 8.14). The only difference now is that we are rotating about a line 1 unit away from the nearest edge of the region. Each cross-section is a washer having outer radius $R(x) = x^2 + 1$ and inner radius $r(x) = 1$ for $1 \leq x \leq 2$. The area of each washer is

$$A(x) = \pi[R(x)^2 - r(x)^2] = \pi[(x^4 + 2x^2 + 1) - 1^2] = \pi(x^4 + 2x^2).$$

The volume of the solid of revolution can be computed by integrating:

$$\int_1^2 \pi(x^4 + 2x^2) \, dx = \pi \left(\frac{x^5}{5} + \frac{2x^3}{3}\right)\Big]_1^2 = \pi\left[\left(\frac{32}{5} + \frac{16}{3}\right) - \left(\frac{1}{5} + \frac{2}{3}\right)\right] \approx 34.14 \text{ cubic units.}$$

∎

EXAMPLE 11 Express the volume of the solid of revolution obtained by rotating the region bounded by $y = \sin(x)$ and $y = \cos(x)$ over the interval $[\pi/4, 5\pi/4]$ about the x-axis.

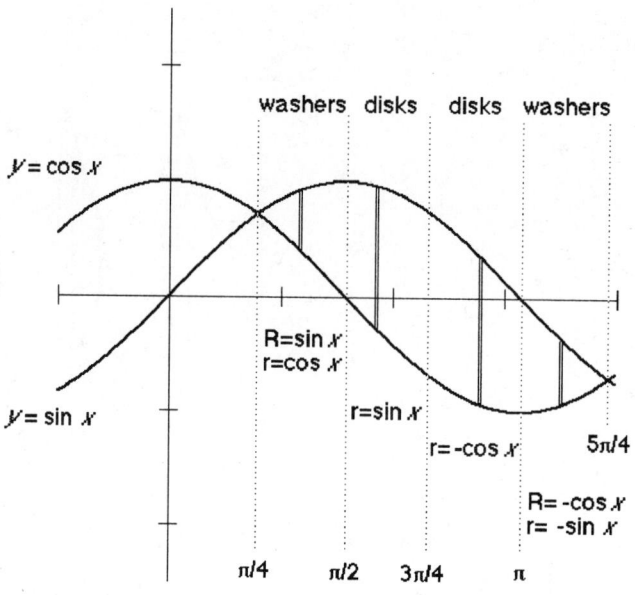

Figure 8.15 Analyzing cross-sections for a solid of revolution.

Solution The region in question is illustrated in Figure 8.15. For any particular value $x \in [\pi/4, 5\pi/4]$, a slice of the region will generate either a disk cross-section or a washer, depending on whether or not the slice is in contact with the axis of revolution. Figure 8.15 shows how the cross-sections vary in form as we move through the interval $[\pi/4, 5\pi/4]$:

For $\pi/4 \leq x \leq \pi/2$, the slices generate washers with an outer radius $R(x) = \sin(x)$ and an inner radius $r(x) = \cos(x)$. Hence, $A(x) = \pi[\sin^2(x) - \cos^2(x)]$, and

$$\int_{\pi/4}^{\pi/2} A(x)\,dx = \pi/2.$$

For $\pi/2 \leq x \leq 3\pi/4$, the slices generate disks with a radius $r(x) = \sin(x)$ because the graph of $y = \sin(x)$ is furthest away from the axis of revolution over this interval. Hence, $A(x) = \pi \sin^2(x)$ for these values, and

$$\int_{\pi/2}^{3\pi/4} A(x)\,dx = \pi^2/8 + \pi/4.$$

For $3\pi/4 \leq x \leq \pi$, the slices generate disks with a radius $r(x) = -\cos(x)$ because the graph of $y = \cos(x)$ is furthest away from the axis of revolution over this interval, and $\cos(x) < 0$. Hence, $A(x) = \pi \cos^2(x)$ for these values, and

$$\int_{3\pi/4}^{\pi} A(x)\,dx = \pi^2/8 + \pi/4.$$

For $\pi \le x \le 5\pi/4$, the slices generate washers with an outer radius $R(x) = -\cos(x)$ and an inner radius $r(x) = -\sin(x)$. Hence, $A(x) = \pi[\cos^2(x) - \sin^2(x)]$ for these values, and

$$\int_{\pi}^{5\pi/4} A(x)\,dx = \pi/2.$$

Adding the results over the subintervals, we find that the total volume of this solid of revolution is $\pi^2/4 + 3\pi/2 \approx 7.17979$ cubic units. ∎

Calculating volume using cylindrical shells

For the special case of a solid of revolution, there is an alternative method of calculating the volume that is quite different from the general cross-sectional area method (which is valid for *any* solid whose cross-sections vary continuously along the axis). This method is sometimes preferred because the resulting definite integral may be easier to calculate using the Second Fundamental Theorem of Calculus (in other words, the antiderivative of the integrand may be easier to recognize). Of course, in the presence of computing machines capable of approximating definite integrals to great precision, this concern may justifiably take a back seat to the concern of correctly modelling the volume of the object in some way as the value of a definite integral.

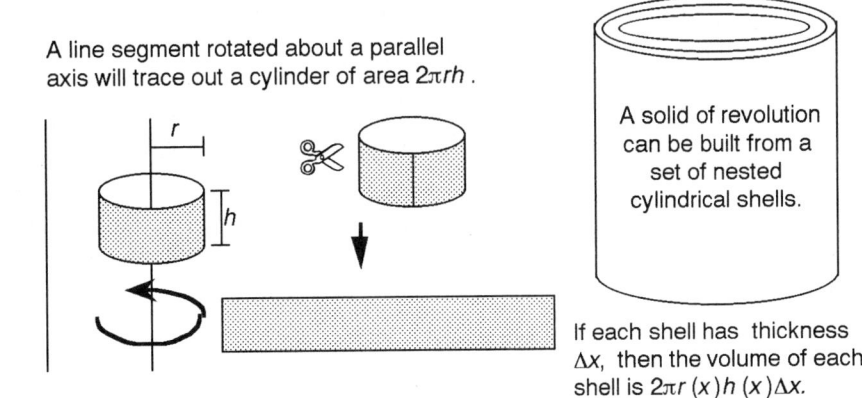

Figure 8.16 Rotation about a parallel axis.

The idea behind the method is to take slices of the rotated region that are *parallel* to the axis of rotation. When rotated about the axis, such a slice traces out a *cylindrical shell* centered about the axis of rotation. If the slice is thin enough, then a good approximation for the volume of this shell is the inner surface area of the shell times the thickness of the slice (see Figure 8.16). Hence, the volume of a thin cylindrical shell is $2\pi r(x)h(x)\Delta x$, where $r(x)$ is the distance from the parallel axis of rotation, $h(x)$ is the height of the shell, and Δx is the thickness of the shell.

If we start at the axis of rotation and sum up the volumes of these nested shells as we move out to the extreme edge of the solid, then we have an approximation for the total volume of the solid. Taking the limit of these approximations as the thickness of the slices approaches zero, we obtain a definite integral representing the exact volume of the solid:

$$\text{Volume} = \int_a^b 2\pi r(x)h(x)\,dx$$

where $x = a$ is the location of the innermost shell, and $x = b$ is the location of the outermost shell.

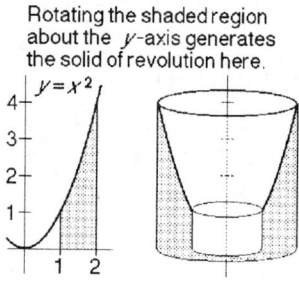

Figure 8.17 Another solid of revolution.

EXAMPLE 12 Find the volume of the solid of revolution obtained by rotating the region bounded by $y = x^2$ the x-axis, the vertical lines $x = 1$ and $x = 2$ about the y-axis.

Solution The region is shown in Figure 8.17. Each vertical slice of the region corresponding to a particular value $x \in [1, 2]$ will trace out a cylinder whose height is $h(x) = x^2$, and whose radius from the y-axis is $r(x) = x$. Hence, the volume of the solid is given by the definite integral

$$\int_1^2 2\pi(x^2)x\,dx = 2\pi \int_1^2 x^3\,dx = \pi x^4/2 \Big]_1^2 = 15\pi/2 \text{ cubic units.}$$

8.2 USING DEFINITE INTEGRALS TO MEASURE VOLUME

EXERCISES

In exercises 1-6, find the volume of an object whose cross-section at x is a square whose side-length is given by the indicated function. Some of the exercises may require the help of a machine.

1. $f : x \mapsto x$ between $x = 0$ and $x = 5$.
2. $f : x \mapsto 9 - x^2$ between $x = -3$ and $x = 3$
3. $f : x \mapsto \frac{1}{2} \sin x$ between $x = 0$ and $x = \pi$.
4. $f : x \mapsto e^{x-1}$ between $x = 0$ and $x = 3$.
5. $f : x \mapsto \frac{|x|}{4}$ between $x = -4$ and $x = 2$.
6. $f : x \mapsto e^{3x} - 1$ between $x = 0$ and $x = 1$.

In exercises 7-15, find the volume of the solid of revolution generated by rotating the described region about a) the x-axis, and b) the y-axis. For each problem, you should graph the region and determine carefully the cross-sectional area function. Then set up the required definite integral or integrals to calculate the volume. A machine root-finder and numerical integrator may be necessary to compute some of the resulting integrals.

7. Region enclosed between the graph of $f : x \mapsto |2 - x|$ and the line $y = 1$.
8. Region bounded by $y = \arctan(x)$, $y = \text{arccot}(x)$ and the y-axis.
9. Region bounded by $y = \dfrac{3}{1 + x^2}$ and $y = e^{x^2}$.
10. Region bounded by $y = \cos(x)$ and $y = (x + 1)/3$.
11. Region enclosed between $y = \ln(x)$, the vertical line $x = 2$, and the horizontal line $y = -2$.
12. Region bounded by $x = 8 - y^2$ and $x = -y + 2$.
13. Region bounded by $x = y^3 - 3$, $x = 1$, and the coordinate axes.
14. Region bounded by $y = \arcsin(x)$ and $y = \arccos(x)$ between $y = \pi/6$ and $y = \pi/3$ is rotated about the y-axis. (Integrate in terms of y.)
15. Region bounded by $y = \sin \frac{1}{2}x$ and $y = \frac{1}{2} \sin x$ between $x = 0$ and $x = \pi$.

---

16. The handle of a screwdriver is 4 inches long, and the tip of the blade is 4 inches from the handle. The handle has cross-sections in the shape of regular hexagons. At the fat end, the handle is 1 inch in diameter. At the narrow end, the handle is 1/2 inch in diameter. The diameter at any point in between is 1/8 the distance from the blade tip. Find the volume of material in the handle.

17. Find a linear function f and an interval $[a, b]$ so that the region between the graph of f and the x-axis between $x = a$ and $x = b$ generates a cone

(pointing left) of length 10 and radius 4 when rotated about the x-axis. Find the volume of this cone.

18. A cone of base radius R and height H can be thought of as the solid of revolution obtained by rotating the region bounded by $y = (R/H)x$, the x-axis, and the line $x = H$ about the $x-axis$. Use this to derive the volume formula for a cone by definite integration.

19. A *torus* is a doughnut-shaped region obtained by rotating a circle about an axis not intersecting the circle. If R is the distance from the center of the circle to the axis, and r is the radius of the circle, then the volume of the torus is known to be

$$V = 2\pi^2 R r^2.$$

Establish this formula by calculating the volume of the solid of revolution obtained by rotating the region between the graphs of $y = R + \sqrt{r^2 - x^2}$ and $y = R - \sqrt{r^2 - x^2}$ about the x-axis.

20. Suppose a container is in the shape of the solid of revolution obtained when the region bounded by $y = -x^2 + 49$, the x-axis, and y-axis is rotated about the y-axis. Assuming that the units along the axes represent cm, how many liters of water would it take to *fill* the solid?

Using formulas for the volumes of cones and cylinders, determine the volumes of the solids of revolution described in exercises 21-24, given that f and g have the graphs shown below.

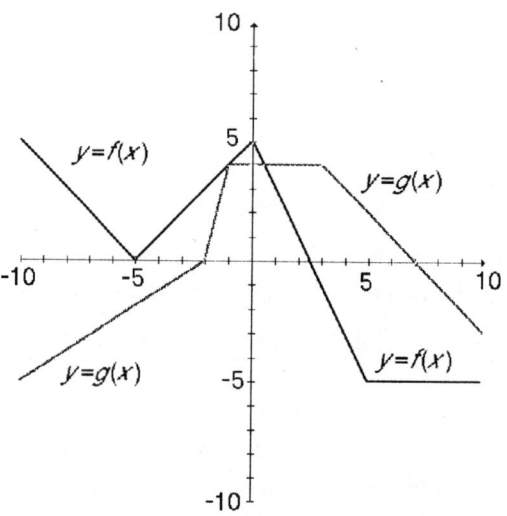

21. Region between the graph of f and the x-axis between $x = 5$ and $x = 10$ rotated about the x-axis.

22. Region between the graph of g and the x-axis between $x = 0$ and $x = 7$ rotated about the x-axis.

23. Region between the graph of f and the x-axis between $x = -5$ and $x = 0$ rotated about the y-axis.

24. Region between the graph of g and the x-axis between $x = -10$ and $x = -2$ rotated about the y-axis.

8.3 USING DEFINITE INTEGRALS TO MEASURE ARC LENGTH

When a curve or path is polygonal (that is, consisting of straight line segments laid end-to-end), then it is a simple matter to compute its total length by simply summing the lengths of the individual segments. In this section, we turn to the measurement problem of calculating the length of a more general curve.

The main idea is simple enough: if we subdivide a smooth curve into several small pieces, then each piece may be approximated by a straight line segment. Summing the lengths of these segments may provide a good approximation to the total length of the curve (see Figure 8.18).

Figure 8.18 Approximating a curve with a polygonal path.

The closer together we choose the subdivision points, the better we expect our approximation to be. Since the shortest distance between two points is along a straight line, our polygonal approximations will always be underestimates of the total length of the curve. Indeed, we *define* the **arc length** of a curve to be the single limiting value (if it exists) of these approximations as we increase the number of subdivision points to be arbitrarily large. (It is possible for a curve to have an infinite or undefined arclength if it "zigs" and "zags" infinitely often between its endpoints.)

In this section we discuss how arc length can be determined by means of definite integration.

Arc length of a function graph

Suppose the curve or path of interest can be described as the graph of a differentiable function $y = f(x)$ over an interval $[a, b]$. If we partition the interval $[a, b]$ into small subintervals of equal length Δx, and connect the points on the graph corresponding to the partition points, then we have a polygonal approximation to the graph. If we zoom in on the graph over one of these subintervals then the graph will appear approximately straight (since a differentiable function is approximately locally linear). Figure 8.19 shows a magnified view of the graph of such a function over one of these tiny intervals.

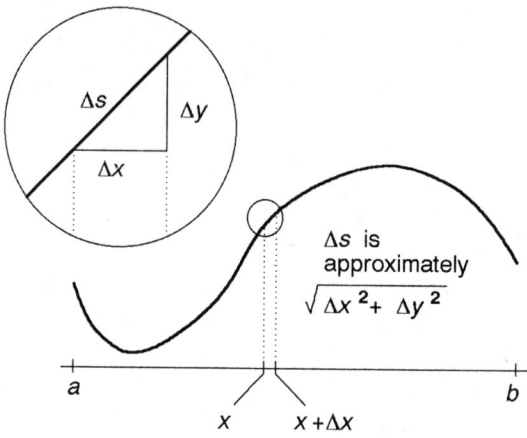

Figure 8.19 Approximating arc length.

We can see that the straight line segment approximation to the curve can be thought of as the hypotenuse of a right triangle with a horizontal leg of length Δx and a vertical leg of length Δy. If we denote the length of this hypotenuse Δs, then the length of the line segment is given by the Pythagorean Theorem:

$$\Delta s \approx \sqrt{(\Delta x)^2 + (\Delta y)^2}.$$

If we factor out $\Delta x = \sqrt{(\Delta x)^2}$, we can rewrite this as

$$\Delta s \approx \Delta x \sqrt{1 + \left(\frac{\Delta y}{\Delta x}\right)^2}.$$

Since f is a differentiable function, the slope of the line segment is

$$\frac{\Delta y}{\Delta x} \approx f'(x)$$

for a point x in the subinterval.

8.3 USING DEFINITE INTEGRALS TO MEASURE ARC LENGTH

Using this approximation, we have

$$\Delta s \approx \sqrt{1 + f'(x)^2}\,\Delta x.$$

Now we sum these lengths Δs for each subinterval from a to b, and take the limit as the subinterval length $\Delta x \to 0$. If the limit exists, we have

$$\text{total arc length} = \lim_{\Delta x \to 0} \sum_a^b \Delta s = \int_a^b \sqrt{1 + [f'(x)]^2}\,dx.$$

EXAMPLE 13 Find the total arc length of the graph of $y = \sin(x)$ over one period.

Solution The function $f : x \mapsto \sin(x)$ has a period of length 2π, so any interval of length 2π will serve our purposes. If we choose $[0, 2\pi]$ as the interval, then

$$\text{total arc length} = \int_0^{2\pi} \sqrt{1 + [f'(x)]^2}\,dx = \int_0^{2\pi} \sqrt{1 + \cos^2 x}\,dx \approx 7.640.$$

∎

EXAMPLE 14 Find the total arc length of the graph of $y = x^3$ between $x = -1$ and $x = 1$.

Solution Since $\dfrac{d}{dx}(x^3) = 3x^2$, we can calculate the arc length of the graph as

$$\int_{-1}^1 \sqrt{1 + (3x^2)^2}\,dx = \int_{-1}^1 \sqrt{1 + 9x^4}\,dx \approx 3.096.$$

∎

In both of these examples, we used a calculator to approximate the value of the definite integral. Definite integrals for the arc length of even very common function graphs are often difficult or impossible to calculate using the Second Fundamental Theorem of Calculus.

Calculating the arc length of a parametrized curve

Recall that the coordinates (x, y) of the points on a curve in the plane are sometimes described as two separate functions of a parameter t. If we think of t as representing time, then the curve may be thought of as the path traced out by the points

$$(x(t), y(t)) \qquad \text{for } a \leq t \leq b.$$

In this situation, the derivatives dx/dt and dy/dt are the instantaneous rates of change in the x and y coordinates of the curve with respect to t. That is, over a very tiny time interval of length Δt, the change in position on the curve will be approximately $\Delta x = \dfrac{dx}{dt}\Delta t$ in the horizontal direction and $\Delta y = \dfrac{dy}{dt}\Delta t$ in the vertical direction. We can see that the actual distance Δs travelled along the curve is approximately

$$\Delta s \approx \sqrt{(\frac{dx}{dt}\Delta t)^2 + (\frac{dy}{dt}\Delta t)^2} = \sqrt{(\frac{dx}{dt})^2 + (\frac{dy}{dt})^2}\Delta t.$$

If we sum the lengths Δs for each subinterval of time, we have

$$\text{total arclength} = \lim_{\Delta t \to 0} \sum_a^b \Delta s = \int_a^b \sqrt{(\frac{dx}{dt})^2 + (\frac{dy}{dt})^2}\, dt.$$

EXAMPLE 15 Find the arc length of the curve parametrized by

$$x(t) = t^3 \qquad y(t) = t^2 \qquad \text{for } 0 \leq t \leq 2.$$

Solution $dx/dt = 3t^2$ and $dy/dt = 2t$, so the arc length is

$$\int_0^2 \sqrt{(3t^2)^2 + (2t)^2}\, dt = \int_0^2 \sqrt{9t^4 + 4t^2}\, dt \approx 9.073.$$

∎

A comment on notation for arc length

A notation sometimes used for the arc length of a curve C is

$$\int_C ds$$

where ds is a shorthand for the appropriate integrand and depends on how the curve is described. For example, if the curve C is the graph $y = f(x)$ for $a \leq x \leq b$, then $ds = \sqrt{1 + [f'(x)]^2}\, dx$ and

$$\int_C ds = \int_a^b \sqrt{1 + [f'(x)]^2}\, dx.$$

If the curve C is described parametrically by $x(t)$ and $y(t)$ for $a \leq t \leq b$, then $ds = \sqrt{(\dfrac{dx}{dt})^2 + (\dfrac{dy}{dt})^2}\, dt$ and

$$\int_C ds = \int_a^b \sqrt{(\frac{dx}{dt})^2 + (\frac{dy}{dt})^2}\, dt.$$

Another possibility can be mentioned here. If we find it convenient to describe a curve C as the locus of points satisfying $x = g(y)$ for $a \leq y \leq b$ and g is differentiable with respect to y, then we can write $ds = \sqrt{1 + [g'(y)]^2}\, dy$ and

$$\int_C ds = \int_a^b \sqrt{1 + [g'(y)]^2}\, dy.$$

EXAMPLE 16 Find the arc length of the graph of $x = \ln(y)$ for $1 \leq y \leq e$.

Solution Here $g(y) = \ln(y)$, so $g'(y) = 1/y$ and

$$\int_1^e \sqrt{1 + (1/y)^2}\, dy \approx 2.003.$$

In this example, we could also describe the curve as the graph of $y = e^x$ for $0 \leq x \leq 1$. Here $f(x) = e^x$, so $f'(x) = e^x$ and

$$\int_0^1 \sqrt{1 + (e^x)^2}\, dx \approx 2.003.$$

Surface area of a surface of revolution

If a line segment is rotated about an axis, it traces out a band whose area is the length Δs of the segment times the average distance the two endpoints travel around the axis. If the average distance from the axis is r, then the average distance travelled around the axis is $2\pi r$ (the path is circular, and r is the radius). Hence, the area of the band is

$$2\pi r \Delta s.$$

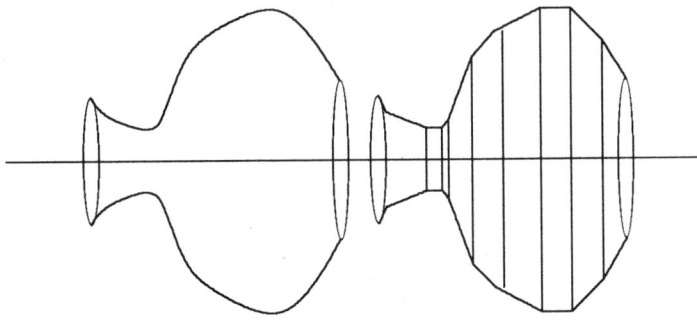

Figure 8.20 Surfaces of revolution.

Now, suppose we rotate a curve around an axis over an interval $[a, b]$. This curve traces out a surface known as a **surface of revolution**. If we first approximate the curve with a polygonal path, then we can approximate the area of this surface by rotating the polygonal path about the same axis. Since the polygonal path is made up of line segments, its surface area will be the sum of the areas of several thin bands of the type we just described (see Figure 8.20).

Of course, the radius of each band may vary, so we write the sum of the areas of the bands as

$$\sum_{a}^{b} 2\pi r(x) \Delta s$$

to denote that $r(x)$ depends on the location of the band over the interval (the length of the band Δs also depends on the location of the band). As we take finer and finer partitions of the interval $[a, b]$ corresponding to better and better polygonal paths, this sum may converge on a single limiting value. Indeed, if the curve is the graph of a differentiable function or can be parametrized by differentiable functions, then we can define the surface area of the surface of revolution to be the definite integral

$$\int_{a}^{b} 2\pi r(x) \, ds.$$

EXAMPLE 17 Find the surface area of the surface of revolution generated by the graph of $y = \sin x$ as it is rotated about the x-axis over the interval $[0, 2\pi]$.

Solution The radius of the surface is $r(x) = |\sin(x)|$ at any point $x \in [0, 2\pi]$, since it is the distance between the x-axis and the graph of the function. We can take $ds = \sqrt{1 + \cos^2(x)} \, dx$ and set up the definite integral for surface area as

$$2\pi \int_{0}^{2\pi} |\sin(x)| \sqrt{1 + \cos^2(x)} \, dx \approx 28.85 \text{ sq. units.}$$

Here we used a machine to carry out the computation of the definite integral. ∎

EXERCISES

In exercises 1-6, find the arc length of the graph of the given function over the interval $[\pi/4, \pi/3]$.

1. $y = \sin(x)$
2. $y = \cos(x)$
3. $y = \tan(x)$
4. $y = \sec(x)$
5. $y = \csc(x)$
6. $y = \cot(x)$

8.4 INTEGRATION USING POLAR COORDINATES

In exercises 7-10, find the arc length of the graph of the given function over the interval $[1/4, 1/3]$.

7. $y = \arcsin(x)$
8. $y = \arccos(x)$
9. $y = \arctan(x)$
10. $y = \text{arccot}(x)$

The hyperbolic sine and cosine functions are defined as follows:

$$\sinh(x) = \frac{e^x - e^{-x}}{2} \quad \text{and} \quad \cosh(x) = \frac{e^x + e^{-x}}{2}.$$

The rest of the hyperbolic functions are defined in a way analagous to the corresponding trigonometric functions:

$$\tanh(x) = \frac{\sinh(x)}{\cosh(x)} \qquad \coth(x) = \frac{\cosh(x)}{\sinh(x)}$$

$$\text{sech}(x) = \frac{1}{\cosh(x)} \qquad \text{csch}(x) = \frac{1}{\sinh(x)}.$$

In exercises 11-16, find the length of the graph of the indicated hyperbolic function over the interval $[1, 4]$.

11. $\sinh(x)$
12. $\cosh(x)$
13. $\tanh(x)$
14. $\coth(x)$
15. $\text{sech}(x)$
16. $\text{csch}(x)$

For each pair of parametric equations in exercises 17-24, find the arc length of the curve over the time interval $0 \le t \le 4$.

17. $x = 2\cos(t), y = 3\sin(t)$
18. $x = -3\cos(t), y = 2\sin(t)$
19. $x = 5 - t, y = 25 - 10t + t^2$
20. $x = 2t + 3, y = 3t + 2$
21. $x = t\cos(t), y = t\sin(t)$
22. $x = \arctan(t), y = t$
23. $x = |5 - t|, y = t$
24. $x = \sqrt{25 - t^2}, y = t$

For exercises 25-40, take the functions and intervals indicated in exercises 1-16 and compute the surface area of the surface of revolution generated by rotating the curve about the x-axis.

8.4 INTEGRATION USING POLAR COORDINATES

In this section, we'll examine what adjustments are necessary to extend integration to curves and regions defined in terms of polar graphs.

To see how we can extend definite integration to regions defined in terms of polar coordinates, let's first review the situation for rectangular coordinates. The area of a region bounded by the graph of $y = f(x)$ and the x-axis between $x = a$ and $x = b$ is given by the definite integral

$$\int_a^b |f(x)|\, dx.$$

The absolute value signs are necessary to guarantee that area both above and below the x-axis is counted as positive contributions to the total area of the region. The value $|f(x)|\,\Delta x$ is the area of a small rectangle of width Δx and height $|f(x)|$, and it represents the approximate area of the region swept out by the graph over a small change in the independent variable Δx. In the limiting process associated with definite integration, this value corresponds to the differential $|f(x)|\,dx$.

Now, let's use that same idea for polar graphs. Suppose $r = r(\theta)$ is given and we want to approximate the area swept out by the graph for a small change in the independent variable $\Delta \theta$. Figure 8.21 illustrates the area swept out by such a graph.

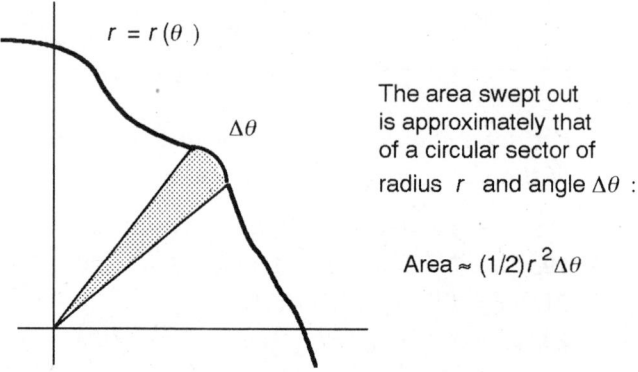

Figure 8.21 Measuring area bounded by a polar graph.

The shaded region in the figure has approximately the area of a circular sector of angle $\Delta \theta$ and radius $r = f(\theta)$, for some θ in the tiny angle interval. Since a circle of radius r has area πr^2, the area of a circular sector of angle $\Delta \theta$ is given by

$$\frac{\Delta \theta}{2\pi} \pi r^2 = \frac{1}{2} r^2 \Delta \theta.$$

8.4 INTEGRATION USING POLAR COORDINATES

To approximate the total area enclosed by the graph between $\theta = a$ and $\theta = b$, we can partition the angle interval into many such tiny subintervals and add the areas of the resulting circular sector area approximations:

$$\text{total area enclosed} \approx \sum_{a}^{b} \frac{1}{2} r^2(\theta) \Delta\theta.$$

The limiting value of these approximations as $\Delta\theta \to 0$ is given by a definite integral:

$$\text{total area enclosed} = \frac{1}{2} \int_{a}^{b} r^2(\theta) \, d\theta.$$

EXAMPLE 18 Using definite integration, find the total area bounded by the curve $r = 2\sin(\theta)$.

Solution The graph of this polar function is shown in Figure 8.22.

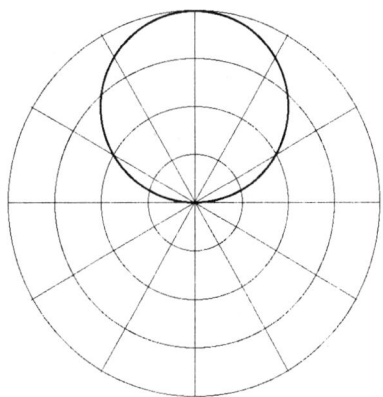

Figure 8.22 Polar graph of $r = 2\sin(\theta)$.

In this case, we know that the graph is a circle of radius 1, so the area bounded by the graph certainly must be $\pi \cdot 1^2 = \pi$. To find this area by definite integration, we need to first determine the interval of integration. Although the period of $2\sin(\theta)$ is 2π, the entire graph is traced out over the interval $[0, \pi]$. Hence, the area is given by the integral

$$\frac{1}{2} \int_{0}^{\pi} r^2 \, d\theta = \frac{1}{2} \int_{0}^{\pi} 4\sin^2(\theta) \, d\theta = 2 \int_{0}^{\pi} \sin^2(\theta) \, d\theta \approx 2 \cdot (1.5708) = 3.1416.$$

(The integral was computed by machine.) ∎

A polar graphing utility on a computer or calculator can be a great aid in checking that the interval of integration is the one you want. In the previous example, we would have obtained an erroneous area of 2π if we had used the interval $[0, 2\pi]$, because the circle is actually traced over *twice* over this interval.

EXAMPLE 19 Find the total area of the cardioid $r = \pi(cos(\theta) - 1)$.

Solution In this case, the entire cardioid is traced out over an interval $[0, 2\pi]$. Hence, the total area is given by

$$\frac{1}{2}\int_0^{2\pi} r^2\,d\theta = \frac{1}{2}\int_0^{2\pi} \pi^2(cos(\theta)-1)^2\,d\theta = \frac{\pi^2}{2}\int_0^{2\pi} (cos^2(\theta) - 2\cos(\theta) + 1)\,d\theta = \frac{3\pi^3}{2}.$$

To judge the reasonableness of this answer, we can estimate the area from the graph (see Figure 8.23).

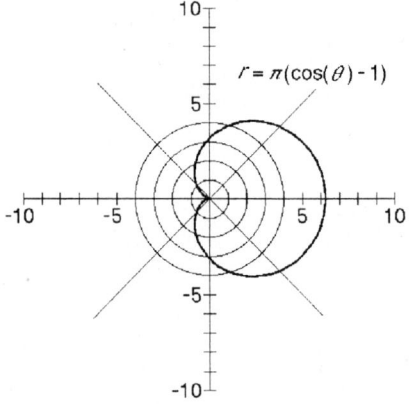

Figure 8.23 Find the area of the cardioid.

The graph is roughly approximated by a circle of radius 4, whose area is $\pi \cdot 4^2 \approx 50$. On the other hand, the value we obtained for the area of the cardioid is

$$\frac{3\pi^3}{2} \approx 46.5.$$

■

More generally, the area bounded between the two graphs $y = f(x)$ and $y = g(x)$ between $x = a$ and $x = b$ is given by the definite integral

$$\int_a^b |f(x) - g(x)|\,dx.$$

For the region bounded between two polar graphs $r = r_1(\theta)$ and $r = r_2(\theta)$ between $\theta = a$ and $\theta = b$, we have the integral

$$\frac{1}{2}\int_a^b |r_1^2(\theta) - r_2^2(\theta)|\,d\theta.$$

8.4 INTEGRATION USING POLAR COORDINATES

The absolute value signs here simply indicate that we should always subtract the smaller squared radius from the larger in our integral.

EXERCISES

For exercises 1-10, find the area of region swept out by the graph over the interval $\pi/6 \leq \theta \leq \pi/4$. You will need to use a machine numerical integrator.

1. $r = 6\cos(\theta)$
2. $r = 3\sin(\theta)\cos^2(\theta)$
3. $r = -2(\cos(\theta) + 1)$
4. $r = 5\sin(\theta)\tan(\theta)$
5. $r = 4\sec(\theta) + 1$
6. $r = \sec(\theta) - 7$
7. $r = 1 + \frac{1}{3}\cos(\theta)$
8. $r = -3 + \cos(\theta)$
9. $r = \cos(2\theta)\sec(\theta)$
10. $r = 4\theta/\pi$

In exercises 11-18, find the total area enclosed by the indicated polar graph.

11. $r = 1 - \sin(\theta)$
12. $r = 1 + \cos(\theta)$
13. $r = \sin(\theta) - 1$
14. $r = \cos(\theta) - 1$
15. $r = 1 + \cos(2\theta)$
16. $r = 1 + \sin(2\theta)$
17. $r = 1 - \sin(2\theta)$
18. $r = 1 - \cos(2\theta)$

In exercises 19-22, find the area enclosed by the inner loop of the indicated polar graph.

19. $r = 1 + 2\cos(\theta)$
20. $r = 1 + 2\sin(\theta)$
21. $r = 1 - 2\sin(\theta)$
22. $r = 1 - 2\cos(\theta)$

In exercises 23-28, find the total area enclosed by one "leaf" of the indicated polar graph.

23. $r = \cos(2\theta)$
24. $r = \sin(2\theta)$
25. $r = \cos(3\theta)$
26. $r = \sin(3\theta)$
27. $r = -\cos(3\theta)$
28. $r = -\sin(3\theta)$

A "rose" is obtained by graphing either of the polar equations

$$r = a\cos(k\theta) \quad \text{or} \quad r = a\sin(k\theta),$$

where a is a positive real number and k is a positive integer.

29. Find the area of one leaf of the rose $r = a\cos(k\theta)$ in terms of a and k.
30. Find the area of one leaf of the rose $r = a\sin(k\theta)$ in terms of a and k.

CHAPTER 9

Applications of the Integral

In this chapter we continue our discussion of applications of integration to measurement. We'll start with the notion of how definite integrals can be used to measure averages. Then we'll examine several physical applications of integration to velocity and distance, force and work, density and mass.

The hypotheses of both Fundamental Theorems of Calculus require that function integrand f must be continuous over a closed bounded interval $[a, b]$. However, it is sometimes possible to integrate functions whose graphs have vertical asymptotes or to integrate functions over infinitely long intervals. These activities are commonly referred to as *improper integration*, not because there is anything "wrong," but rather to emphasize that they represent an extension beyond the "proper" setting of the Fundamental Theorems of Calculus.

We will conclude this chapter with a look at some applications of integration to probability. One application has to do with problems that can be modelled in geometric terms of area measurement. The potential of definite integration to help us here is clear. More generally, integration is a key tool in the study of continuous probability density functions.

9.1 USING DEFINITE INTEGRALS TO MEASURE AVERAGES

Another use of definite integrals is in determining *average* or *mean* values of continuously varying quantities. The computation of an average is quite familiar in the case of a set of finitely many values. We simply sum the values and divide by the number of values. For example, if six students have test scores of 75, 98, 82, 75, 69 and 84, then the average score is

$$\frac{75 + 98 + 82 + 75 + 69 + 84}{6} = \frac{483}{6} = 80.5.$$

Notice that if all six students had the same score of 80.5, then the total of their scores would be the same, since $6(80.5) = 483$. We might think of this

9.1 USING DEFINITE INTEGRALS TO MEASURE AVERAGES

as the "leveling" interpretation of average. That is, if all our values were at the same common level, the average is the common value yielding the same sum.

Now, suppose we have a function f defined over an interval $[a,b]$, so that we have an output value $f(x)$ for each input $x \in [a,b]$. Is there a way of computing the average output value $f(x)$? It makes no sense to try to sum up all the values $f(x)$ for each x, because there are infinitely many of them. However, if we think of the leveling interpretation of average, we could ask a meaningful question: What common level value over the interval $[a,b]$ would produce the same net effect as f? In other words, what *constant* function $x \longmapsto h$ would give us the same definite integral as $x \longmapsto f(x)$ over the interval $[a,b]$?

We can determine h by stating this requirement as an equation to solve:

$$\int_a^b h\, dx = \int_a^b f(x)\, dx.$$

Since $\int_a^b h\, dx$ is simply $h(b-a)$, we can define the average value of an integrable function by solving for h.

Definition 1 The **average value of f over the interval** $[a,b]$ is

$$\frac{\int_a^b f(x)\, dx}{b-a}$$

provided the integral exists.

EXAMPLE 1 Find the average value of $f : x \longmapsto x^2$ over the interval $[1,3]$.

Solution Here $a = 1$, $b = 3$, and $f(x) = x^2$, so

$$\text{avg value of } f = \frac{\int_a^b f(x)\, dx}{b-a} = \frac{\int_1^3 x^2\, dx}{3-1} = \frac{26/3}{2} = \frac{13}{3}.$$

∎

Graphically, if $f(x)$ is positive over the interval $[a,b]$, then its average value is the height h of a rectangle of length $b-a$ having the same area as $\int_a^b f(x)\, dx$. In fact, if f is continuous over $[a,b]$, then this height must be achieved at least once by the function over the interval. In other words, there exists $a \leq x_0 \leq b$ such that $f(x_0) = h$. The *Mean Value Theorem for Integrals* states that any continuous function must achieve its average value at least once over a closed interval.

Theorem 9.1

> **Mean Value Theorem for Integrals.**
> **Hypothesis:** Suppose f is a continuous function on $[a, b]$.
> **Conclusion:** $f(x_0)(b - a) = \int_a^b f(x)\, dx$ for at least one value $a \leq x_0 \leq b$.

Reasoning Assume f is a continuous function on $[a, b]$. Figure 9.1 illustrates the situation for a function f that happens to be positive over the interval $[a, b]$, but all our observations will hold for continuous functions in general.

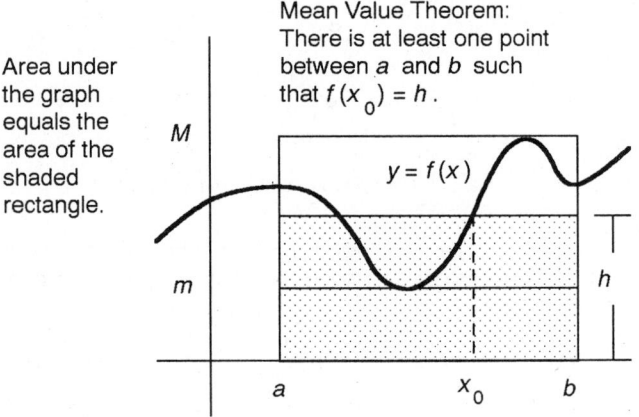

Figure 9.1 Illustration of the Mean Value Theorem for Integrals.

Suppose $M = $ maximum value $f(x)$ attains on $[a, b]$ and $m = $ minimum value $f(x)$ attains on $[a, b]$. (Since f is continuous, we know f must attain these values by the Extreme Value Theorem.) This means that the graph $y = f(x)$ must lie between the horizontal lines $y = M$ and $y = m$ as shown in Figure 9.1, and we must have

$$m(b - a) \leq \int_a^b f(x)\, dx \leq M(b - a).$$

For the positive function pictured, this simply means that the area under the graph of $y = f(x)$ must be between the areas of the rectangles of heights m and M. Geometrically, this means that there is some height $m \leq h \leq M$ such that the rectangle of height h and length $b - a$ has exactly the same area $h(b - a)$ as that under the graph of $y = f(x)$. If we divide through by the positive number $b - a$, the inequality becomes

$$m \leq \frac{\int_a^b f(x)\, dx}{b - a} \leq M.$$

9.1 USING DEFINITE INTEGRALS TO MEASURE AVERAGES

Since the value in the middle of this inequality lies between the extreme values m and M of the function, the Intermediate Value Theorem for continuous functions guarantees that there exists at least one value x_0 such that

$$f(x_0) = \frac{\int_a^b f(x)\, dx}{b-a}.$$

Multiplying both sides by $(b-a)$ gives us the conclusion,

$$f(x_0)(b-a) = \int_a^b f(x)\, dx.$$

□

Think of the graph in Figure 9.1 as representing the elevation of the terrain faced by a road-building crew. If the crew wants to make a road at a certain level elevation, then they must fill in the places below that level and cut down places above that level. The crew can use some of the material from the high places to fill in the low places. The mean value of the function represents an optimal level to choose, in the sense that no extra material would need to be trucked in or out.

Center of mass

Have you ever noticed a parent and child on a see-saw (teeter-totter)? In order to balance, the parent must typically ride much closer to the fulcrum (pivoting point) than the child. In fact, to balance, the product of the parent's weight times the parent's distance from the fulcrum must be equal to the child's weight times the child's distance from the fulcrum. For example, if the child weighs 30 kilograms and the parent weighs 60 kilograms, then the child must ride twice as far out as the parent, so that the "weighted" distances from the fulcrum are equal (see Figure 9.2).

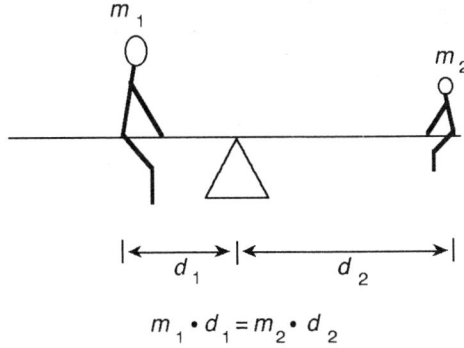

Figure 9.2 Center of mass is at the balance point.

The location of the balance point for a set of weights along a straight line is called the **center of mass**, and we can think of it as a *weighted average*. To balance on a see-saw, two people must position themselves so that their center of mass coincides with the fulcrum point.

To find the center of mass, we first need to establish a coordinate system to describe the positions of the weights along a line. Let's walk through a particular example to get a feel for how the center of mass can be located in general.

EXAMPLE 2   Suppose we choose a coordinate system so that the parent's position is $x = 3$ and the child's is $x = 7$, as depicted in Figure 9.3. Find the coordinate $\bar{x}$ of the center of mass.

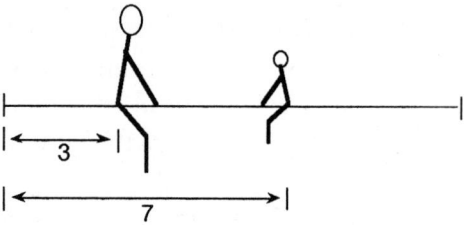

Figure 9.3 Finding the location of the center of mass.

Solution   Since the parent weighs 60 kilograms and the child weighs 30 kilograms, we need to place the fulcrum at a point $\bar{x}$ between them so that

$$60(\bar{x} - 3) = 30(7 - \bar{x}).$$

If we simplify, we find

$$60\bar{x} - 60 \cdot 3 = 30 \cdot 7 - 30\bar{x},$$

and solving for $\bar{x}$:

$$\bar{x} = \frac{30 \cdot 7 + 60 \cdot 3}{30 + 60} = \frac{390}{90} = \frac{13}{3}.$$

We can check that the the child is twice as far from the balance point $\bar{x} = 13/3$ as the adult: $7 - \frac{13}{3} = \frac{8}{3} = (2)\frac{4}{3} = 2(\frac{13}{3} - 3)$.

∎

9.1 USING DEFINITE INTEGRALS TO MEASURE AVERAGES

 We have been using the everyday reference to kilogram as a unit of *weight*, but a kilogram is actually a unit of *mass*. The mass of an object can be thought of as the amount of material in the object, while the weight of an object is the force that gravity exerts on the object. (A 60 kilogram person has the same mass on the moon as on earth, but only 1/6 the weight.)

In general, we can see that if two objects having masses m_1 and m_2 are located at coordinates x_1 and x_2, respectively, then their center of mass is located at

$$\overline{x} = \frac{x_1 m_1 + x_2 m_2}{m_1 + m_2}.$$

If we add a third mass m_3 at position x_3, we can use the same kind of analysis to find that

$$\overline{x} = \frac{x_1 m_1 + x_2 m_2 + x_3 m_3}{m_1 + m_2 + m_3}.$$

In fact, if we have any number of weights $m_1, m_2, \ldots, m_n$, at positions $x_1, x_2, \ldots, x_n$, then the center of mass has position

$$\overline{x} = \frac{x_1 m_1 + x_2 m_2 + \cdots + x_n m_n}{m_1 + m_2 + \cdots + m_n} = \frac{\sum_{i=1}^{n} x_i m_i}{\sum_{i=1}^{n} m_i}.$$

Computationally, the center of mass is the **moment** divided by the total mass. The moment is the weighted sum of the products of distance and mass (the $x_i m_i's$), and the total mass is the sum of the $m_i's$.

The mass **density** of an object is a measure of the ratio of its mass to one or more of its dimensions. Hence, units like kg/m^3 (mass per unit of volume) are appropriate for density, but it is also common to use units like kg/m (mass per unit of length) or kg/m^2 (mass per unit of area) depending on the application. For example, if a thin rod 12 meters long has a constant density of 3 kilograms per meter (kg/m), then the mass of the rod is

$$\text{mass of rod} \;=\; 3\frac{\text{kg}}{\text{m}} \cdot 12 \text{ m} \;=\; 36 \text{ kg}.$$

Now, suppose we have a thin rod whose mass density is variable along its length. If we set up a coordinate axis along the length of the rod so that it lies over an interval $[a, b]$, then it has a mass density value $m(x)$ at each point in the interval. If we sliced the rod up into sufficiently small equal-sized pieces of length Δx, then each segment has approximately constant density, and a mass approximately equal to $m(x)\Delta x$ (see Figure 9.4).

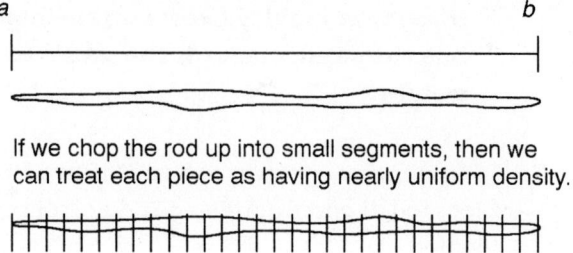

Figure 9.4 Finding the mass of a rod of non-uniform density.

To find the approximate total mass of the rod, we can sum the masses of each of the tiny segments:

$$\text{total mass of the rod} \approx \sum_{a}^{b} m(x)\Delta x.$$

As the size of the segments is taken smaller and smaller, the better the approximation. The limiting value of this sum of masses gives us the exact mass of the rod:

$$\text{total mass of the rod} = \int_{a}^{b} m(x)\,dx.$$

We can use this same type of analysis to compute the center of mass of the rod. We multiply the mass of each tiny segment by its x-coordinate and sum the results to get the approximate moment of the rod, and then divide by the approximate total mass:

$$\text{center of mass} \approx \frac{\sum_{a}^{b} x m(x)\Delta x}{\sum_{a}^{b} m(x)\Delta x}.$$

Passing to the limiting value of these approximations as $\Delta x \to 0$:

$$\text{center of mass} = \frac{\int_{a}^{b} x m(x)\,dx}{\int_{a}^{b} m(x)\,dx}.$$

9.1 USING DEFINITE INTEGRALS TO MEASURE AVERAGES

EXAMPLE 3 A right triangle made of a uniform material has a base of 4 inches and a height of 3 inches. If stood on edge, where along the base would it balance?

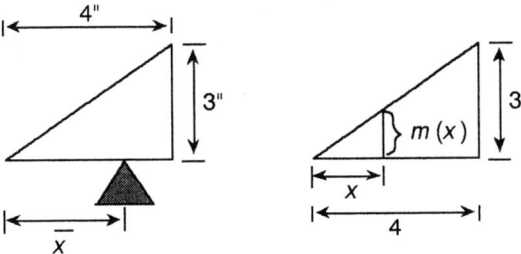

Figure 9.5 Finding the balance point for a triangle.

Solution Taking our reference point to be the left hand corner of the base, so that x refers to the distance along the base from the left, the density, $m(x)$, will be proportional to the height of the triangle over that point. By using the similar triangle property (see Figure 9.5),

$$m(x) = \frac{3x}{4} \text{ for } 0 \leq x \leq 4$$

so that

$$\bar{x} = \frac{\int_0^4 x \cdot m(x)\,dx}{\int_0^4 m(x)\,dx} = \frac{\frac{3}{4}\int_0^4 x^2\,dx}{\frac{3}{4}\int_0^4 x\,dx}$$

$$= \frac{\left.\frac{x^3}{3}\right]_0^4}{\left.\frac{x^2}{2}\right]_0^4} = \frac{64/3}{8} = \frac{8}{3}.$$

The balance point is located $2\frac{2}{3}$ inches from the left corner. ∎

In general, considering a region R in the plane as a flat plate of uniform density, we can use this integration technique to find the horizontal balance point $\bar{x}$:

$$\bar{x} = \frac{\int_a^b x m(x)\,dx}{\int_a^b m(x)\,dx}$$

where the region R lies over the interval $[a, b]$ and $m(x)$ is the height of the region at x.

EXAMPLE 4 Suppose a flat plate of uniform density has the shape of the region bounded by $y = x^2$ and $y = x^3$. Find $\bar{x}$.

Solution The region in question is shown in Figure 9.6.

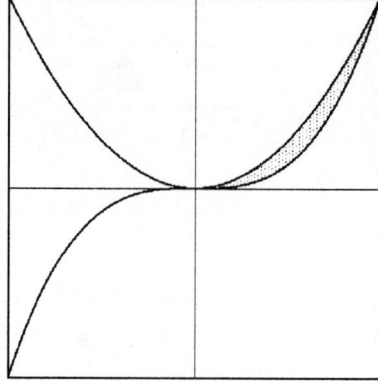

Figure 9.6 What is $\bar{x}$ for the shaded region?

$$\bar{x} = \frac{\int_0^1 x(x^2 - x^3)\, dx}{\int_0^1 (x^2 - x^3)\, dx} = \frac{1/20}{1/12} = 0.6.$$

■

Moving averages

Another kind of weighted average is useful in situations where the data is "noisy," that is, has lots of short-term ups and downs that can hide what is happening over a longer term. For example, this *moving average* is used in analyzing stock market trends and consists of replacing each day's stock price by its average over some fixed previous period. For example, instead of daily charting the price of a particular stock, we might daily chart the average of the previous 5 days, or previous 30 days, or whatever. This tends to smooth out the day-to-day fluctuations making it easier to spot longer term trends.

9.1 USING DEFINITE INTEGRALS TO MEASURE AVERAGES

Definition 2 If $x \longmapsto f(x)$ is a continuous function, and A is a positive real number, then the **moving average of f over period** A is the new function

$$x \longmapsto \frac{\int_{x-A}^{x} f(u)\,du}{A}.$$

Note that the value of the moving average of f over period A at a point x is the average of f over the "previous" interval of the form $[x - A, x]$.

EXAMPLE 5 Compute the moving average over a period 1 of the function $x \longmapsto \sin(4x)$ and compare the amplitude (half the distance from the maximum to the minimum values) of the moving average with the amplitude of the original.

Solution The moving average is given by

$$x \longmapsto \frac{\int_{x-1}^{x} \sin(4u)\,du}{1} = \frac{1}{4}\cos(4u)\Big]_{x-1}^{x} = \frac{1}{4}(\cos(4x) - \cos(4x-1)).$$

Graphing both the original and moving average functions yield those shown in Figure 9.7.

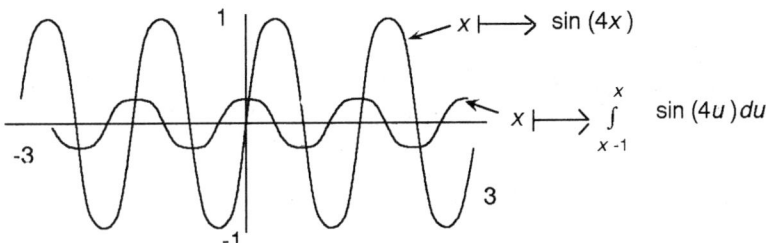

Figure 9.7 Graph of $y = \sin(4x)$ and its moving average over period 1.

The maximum and minimum values of the original are 1 and -1, that of the moving average are $\approx .24$ and $-.24$. Using the moving average reduced the variability to somewhat less than a quarter of the original. ∎

EXERCISES

1. Find the average value of the function $f(x) = x$ over $[2, 4]$.
2. Find the average value of the function $f(x) = x^2$ over $[0, 3]$.
3. Find the average value of the function $f(x) = \sin 2x + \frac{1}{2}\cos x$ over $[0, \frac{\pi}{4}]$.
4. Find the average value of the function $f(x) = e^x$ over $[1, 4]$.
5. What is the average area of all possible circles with radii between 3 and 6?
6. The president of Platypus would like to know the average profit the company makes per person-hour. You know that there are fixed costs of being an employer of $100 per hour, no matter what the number of employees (for administration costs), and the cost of x person-hours is $28x$. The company brings in $(132 + .01x)$ for the xth person-hour worked in a month. Last month the company employees worked 10000 person-hours. What is the average value of the profit per person-hour?

For exercises 7-10, suppose the parent weighs 60 *kilograms and the child weighs* 30 *kilograms.*

7. If the parent's position is given as $x = 18$ and the child's position is given as $x = 22$, what is the location of their center of mass $\bar{x}$?
8. The parent and child in the previous exercise have the same masses and are the same distance apart as the parent and child in the earlier example. Compare your result with that of the example, noting the distance of each person to the center of mass.
9. Suppose each person is handed a 10 kg weight. Will the location of the center of mass change?
10. Now suppose the parent is handed a 10 kg and the child a 5 kg weight. Will the location of the center of mass change?

Consider each of the regions in exercises 11-20 as a flat plate of uniform density and find $\bar{x}$.

11. $y = x^3$ and $y = \sqrt{x}$ for $0 \leq x \leq 1$.
12. $y = 3x^2 - 6 + 4$, the coordinate axes, and the line $x = 2$.
13. $y = e^x$, $x = 0$, $x = 1$, and $y = 0$.
14. $y = \sqrt{1 - x^2}$ and the x-axis for $0 \leq x \leq 1$.
15. $y = \sin(x)$ and $y = 1/2$ for the interval $\pi/6 \leq x \leq 5\pi/6$.
16. $y = \tan(x)$ and $y = \sec(x)$ for $-\pi/2 \leq x \leq \pi/3$.
17. $x^2 = 4 - y$ and $x + y = 2$ over the interval $-1 \leq x \leq 2$.

18. $y = e^x$ and $y = \sin(x)$ for $0 \le x \le 3$.
19. $y = \ln(x)$ and $y = e^x$ for $1/2 \le x \le 10$.
20. $y = 2\sin(x)$ and $y = 3\cos(x)$ over the interval $-1 \le x \le 1$.

---

21. Suppose a right triangle has legs of 3 inches and 4 inches. If stood up so the short leg is horizontal, where would the triangle balance on a point? Note: if you draw a perpendicular line from each balance point on the two legs of this triangle, the point where they cross is the center of mass for the triangle. This means that triangle laid *flat* should balance here on a point (like the point of a pencil). Try it by cutting out a cardboard triangle having these dimensions.

22. Suppose a right triangle has legs of 3 inches and 4 inches. If stood up so the hypotenuse is horizontal, where would the triangle balance on a point?

23. Suppose a right triangle has legs of b inches and h inches. If stood up so the leg of b inches is horizontal, where would the triangle balance on a point?

24. Suppose a right triangle has legs of b inches and h inches. If stood up on the leg of h inches, where would the triangle balance?

25. Suppose a right triangle has legs of b inches and h inches. If stood up on the hypotenuse, where would the triangle balance?

For each of the exercises 26-30, find the moving average function over the indicated period. Compare the graphs of the original function and the moving average function over the interval $[-3, 3]$.

26. $x \longmapsto \cos(x)$ over a period 1
27. $x \longmapsto x$ over a period 2
28. $x \longmapsto x$ over a period 3
29. $x \longmapsto x^2 \sin(x^3)$ over a period 0.5
30. $x \longmapsto x \cos(x^2)$ over a period 1

---

31. Find data on the daily high temperature in your area over the last thirty days. Starting with day 5, plot a line graph of these temperatures over the interval $[5, 30]$ (in other words, connect the dots with a polygonal path). Now compute the average high temperature for days 1-5, 2-6, 3-7, and so on up to 26-30. Plot these moving averages over the same interval $[5, 30]$ and note the effect on the fluctuation in the graph.

9.2 USING DEFINITE INTEGRALS TO MAKE PHYSICAL MEASUREMENTS

In this section, we discuss some of the different physical applications which require the use of definite integration as a measurement tool.

Net and total distance travelled

Let's review our example of the distance travelled by a car. If $s(t)$ represents the position of the car (or any other object) at time t, we know that the velocity v and acceleration a represent the first and second derivatives:

$$v(t) = s'(t) \quad \text{and} \quad a(t) = s''(t).$$

We can recover the position function from a knowledge of the velocity function and an initial position, or we can even recover the position function from a knowledge of the acceleration function and an initial velocity and position.

EXAMPLE 6 Suppose an object moves with an acceleration $a(t) = -16t$ ft/sec^2 with an initial velocity $v(0) = 32$ ft/sec and an initial position $s(0) = 8$ ft. Find the velocity and position functions.

Solution We obtain the velocity by antidifferentiating the acceleration:

$$v(t) = \int a(t)\, dt = \int -16t\, dt = -8t^2 + C_1$$

where the constant C_1 is determined by the initial velocity

$$32 = v(0) = -8(0)^2 + C_1 = C_1.$$

Therefore, $v(t) = -8t^2 + 32$ ft/sec. Now we can antidifferentiate the velocity function to find the position function:

$$s(t) = \int (-8t^2 + 32)\, dt = -\frac{8}{3}t^3 + 32t + C_2,$$

and the second constant is determined by the initial position

$$8 = s(0) = -\frac{8}{3}0^3 + 32(0) + C_2 = C_2.$$

Thus, $s(t) = -\frac{8}{3}t^3 + 32t + 8$ feet. ∎

Since position s is the antiderivative of velocity v, we have

$$\int_a^b v(t)\, dt = s(b) - s(a).$$

In other words, integrating the velocity $v(t)$ over a time interval $[a, b]$ yields the *change in position of the object* over that time interval. We should think of this as the *net distance* travelled, and distinguish it carefully from the notion of *total distance* travelled. For example, if you walk 3 steps forward and then 2 steps backward, the net change in your position is 1 step forward, even though you have travelled a total of 5 steps.

The direction of movement is indicated by the sign of the velocity. A negative velocity indicates movement in the opposite direction of a positive velocity. The magnitude (absolute value) of the velocity indicates the *speed*, regardless of the direction. In terms of using integration to determine distance travelled, we have

$$\text{net distance travelled} = \int_a^b v(t)\,dt,$$

and

$$\text{total distance travelled} = \int_a^b |v(t)|\,dt.$$

EXAMPLE 7 An object moves horizontally with distance to the right considered positive. Find the net and total distance travelled by the object over the time interval $1 \leq t \leq 3$ if it has velocity $v : t \longmapsto -8t^2 + 32$.

Solution Figure 9.8 shows the graph of this velocity v as a function of time t.

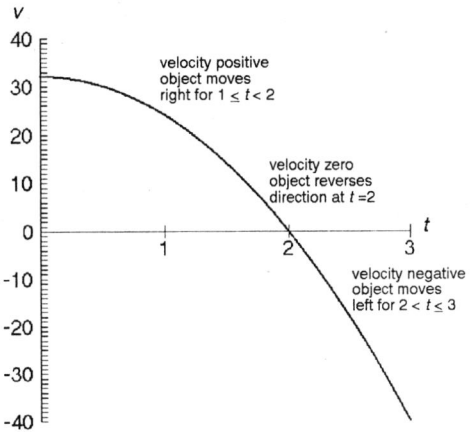

Figure 9.8 Graph of the velocity function.

We can tell what direction the object is moving at any particular time t by noting the sign of the velocity $v(t)$. In this case, the velocity is positive for $1 \leq t < 2$ and the object moves to the right during this time. The velocity is negative for $2 < t \leq 3$ and the object moves to the left during this time.

The object reverses direction at the instant $t = 2$ when the velocity is zero. The net distance travelled by the object is measured by the net area above the t-axis over $[1,3]$:

$$\int_1^3 (-8t^2 + 32)\, dt = -\frac{8}{3}t^3 + 32t \Big]_1^3 = 24 - \frac{88}{3} = -\frac{16}{3} \text{ ft.}$$

This means the object moves a net distance of $16/3$ units to the **left**.

The speed $|v(t)|$ of the object is plotted in Figure 9.9.

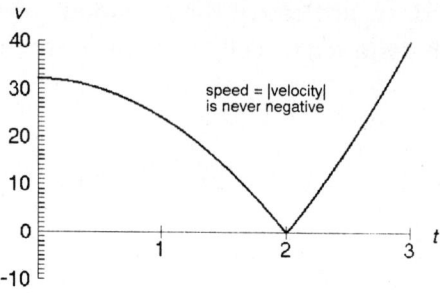

Figure 9.9 Graph of the speed function.

The total distance travelled is

$$\int_1^3 |-8t^2 + 32|\, dt = \int_1^2 (-8t^2 + 32)\, dt + \int_2^3 (8t^2 - 32)\, dt$$

$$= -\frac{8}{3}t^3 + 32t \Big]_1^2 + \frac{8}{3}t^3 - 32t \Big]_2^3 = \frac{40}{3} + \frac{56}{3} = \frac{96}{3} = 32 \text{ ft.}$$

From this computation, we can see that the object moved $40/3$ units to the right, reversed direction, and then moved $56/3$ units to the left for the total distance of 32 units and the net distance of $16/3$ units to the left. ∎

In the previous example, the integral of the absolute value of the velocity

$$\int_1^3 |v(t)|\, dt = 32$$

gives us the correct total distance travelled. Note that we get an entirely different result if we simply take the absolute value of the integral of velocity:

$$\left| \int_1^3 v(t)\, dt \right| = |-16/3| = 16/3.$$

9.2 USING DEFINITE INTEGRALS TO MAKE PHYSICAL MEASUREMENTS

 Be careful with integrals involving absolute value. In general,

$$\int_a^b |f(x)|\,dx \neq \left| \int_a^b f(x)\,dx \right|$$

unless f never changes sign over $[a, b]$.

Work

One application we will describe in detail concerns the notion of **work** as it is defined in physics. Work can be thought of roughly as a "force times distance." In the simplest case, we might imagine lifting a weight F (force due to gravity) a distance Δx (see Figure 9.10).

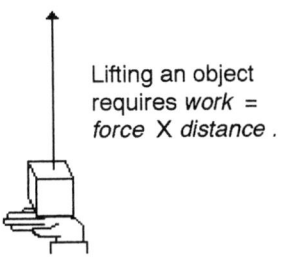

Lifting an object requires *work* = force × *distance*.

Figure 9.10 Work = force × distance.

The work performed in this instance is equal to the weight F times the distance Δx:

$$W = F\Delta x.$$

From this definition we can deduce appropriate units of measurement for work. In the English system, force can be measured in pounds (lbs) and distance in feet (ft), so an appropriate unit for work would be the foot-pound (ft lb). In the metric system, force can be measured in *newtons* (1 newton = 1 kg m/sec^2) where mass is measured in kilograms (kg), distance in meters (m), and time in seconds (sec). In this case, an appropriate unit for work would be the newton-meter, or equivalently, the *joule*. (If force is measured in dynes (g cm/sec^2 with mass measured in grams and distance in centimeters, then work is measured in dyne-centimeters, or equivalently, *ergs*.)

Now, suppose that the force exerted varies as we move the object. For example, suppose we push an object along an axis from $x = a$ to $x = b$, but the applied force $F(x)$ depends on the specific point x along the path. If the force $F(x)$ varies continuously over the interval $[a, b]$ and we restrict our attention to a very small subinterval $[x, x + \Delta x]$, then over this subinterval the work performed is approximately

$$F(x)\Delta x.$$

If we compute the work performed over subintervals of length Δx from a to b (adjusting $F(x)$ for the computation over each subinterval) and add the results, we get an approximation to the total work performed:

$$\text{total work } W \approx \sum_a^b F(x)\Delta x.$$

Again we see the Riemann sum model, this time for work. Taking the limiting value of this sum as $\Delta x \to 0$, we have

$$\text{total work } W = \int_a^b F(x)\,dx.$$

EXAMPLE 8 Find the work done by moving a particle along the x-axis from $x = 3$ to $x = 6$, if the force exerted on the particle at x is $x^2 + 3$ newtons and x is measured in units of meters.

Solution Here $F(x) = x^2 + 3$, so the work performed is

$$W = \int_3^6 (x^2 + 3)\,dx = \frac{x^3}{3} + 3x \Big]_3^6 = 90 - 18 = 72 \text{ joules.}$$

■

Examples abound of forces that vary depending on distances or location including gravitational and magnetic forces. In the remainder of this section, we examine several examples involving the calculation of work.

Work performed in stretching or compressing a spring

One example common to most people's experience has to do with compression and stretching of springs.

If we compress a spring, its tendency is to resist in an attempt to return to its natural length. The more we compress the spring, the harder the spring pushes back on our hand. Indeed, Hooke's law states that the force F required to compress (or stretch) a spring is proportional to the displacement x from its natural length. Figure 9.11 shows a vertically hung spring before and after a weight F has been attached.

9.2 USING DEFINITE INTEGRALS TO MAKE PHYSICAL MEASUREMENTS

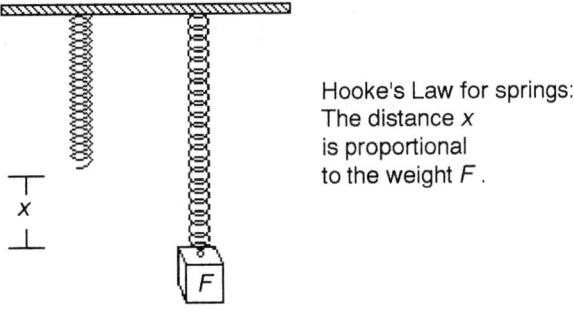

Hooke's Law for springs: The distance x is proportional to the weight F.

Figure 9.11 Hooke's law for springs.

For example, if a weight of 10 pounds stretches the spring 4 inches from its natural length, then a weight of 30 pounds will stretch the spring 12 inches from its natural length. If x is the displacement from the natural length, then

$$F(x) = kx,$$

for some positive constant k. The magnitude of this *spring constant* k depends on the size, material, and construction of the spring. The spring on a car's shock absorber has a large k value; the spring in a ball point pen has a small k value.

The spring constant for a given spring can be determined empirically by observing the force required to compress or stretch the spring a set distance. For the spring just mentioned, the spring constant is

$$k = \frac{10 \text{ pounds}}{4 \text{ inches}} = 2.5 \text{ pounds per inch}.$$

If we know the spring constant k, then we can calculate the work performed in stretching or compressing the spring from displacement $x = a$ to $x = b$:

$$W = \int_a^b kx \, dx.$$

EXAMPLE 9 If a spring has a natural length of 18 inches, and a force of 10 pounds is sufficient to compress it to a length of 16 inches, what is the spring constant k and how much work would be done in compressing it from a length of 16 inches to a length of 12 inches?

Solution Since a force of 10 pounds compresses the spring 2 inches from its natural length, we have

$$k = \frac{10 \text{ pounds}}{2 \text{ inches}} = 5 \text{ pounds per inch.}$$

Compressing the spring from a length of 16 inches to a length of 12 inches represents a displacement from $x = 2$ to $x = 6$ inches from its natural length. Hence,

$$W = \int_2^6 5x \, dx = \frac{5x^2}{2} \Big]_2^6 = 90 - 10 = 80 \text{ inch-pounds.}$$

■

Work lifting and pumping

Anytime we move an object up vertically, we have performed work against the force that gravity exerts on the object. For example, if we lift a 15 pound weight 8 inches, the work performed is

$$W = 15 \text{ pounds} \times \frac{2}{3} \text{ feet} = 10 \text{ foot pounds.}$$

Integration becomes necessary if different parts of the object must be lifted different distances, as is illustrated in the next two examples.

EXAMPLE 10 A uniform cable 40 feet long and weighing 60 pounds hangs vertically from the top of a building. If a 500 pound weight is attached to the end of the cable, what work is required to pull it to the top?

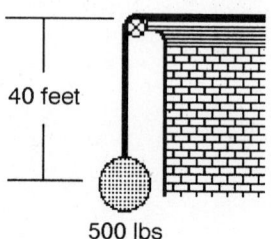

Figure 9.12 A weight hanging from a uniform cable.

Solution A picture of the situation is shown in Figure 9.12. It is easy to calculate the work required just to lift the weight itself to the top of the building:

$$W_{weight} = 500 \text{ lbs} \times 40 \text{ feet} = 20000 \text{ ft lbs}.$$

The difficulty lies with the cable, for different parts of the cable must be lifted different distances. (The bottom of the cable must be lifted the entire 40 feet, but the middle of the cable must be lifted only 20 feet.)

Imagine a very small segment of cable of length Δx feet. Since the entire cable weighs 60 pounds and is 40 feet long, this small piece of cable weighs

$$\frac{60 \text{ lbs}}{40 \text{ feet}} \cdot \Delta x \text{ feet} = 1.5 \Delta x \text{ lbs}.$$

If this piece of cable is at a distance x feet below the top of the building, then the work required to lift it is approximately $1.5 x \Delta x$ ft lbs, and the total work to lift all the tiny segments is

$$W_{cable} \approx \sum_{0}^{40} 1.5 x \Delta x.$$

Taking the limiting value as $\Delta x \to 0$, the exact total work required to pull the cable to the top of the building is

$$W_{cable} = \int_{0}^{40} 1.5 x \, dx = 0.75 x^2 \Big]_{0}^{40} = 1200 \text{ ft lbs}.$$

Adding the results for the weight and the cable gives us

$$W_{total} = W_{weight} + W_{cable} = 21200 \text{ ft lbs}.$$

■

Pumping fluid to a higher elevation is essentially a lifting problem. The next example illustrates how integration can be useful in measuring the work performed in this way.

EXAMPLE 11 Suppose a tank has a rectangular base of 2 by 4 feet and a vertical height of 3 feet. If the tank is filled with water (62.5 pounds per cubic foot), find the work required to pump the water 2 feet above the top of the tank.

Solution Figure 9.13 illustrates the tank in question.

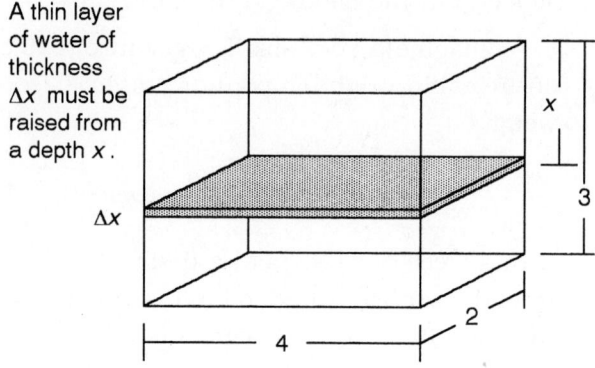

Figure 9.13 Pumping water from a rectangular tank.

If we think of the water as consisting of a stack of very thin layers of water, each of depth Δx feet, then we can approximate the work required to lift a single layer to the desired height:

$$\text{weight of layer} \times \text{distance lifted}.$$

The volume of water in a layer of thickness Δx is

$$2 \text{ feet} \times 4 \text{ feet} \times \Delta x \text{ feet} = 8\Delta x \text{ ft}^3$$

and thus, the weight of the layer is

$$(62.5 \text{ lbs per cu ft})(8\Delta x \text{ cu ft}) = 500\Delta x \text{ lbs}.$$

If the layer is located x feet from the top, then the layer must be lifted $x + 2$ feet. The work required to lift the layer is

$$500(x + 2)\Delta x \text{ ft lbs}.$$

The total work in lifting all the layers is approximately

$$\sum_0^3 500(x + 2)\Delta x.$$

The exact work is given by

$$\int_0^3 500(x+2)\,dx = 250x^2 + 1000x \Big]_0^3 = 5250 \text{ ft lbs}.$$ ∎

9.2 USING DEFINITE INTEGRALS TO MAKE PHYSICAL MEASUREMENTS

In this example, the layers of water were of uniform dimensions because the tank was a rectangular box. In other cases, the size of the layer may depend on the depth it was taken from.

EXAMPLE 12 Suppose oil with a density of 50 lbs per cubic foot half-fills a hemispherical reservoir of radius 4 feet. How much work does it take to pump the oil to the top of the

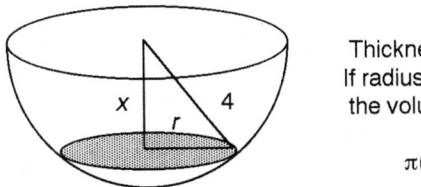

Thickness of layer is Δx. If radius of layer is r, then the volume of the layer is

$$\pi(16 - x^2)\Delta x.$$

Figure 9.14 Pumping oil out of a hemispherical tank.

Solution Figure 9.14 illustrates the reservoir. If we use x to denote the depth to a thin layer of oil in the tank, then the radius r of the circular layer satisfies the Pythagorean relationship:

$$x^2 + r^2 = 4^2.$$

The volume of this layer is approximately $\pi r^2 \Delta x = \pi(16 - x^2)\Delta x$ cubic feet, and the weight of the layer is $50\pi(16 - x^2)\Delta x$ pounds. The work required to lift it the x feet to the top of the tank is approximately

$$50\pi x(16 - x^2)\Delta x.$$

Since the tank is only half-full, the layers range from a depth of $x = 2$ feet to $x = 4$ feet. The total work performed in pumping out this tank is thus

$$\int_2^4 \pi x(16 - x^2)\,dx = \pi(8x^2 - x^4/4)\Big]_2^4 = 36\pi \text{ ft lbs.}$$

∎

EXERCISES

In exercises 1-5, you are given an acceleration function for a moving object and an initial velocity and position. Use this information to find the velocity and position functions.

1. $a: t \longmapsto t - t^2 \quad v(0) = -2 \quad s(0) = 3.$
2. $a: t \longmapsto e^{-t} \quad v(0) = 0 \quad s(0) = 4.$

3. $a: t \longmapsto \sin(t)$ $\quad v(0) = -1$ $\quad s(0) = 1.$

4. $a: t \longmapsto \dfrac{3}{1+t^2}$ $\quad v(0) = 3.5$ $\quad s(0) = 1.25.$

5. $a: t \longmapsto \sqrt{t}$ $\quad v(0) = -4$ $\quad s(0) = 2.$

In exercises 6-10, you are given a velocity function. Use it to calculate the net and total distance travelled during the indicated time interval.

6. $v: t \longmapsto \ln(2-t)$ $\quad 0.5 \leq t \leq 1.5.$

7. $v: t \longmapsto e^{-t}$ $\quad -1 \leq t \leq 1.$

8. $v: t \longmapsto \dfrac{2}{t^3}$ $\quad 0.5 \leq t \leq 2.35.$

9. $v: t \longmapsto \sin(2\pi t)$ $\quad 0 \leq t \leq 1.$

10. $v: t \longmapsto \sqrt{4-t^2}$ $\quad 0 \leq t \leq 2.$

---

11. A ball is thrown straight upwards at 64ft/sec at time $t=0$. If it is being attracted by the gravity of the earth so that it falls toward the earth with an acceleration of 32ft/sec^2, how far does it travel in 3 seconds?

12. Assume the brakes of an automobile produce a constant deceleration of k ft/sec^2. What is k if a car that is initially traveling 60 mph (88 ft/sec) is brought to a full stop 150 ft after braking begins? How far will the same automobile travel during the time it takes to slow from 60 mph to 30 mph by braking?

---

13. Find the work done by a force of $3\sqrt{x}$ moving a particle along the x-axis from $x=0$ to $x=4$.

14. Find the work done by a force $F(x) = \sin \pi x$ from $x=-1$ to $x=1$. Does this answer make sense? Explain.

15. A spring of natural length 13 inches stretches to a length of 18 inches under a weight of 8 pounds. Find the work done in stretching the spring from a length of 16 inches to a length of 20 inches.

16. A spring of natural length 15 inches stretches to a length of 16 inches under a weight of 4 pounds. Find the work done in stretching the spring from a length of 18 inches to a length of 23 inches.

17. A spring of natural length 13 inches compresses to a length of 6 inches under a weight of 6 pounds. Find the work done in compressing the spring from a length of 9 inches to a length of 5 inches.

18. A spring of natural length 12 inches compresses to a length of 5 inches under a weight of 8 pounds. Find the work done in compressing the spring from a length of 10 inches to a length of 6 inches.

9.2 USING DEFINITE INTEGRALS TO MAKE PHYSICAL MEASUREMENTS

19. A spring has a natural length of 1 foot and a force of 15 pounds is required to hold it stretched to a total length of 2 feet. How much work is done in compressing this spring to a length of 6 inches?

20. A spring has a natural length of 6 inches. A 12,000 pound force compresses it to $5\frac{1}{2}$ inches. Find the work done in compressing it from 6 inches to 5 inches.

21. A spring, which has a natural length of 4 feet and is stretched one foot by a force of 10 pounds, is attached to a wall. Another spring, which has a natural length of 3 feet and is stretched $\frac{1}{2}$ feet by a force of 2 pounds, is attached to another wall 10 feet away. Find the force on the second spring if the two loose ends are connected. What is the length of each of the two springs when connected?

22. One observes that a force of 10 pounds stretches a spring .87 inches. How much work is required to stretch the spring one foot?

23. If one hangs a spring (of constant 10 lbs./ft) vertically and attaches a 10 lb. weight to the spring, how much work is done in raising the weight 6 inches from where it hangs naturally? (Note that we have two forces here, the spring and gravity.)

In exercises 24-26, suppose a tank has a rectangular base of 2 by 4 feet and a vertical height of 3 feet.

24. If the tank is filled with water (62.5 pounds per cubic foot), find the work required to pump the water out of the top of the tank.

25. If the tank is half-filled with water (62.5 pounds per cubic foot), find the work required to pump the water 2 feet above the top of the tank.

26. If the tank is half-filled with oil (50 pounds per cubic foot), find the work required to pump the water out of the top of the tank.

In exercises 27-30 suppose a vertical cylindrical can has diameter 30 cm and a height of 60 cm.

27. If the can is filled with water, find the work done to pump the water out of the top of the tank.

28. If the can is filled with water, find the work done to pump the water 20 cm above the top of the tank.

29. If the can is half-filled with water, find the work done to pump the water out of the top of the tank.

30. If the can is half-filled with water, find the work done to pump the water 20 cm above the top of the tank.

In exercises 31-34 suppose a vertical cone (point down) has radius 30 *inches and a height of* 40 *inches.*

31. If the cone is filled with water, find the work done to pump the water out of the top of the cone.

32. If the cone is filled with water, find the work done to pump the water 20 inches above the top of the cone.

33. If the cone is filled to a depth of 20 inches with water, find the work done to pump the water out of the top of the cone.

34. If the cone is half-filled with water, find the work done to pump the water to the top of the cone. (Note: "half-filled" means in terms of volume. This problem is different than the preceding one.)

---

35. A uniform cable 50 feet long and weighing 75 pounds hangs vertically from the top of a building. If a 400 pound weight is attached to the end of the cable, what work is required to pull it to the top?

36. A bucket weighing 20 lbs containing 60 lbs of sand is attached to a rope 100 ft long and weighing 10 lbs and is hanging in a well 150 ft deep. Find the work done in lifting the bucket to the top of the well.

37. A swimming pool full of water is in the form of a rectangular parallelepiped 5 ft deep, 15 ft wide, and 25 ft long. Find the work required to pump the water out of the pool up to a level 1 ft above the surface of the pool.

38. Another swimming pool full of water has the dimensions shown below. Find the work required to pump the water out of the pool.

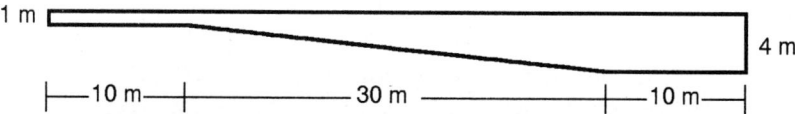

39. The magnitude of the repulsion of two electric charges Q_1 and Q_2 is $F = \dfrac{kQ_1Q_2}{R^2}$ where k is a constant and R is the distance between Q_1 and Q_2. How much work is required to move the charges from 10 centimeters apart to 10^{-6} cm. apart?

40. Newton's law of gravitation says that the force of attraction of two masses is $F = \dfrac{GM_1M_2}{R^2}$ where G is a constant and R is the distance between the centers of the two masses. For objects near the earth this works out approximately to $F = \dfrac{4000^2 M}{R^2}$ where M is the mass of the object and R is the distance to the center of the earth. How much work is required to lift a 1000 pound rocket from the surface of the earth to a 500 mile high orbit? How much more work to lift it to a 1000 mile high orbit?

9.3 IMPROPER INTEGRALS

Normally, a definite integral $\int_a^b f(x)\,dx$ is defined over a *closed* and *bounded* interval $[a,b]$. In this section, we want to extend the idea of integration beyond these restrictions in two different ways. First, we'll investigate when it is possible to assign a reasonable value to definite integrals over *unbounded* intervals like $[a,\infty)$, $(-\infty, b]$, or even to the whole real line $(-\infty, \infty)$. Then we'll examine definite integrals that are not continuous over the given interval. In both cases, we refer to such integrals as **improper integrals**, since they do not satisfy the usual hypotheses of the Fundamental Theorems of Calculus. However, we still may be able to assign a sensible value to the integral in some of these situations.

Improper integrals—horizontal type

First, let's consider functions defined over unbounded intervals. Figure 9.15 illustrates two possibilities.

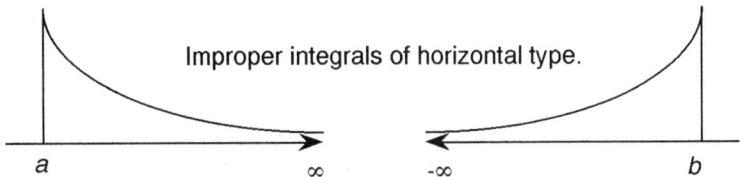

Figure 9.15 Improper integrals over unbounded intervals.

If the function values $f(x)$ are positive over the entire unbounded interval, then we can imagine the integration problem as one of measuring the area of an infinitely long region. At first thought, it might appear that an integral like

$$\int_a^\infty f(x)\,dx \quad \text{or} \quad \int_{-\infty}^b f(x)\,dx$$

could not help but represent an infinite quantity.

Nevertheless, there are cases where it makes sense to assign a finite value to such integrals. A physical example to keep in mind is the case of radioactive decay. Suppose f represents the *rate of decay* of some piece of radioactive material. Such material never decays completely in any finite amount of time. On the other hand, the rate of decay becomes slower and slower as time passes. If we have 100 grams of the material present at time $t=0$, then $\int_0^b f(t)\,dt$ yields the amount of decay at time $t=b$. Theoretically,

$$\int_0^\infty f(t)\,dt = 100 \text{ grams}.$$

In other words, given an infinitely long period of time, the finite amount of material would decay completely.

Using the language of limits, what we are really saying is that the value of

$$\int_0^b f(t)\,dt$$

approaches 100 grams as $b \to \infty$.

Terminology of improper integrals

We will call integrals over unbounded intervals *horizontal type improper integrals*. A horizontal type improper integral can be thought of as the limiting value of a "proper" definite integral as one endpoint approaches ∞ or $-\infty$.

Definition 3 | **Horizontal type improper integrals.**

$$\int_a^\infty f(x)\,dx \quad \text{is shorthand for} \quad \lim_{b\to\infty} \int_a^b f(x)\,dx$$

and

$$\int_{-\infty}^b f(x)\,dx \quad \text{is shorthand for} \quad \lim_{a\to-\infty} \int_a^b f(x)\,dx.$$

If the limiting value exists, then we say the improper integral **converges**. Otherwise, we say the improper integral **diverges**.

EXAMPLE 13 Determine whether $\int_1^\infty \frac{1}{x^2}\,dx$ converges or diverges. If it converges, find its value.

Solution First, we calculate $\int_1^b \frac{1}{x^2}\,dx$, using the parameter b as the upper limit of integration:

$$\int_1^b \frac{1}{x^2}\,dx = \int_1^b x^{-2}\,dx = -\frac{1}{x}\Big]_1^b = -\frac{1}{b} - \left(-\frac{1}{1}\right) = 1 - \frac{1}{b}.$$

Now, we take the limit as $b \to \infty$.

$$\int_1^\infty \frac{1}{x^2}\,dx = \lim_{b\to\infty} \int_1^b \frac{1}{x^2}\,dx = \lim_{b\to\infty} 1 - \frac{1}{b} = 1,$$

9.3 IMPROPER INTEGRALS

since $1/b \to 0$ as $b \to \infty$. We say that this improper integral *converges* and has the value 1. ∎

EXAMPLE 14 Determine whether $\int_4^\infty \frac{1}{\sqrt{x}} \, dx$ converges or diverges. If it converges, find its value.

Solution Using b as the upper limit of integration, we have

$$\int_4^b \frac{1}{\sqrt{x}} \, dx = \int_4^b x^{-1/2} \, dx = 2x^{1/2} \Big]_4^b = 2\sqrt{b} - 4.$$

Now,

$$\int_4^\infty \frac{1}{\sqrt{x}} \, dx = \lim_{b \to \infty} \int_4^b \frac{1}{\sqrt{x}} \, dx = \lim_{b \to \infty} 2\sqrt{b} - 4 = \infty,$$

since $2\sqrt{b}$ grows without bound as $b \to \infty$. We say that this improper integral *diverges*. ∎

Both of the functions $x \longmapsto 1/x^2$ and $x \longmapsto 1/\sqrt{x}$ have output values that approach 0 as $x \to \infty$. What the last two examples show is that this is not a strong enough condition for the improper integral to converge. Evidently, $1/x^2$ approaches 0 much faster than $1/\sqrt{x}$ as $x \to \infty$ (substituting large positive values x into both expressions supports this).

EXAMPLE 15 Determine whether $\int_{-\infty}^{-2} e^x \, dx$ converges or diverges. If it converges, find its value.

Solution First, we replace the lower limit of integration by a parameter a:

$$\int_a^{-2} e^x \, dx = e^x \Big]_a^{-2} = e^{-2} - e^a.$$

Now we take the limit as $a \to -\infty$:

$$\int_{-\infty}^{-2} e^x \, dx = \lim_{a \to -\infty} \int_a^{-2} e^x \, dx = \lim_{a \to -\infty} e^{-2} - e^a = e^{-2}$$

since $e^a \to 0$ as $a \to -\infty$. We say the improper integral converges to the value $1/e^2$. ∎

EXAMPLE 16 Determine whether $\int_{-\infty}^{0} \cos(x)\,dx$ converges or diverges. If it converges, find its value.

Solution $\int_{-\infty}^{0} \cos(x)\,dx = \lim_{a \to -\infty} \int_{a}^{0} \cos(x)\,dx = \lim_{a \to -\infty} \sin(x)\Big]_{a}^{0} = \lim_{a \to -\infty} -\sin(a).$

This limit does not exist, since $\sin(a)$ oscillates between -1 and 1 periodically as $a \to -\infty$. We say the improper integral *diverges*. ∎

If both limits of integration are infinite, then we consider the integral to be a sum of two improper integrals of the type we've been discussing:

$$\int_{-\infty}^{\infty} f(x)\,dx = \int_{-\infty}^{c} f(x)\,dx + \int_{c}^{\infty} f(x)\,dx$$

where c can be any real number we choose. If the graph $y = f(x)$ has symmetry about the vertical line $x = c$, then that would be a good choice. If there is no particular symmetry to the graph, then $c = 0$ is as good a choice as any. If *either* or both of the two integrals

$$\int_{-\infty}^{c} f(x)\,dx \quad \text{and} \quad \int_{c}^{\infty} f(x)\,dx$$

diverge, then we say that $\int_{-\infty}^{\infty} f(x)\,dx$ diverges.

EXAMPLE 17 Determine whether $\int_{-\infty}^{\infty} \frac{1}{1+x^2}\,dx$ converges or diverges. If it converges, find its value.

Solution The graph $y = \frac{1}{1+x^2}$ is a bell-shaped curve having symmetry about the vertical line $x = 0$ (see Figure 9.16).

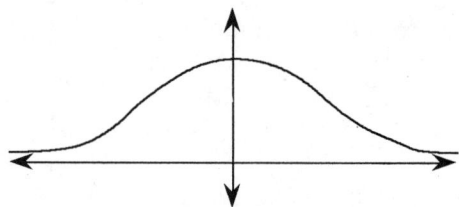

Figure 9.16 Graph of $y = 1/(1+x^2)$.

In other words, $x \longmapsto 1 + x^2$ is an even function and we can write

$$\int_{-\infty}^{\infty} \frac{1}{1+x^2}\,dx = \int_{-\infty}^{0} \frac{1}{1+x^2}\,dx + \int_{0}^{\infty} \frac{1}{1+x^2}\,dx = 2\int_{0}^{\infty} \frac{1}{1+x^2}\,dx.$$

9.3 IMPROPER INTEGRALS

The improper integral $\int_0^\infty \frac{1}{1+x^2}\,dx$ can be evaluated as

$$\lim_{b\to\infty}\int_0^b \frac{1}{1+x^2}\,dx = \lim_{b\to\infty}\arctan(x)\Big]_0^b = \lim_{b\to\infty}\arctan(b) = \pi/2.$$

Hence, the improper integral converges, and

$$\int_{-\infty}^\infty \frac{1}{1+x^2}\,dx = 2(\pi/2) = \pi.$$

∎

In this example, we were able to exploit the symmetry of the function to evaluate only one of the two "halves" of the integral.

Beware.

$$\int_{-\infty}^\infty f(x)\,dx \quad \textbf{may or may not equal} \quad \lim_{a\to\infty}\int_{-a}^a f(x)\,dx.$$

Use of this shortcut may be injurious to the correctness of your result.

Improper integrals—vertical type

The Second Fundamental Theorem of Calculus guarantees that the definite integral

$$\int_a^b f(x)\,dx$$

exists provided the function f is continuous over the closed bounded interval $[a,b]$. We have just seen how it is sometimes possible to assign a finite value to improper integrals such as

$$\int_a^\infty f(x)\,dx \quad \text{or} \quad \int_{-\infty}^b f(x)\,dx$$

where the interval is not bounded. We call these *horizontal type* improper integrals.

Another way to relax the hypotheses of the Second Fundamental Theorem of Calculus is to not insist on the function being continuous over the interval. In an earlier discussion of integration, we pointed out that discontinuities such as holes, skips, and jumps really present no problem, provided there are only finitely many of them in the interval. For these types of discontinuities, one can simply split the interval up at these points, and then redefine the function at all the newly created endpoints so that the Second Fundamental Theorem applies over each.

In this section we take a look at a more serious type of discontinuity. *Vertical type* improper integrals are integrals of the form

$$\int_a^b f(x)\,dx$$

where there is a *vertical asymptote* somewhere in the interval. Figure 9.17 illustrates two possibilities.

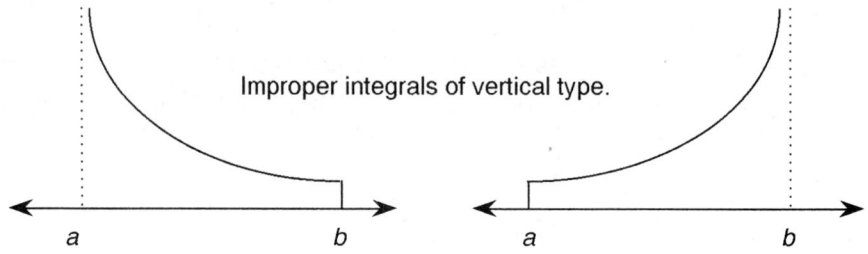

Figure 9.17 Improper integrals over intervals with a vertical asymptote.

If the function values $f(x)$ are all positive except at the asymptote, then we can think of the improper integral as the measure of the area of an infinitely tall region. Whether or not this area is finite will depend on how fast the graph $y = f(x)$ approaches the vertical asymptote as x approaches the location of the asymptote. The strategy for treating such an integral is very similar to the strategy for evaluating horizontal type improper integrals. If the vertical asymptote occurs at one end of the interval, then we first replace that endpoint with a parameter (placeholder) and then study the limiting behavior of the definite integral as the parameter's value approaches the "bad" end.

Definition 4

Vertical type improper integrals. If f is continuous over $(c, b]$, with a vertical asymptote at $x = c$, then

$$\int_c^b f(x)\,dx \quad \text{is shorthand for} \quad \lim_{a \to c^+} \int_a^b f(x)\,dx.$$

If f is continuous over $[a, c)$, with a vertical asymptote at $x = c$, then

$$\int_a^c f(x)\,dx \quad \text{is shorthand for} \quad \lim_{b \to c^-} \int_a^b f(x)\,dx.$$

In either case, if the limiting value exists, then we say the improper integral **converges**. Otherwise, we say the improper integral **diverges**.

9.3 IMPROPER INTEGRALS

We'll illustrate with several examples.

EXAMPLE 18 Determine whether $\int_0^1 \frac{1}{\sqrt{x}}\, dx$ converges or diverges. If the improper integral converges, find its value.

Solution Note that $1/\sqrt{x}$ is not defined at $x = 0$. However, $\int_a^1 \frac{1}{\sqrt{x}}\, dx$ is defined if $a > 0$, and we can compute

$$\int_a^1 \frac{1}{\sqrt{x}}\, dx = \int_a^1 x^{-1/2}\, dx = \left.\frac{x^{1/2}}{1/2}\right]_a^1 = 2 - 2\sqrt{a}.$$

Hence,

$$\int_0^1 \frac{1}{\sqrt{x}}\, dx = \lim_{a \to 0^+} \int_a^1 \frac{1}{\sqrt{x}}\, dx = \lim_{a \to 0^+} 2 - 2\sqrt{a} = 2.$$

We say the integral *converges* to 2. ∎

EXAMPLE 19 Determine whether $\int_0^e \frac{1}{x}\, dx$ converges or diverges. If the improper integral converges, find its value.

Solution

$$\lim_{a \to 0^+} \int_0^e \frac{1}{x}\, dx = \lim_{a \to 0^+} \left(\ln|x|\Big]_a^e\right) = \lim_{a \to 0^+} (\ln e - \ln a) = \lim_{a \to 0^+} 1 - \ln a = +\infty.$$

We say the integral *diverges*. ∎

EXAMPLE 20 Determine whether $\int_1^2 (x - 2)^{-2/3}\, dx$ converges or diverges. If the improper integral converges, find its value.

Solution This time the vertical asymptote occurs at the right endpoint. So,

$$\int_1^2 (x - 2)^{-2/3}\, dx = \lim_{b \to 2^-} \int_1^b (x - 2)^{-2/3}\, dx = \lim_{b \to 2^-} 3(x - 2)^{1/3}\Big]_1^b$$

$$= \lim_{b \to 2^-} 3(b - 2)^{1/3} - 3(1 - 2)^{1/3} = \lim_{b \to 2^-} 3(b - 2)^{1/3} + 3 = 3.$$

The improper integral converges to 3. ∎

If a vertical asymptote appears between the endpoints then we agree to split the interval into two pieces at the vertical asymptote. Again, if *either* or both of these improper integrals diverges, we say the entire improper integral diverges.

EXAMPLE 21 Determine whether $\int_{-1}^{1} \frac{1}{x} dx$ converges or diverges. If the improper integral converges, find its value.

Solution A glance at the graph of $y = 1/x$ might suggest to us that this integral should be 0, since the region below the x-axis is symmetrical to the region above. However, we cannot say that $\int_{-1}^{1} \frac{1}{x} dx$ converges unless each of

$$\int_{-1}^{0} \frac{1}{x} dx \quad \text{and} \quad \int_{0}^{1} \frac{1}{x} dx$$

converges.

$$\int_{a}^{1} \frac{1}{x} dx = \lim_{a \to 0^+} \ln|x| \Big]_{a}^{1} = \lim_{a \to 0^+} \ln|1| - \ln|a| = \lim_{a \to 0^+} 0 - \ln|a| = \infty.$$

So, this improper integral diverges. ∎

EXAMPLE 22 Determine whether $\int_{1}^{4} (x-2)^{-2/3} dx$ converges or diverges. If the improper integral converges, find its value.

Solution This is similar to a previous example, except now the vertical asymptote $x = 2$ occurs in the interior of the interval.

$$\int_{1}^{3} (x-2)^{-2/3} dx = \lim_{b \to 2^-} \int_{1}^{b} (x-2)^{-2/3} dx + \lim_{a \to 2^+} \int_{a}^{4} (x-2)^{-2/3} dx.$$

We have already computed the first of these improper integrals, and found it to converge to 3. As for the second improper integral,

$$\lim_{a \to 2^+} \int_{a}^{4} (x-2)^{-2/3} = \lim_{a \to 2^+} \frac{3}{5}(x-2)^{1/3} \Big]_{a}^{4} = 3\sqrt[3]{2}.$$

Since both integrals converge, we can conclude that the entire improper integral converges to $3 + 3\sqrt[3]{2}$. ∎

EXERCISES

In exercises 1-10, determine whether the improper integral converges or diverges. If it converges, find its value.

1. $\int_{1}^{\infty} \frac{1}{x^3} dx$
2. $\int_{3}^{\infty} \frac{1}{x^{2/3}} dx$
3. $\int_{0.5}^{\infty} \frac{1}{x^2} dx$
4. $\int_{8}^{\infty} \frac{1}{x^{1/3}} dx$
5. $\int_{1}^{\infty} \frac{1}{x^{3/2}} dx$
6. $\int_{0.25}^{\infty} \frac{1}{x^{1/2}} dx$

9.3 IMPROPER INTEGRALS

7. $\int_1^\infty \dfrac{1}{x^{1.01}}\, dx$

8. $\int_1^\infty \dfrac{1}{x^{0.99}}\, dx$

9. $\int_1^\infty \dfrac{1}{x}\, dx$

10. $\int_1^\infty \dfrac{1}{x-1}\, dx$

---

11. On the basis of your answers to exercises 1-10, formulate a conjecture regarding the convergence or divergence of

$$\int_1^\infty \dfrac{1}{x^p}\, dx.$$

Specifically, does the integral converge or diverge when $p < 1$, when $p = 1$, and when $p > 1$? Then test your conjecture by evaluating the integral in terms of p.

12. For what values p does the improper integral $\int_{-\infty}^1 \dfrac{1}{x^p}\, dx$ converge?

In exercises 13-20, determine whether the improper integral converges or diverges. If it converges, find its value.

13. $\int_0^\infty e^{2x}\, dx$

14. $\int_0^\infty e^{-2x}\, dx$

15. $\int_{-\infty}^0 e^{2x}\, dx$

16. $\int_{-\infty}^0 e^{-2x}\, dx$

17. $\int_0^\infty e^{x/2}\, dx$

18. $\int_0^\infty e^{-x/2}\, dx$

19. $\int_{-\infty}^0 e^{x/2}\, dx$

20. $\int_{-\infty}^0 e^{-x/2}\, dx$

---

21. On the basis of your answers to exercises 13-20, formulate a conjecture regarding the convergence or divergence of

$$\int_0^\infty e^{px}\, dx \quad \text{and} \quad \int_{-\infty}^0 e^{px}\, dx.$$

Specifically, what happens when $p < 0$ and when $p > 0$? In the cases when the integral converges, find its value in terms of p.

22. For the same integral as in exercise 21, what happens when $p = 0$?

In exercises 23-26, determine whether the improper integral converges or diverges. If it converges, find its value.

23. $\int_{-\infty}^\infty \dfrac{e^x}{1+e^{2x}}\, dx$

24. $\int_{-\infty}^\infty e^{-|x|}\, dx$

25. $\int_{-\infty}^{\infty} x \, dx$

26. $\int_{2}^{\infty} \dfrac{1}{x\sqrt{x^2 - 1}} \, dx$

In exercises 27-36, determine whether the improper integral converges or diverges. If it converges, find its value.

27. $\int_{-1}^{0} \dfrac{1}{x^3} \, dx$

28. $\int_{-3}^{0} \dfrac{1}{x^{2/3}} \, dx$

29. $\int_{-0.5}^{0} \dfrac{1}{x^2} \, dx$

30. $\int_{-8}^{0} \dfrac{1}{x^{1/3}} \, dx$

31. $\int_{0}^{2} \dfrac{1}{x^{3/2}} \, dx$

32. $\int_{0}^{0.25} \dfrac{1}{x^{1/2}} \, dx$

33. $\int_{0}^{1} \dfrac{1}{x^{1.01}} \, dx$

34. $\int_{0}^{1} \dfrac{1}{x^{0.99}} \, dx$

35. $\int_{0}^{1} \dfrac{1}{x} \, dx$

36. $\int_{0}^{1} \dfrac{1}{x^{-1}} \, dx$

---

37. On the basis of your answers to exercises 27-36, formulate a conjecture regarding the convergence or divergence of

$$\int_{0}^{1} \dfrac{1}{x^p} \, dx.$$

Specifically, does the integral converge or diverge when $p < 1$, when $p = 1$, and when $p > 1$? Then test your conjecture by evaluating the integral in terms of p.

38. For what values p does the improper integral $\int_{-1}^{0} \dfrac{1}{x^p} \, dx$ converge?

---

39. Determine whether $\int_{0}^{4} \dfrac{1}{(x-1)^{\frac{1}{3}}} \, dx$ converges, and if it does, compute its value.

40. Determine whether $\int_{0}^{1} \ln(x) \, dx$ converges, and if it does, compute its value.

41. Analyze the following problem:

$$\int_{0}^{\pi} \sec^2(x) \, dx = \tan x \Big]_{0}^{\pi} = \tan \pi - \tan 0 = 0.$$

If there is anything wrong with the solution, clearly state what is wrong and evaluate the integral correctly.

42. Determine whether $\int_{1}^{2} \dfrac{x}{x^2 - 1} \, dx$ converges, and if it does, compute its value.

9.3 IMPROPER INTEGRALS

43. Determine whether $\int_1^3 \ln x \, dx$ converges, and if it does, determine its value.

44. A student looks at the graph of $y = 1/x$ over the interval $[-2, 2]$ and notes that the graph is symmetric with respect to the origin. The student concludes that

$$\int_{-2}^0 \frac{1}{x} \, dx = -\int_0^2 \frac{1}{x} \, dx,$$

and therefore, $\int_{-2}^0 \frac{1}{x} \, dx = 0$. Why is the student wrong?

---

45. Verify that integral $\int_{-a}^a x \, dx = 0$ for any positive number a. This of course means that

$$\lim_{a \to \infty} \int_{-a}^a x \, dx = 0.$$

Compare this with the answer you obtained in exercise 25. (This example shows why this method of evaluating an improper integral over $(-\infty, \infty)$ is unreliable.)

46. A nonnegative function $p(x)$ can be a probability density function provided that the integral $\int_{-\infty}^{\infty} p(x) \, dx = 1$. Let

$$p(x) = \frac{1}{\sqrt{2\pi}} e^{-(x^2/2)}.$$

(This function p is known as the standard normal probability density function.) The graph of $y = p(x)$ is a bell-shaped curve symmetric about $x = 0$, so

$$\int_0^{\infty} p(x) \, dx = \int_{-\infty}^0 p(x) \, dx.$$

Since $p(x)$ has no closed-form antiderivative, we cannot use the Second Fundamental Theorem of Calculus to evaluate it. Using a numerical integrator, approximate

$$\int_0^b p(x) \, dx$$

for $b = 1, 2, 3, 4, 5,$ and 10. Do you believe that $\int_{-\infty}^{\infty} p(x) \, dx = 1$?

47. In the theory of differential equations, if f is a function, then the **Laplace Transform** $L[s]$ of f is defined by

$$L[s] = \int_0^{\infty} e^{-sx} f(x) \, dx,$$

for every real number s for which the improper integral converges. Find the Laplace Transform of $f : x \longmapsto \sin(x)$. In other words evaluate

$$\int_0^\infty e^{-sx} \sin x \, dx.$$

Your answer should be in terms of s. (Treat s as a constant when integrating.)

48. The *Horn of Gabriel* is obtained by rotating the graph of $y = 1/x$ about the x-axis over the interval $[1, \infty)$. The surface area of Gabriel's Horn and the volume within Gabriel's Horn are both given by improper integrals, since the interval is infinite. Which of the two improper integrals converges? (It's been said that the only way to paint Gabriel's Horn is by filling it with paint.)

9.4 USING DEFINITE INTEGRALS TO MEASURE PROBABILITIES

Integration plays an especially important role in analyzing probabilities. There are two main branches of probability— discrete and continuous. Discrete probability analysis essentially is concerned with "counting" problems. Here are some examples of discrete probablility questions:

If I flip a fair coin 5 times, what is the probability that I get exactly 2 heads?

What is the probability that I get at least 3 heads?

What is the probability that I get 7 heads?

The answers to these three questions are respectively 5/16, 1/2, and 0. (In discrete probability analysis, a probability of 0 represents an impossible event.)

Calculus is particularly important in solving problems in *continuous* probability analysis. Many of these problems relate directly to area measurement, and so it is natural that definite integration is useful. Here is an example of a continuous probability problem directly solved in terms of area measurement:

If a dart hits a circular dartboard of radius 9 inches at random, what is the probability of hitting the bullseye, which is a circle of radius 1 inches (see Figure 9.18)?

9.4 USING DEFINITE INTEGRALS TO MEASURE PROBABILITIES

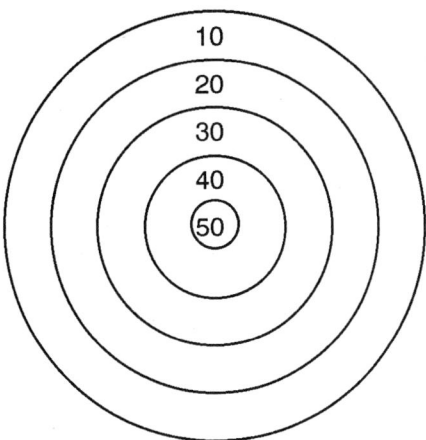

Figure 9.18 What is the probability of hitting a bullseye with a random dart?

The solution to the dartboard problem requires computing the ratio of the area of the bullseye to the area of the dartboard:

$$\text{probability of bullseye} = \frac{\pi \cdot 1^2}{\pi \cdot 9^2} = \frac{1}{81}.$$

The words "at random" are key here, for this means that we are just as likely to hit any point as any other on the dartboard. We would expect an experienced dart thrower to have a much higher than $1/81$ chance of a bullseye because the throws would not be distributed at random.

What is the probability of the dart hitting the *exact* center of the dartboard, if it hits the dartboard at random? (By this, we mean that the exact center of the dart's point coincides with the exact center of the dartboard.) Even though this is a theoretical possibility, we are forced to assign a 0 probability to this event, because the area of a single point is 0. That is one of the key distinctions between discrete and continuous probability analysis— a probability of 0 in discrete analysis means that the event cannot possibly happen, while in continuous analysis, the event is not necessarily impossible (but it surely is *extremely* unlikely).

EXAMPLE 23 What is the probability that a point (x, y), randomly chosen from $[-1, 1] \times [0, 1]$ satisfies $y \leq 1 - x^2$?

Solution The rectangle $[-1, 1] \times [0, 1]$ is a rectangle of length $1 - (-1) = 2$ and height $1 - 0 = 1$. To satisfy the inequality, a point (x, y) in this rectangle must fall on or below the graph of $y = 1 - x^2$. This region is shown in Figure 9.19.

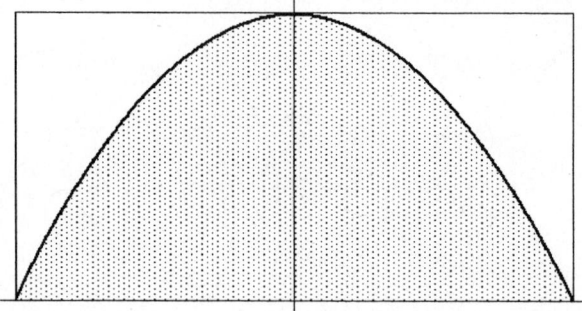

Figure 9.19 What fraction of the rectangle is shaded?

The total area is

$$2 \cdot 1 = 2 \text{ square units.}$$

Now we must calculate the area of the subregion of interest. The points in the rectangle satisfying $y \leq 1 - x^2$ are those lying on or under the graph of $y = 1 - x^2$ over the interval $[-1, 1]$. This area is given by the definite integral

$$\int_{-1}^{1} 1 - x^2 \, dx = \frac{4}{3}.$$

Thus, the probability that a randomly chosen point from the rectangle satisfies the inequality is $(4/3)/2 = 2/3$. ∎

Whenever a real number x is chosen randomly from an interval $[a, b]$, the probability that x falls in a subinterval of length Δx is

$$\frac{\Delta x}{b - a}.$$

EXAMPLE 24   If a real number x is randomly chosen from the interval $[-3, 5]$, what is the probability that $x > 2$? What is the probability that $-1 \leq x \leq 1$?

Solution   The total length of the interval is $5 - (-3) = 8$.

To have $x > 2$, we must have $x \in (2, 5]$. The length of this interval is $\Delta x = 5 - 2 = 3$. Hence, the probability that $x > 2$ is $3/8$.

The length of the interval $[-1, 1]$ is $\Delta x = 1 - (-1) = 2$. Hence, the probability that $-1 \leq x \leq 1$ is $2/8 = 1/4$.

Notice that in determining the length of an interval, it does not matter whether the endpoints are included or not. ∎

Whenever a point (x, y) is chosen randomly from a region R, then the probability that (x, y) falls in a given subregion S is (area of S)/(area of R).

9.4 USING DEFINITE INTEGRALS TO MEASURE PROBABILITIES

EXAMPLE 25 Suppose a point (x, y) is randomly chosen from $[-1, 1] \times [-1, 1]$. What is the probability that $x^3 < y < x^2$?

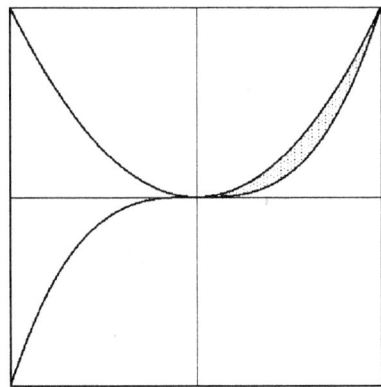

Figure 9.20 What fraction of the square is shaded?

Solution The region R is a square with side length 2, hence the area of R is 4. To satisfy the inequalities, the point (x, y) must fall in the subregion S above the graph of $y = x^3$ and below the graph of $y = x^2$. The region R with subregion S is shown in Figure 9.20.

We can find the area of S by integrating

$$\int_0^1 (x^2 - x^3) = 1/3 - 1/4 = 1/12.$$

The probability that a randomly chosen point from R falls in S is

$$\frac{1/12}{4} = \frac{1}{48} \approx 0.02083.$$

∎

Exactly what does the probability measurement tell us? One natural interpretation of probability measure is that of *expected frequency ratio* of repeated trials. For example, if we choose several points at random from the region R of the previous example, and we count the number of chosen points that fall in S, then we should expect the ratio

$$\frac{\text{number of points in } S}{\text{total number of points}} \approx \frac{1}{48}.$$

In practice, this means that we expect a frequency of approximately 2 out of every 100 points falling in the region. In actual experimentation, it is quite possible to get a ratio that is very far from the value of the theoretical probability, particularly if we choose only a few points. But, as the number of trials becomes very large, it is less and less likely that this frequency ratio will vary drastically from the probability. In fact, one way to check a computed probability is by comparing it to the frequency ratio results of several repeated actual or simulated trials.

Using a random number generator to estimate π

Many calculators and computers have a random number generator that can provide (almost) random numbers, usually between 0 and 1. If we use such a random number generator to choose an ordered pair (x, y) by first using it to choose x then to choose y, then we can think of this as a point (x, y) randomly chosen from the unit square $[0, 1] \times [0, 1]$. If we generate sufficiently many random points in this way, we can actually use their distribution to make approximate area measurements.

EXAMPLE 26 What is the probability that a randomly chosen point (x, y) from the unit square $[0, 1] \times [0, 1]$ is within 1 unit of the origin?

Solution   In terms of area measurement, we can see that this probability is the ratio of the area of a quarter circle to the area of the unit square (see Figure 9.21).

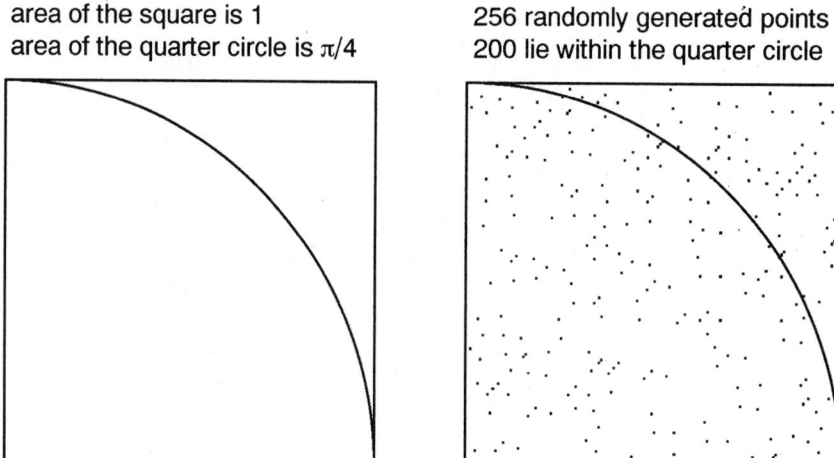

Figure 9.21 Estimating the area of the quarter circle with random points.

The area of the quarter circle of radius 1 is

$$\frac{\pi 1^2}{4} = \frac{\pi}{4}.$$

The unit square $[0, 1] \times [0, 1]$ has area 1, so the probability that the randomly chosen point (x, y) falls in the quarter circle is $\dfrac{\pi/4}{1} = \pi/4$. ∎

In terms of the frequency ratio interpretation of probability, this suggests that we can estimate $\pi/4$ by repeatedly choosing points at random in the square and finding the proportion of those falling within the quarter circle. Equivalently, we need to compute the proportion of points (x, y) in $[0, 1] \times [0, 1]$ satisfying the inequality

$$x^2 + y^2 \leq 1.$$

Figure 9.21 shows the results of 256 points generated at random. In 200 trials out of 256, the inequality was satisfied. This leads to the estimate $\pi/4 \approx 200/256 = 0.78125$. If we multiply the estimate by 4, we have

$$\pi \approx 3.125.$$

While we know that this is not a particularly accurate estimate of π, it illustrates the idea. With many more trials, we could expect a better approximation.

Probability mass

In discrete probability, the notion of **probability mass** is a useful one. If a fair coin is flipped, then heads and tails are equally likely and we say that each has a *probability mass* of $1/2$. If the coin is weighted in such a way that heads comes up on average two times for every one time tails comes up, then we say that heads has a probability mass of $2/3$ and tails has a probability mass of $1/3$. In the case of the fair coin, the probability masses are distributed *uniformly*, while they are not for the weighted coin.

No matter how the probability masses are distributed, notice that none can be a negative number, and they must sum up to a total of 1.

If a real number X is assigned to each possible event, then we say that X is a **random variable**, regardless of whether or not the probability masses are distributed uniformly. For example, we could assign the values $X = 1$ for heads and $X = 2$ for tails in flipping a coin. The **expected value** $E(X)$ of a discrete random variable X is computed by multiplying each possible value X by its probability mass and adding the results.

EXAMPLE 27 For the fair coin we have

$$E(X) = 1 \cdot (1/2) + 2 \cdot (1/2) = 1/2 + 1 = 3/2.$$

For the weighted coin we have

$$E(X) = 1 \cdot (2/3) + 2 \cdot (1/3) = 2/3 + 2/3 = 4/3.$$

Notice that the expected value is closer to 1 for the coin weighted in favor of heads. ∎

The expected value has an interpretation closely tied to averages and centers of mass: If we flipped a coin many times, recording $X = 1$ for heads and $X = 2$ for tails, the average of these values should be approximately the expected value $E(X)$. If we think of the values of X as coordinate values along an axis, then the expected value $E(X)$ is just the center of probability mass $\overline{X}$. (This is because the total probability mass is 1.) In fact, the notation $\overline{X}$ is often used for the mean value of a random variable X.

To compute the expected value of a function f of X, you simply multiply each function value $f(X)$ by the corresponding probability mass of X and add the results.

EXAMPLE 28 When a fair die is rolled, each of the faces 1 through 6 are equally likely, so the probability masses are uniformly distributed and each has a probability mass of $1/6$. If X is the number rolled, find $E(X)$ and $E(X^2)$.

Solution $E(X) = (1/6)(1 + 2 + 3 + 4 + 5 + 6) = 21/6 = 7/2 = 3.5$, and $E(X^2) = (1/6)(1^2 + 2^2 + 3^2 + 4^2 + 5^2 + 6^2) = 91/6 \approx 15.167$. ∎

Mean, moments, variance, and standard deviation

The **mean** μ of a random variable X is simply its expected value $E(X)$ (μ is the Greek letter corresponding to "m" as in *middle*).

The **variance** σ^2 of a random variable X is

$$\text{variance} = \sigma^2 = E(X^2) - E(X)^2.$$

Computationally, the variance is the average of the squares minus the square of the average. Equivalently, it gives the expected value of the squared distance from the mean. That is,

$$\sigma^2 = E((X - \mu)^2).$$

The square root of the variance is called the standard deviation:

$$\text{standard deviation} = \sigma = \sqrt{\text{variance}}.$$

The variance and standard deviation measure how spread out the probability density is (σ is the Greek letter corresponding to "s" as in *spread*).

9.4 USING DEFINITE INTEGRALS TO MEASURE PROBABILITIES

EXAMPLE 29 If X is the number rolled on a fair die, what is the variance and standard deviation?

Solution Using the computations from the previous example, the variance is

$$\sigma^2 = E(X^2) - E(X)^2 = 91/6 - (7/2)^2 = 35/12 \approx 2.917$$

and the standard deviation is $\sigma = \sqrt{\sigma^2} = \sqrt{35/12} \approx 1.708$. ∎

The nth **moment** of X is $E(X^n)$. For example, the first moment is $E(X) = \mu$. The second moment is $E(X^2)$, the third moment is $E(X^3)$, and so on.

Probability density

In continuous probability analysis, it does not make sense to talk about a single value of a random variable X as having a positive probability mass. Much like mass density functions, we can talk of **probability density functions.** The simplest case of a probability density function is for the random choice of a real number X from an interval $[a, b]$. In this case, each value of X in the interval has a density

$$p(X) = \frac{1}{b-a}.$$

Otherwise, $p(X) = 0$.

EXAMPLE 30 If X is a random variable representing a real number chosen at random from the interval $[1, 4]$, find the probability density function for X. What is the probability that $2 < X < 2.5$?

Solution We have a probability density of $p(X) = 1/(4-1) = 1/3$ for each value $1 \leq X \leq 4$, and $p(X) = 0$ for all other values X. We can plot the probability density function as shown in Figure 9.22.

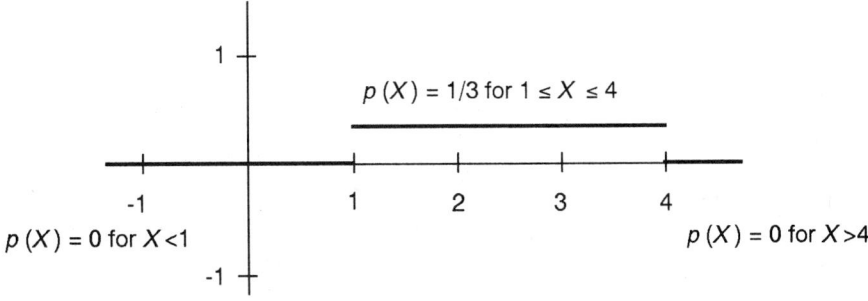

Figure 9.22 Uniform probability density function for the interval $[1, 4]$.

The probability that $2 < X < 2.5$ is

$$(2.5 - 2)/(4 - 1) = 1/6 \approx 0.1667.$$

Notice that this probability value is precisely the area under the graph of $y = p(X)$ over the interval $[2, 2.5]$. A probability density function allows us to calculate probabilities directly by integration. In this case the probability density function was uniform (constant over the interval). More generally, the values of a probability density function could vary. The only two specific requirements that a valid probability density function p must satisfy are:

1) $p(X) \geq 0$ for all X

2) the total area under the graph of $y = p(X)$ must be exactly 1.

EXAMPLE 31 Which of the following are valid probability density functions?

$$p : X \longmapsto .5 \sin(X) \text{ for } 0 \leq X \leq \pi, \text{ and } p(X) = 0, \text{ otherwise.}$$

$$q : X \longmapsto X^2 \text{ for } -1 \leq X \leq 1, \text{ and } q(X) = 0, \text{ otherwise.}$$

$$g : X \longmapsto \ln(X) \text{ for } .5 \leq X \leq 2, \text{ and } g(X) = 0, \text{ otherwise.}$$

Solution For each function, we need to check

1) that all its outputs are nonnegative, and

2) the total area under the graph is 1.

First, $p(X) \geq 0$ for all $X \in [0, \pi]$, because $\sin(X) \geq 0$ for these values X. Checking the total area under the graph, we find:

$$\int_0^\pi p(X)\, dX = \int_0^\pi .5 \sin(X)\, dX = -.5 \cos(X) \Big]_0^\pi = (-.5)(-1) - (-.5)(1) = 1.$$

Hence, p is a valid probability density function.

Second, $q(X) \geq 0$ for all $X \in [-1, 1]$, but

$$\int_{-1}^1 q(X)\, dX = \int_{-1}^1 X^2\, dX = \frac{X^3}{3} \Big]_{-1}^1 = 1/3 - (-1/3) = 2/3 \neq 1.$$

So, q is *not* a valid probability density function.

Third, g is not a valid probability density function because $\ln(X) < 0$ for $.5 \leq X < 1$. ∎

If a function p is the probability density function for a random variable X, then the probability that $a \leq X \leq b$ is simply

$$\int_a^b p(X)\, dX.$$

9.4 USING DEFINITE INTEGRALS TO MEASURE PROBABILITIES

The same integral gives the probability that $a < X \leq b$, $a \leq X < b$, or $a < X < b$, since the probability mass at a single point is 0.

EXAMPLE 32 Suppose X is a random variable with probability density

$$p : X \longmapsto .5\sin(X) \text{ for } 0 \leq X \leq \pi, \text{ and } p(X) = 0, \text{ otherwise.}$$

Find the probability that $2.5 < X \leq 3.5$.

Solution The probability is given by

$$\int_{2.5}^{3.5} p(X)\,dX.$$

Since $p(X) = 0$ for $X > \pi$, we have

$$\int_{2.5}^{3.5} p(X)\,dX = \int_{2.5}^{\pi} .5\sin(X)\,dX \approx 0.1.$$

■

To compute the expected value $E(X)$ of X, we multiply each value of X by its probability density $p(X)$ and integrate over all values of X (those for which $p(X) > 0$ are the only values that contribute). To find the expected output value $f(X)$, where f is a function, we multiply each function value $f(X)$ by the corresponding probability density $p(X)$ and integrate over all values of X.

EXAMPLE 33 For the same random variable as in the previous example, find the first and second moments.

Solution
$$E(X) = \int_0^{\pi} X(.5)\sin(X)\,dX = \pi/2 \approx 1.571.$$

$$E(X^2) = \int_0^{\pi} X^2(.5)\sin(X)\,dX \approx 2.935.$$

■

EXAMPLE 34 For the same variable as in the previous example, find the variance and standard deviation of X.

Solution Using the computations of $E(X^2)$ and $E(X)$ from before, the variance is

$$\sigma^2 = E(X^2) - (E(X))^2 \approx 2.935 - (1.571)^2 \approx 0.467.$$

The standard deviation is $\sigma \approx \sqrt{2.935 - (1.571)^2} \approx 0.683$.

■

The normal probablity density function

One of the most useful and frequently used probability density functions is the **normal** or **Gaussian**) density. The graph of this density function has the shape of the "bell-shaped curve" familiar from graphs of test scores, population height, manufacturing tolerances, and a host of other examples of real number measurements.

The *standard normal* density function has the formula

$$p(X) = \frac{1}{\sqrt{2\pi}} e^{-X^2/2}$$

for all real values of X. Notice that $p(X) > 0$ for all values of X. The constant $1/\sqrt{2\pi}$ guarantees that the total area under the graph is 1. The graph of the standard normal density function is shown in Figure 9.23.

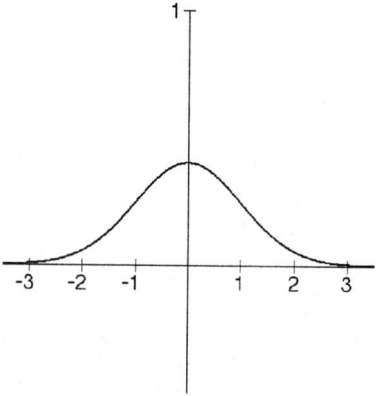

Figure 9.23 Graph of the standard normal density function $p(X) = \frac{1}{\sqrt{2\pi}} e^{-X^2/2}$.

The center peak of the symmetric bell is located at $X = 0$, and this is the center of probability mass. In other words, $E(X) = 0$.

EXAMPLE 35 Suppose X is a random variable with a standard normal probability density. What is the probability that $-1 \leq X \leq 1$?

Solution The probability is given by

$$\int_{-1}^{1} \frac{1}{\sqrt{2\pi}} e^{-X^2/2} \, dX \approx 0.6827.$$

To approximate this integral, we used a machine, because the standard normal probability density function has *no closed-form antiderivative*. For this reason, large tables of approximate values for these integrals have been compiled. ∎

This example shows that over $2/3$ of the area under the graph falls between $X = -1$ and $X = 1$. These values $X = -1$ and $X = 1$ are special for another reason: they are the locations of the two inflection points for the standard normal density function.

There is, in fact, a whole family of normal density functions. Each has the characteristic bell-shaped graph, but the center and the spread of the bell vary. If the center peak of the bell occurs at $X = \mu$ and the two inflection points are located at $X = \mu - \sigma$ and $X = \mu + \sigma$, then the normal probability density function must have the formula

$$p(X) = \frac{1}{\sqrt{2\pi}\sigma} \exp\left(-\frac{(X-\mu)^2}{2\sigma^2}\right).$$

As is suggested by the choice of symbols, the *parameters* μ and σ are the mean and standard deviation of the density function. If X is a random variable with this probability density function, we say that X is **normally distributed with mean μ and standard deviation** σ.

EXAMPLE 36 What is the probability density function for a normally distributed random variable X with mean $\mu = -2$ and standard deviation $\sigma = 3$?

Solution The probability density function is $p : X \longmapsto \frac{1}{3\sqrt{2\pi}} \exp\left(-\frac{(X+2)^2}{18}\right)$. ∎

EXERCISES

Suppose the dartboard of Figure 9.18 consists of concentric circles of radii 9 inches, 7 inches, 5 inches, 3 inches, and a bullseye of radius 1 inch. Suppose X is the score of a dart hitting the dartboard at random. Exercises 1-4 refer to this dartboard.

1. What is the probability of getting $X = 10$? $X = 20$? $X = 30$? $X = 40$? $X = 50$? What is the probability of the dart hitting the border between two regions?
2. What is the mean of X?
3. What are the second and third moments of X?
4. What are the variance and standard deviation of X?

Exercises 5-15 refer to a point (x, y) that is chosen at random from the rectangle $[-1, 2] \times [-1, 2]$. For each exercise, calculate the probability or probabilities indicated. If you have a random number generator that produces random numbers R between 0 and 1, then $3R - 1$ will be a random number in the interval $[-1, 2]$. You can use this to produce 100 random points in the rectangle and empirically test your probabilities by counting the number of points satisfying the requirements in each of the exercises.

5. What is the probability that the point is in the first quadrant? second quadrant? third quadrant? fourth quadrant?

6. What is the probability that the point is on one of the coordinate axes?

7. What is the probability that the point is within the unit circle?

8. What is the probability that $y \geq 2x - 1$?

9. What is the probability that $x^3 < y$?

10. What is the probability that $y < x^3$?

11. What is the probability that $x^2 < y$?

12. What is the probability that $y < x^2$?

13. What is the probability that $x^2 < y < x^3$?

14. What is the probability that $x^3 < y < x^2$?

15. What is the probability that $|x| > |y|$?

Which of the functions in exercises 16-20 are valid probability density functions? For each valid probability density function, find its mean and its variance. For each function that is not a valid probability density function, explain what is wrong.

16. $p : X \longmapsto \dfrac{\cos(X)}{2}$ for $-\pi/2 \leq X \leq \pi/2$, $p(X) = 0$ otherwise

17. $p : X \longmapsto 1/X$ for $1 \leq X \leq e$, $p(X) = 0$ otherwise

18. $p : X \longmapsto \cos(X)$ for $0 \leq X \leq \pi$, $p(X) = 0$ otherwise

19. $p : X \longmapsto \dfrac{X^2}{3}$ for $-1 \leq X \leq 2$, $p(X) = 0$ otherwise

20. $p : X \longmapsto \ln(X)$ for $1 \leq X \leq e$, $p(X) = 0$ otherwise

Exercises 21-25 refer to the standard normal probability density function

$$p : X \longmapsto = \dfrac{e^{-X^2/2}}{\sqrt{2\pi}}.$$

21. Find the probability that $0.5 < X < 2$ accurate to three decimal places (you will definitely need a machine for this).

22. Find the probability that $-3 \leq X \leq 3$ accurate to three decimal places. Do you agree with the statement: "Over 99% of the area under the bell curve falls within three standard deviations of the mean" ?

23. Show that the inflection points of the graph of $y = p(X)$ occur at $X = \pm 1$.

24. Find the maximum value of $p(X)$.

25. Using a numerical integrator, find a value a so that the probability $-a \leq X \leq a$ is $1/2$ accurate to three decimal places.

Platypus finds that the metal rods coming off one of their manufacturing machines have diameters X, which are normally distributed with a mean $\mu = 0.5$ cm and a standard deviation $\sigma = .01$ cm. Exercises 26-30 concern these rods.

26. What is the probability density function for the diameter X of the rods?

27. A rod is unacceptable if its diameter is outside the range $.48 \leq X \leq .52$ cm. What is the probability that a rod is unacceptable? (Find the probability that it is acceptable, and subtract this value from the total probability 1.)

28. One of the machine settings slips, so that the mean diamater is now $\mu = 0.51$ cm with the same standard deviation $\sigma = 0.1$ cm. What is the new probability density function?

29. If the acceptable range is still $.48 \leq X \leq .52$ cm, what is the probability that a rod is unacceptable after the machine setting slips?

30. The machine setting is corrected so the $\mu = 0.5$, but the age of the machine has increased the standard deviation to $\sigma = .02$ cm. What is the new probability density function, and what is the probability that a rod is unacceptable?

---

31. Suppose a coaster of diameter 4 inches is thrown onto a floor of 8 inch by 8 inch square tiles. Assuming that the floor is large enough that there is no chance of the coaster leaning up against a wall, what is the probability that the coaster lands entirely within one of the square tiles (with no part touching a tile border)? Hint: The center of the coaster must land somewhere within one of the square tiles. Where must the center of the coaster be within the square so that the entire coaster is within the square?

32. Suppose X is a random number chosen from the interval $[a, b]$. What is the mean and variance?

One of the more famous problems in probability is Buffon's Needle Problem. We can state it in the following way: suppose that a needle of length $\ell = 1$ unit is thrown randomly onto a wooden floor with parallel boards of width $d = 2$ units (see picture below). What is the probability that the needle will land lying across one of the board lines? In analyzing this problem, a couple of simplifying assumptions are normally made. First, the horizontal position of the needle can be ignored. Second, throwing the needle randomly can be accurately modelled by randomly choosing a pair of numbers (x, y), where x is the distance between the center of the needle and the lower board edge and y is the angle in radians that the needle makes with a line parallel to the board edges. In other words, (x, y) is an ordered pair randomly chosen from $[0, 2] \times [0, \pi]$.

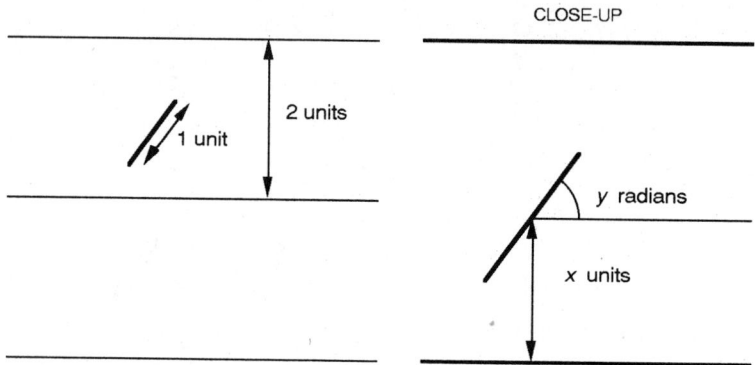

The boards, the needle, and the dimensions of Buffon's needle problem.

33. Show that the vertical distance from a tip of the needle to the horizontal line through its center is $\frac{1}{2}\sin(y)$.

34. Write down an inequality describing what pairs (x, y) represent needles that touch the board edges. (When does the vertical distance of the needle tip to its center exceed the distance to either the lower or upper board edge?)

35. Find the probability that the needle touches a board edge. (What "points" in $[0, 2] \times [0, \pi]$ satisfy the inequality you obtained in the previous exercise?)

36. Suppose, in general, that the needle has a length ℓ and the distance between boards is d (and $d > \ell$). Find the probability (in terms of ℓ and d) that the needle lands across a board border.

To get some feeling about how different, perhaps equally natural, interpretations of "random" can lead to different predictions, consider choosing a chord of a unit circle "at random" and finding the probability that the length of the chord will be less than 1. We will assume that the probability is symmetric around the circle so that we can assume that one of the endpoints of the chosen chord is at the bottom of the circle, or that the chord is parallel to the horizontal axis.

37. In this case, by choosing a chord "randomly" we mean that we randomly choose the angle $\theta \in [0, \pi]$ that the chord makes with the horizontal (see picture below). By considering right triangles we can see that the length of the chord is $2\sin(\theta)$. What is the probability that the chord length is less than 1?

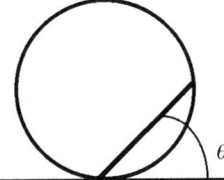 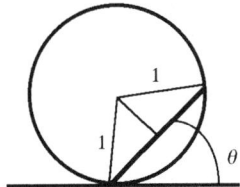

38. With the same situation as the previous exercise, what is the expected value of the chord length?

39. In this version of choosing a random chord, we randomly choose a height $h \in [-1, 1]$ at which the chord (assumed parallel to the horizontal axis) cuts the circle (see picture below). What is the probability that the chord has length less than 1 in this case?

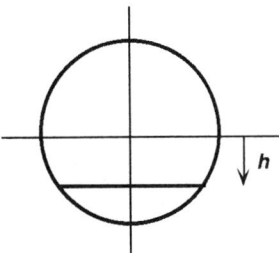

40. With the same situation as the previous exercise, what is the expected value of the chord length?

---

41. The lifetime t of a particular kind of particle is a random variable with probability density function $t \longmapsto 0.27\exp(-0.27t)$ for $t \geq 0$, t in seconds. For what value of t is there a 50% chance that a randomly chosen particle has decayed?

CHAPTER 10

Sequences and Approximations

Approximations and infinite processes are two of the distinguishing features of the mathematics of calculus, and in many ways the ideas of approximation and infinite processes go hand in hand. To get some feel for these ideas, let's look at a familiar example.

Suppose you want to express the fraction 22/7 as a decimal. As we perform the long division of 22 by 7 as shown in Figure 10.1, we get a sequence of partial results

$$a_1 = 3, \ a_2 = 3.1, \ a_3 = 3.14, \ a_4 = 3.142, \ a_5 = 3.1428, \ldots$$

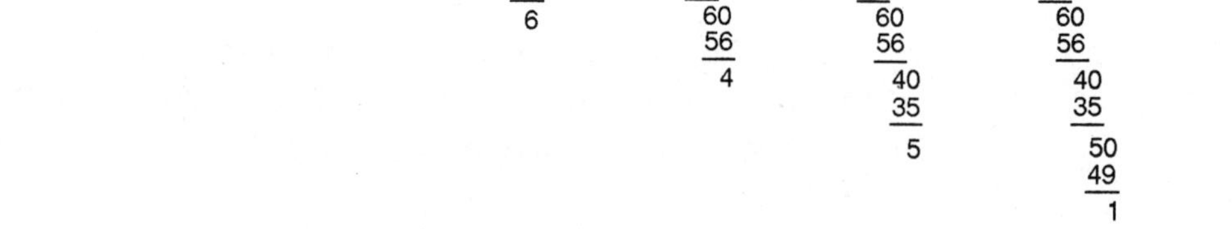

Figure 10.1 Long division as an infinite process.

Since 22/7 is reduced to lowest terms and its denominator 7 is not a factor of any power of 10, we know that this long division process will never end. We say the decimal representation of 22/7 does not terminate. We can think of the partial results as forming an infinite sequence of decimal approximations, each one closer to the exact value 22/7.

When you first encountered such a long division problem years ago, the realization that the process would never end may have troubled you, since it means that it is physically impossible to write down the "complete" decimal expansion of 22/7. Nevertheless, the infinite sequence of results we get in the process of dividing somehow does have a definite "end," in the sense that it represents 22/7. We could say that the real number 22/7 is the **limit** of the sequence. Similarly, the real number $\sqrt{2}$ is the limit of the sequence

$$b_1 = 1, \; b_2 = 1.4, \; b_3 = 1.41, \; b_4 = 1.414, \; b_5 = 1.4142, \ldots$$

These infinite processes each return a sequence of *successive approximations* to particular real numbers. Keeping that in mind raises several important questions when we examine other infinite processes that produce sequences of numerical values:

Do the numbers in the sequence *converge* to a particular real number?

If so, what is that limit?

How many steps does it take to achieve a given level of accuracy?

This chapter is devoted to examining these questions. We start by looking at several examples of sequences of numbers and how they behave, and then get down to the important questions of convergence, speed of convergence, and computation of limits and some valuable techniques for addressing these questions. We'll then turn to some approximation techniques for finding solutions or *roots* of equations.

10.1 EXAMPLES OF SEQUENCES

You can generally think of a *sequence* as an *infinitely long list* of numbers (or any other objects, for that matter). We call the individual numbers in a sequence the **terms** and we identify them by their position in the list. For example, in the sequence

$$a_1 = 3, \, a_2 = 3.1, \, a_3 = 3.14, \, a_4 = 3.142, \, a_5 = 3.1428, \ldots$$

we call a_1 the first term, a_2 the second term, a_3 the third term, and so on. In general, a_n is called the nth term of the sequence. The subscript n used to label the terms of the sequence is sometimes called the **index**.

Technically speaking, a sequence of real numbers can be considered a function from the set of positive integers $\mathbb{N} = \{1, 2, 3, \ldots\}$ to the set of real numbers $\mathbb{R}$. In other words, the numbers in the list

$$a_1, a_2, a_3, \ldots, a_n, \ldots$$

can be viewed as the outputs

$$f(1), f(2), f(3), \ldots, f(n), \ldots$$

of a function

$$f : \mathbb{N} \to \mathbb{R}.$$

Here are several examples of sequences.

EXAMPLE 1   The **harmonic sequence** consists of the reciprocals of the positive integers in order:

$$1, \frac{1}{2}, \frac{1}{3}, \frac{1}{4}, \ldots, \frac{1}{n}, \ldots$$

■

EXAMPLE 2   The positive integers themselves form a sequence:

$$1, 2, 3, 4, \ldots$$

■

EXAMPLE 3   The values appearing as terms in a sequence need not all be distinct. The sequence

$$0, 1, 2, 0, 1, 2, 0, 1, 2, \ldots$$

has infinitely many terms (as does any sequence), but only two distinct values. It's quite possible to have a **constant sequence** such as

$$-\frac{5}{7}, -\frac{5}{7}, -\frac{5}{7}, -\frac{5}{7}, \ldots$$

with only one value appearing.

■

10.1 EXAMPLES OF SEQUENCES

EXAMPLE 4 If the signs of the terms in a sequence alternate back and forth between positive and negative, then we call it an **alternating sequence**. The *alternating harmonic sequence* is

$$1, -\frac{1}{2}, \frac{1}{3}, -\frac{1}{4}, \frac{1}{5}, -\frac{1}{6}, \ldots$$

∎

The shorthand notation for a sequence is

$$\{a_n\}_{n=1}^{\infty}$$

or simply

$$\{a_n\}.$$

Occasionally, it is convenient to have the index n start out with a value other than $n = 1$. In particular, $n = 0$ is often a starting index value for a sequence.

EXAMPLE 5 The sequence of powers of two:

$$1, 2, 4, 8, 16, 32, \ldots$$

can be written as

$$\{2^n\}_{n=0}^{\infty}$$

since $2^0 = 1, 2^1 = 2, 2^2 = 4$, and so on. ∎

 We should be careful to distinguish between a sequence and an infinite set. A sequence has a definite ordering of its terms, while a set is simply a collection of objects with no particular ordering.

Closed-form sequences

As we mentioned above, a sequence of real numbers can be thought of as a function assigning a real number a_n to each positive integer index n. Certainly, it's physically impossible to write down all the terms in a sequence. However, we could say that we "know" a sequence if we have some means by which to find (predict) the value of any term in the sequence.

The most convenient way this could be accomplished is if we had an *explicit formula*

$$a_n = f(n)$$

providing the value a_n in terms of n. When this is possible, we say we have a **closed-form** description of the sequence.

EXAMPLE 6 The closed-form description of the harmonic sequence

$$1, \frac{1}{2}, \frac{1}{3}, \frac{1}{4}, \ldots$$

is

$$\{\frac{1}{n}\}_{n=1}^{\infty}.$$

∎

EXAMPLE 7 The closed-form description of the alternating harmonic sequence

$$1, -\frac{1}{2}, \frac{1}{3}, -\frac{1}{4}, \frac{1}{5}, -\frac{1}{6}, \ldots$$

is

$$\{\frac{(-1)^{n+1}}{n}\}_{n=1}^{\infty}.$$

The factor $(-1)^{n+1}$ provides the alternating signs: when n is odd ($n = 1, 3, 5, 7, \ldots$), then $n+1$ is even and $(-1)^{n+1} = 1$. When n is even ($n = 2, 4, 6, 8, \ldots$), then $n+1$ is odd and $(-1)^{n+1} = -1$. ∎

EXAMPLE 8 The closed-form description of the constant sequence

$$-\frac{5}{7}, -\frac{5}{7}, -\frac{5}{7}, \ldots$$

is

$$\{-\frac{5}{7}\}_{n=1}^{\infty}.$$

∎

If you are trying to find a closed-form formula for a sequence, then you certainly should check that the formula generates the correct values for the some specific index values n.

10.1 EXAMPLES OF SEQUENCES

Arithmetic sequences

One important type of closed-form sequence is an **arithmetic sequence**. An arithmetic sequence is distinguished by the fact that the *difference of any two successive terms is constant*. In other words, given any term, we obtain the next term by adding a specific constant.

EXAMPLE 9 The following are arithmetic sequences:

$$1, 2, 3, 4, 5, \ldots \qquad \text{(constant addend} = 1)$$
$$1, 3, 5, 7, 9, \ldots \qquad \text{(constant addend} = 2)$$
$$-17, -22, -27, -32, \ldots \qquad \text{(constant addend} = -5)$$

■

We can find a closed-form formula for any arithmetic sequence, provided we know the value of the initial term and the constant addend. It is convenient to start with the index $n = 0$ so that a_0 is the initial term. If b is the constant difference, then the closed-form formula for the arithmetic sequence is

$$\{a_0 + bn\}_{n=0}^{\infty}.$$

The formula $a_0 + bn$ generates

$$a_0, a_0 + b, a_2 b, a_0 + 3b, \ldots$$

for $n = 0, 1, 2, 3, \ldots$.

EXAMPLE 10 Find the closed-form formulas for the sequences in the previous example.

Solution For the first sequence $1, 2, 3, \ldots$, we have $a_0 = 1$ and $b = 1$, so

$$\{1 + n\}_{n=0}^{\infty}$$

describes the sequence.

For the second sequence $1, 3, 5, 7, 9, \ldots$, we have $a_0 = 1$ and $b = 2$, so

$$\{1 + 2n\}_{n=0}^{\infty}$$

describes the sequence.

For the third sequence $-17, -22, -27, -32, \ldots$, we have $a_0 = -17$ and $b = -5$, so

$$\{-17 - 5n\}_{n=0}^{\infty}$$

describes the sequence. ∎

An arithmetic sequence can be spotted by examining the difference between successive terms to see if it is constant.

Geometric sequences

A **geometric sequence** is distinguished by the fact that the *ratio of any two successive terms is constant*. In other words, given any term, we obtain the next term by multiplying by a specific constant.

EXAMPLE 11 The following are geometric sequences:

$$1, 2, 4, 8, 16, \ldots \qquad \text{(constant multiple} = 2)$$

$$3, \frac{3}{2}, \frac{3}{4}, \frac{3}{8}, \frac{3}{16}, \ldots \qquad \text{(constant multiple} = \frac{1}{2})$$

$$-7, \frac{7}{3}, -\frac{7}{9}, \frac{7}{27}, -\frac{7}{81}, \ldots \qquad \text{(constant multiple} = -\frac{1}{3})$$

∎

We can find a closed-form formula for any geometric sequence provided we know the value of the initial term and the constant multiple. Again, it is convenient to start with the index $n = 0$ so that a_0 is the initial term. If r is the constant multiple, then the closed-form formula for the geometric sequence is

$$\{a_0 r^n\}_{n=0}^{\infty}.$$

The formula $a_0 r^n$ generates

$$a_0, a_0 r, a_0 r^2, a_0 r^3, \ldots$$

for $n = 0, 1, 2, 3, \ldots$.

10.1 EXAMPLES OF SEQUENCES

EXAMPLE 12 Find closed-form formulas for the sequences in the previous example.

Solution For the first sequence $1, 2, 4, 8, 16, \ldots$, we have $a_0 = 1$ and $r = 2$, so

$$\{2^n\}_{n=0}^{\infty}$$

describes the sequence.

For the second sequence $3, \frac{3}{2}, \frac{3}{4}, \frac{3}{8}, \frac{3}{16}, \ldots$ we have $a_0 = 3$ and $r = \frac{1}{2}$, so

$$\{\frac{3}{2^n}\}_{n=0}^{\infty}$$

describes the sequence.

For the third sequence $-7, \frac{7}{3}, -\frac{7}{9}, \frac{7}{27}, -\frac{7}{81}, \ldots$ we have $a_0 = -7$ and $r = -\frac{1}{3}$, so

$$\{\frac{-7}{(-3)^n}\}_{n=0}^{\infty}$$

describes the sequence. ∎

A geometric sequence can be spotted by examining the ratio of successive terms to see if it is constant.

EXAMPLE 13 Is the harmonic sequence arithmetic, geometric, or neither?

Solution The harmonic sequence is

$$1, \frac{1}{2}, \frac{1}{3}, \frac{1}{4}, \frac{1}{5}, \ldots.$$

To check whether this is an arithmetic sequence, we note that $1/2 - 1 = -1/2$, but $1/3 - 1/2 = -1/6$, so the difference between successive terms is not constant, and the sequence cannot be arithmetic.

To check whether this is a geometric sequence, we note that $\frac{1/2}{1} = 1/2$, but $\frac{1/3}{1/2} = 2/3$, so the ratio of successive terms is not constant, and the sequence cannot be geometric.

We can conclude that the harmonic sequence is neither arithmetic nor geometric. ∎

Iterative and recursive sequences

When we say a sequence is described in closed form, we mean that we have an explicit formula for a_n written in terms of n. For example, the geometric sequence

$$a_0 = 1, a_1 = \frac{1}{2}, a_2 = \frac{1}{4}, a_3 = \frac{1}{8}, a_4 = \frac{1}{16} \ldots$$

can be described by the formula

$$a_n = \frac{1}{2^n} \quad \text{for} \quad n = 0, 1, 2, \ldots$$

In fact, we could write the sequence as

$$\{\frac{1}{2^n}\}_{n=0}^{\infty}.$$

Sometimes the terms in a sequence depend directly on one or more of the preceding terms in the sequence. When this is the case, it may be preferable or more convenient to describe the sequence using this relationship.

A sequence is said to be defined **recursively** or **inductively** if the first term is given (or the first few terms are given) and we have explicit instructions on how to obtain the subsequent terms using the values of preceding terms.

EXAMPLE 14 Describe the geometric sequence $1, \frac{1}{2}, \frac{1}{4}, \frac{1}{8}, \ldots$ recursively.

Solution

$$a_0 = 1$$
$$a_n = \frac{1}{2} a_{n-1} \quad \text{for} \quad n = 1, 2, 3, \ldots$$

∎

A recursive description of a sequence allows us to "build" up the terms much like a stack of blocks. In this example, we are given

$$a_0 = 1.$$

The **recursion formula** $a_n = \frac{1}{2} a_{n-1}$ is used over and over again to obtain the subsequent terms:

$$a_1 = \frac{1}{2} a_{1-1} = \frac{1}{2} a_0 = \frac{1}{2} \cdot 1 = \frac{1}{2}$$

10.1 EXAMPLES OF SEQUENCES

$$a_2 = \frac{1}{2}a_{2-1} = \frac{1}{2}a_1 = \frac{1}{2} \cdot \frac{1}{2} = \frac{1}{4}$$

$$a_3 = \frac{1}{2}a_{3-1} = \frac{1}{2}a_2 = \frac{1}{2} \cdot \frac{1}{4} = \frac{1}{8}$$

$$\vdots$$

We can see that any geometric sequence can be described recursively in a similar way. If a is the first term in the sequence and r is the constant factor, then the sequence

$$a, ar, ar^2, ar^3, \ldots$$

can be described recursively:

$$a_0 = a$$

$$a_n = ra_{n-1}.$$

Similarly, any arithmetic sequence can be described recursively.

EXAMPLE 15 Describe the arithmetic sequence 3, 7, 11, 15, 19, ... recursively.

Solution

$$a_0 = 3$$

$$a_n = a_{n-1} + 4 \quad \text{for } n = 1, 2, \ldots$$

∎

In general, if a is the first term of an arithmetic sequence, and b is the constant addend, then the sequence

$$a, a+b, a+2b, a+3b, \ldots$$

can be described recursively:

$$a_0 = a$$

$$a_n = a_{n-1} + b.$$

Perhaps the most famous of all sequences is the so-called **Fibonacci sequence**. Fibonacci, also known as Leonardo de Pisa, was a thirteenth century Italian mathematician. The sequence named after him is usually described recursively as follows:

$$a_0 = 1$$

$$a_1 = 1$$

$$a_n = a_{n-1} + a_{n-2} \quad \text{for } n = 2, 3, 4, \ldots.$$

In other words, the first two terms of the Fibonacci sequence are both 1, and each subsequent term is determined by adding the previous two terms. The Fibonacci sequence starts out

$$1, 1, 2, 3, 5, 8, 13, 21, 34, 55, 89, \ldots$$

The Fibonacci sequence appears quite often in natural number patterns, and it has a multitude of interesting properties. In fact, an entire journal (*The Fibonacci Quarterly*) is devoted to it.

The recursive descriptions for arithmetic and geometric sequences are examples of **iteration**. An **iterative process** is one in which we start with an initial input x_0 and then successively apply a function f to it by using the output at each stage as the input to the next.

For example, if

$$f(x) = x + 4$$

and our initial input is $x_0 = 3$, then

$$x_1 = f(x_0) = f(3) = 7$$

$$x_2 = f(x_1) = f(7) = 11$$

$$x_3 = f(x_2) = f(11) = 15$$

$$x_4 = f(x_3) = f(15) = 19$$

$$\vdots$$

We can see that this function f, along with the initial input $x_0 = 3$, generates the arithmetic sequence of the last example.

☞ **Any function f and initial input x_0 can be used to define an iterative sequence, provided the output at each step is still in the domain of the function.**

10.1 EXAMPLES OF SEQUENCES

EXAMPLE 16 Find the first 5 terms of the iterative sequence defined by

$$f(x) = \frac{x^2 + 2}{2x}$$

with initial input $x_0 = 1$.

Solution With $x_0 = 1$, we have

$$x_1 = f(x_0) = f(1) = \frac{1^2 + 2}{2} = \frac{3}{2} = 1.5$$

$$x_2 = f(x_1) = f(1.5) = \frac{(1.5)^2 + 2}{3} \approx 1.41666666667$$

$$x_3 = f(x_2) \approx f(1.41666666667) \approx 1.41421568627$$

$$x_4 = f(x_3) \approx f(1.41421568627) \approx 1.41421356237$$

$$x_5 = f(x_4) \approx f(1.41421356237) \approx 1.41421356237.$$

The values listed for x_2, x_3, x_4, and x_5 are rounded to eleven decimal places, and from the result of x_5 we would anticipate that the subsequent terms x_6, x_7, and so on will stay "locked on" to the value 1.41421356237. ∎

Many approximation methods are essentially iterative processes— an initial guess for a solution is tested and then adjusted accordingly. This yields an infinite sequence of approximations, each better (one hopes) than the one preceding it.

EXERCISES

For exercises 1-4, describe the arithmetic sequence with given initial term and constant addend with a closed-form formula and with a recursive formula.

1. $a_0 = 3$, $b = -5$
2. $a_0 = -4$, $b = 0.5$
3. $a_0 = \pi$, $b = 0$
4. $a_0 = 0$, $b = \pi$

For exercises 5-8, describe the geometric sequence with given initial term and constant factor with a closed-form formula and with a recursive formula.

5. $a_0 = 3$, $r = -5$
6. $a_0 = -4$, $r = 0.5$
7. $a_0 = \pi$, $r = 0$
8. $a_0 = 0$, $r = \pi$

The following sequence of numbers approaches the golden ratio $\frac{1+\sqrt{5}}{2}$:

$$\frac{1}{1}, \frac{2}{1}, \frac{3}{2}, \frac{5}{3}, \frac{8}{5}, \frac{13}{8}, \ldots$$

Exercises 9-14 refer to this sequence.

9. Describe the pattern you see in the numerators. Describe the pattern you see in the denominators.

10. What is the first term in the sequence to be within .001 less than the golden ratio?

11. What is the first term in the sequence to be within .001 more than the golden ratio?

12. What are the first two consecutive terms in the sequence that are within .000001 of each other?

13. Find the first 10 terms of the sequence whose nth term is given by

$$a_n = \frac{1}{\sqrt{5}}\left[\left(\frac{1+\sqrt{5}}{2}\right)^n - \left(\frac{1-\sqrt{5}}{2}\right)^n\right].$$

This is the closed-form formula for what sequence?

14. Use the previous exercise to find a closed-form formula for the sequence

$$\frac{1}{1}, \frac{2}{1}, \frac{3}{2}, \frac{5}{3}, \frac{8}{5}, \frac{13}{8}, \ldots$$

The nth term of a sequence is given in each of exercises 15-18. Find the first five terms of each sequence (starting with $n = 1$).

15. $a_n = \frac{(-1)^n n}{n+1}$

16. $a_n = n \cos \frac{n\pi}{2}$

17. $a_1 = \sqrt{3}$ and $a_n = \sqrt{3a_{n-1}}$ for $n \geq 2$.

18. $a_n = -\sqrt{n}$

The initial term and a function are given in each of exercises 19-22. Find the first five terms of the iterative sequence based on this information.

19. $x_0 = 0$, $f(x) = x - \cos(x)$

20. $x_0 = 1$, $f(x) = \frac{x+1}{2}$

21. $x_0 = 0.5$, $f(x) = x^2$

22. $x_0 = 10$, $f(x) = 1/x$

10.2 CONVERGENCE AND DIVERGENCE OF SEQUENCES

A sequence of real numbers $\{a_n\}_{n=1}^{\infty}$ is said to *converge* if the terms in the sequence eventually "stabilize" toward some single limiting value. Otherwise, we say the sequence *diverges*. For example, in the last section, we looked at an iterative sequence that can be described using the following recursion formula:

$$a_0 = 1$$
$$a_n = \frac{a_{n-1}^2 + 2}{2a_{n-1}}.$$

The first few terms of this sequence are (rounded to 11 decimal places):

$$a_0 = 1$$
$$a_1 = 1.5$$
$$a_2 = 1.41666666667$$
$$a_3 = 1.41421568627$$
$$a_4 = 1.41421356237$$
$$a_5 = 1.41421356237$$
$$a_6 = 1.41421356237$$
$$\vdots$$

Now, the terms $a_4, a_5, a_6, \ldots$ are not equal, but rounded to 11 decimal places, they are indistinguishable. This sequence appears to be stabilizing (very quickly) and the numerical evidence suggests that the sequence converges. It is known that this sequence has a limiting value of exactly $\sqrt{2}$, which approximated to 11 decimal places is 1.41421356237. In fact, if you take any positive number as the initial term a_0 and use the same recursion formula, the terms a_n will stabilize to a limiting value of

$$\sqrt{2} \approx 1.41421356237.$$

(Try it with starting values such as $a_0 = 1000$ or $a_0 = .0003$.)

In general, if $a > 0$, then the iterative sequence defined by

$$f(x) = \frac{x^2 + a}{2x}$$

can be used to approximate $\sqrt{a}$.

Two questions we'll explore in this section are:

How can you tell whether a sequence converges or diverges?

If a sequence converges, how can you determine the limiting value?

When does a sequence have a limit?

First, let's make precise what we mean by convergence.

Definition 1

> A sequence $\{a_n\}_{n=1}^{\infty}$ **converges to a limit** L provided that for any given tolerance $\epsilon > 0$, there is a specific index N for which
> $$L - \epsilon < a_n < L + \epsilon$$
> whenever $n \geq N$. If there is a certain tolerance $\epsilon > 0$ for which no such N exists, then we say the sequence **diverges**.

You can think of N as the position at which the "tail" of the sequence has only terms that are within ϵ of L.

$$a_1, a_2, a_3, \ldots, \underbrace{a_N, a_{N+1}, a_{N+2}, \ldots}_{\text{all these terms are within } \varepsilon \text{ of the limit } L.}$$

No matter how small a positive tolerance ϵ we are given, we must be able to find a tail of the sequence with all the terms within that tolerance of L.

Since we use decimal numbers so often to represent real numbers, you may find it convenient to think of the tolerance ϵ in terms of *decimal places of agreement*: Given any number of decimal places, there must be a position in the sequence after which all the terms are indistinguishable from the limit when rounded to that number of decimal places.

EXAMPLE 17 Let's illustrate this idea using the iterative sequence we just discussed. Here, $L = \sqrt{2}$. If we were to round all the terms to two decimal places, then we have

$$a_0 = 1, a_1 = 1.5, a_2 \approx 1.42, a_3 \approx 1.41, a_4 \approx 1.41, a_5 \approx 1.41, a_6 \approx 1.41, \ldots$$

Note that starting with a_3, all the terms match the two-decimal approximation of $\sqrt{2} \approx 1.41$ (in other words, $N = 3$ for two decimal places of agreement). If we round all the terms to eleven decimal places, then $a_4, a_5, a_6, \ldots$ all agree with $\sqrt{2}$ (so $N = 4$ for eleven decimal places of agreement). ∎

The number N gives us a measure of the *speed of convergence*. If we need only a small value of N to achieve a large number of decimal places of agreement between the limit and all subsequent terms $a_N, a_{N+1}, \ldots$, then our sequence converges quickly (like the example above). If very large

10.2 CONVERGENCE AND DIVERGENCE OF SEQUENCES

values of N are required to achieve a small number of decimal places of agreement, then our sequence converges slowly.

 Speed of convergence can be misleading. If a sequence converges very slowly, the numerical evidence we obtain by examining the first few terms may suggest that the sequence diverges. On the other hand, the first few terms of a divergent sequence might suggest that the sequence is converging.

EXAMPLE 18 The alternating harmonic sequence $\{\frac{(-1)^n}{n}\}_{n=1}^{\infty}$

$$-1, \frac{1}{2}, -\frac{1}{3}, \frac{1}{4}, -\frac{1}{5}, \ldots$$

converges to the limit $L = 0$. Given any positive tolerance ϵ, we can choose N so large that

$$-\epsilon < \pm\frac{1}{n} < \epsilon \text{ for all } n \geq N.$$

(Just pick $N > \frac{1}{\epsilon}$ so that $\frac{1}{N} < \epsilon$.) ∎

Types of divergent behavior–unbounded, oscillatory, chaotic

Convergent behavior of a sequence is marked by the eventual stabilization of its terms to a single limiting value. This stabilization can be visualized if the terms in decimal form are flashed before our eyes in quick succession– we should see a "fixing" of more and more of the leading decimal places. Or, if enough terms are listed vertically in a long column, we can spot this stabilization (warning– *enough* could be a lot).

Divergent behavior, on the other hand, can take on many different forms. Let's look at some examples of divergent sequences.

EXAMPLE 19 The sequence of positive integers $\{n\}_{n=1}^{\infty}$

$$1, 2, 3, 4, 5, \ldots$$

is divergent, for the terms simply increase without bound toward no single limiting value. ∎

EXAMPLE 20 Another example of an unbounded sequence is

$$-2, 4, 6, -8, 10, 12, -14, \ldots$$

(all the terms are even, and every third term is negative). ■

EXAMPLE 21 The geometric sequence $\{(1.01)^n\}_{n=1}^{\infty}$:

$$1.01, 1.0201, 1.030301, \ldots$$

is also an unbounded sequence, though its terms grow slowly. (What's the thousandth term $a_{1000} = 1.01^{1000}$?) ■

An **unbounded** sequence contains *arbitrarily* large positive and/or negative terms. Such a sequence can never converge to a single finite limiting value. A **bounded** sequence has all of its terms within some bounded interval of real numbers. However, a bounded sequence can still diverge.

EXAMPLE 22 The terms of the sequence

$$0, 1, 0, 1, 0, 1, \ldots$$

oscillate forever between 0 and 1. The sequence is *bounded* (certainly all the terms lie within $[0, 1]$) but *diverges* because the terms do not tend toward a single limiting value. ■

This example goes to show that divergent behavior need not be wild, unbounded, or unpredictable. Sequences whose terms oscillate between two values or cycle periodically through some set of values are certainly "well-behaved," but not convergent.

EXAMPLE 23 Find the 1000th term of the sequence $\{a_n\}_{n=1}^{\infty}$ where the terms a_n follow the pattern

$$1, 2, 3, 1, 2, 3, 1, 2, 3, \ldots$$

Solution This sequence cycles through the values $1, 2, 3$ over and over again. We can see that 3 appears every third term (in other words, $3 = a_3, a_6, a_9, \ldots$). So $a_{999} = 3$, and the 1000th term $a_{1000} = 1$. This sequence definitely does not converge. ■

Of course, a sequence whose terms wildly fluctuate in a chaotic manner without ever stabilizing is divergent. We should point out that mathematicians and scientists alike are finding many examples of seemingly chaotic behavior that nevertheless can be described by fairly simple mathematical patterns and relations.

Monotonic and Cauchy sequences

Convergence and divergence are words used to describe whether or not a sequence has a limit. We can also describe how the terms of a sequence behave relative to each other. A **monotonically increasing** sequence $\{a_n\}_{n=1}^{\infty}$ has the property

$$a_1 \leq a_2 \leq a_3 \leq \cdots \leq a_n \leq \cdots,$$

while a **strictly monotonically increasing** sequence $\{a_n\}_{n=1}^{\infty}$ has the even stronger property

$$a_1 < a_2 < a_3 < \cdots < a_n < \cdots$$

Similarly, a **monotonically decreasing** sequence $\{a_n\}_{n=1}^{\infty}$ has the property

$$a_1 \geq a_2 \geq a_3 \geq \cdots \geq a_n \geq \cdots$$

while a **strictly monotonically decreasing** sequence $\{a_n\}_{n=1}^{\infty}$ has the even stronger property

$$a_1 > a_2 > a_3 > \cdots > a_n > \cdots$$

Any sequence that is either monotonically increasing or decreasing may be referred to as **monotonic**.

EXAMPLE 24 The harmonic sequence $\{\frac{1}{n}\}_{n=1}^{\infty}$

$$1, \frac{1}{2}, \frac{1}{3}, \frac{1}{4}, \ldots, \frac{1}{n}, \ldots$$

is monotonically decreasing. In fact, it is strictly monotonically decreasing since $\frac{1}{n} > \frac{1}{n+1}$ for any $n = 1, 2, 3, \ldots$. ∎

EXAMPLE 25 The geometric sequence $\{2^n\}_{n=0}^{\infty}$

$$1, 2, 4, 8, 16, \ldots$$

is monotonically increasing. In fact, it is strictly monotonically increasing since $2^n < 2^{n+1}$ for any $n = 0, 1, 2, 3, \ldots$. ∎

EXAMPLE 26 The oscillating sequence

$$0, 1, 0, 1, 0, 1, \ldots$$

is neither monotonically increasing nor decreasing. ∎

A sequence is called **Cauchy** (named after Augustin-Louis Cauchy, 1789-1857) if the terms get (and stay) arbitrarily close to each other. Here's the precise definition.

Definition 2

> A sequence $\{a_n\}_{n=1}^{\infty}$ is **Cauchy** if it has the property that for any given positive tolerance $\epsilon > 0$, there is a corresponding index value N such that
>
> $$|a_n - a_m| < \epsilon$$
>
> whenever $n, m \geq N$.

Put in other words, given a tolerance $\epsilon > 0$, we can always find some tail of a Cauchy sequence containing terms all within ϵ of each other.

EXAMPLE 27 The harmonic sequence is a Cauchy sequence. Given any $\epsilon > 0$, we can find an index value N (just choose any integer $N > \frac{1}{\epsilon}$) such that

$$|\frac{1}{n} - \frac{1}{m}| < \epsilon \text{ whenever } n, m \geq N.$$

∎

If we think about it a bit, we can realize that any convergent sequence must also be Cauchy. The reasoning is as follows: all the terms in the tail of a convergent sequence must be close to some single limit value L. This forces all the terms in the tail to be close to each other (Cauchy).

Conversely, a Cauchy sequence of real numbers must converge to a single limiting value. This is perhaps not as obvious, but the general idea is that the only way for the terms in a sequence to become arbitrarily close to each other is for them to become arbitrarily close to a single value L.

We can also see that a convergent sequence must be bounded. The reasoning here is that all the terms in a tail of the sequence must be close to a single value L. That accounts for all but the finitely many terms preceding the tail. This means we can find a bounded interval big enough to include the tail of the sequence as well as all terms preceding the tail.

Even though a convergent sequence must be bounded, the converse is not true— there are certainly bounded sequences that don't converge (remember $0, 1, 0, 1, 0, 1, \ldots$). However, if a sequence is both bounded *and* monotonic, then it must converge. The reasoning makes use of a fundamental property of real numbers: given any bounded set of real numbers, there is a *smallest* closed interval containing all the numbers in the set. For example, the *smallest* closed interval containing all the numbers in the harmonic sequence $\{\frac{1}{n}\}_{n=1}^{\infty}$ is $[0, 1]$. Now, if the sequence is monotonic (like the harmonic sequence is), then the terms in the sequence will have to cluster up to one end or the other of this closed interval, and that endpoint is the limit of the sequence. (The harmonic sequence is monotonically decreasing, so its terms approach 0, the left endpoint of the interval.)

10.2 CONVERGENCE AND DIVERGENCE OF SEQUENCES

We summarize all these observations below:

Theorem 10.1

> **Hypothesis:** $\{a_n\}$ is a sequence of real numbers.
> **Conclusion 1:** $\{a_n\}$ is convergent if and only if $\{a_n\}$ is Cauchy.
> **Conclusion 2:** If $\{a_n\}$ is convergent, then $\{a_n\}$ is bounded (but the converse need not hold). **Conclusion 3:** If $\{a_n\}$ is bounded *and* monotonic, then $\{a_n\}$ is convergent.

□

Limit properties of sequences

For shorthand, we'll use $a_n \to L$ to represent $\lim_{n \to \infty} a_n = L$.

If the nth term of a sequence is given by an explicit formula in n,

$$a_n = f(n)$$

then we can investigate the behavior of the sequence both numerically and graphically.

Numerically, we can simply substitute the values $n = 1, 2, 3, \ldots$ and so on into the formula to get a feel for the sequence. You can "see" convergent behavior numerically in the stabilization of the digits in the decimal form of each term. That is, if $a_1, a_2, a_3, \ldots$, are flashed before our eyes quickly in sequence (with the position of the decimal point fixed in place), then the result is like a movie where the action becomes still from left to right. How long the movie must run for a convergent sequence to show stabilization of each digit is an indication of the speed of convergence of the sequence. Substituting large values of n may (or may not) give us a good approximation to the limit of a convergent sequence. But, if you use a machine to compute the values a_n for large n, be particularly sensitive to round-off and cancellation errors that might occur.

Graphically, we can plot

$$y = f(x)$$

and examine the graph for large values of x. The y-coordinates of the points

$$(n, f(n))$$

that lie on this graph correspond to terms in the original sequence. A limiting value L to the sequence will correspond to a horizontal asymptote $y = L$ to the graph of $y = f(x)$ (see Figure 10.2).

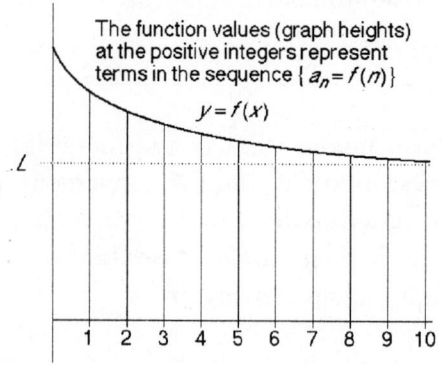

Figure 10.2 Investigating a sequence graphically.

If you use a machine grapher to explore the behavior of a sequence, beware that it is subject to the same limitations as the machine's numerical computations.

These numerical and graphical explorations can help give you a better feel for the behavior of the sequence, but ultimately neither can provide a definitive answer to the question: What is $\lim_{n\to\infty} a_n$?

If the formula describing a_n is composed or built up algebraically from familiar functions, then the same limit properties enjoyed by functions as described in chapter 2 are shared by sequences. Put simply, if we can recognize a given sequence as being built term-by-term out of component sequences with known limits, then we can build the limit of the sequence in the same way.

More precisely, if $a_n \to L_1$ and $b_n \to L_2$, then all of the following are true:

$$a_n + b_n \to L_1 + L_2$$

$$a_n - b_n \to L_1 - L_2$$

$$a_n b_n \to L_1 L_2$$

$$a_n/b_n \to L_1/L_2 \quad \text{(provided } b_n\text{'s and } L_2 \neq 0)$$

$$C a_n \to C L_1 \quad (C \text{ a constant})$$

$$f(a_n) \to f(L_1) \quad (f \text{ a continuous function}).$$

We might call these *determinate forms* for sequences, since we can determine their limits from their component parts. In the next section we'll examine some useful techniques for finding limits of both functions and sequences when they have *indeterminate forms*.

10.2 CONVERGENCE AND DIVERGENCE OF SEQUENCES

EXERCISES

Indicate whether each of the statements in exercises 1-4 is true or false. If false, give a counterexample.

1. All bounded, monotonic sequences converge.
2. All non-monotonic sequences diverge.
3. All Cauchy sequences are bounded and monotonic.
4. All unbounded sequences diverge.

For exercises 5-9 list the first 5 terms of each sequence $\{a_n\}_1^\infty$. For each, classify as bounded or not, monotonic or not, convergent or not, Cauchy or not.

5. $a_n = 2^{\cos n\pi}$
6. $a_n = \left(\frac{n-1}{n+1}\right)^n$
7. $-2, -3.1, -4.01, -5.001, \ldots$ (continues with same pattern)
8. sequence pictured below continues with same pattern

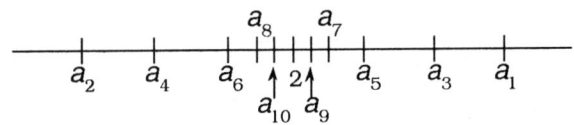

9. $a_{n+1} = \frac{1}{1+a_n}$

Exercises 10-13 refer to the sequence with closed-form $a_n = 1 + (-1)^n \left(\frac{1}{n}\right)$. Note that $\lim_{n \to \infty} a_n = 1$.

10. Plot and label the first 7 terms of this sequence $\{a_n\}_1^\infty$.
11. Find the first term in the sequence to be within .1 less than the limit.
12. Find the first term in the sequence to be within .1 more than the limit.
13. Find the first 2 terms in the sequence to be within .1 of each other.

For each of exercises 14-18, create three examples of closed-form sequences satisfying the given descriptions.

14. bounded, monotonic.
15. bounded, non-monotonic, convergent.
16. unbounded, monotonic.
17. unbounded, non-monotonic.
18. bounded, non-monotonic, divergent.

---

19. Let $a_n = (1 + \frac{1}{n})^n$. Find $a_1, a_2, a_3, a_{1000}, a_{2000}, a_{3000}$. Is the sequence $\{a_n\}$ bounded and/or monotonic? If so, what does the limit appear to be? Approximately, how close is a_{3000} to the limit?

The nth term of a sequence $\{a_n\}$ is given in each of exercises 20-21. Find the limit of each sequence, and then find the smallest value N such that all the terms after a_N in the sequence $\{a_n\}$ are within .001 of the limit.

20. $a_n = \dfrac{1}{n!}$ (Note: $n! = 1 \cdot 2 \cdot 3 \cdots\cdots n$.)

21. $a_n = \dfrac{1}{2^n}$

22. Make a statement comparing the rates of convergence of the above 2 sequences.

The nth term of a sequence is given in each of exercises 23-26. Find the first five terms of each sequence, and classify each as bounded or not, monotonic or not, and determine whether or not the sequence converges without finding the limit.

23. $a_n = \dfrac{(-1)^n n}{n+1}$
24. $a_n = n \cos(\frac{n\pi}{2})$
25. $a_1 = \sqrt{3}$ and $a_n = \sqrt{3 a_{n-1}}$ for $n \geq 2$.
26. $a_n = -\sqrt{n}$

10.3 FINDING LIMITS – INDETERMINATE FORMS

As you can see, the properties of limits for both sequences and functions are very similar. Indeed, in the case of a closed-form sequence $\{a_n\}$ where

$$a_n = f(n),$$

we can say that if

$$\lim_{x \to \infty} f(x) = L$$

where x is a real variable, then we must also have

$$\lim_{n \to \infty} a_n = L.$$

The properties of limits allow us to break down a closed-form formula into component parts and then study the limiting behavior of each. In this section we'll examine some techniques for determining limits of functions (and sequences) whose forms require closer examination.

Indeterminate quotients—L'Hôpital's Rule

Whether we are studying sequences or functions, the limiting value of a quotient can pose special problems. Let's look at the case of functions first. Suppose we have two functions f and g, and we know that

$$\lim_{x \to a} f(x) = L$$

and

$$\lim_{x \to a} g(x) = M.$$

Then,

$$\lim_{x \to a} \frac{f(x)}{g(x)} = \frac{L}{M} \qquad \text{(provided } M \neq 0\text{)}.$$

Now, what happens if $\lim_{x \to a} g(x) = 0$?

Case 1. If $\lim_{x \to a} f(x) = L \neq 0$, then *we can definitely say that*

$$\lim_{x \to a} \frac{f(x)}{g(x)} \qquad \text{does not exist.}$$

In this case, we may be able to say a bit more about the behavior of the quotient near $x = a$ depending on the behavior of g and whether the limit of $f(x)$ is positive or negative. Recall, we use the notation

$$g(x) \to 0^+$$

to mean that $g(x)$ approaches 0 through strictly positive values, and

$$g(x) \to 0^-$$

to mean that $g(x)$ approaches 0 through strictly negative values. With that meaning understood, we can form Table 10.1.

$\lim_{x \to a} f(x)$	$\lim_{x \to a} g(x)$	$\lim_{x \to a} \frac{f(x)}{g(x)}$
$L > 0$	0^+	$+\infty$
$L < 0$	0^+	$-\infty$
$L > 0$	0^-	$-\infty$
$L < 0$	0^-	$+\infty$
L = any real number	$M \neq 0$	L/M
0	0	?

Table 10.1 Limiting behavior of a quotient $f(x)/g(x)$.

We need to emphasize that $\lim_{x \to a} \dfrac{f(x)}{g(x)}$ is undefined in the first four lines of the third column of the table. The symbols $+\infty$ and $-\infty$ simply describe the behavior of the quotient $f(x)/g(x)$ as $x \to a$. Each result should make sense to you. For example, in the third line, we say that if the numerator approaches a *positive* value and the *negative* denominator becomes *smaller and smaller in magnitude*, then the quotient will be a *negative* number of larger and larger magnitude.

We've left a question mark for the result of the last line of the table, when both $f(x)$ and $g(x)$ approach 0 as $x \to a$. Depending on "how fast" the numerator and denominator approach 0 respectively, we could obtain almost any result.

EXAMPLE 28 Find the following limits, if they exist.

$$\lim_{x \to 0} \frac{3x}{2x} \qquad \lim_{x \to 0} \frac{3x^2}{2x} \qquad \lim_{x \to 0} \frac{3x}{2x^2}$$

Solution In all three of these quotients, the values of the expressions in the numerator and denominator both approach 0 as $x \to 0$.

In the first quotient, $\dfrac{3x}{2x} = \dfrac{3}{2}$ for every $x \neq 0$, so

$$\lim_{x \to 0} \frac{3x}{2x} = \frac{3}{2}.$$

10.3 FINDING LIMITS – INDETERMINATE FORMS

In the second quotient, $\frac{3x^2}{2x} = \frac{3x}{2}$ for every $x \neq 0$. Since $\frac{3x}{2} \to 0$ as $x \to 0$ we have

$$\lim_{x \to 0} \frac{3x^2}{2x} = 0.$$

In the third quotient, $\frac{3x}{2x^2} = \frac{3}{2x}$ for every $x \neq 0$. Since $\frac{3}{2x} \to \pm\infty$ as $x \to 0^{\pm}$ (in other words, $\frac{3}{2x} \to +\infty$ as $x \to 0$ through positive numbers, and $\frac{3}{2x} \to -\infty$ as $x \to 0$ through negative numbers), there is no limiting value and

$$\lim_{x \to 0} \frac{3x^2}{2x}$$

does not exist. ∎

Quotient limits like these are said to be of indeterminate form $0/0$. The expression $0/0$ is not a real number— it is simply shorthand for

$$\lim_{x \to a} \frac{f(x)}{g(x)}$$

when $f(x), g(x) \to 0$ as $x \to a$. In the example above, we were able to determine the limiting behavior of each quotient by examining an equivalent expression for values $x \neq a$. This is not always possible to do. Recall that we found

$$\lim_{x \to 0} \frac{\sin x}{x} = 1$$

in Chapter 2 by use of a geometric argument. There is no algebraic or trigonometric simplification allowing us to determine this limit.

In fact, whenever we attempt to calculate the derivative of a function at a point

$$f'(a) = \lim_{x \to a} \frac{f(x) - f(a)}{x - a}$$

we are working with an indeterminate form. Interestingly, we can turn the tables and use derivatives to help us calculate the limit of an indeterminate form $0/0$.

L'Hôpital's Rule

To see how derivatives might be useful in finding limits of indeterminate forms, let's first look at a problem involving linear functions.

Suppose we have

$$f(x) = m_1(x - a)$$

and

$$g(x) = m_2(x - a).$$

This means the graphs of f and g are straight lines with slopes m_1 and m_2 respectively, and a common x-intercept at $x = a$. Figure 10.3 shows a typical picture of such a situation.

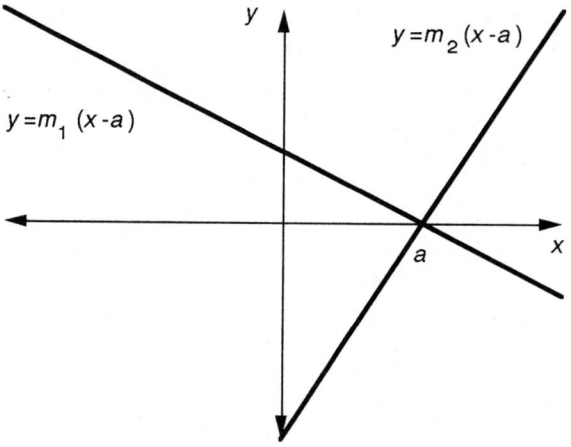

Figure 10.3 Two linear functions with a common x-intercept a.

Now $\lim\limits_{x \to a} \dfrac{f(x)}{g(x)}$ has the indeterminate form $0/0$, but for $x \neq a$,

$$\frac{f(x)}{g(x)} = \frac{m_1(x-a)}{m_2(x-a)} = \frac{m_1}{m_2}.$$

We can conclude that

$$\lim_{x \to a} \frac{f(x)}{g(x)} = \frac{m_1}{m_2}.$$

In other words, the limit of the quotient of the two linear functions at $x = a$ is the *quotient of their slopes*.

10.3 FINDING LIMITS – INDETERMINATE FORMS

Now, imagine that we have two functions f and g that are differentiable except possibly at $x = a$, and both functions have a common limit 0 at $x = a$:

$$\lim_{x \to a} f(x) = \lim_{x \to a} g(x) = 0.$$

A typical situation is pictured in Figure 10.4.

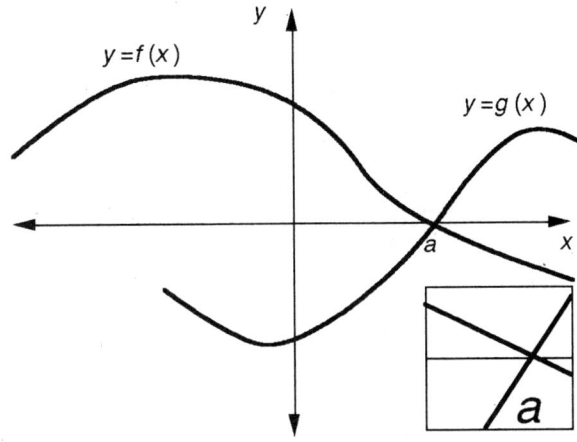

Figure 10.4 Two functions with a common limit $L = 0$ at $x = a$.

Since f and g are differentiable, zooming in on the graph could result in a picture similar to that in Figure 10.3. This suggests that when x is sufficiently close to a, the value of $\dfrac{f(x)}{g(x)}$ will be close to $\dfrac{f'(x)}{g'(x)}$, the ratio of the slopes of f and g. This is indeed the case and the result is known as L'Hôpital's Rule (named after Guillaume Francois Antoine de L'Hôpital (1661-1704), the author of the first calculus textbook!).

Theorem 10.2

L'Hôpital's Rule.
Hypothesis 1: $\lim\limits_{x \to a} f(x) = 0 = \lim\limits_{x \to a} g(x)$.
Hypothesis 2: $\lim\limits_{x \to a} \dfrac{f'(x)}{g'(x)} = L$.
Conclusion: $\lim\limits_{x \to a} \dfrac{f(x)}{g(x)} = L$.

EXAMPLE 29 Find $\lim_{x \to 0} \dfrac{\sin x}{x}$ using L'Hôpital's Rule.

Solution

$$\lim_{x \to 0} \frac{\sin x}{x} = \lim_{x \to 0} \frac{\cos x}{1} = 1.$$

(Graphically, when we zoom in on the graph of $y = \sin x$ it becomes indistinguishable from the line $y = x$; both have slope 1 at $x = 0$.) ∎

EXAMPLE 30 Find $\lim_{x \to \pi/4} \dfrac{\tan x - 1}{x - \pi/4}$.

Solution $\lim_{x \to \pi/4} \dfrac{\tan x - 1}{4x - \pi} = \lim_{x \to \pi/4} \dfrac{\sec^2 x}{4} = \dfrac{2}{4} = \dfrac{1}{2}.$ ∎

It is possible that $\lim_{x \to a} f'(x)/g'(x)$ is also of the indeterminate form $0/0$. (This might be evident graphically by both graphs appearing to have a horizontal tangent line $y = 0$ at $x = a$.) In this situation, L'Hôpital's Rule can simply be applied a second time, a third time, and so on until the form is no longer indeterminate.

EXAMPLE 31 Find $\lim_{x \to 1} \dfrac{\sin(x) - x}{\cos(x) - 1}$.

Solution Applying L'Hôpital's Rule once, we have

$$\lim_{x \to 1} \frac{\sin(x) - x}{\cos(x) - 1} = \lim_{x \to 1} \frac{\cos(x) - 1}{-\sin(x)}.$$

This last limit is still of the indeterminate form $0/0$, so we apply L'Hôpital's Rule once again:

$$\lim_{x \to 1} \frac{\cos(x) - 1}{-\sin(x)} = \lim_{x \to 1} \frac{-\sin(x)}{-\cos(x)} = \frac{0}{-1} = 0.$$

∎

10.3 FINDING LIMITS – INDETERMINATE FORMS

Extending L'Hôpital's Rule

L'Hôpital's Rule is a powerful tool for finding limits of quotients having the indeterminate form $0/0$. Let's summarize it once again:

Suppose $f(x) \to 0$ and $g(x) \to 0$ as $x \to a$.

If $\lim_{x \to a} \dfrac{f'(x)}{g'(x)} = L$, then $\lim_{x \to a} \dfrac{f(x)}{g(x)} = L$.

L'Hôpital's Rule also holds for one-sided limits, where $x \to a^+$ or $x \to a^-$ replaces $x \to a$ above. We can also extend L'Hôpital's Rule to the situation where $x \to \infty$ or $x \to -\infty$:

Suppose $f(x) \to 0$ and $g(x) \to 0$ as $x \to \infty$.

If $\lim_{x \to \infty} \dfrac{f'(x)}{g'(x)} = L$, then $\lim_{x \to \infty} \dfrac{f(x)}{g(x)} = L$.

To see why, we can rewrite these limits in terms of a different variable. If we let

$$h = \frac{1}{x}$$

then $x = \dfrac{1}{h}$ and

$$\lim_{x \to \infty} \frac{f(x)}{g(x)} = \lim_{h \to 0^+} \frac{f(1/h)}{g(1/h)}.$$

If

$$f(x) \to 0 \quad \text{and} \quad g(x) \to 0 \quad \text{as } x \to \infty,$$

then

$$f(1/h) \to 0 \quad \text{and} \quad g(1/h) \to 0 \quad \text{as } h \to 0^+.$$

We can apply L'Hôpital's Rule (with respect to h) to obtain

$$\lim_{h \to 0^+} \frac{f(1/h)}{g(1/h)} = \lim_{h \to 0^+} \frac{f'(\frac{1}{h}) \cdot (-1/h^2)}{g'(1/h) \cdot (-1/h^2)} = \lim_{h \to 0^+} \frac{f'(1/h)}{g'(1/h)}.$$

The last limit is the same as $\lim_{x \to \infty} \dfrac{f'(x)}{g'(x)}$.

(A similar argument can be applied to the case $x \to -\infty$:
if $h = 1/x$, then $h \to 0^-$ as $x \to -\infty$.)

L'Hôpital's Rule also applies to the indeterminate form ∞/∞. We use this to mean any limit of a quotient $\dfrac{f(x)}{g(x)}$ where the magnitude of $f(x)$ and $g(x)$ both increase without bound. That is, $f(x) \to \infty$ or $-\infty$ and $g(x) \to \infty$ or $-\infty$. In this case, if

$$\lim_{x \to a} \frac{f'(x)}{g'(x)} = L$$

then

$$\lim_{x \to a} \frac{f(x)}{g(x)} = L.$$

EXAMPLE 32 Find $\lim_{x \to \infty} \dfrac{e^{-x}}{\ln(1 + \frac{1}{x})}$.

Solution This is of the indeterminate form $0/0$ since $e^{-x} \to 0$ and $\ln(1 + \frac{1}{x}) \to \ln 1 = 0$ as $x \to \infty$. Using L'Hôpital's Rule, we have

$$\lim_{x \to \infty} \frac{e^{-x}}{\ln(1 + \frac{1}{x})} = \lim_{x \to \infty} \frac{-e^{-x}}{\frac{1}{1 + \frac{1}{x}} \cdot \frac{-1}{x^2}} = \lim_{x \to \infty} \frac{e^{-x}}{\frac{1}{x^2 + x}}$$

$$= \lim_{x \to \infty} \frac{x^2 + x}{e^x} = \lim_{x \to \infty} \frac{2x + 1}{e^x} = \lim_{x \to \infty} \frac{2}{e^x} = 0.$$

∎

Note that in the second line of this limit computation, the quotient was rewritten in the indeterminate form ∞/∞.

There are many other indeterminate forms that can pose problems when we attempt to find limits. Fortunately, L'Hôpital's Rule can be applied to many of these forms just by rewriting the function expression so that it takes on either of the indeterminate quotient forms $0/0$ or ∞/∞.

Indeterminate form $0 \cdot \infty$

In the case of a limit of the form

$$\lim_{x \to a} f(x) \cdot g(x)$$

where $f(x) \to 0$ and $g(x) \to \infty$ or $-\infty$, then the limiting value is unclear (one factor grows large while the other grows small). However, we can rewrite this as

$$\lim_{x \to a} \frac{f(x)}{1/g(x)} \qquad \left(\frac{0}{0}\right)$$

or

$$\lim_{x \to a} \frac{g(x)}{1/f(x)} \qquad \left(\frac{\infty}{\infty}\right)$$

so that it takes on one of the indeterminate forms for L'Hôpital's Rule.

10.3 FINDING LIMITS – INDETERMINATE FORMS

EXAMPLE 33

$$\lim_{x \to \infty} x \tan \frac{1}{x} = \lim_{x \to \infty} \frac{\tan \frac{1}{x}}{1/x} = \lim_{x \to \infty} \frac{\sec^2(\frac{1}{x}) \cdot \frac{-1}{x^2}}{-1/x^2} = \sec^2 0 = 1.$$

∎

Indeterminate form $\infty - \infty$

Similarly, in the case of a limit

$$\lim_{x \to a} f(x) - g(x)$$

where $f(x) \to \infty$ and $g(x) \to \infty$, we need to rewrite the expression so that it takes on the form $0/0$ or ∞/∞.

EXAMPLE 34 Find $\lim_{x \to \infty} x - \sqrt{x}$.

Solution We can rewrite this as

$$\lim_{x \to \infty} x - \sqrt{x} = \lim_{x \to \infty} (x - \sqrt{x}) \cdot \frac{x + \sqrt{x}}{x + \sqrt{x}} = \lim_{x \to \infty} \frac{x^2 - x}{x + \sqrt{x}}.$$

Now the limit has the form ∞/∞ and we can apply L'Hôpital's Rule:

$$\lim_{x \to \infty} \frac{x^2 - x}{x + \sqrt{x}} = \lim_{x \to \infty} \frac{2x}{1 - 1/\sqrt{x}} = \infty,$$

because the numerator grows without bound and the denominator $\to 1$. ∎

L'Hôpital's Rule can often be used to evaluate limits of sequences if the nth term a_n has a closed form that is indeterminate as $n \to \infty$. For instance, we can see that the sequence $\{a_n = n - \sqrt{n}\}_{n=1}^{\infty}$ diverges by using the result of the previous example.

Indeterminate exponential forms $1^\infty, 0^0, \infty^0$

These refer to limits of the form

$$\lim_{x \to a} f(x)^{g(x)}.$$

These are indeterminate when any of the following situations are present:

1) $f(x) \to 1$ while $g(x) \to \infty$,
2) $f(x)$ and $g(x)$ both $\to 0$, or
3) $f(x) \to \infty$ while $g(x) \to 0$.

Here's a technique for handling these exponential forms:

Step 1. Take the natural logarithm ln of the limit expression.

Step 1. Compute this new limit.

Step 1. Take the natural exponential exp of the result.

The natural logarithm and exponential functions are continuous, so

$$\exp(\lim_{x \to a} \ln(f(x))) = \lim_{x \to a} f(x).$$

EXAMPLE 35 Find $\lim_{x \to \infty} \left(1 + \frac{1}{x}\right)^x$.

Solution This is of the indeterminate form 1^∞. We first take the natural logarithm of the expression:

$$\lim_{x \to \infty} \ln\left(1 + \frac{1}{x}\right)^x = \lim_{x \to \infty} x \ln\left(1 + \frac{1}{x}\right).$$

We compute this new limit

$$\lim_{x \to \infty} \frac{\ln(1 + 1/x)}{1/x} = \lim_{x \to \infty} \frac{\frac{1}{1+1/x} \cdot \frac{-1}{x^2}}{-1/x^2} = \lim_{x \to \infty} \frac{1}{1 + 1/x} = 1.$$

We take the natural exponential to find the original limit value:

$$\lim_{x \to \infty} \left(1 + \frac{1}{x}\right)^x = e^1 = e.$$

∎

EXAMPLE 36 Find $\lim_{x \to \infty} x^{1/x}$.

Solution This has the indeterminate form ∞^0. We compute the limit of the natural logarithm of the expression:

$$\lim_{x \to \infty} \ln(x^{\frac{1}{x}}) = \lim_{x \to \infty} \frac{1}{x} \ln(x) = \lim_{x \to \infty} \frac{\ln x}{x}.$$

This has the indeterminate form ∞/∞, so we may use L'Hôpital's Rule to find

$$\lim_{x \to \infty} \frac{\ln(x)}{x} = \lim_{x \to \infty} \frac{1/x}{1} = \lim_{x \to \infty} \frac{1}{x} = 0.$$

Taking the natural exponential, we have $\lim_{x \to \infty} x^{1/x} = e^0 = 1$.

∎

10.3 FINDING LIMITS – INDETERMINATE FORMS

EXAMPLE 37 Find $\lim_{x \to 0} x^x$.

Solution This has the indeterminate form 0^0. We compute the limit of the natural logarithm of the expression:

$$\lim_{x \to 0} \ln(x^x) = \lim_{x \to 0} x \ln x = \lim_{x \to 0} \frac{\ln x}{1/x} = \lim_{x \to 0} \frac{1/x}{-1/x^2} = \lim_{x \to 0} -x = 0.$$

So $\lim_{x \to 0} x^x = e^0 = 1$. ∎

EXERCISES

In exercises 1-15, you are given a limit of a function f having the indeterminate form indicated in parentheses. For each, investigate the limiting behavior of the function by graphing $y = f(x)$ over suitable values and by making a table of function values for appropriate inputs x. Then use L'Hôpital's rule to evaluate the limit. Remember to make sure that you still have an appropriate indeterminate form if you use L'Hôpital's rule more than once.

1. $\lim_{x \to \frac{1}{2}} \dfrac{2x - \sin \pi x}{4x^2 - 1}$ $\qquad \left(\dfrac{0}{0}\right)$

2. $\lim_{x \to \infty} \dfrac{2e^{3x} + \ln x}{e^{3x} + x^2}$ $\qquad \left(\dfrac{\infty}{\infty}\right)$

3. $\lim_{x \to 10} \dfrac{\ln(x-9)}{x-10}$ $\qquad \left(\dfrac{0}{0}\right)$

4. $\lim_{x \to \infty} \left(1 + \dfrac{1}{x}\right)^x$ $\qquad (1^\infty), (0 \cdot \infty)$

5. $\lim_{x \to 1^+} (x - 1)^{\ln x}$ $\qquad (0^0)$

6. $\lim_{x \to 0} \dfrac{1}{x} - \dfrac{1}{\ln(x+1)}$ $\qquad (\infty - \infty)$

7. $\lim_{x \to \infty} x^{\frac{1}{x}}$ $\qquad (\infty^0)$

8. $\lim_{x \to \infty} \dfrac{\ln x}{x^{10}}$

9. $\lim_{x \to \infty} \dfrac{\ln x}{x^{\frac{1}{10}}}$

   (Note: Exercises 8 and 9 show how slowly $\ln x$ grows.)

10. $\lim_{x \to \infty} \dfrac{e^x}{x^{10}}$

    (Note: Exercise 10 shows how quickly e^x grows.)

11. $\lim_{x \to 0} \dfrac{2e^x - x^2 - 2x - 2}{x^3}$ $\qquad \left(\dfrac{0}{0}\right)$

12. $\lim_{x \to \infty} \left(\sqrt[5]{x^5 - 3x^4 + 17} - x\right)$ $\qquad (\infty - \infty)$

13. $\lim_{x \to \infty} \dfrac{x + \cos(x)}{x}$

14. $\lim\limits_{x \to 0} \dfrac{\sin(x) - x}{x - \tan(x)}$

15. $\lim\limits_{x \to 0+} \dfrac{e^{-1/x}}{x}$

10.4 ROOT-FINDING APPROXIMATION METHODS

Your experience has shown you that many equations involving a single (real number) unknown x can be expressed in the form

$$f(x) = 0$$

where f is a real-valued function of x. In other words, finding the solution to an equation is often equivalent to finding the **roots** or **zeroes** of a certain function f.

What exactly does it mean to solve such an equation? We are used to thinking in terms of *exact solutions*. That is, we want to find the precise real values x which make $f(x)$ *exactly equal to* 0. However, if f is a function modelling some physical process, then it may well be based on physical measurements, all of which have some bounds on their accuracy. An "exact" solution to an equation derived from imprecise data will therefore be an imprecise solution to the "true" equation. The point we are making is that in many real life applications, what we seek is not so much an exact solution, but a *sufficiently accurate* solution for our purposes.

Furthermore, even if we had an equation based on perfectly accurate data, we would be fortunate indeed if it fell into the category of solvable equations. For example, take a simple polynomial equation of the form

$$p(x) = 0.$$

If $p(x)$ happens to be linear (degree 1) or quadratic (degree 2) then we're in luck, because we have a definite procedure for solving for x:

To solve $mx + b = 0$, with $m \neq 0$, take $x = -b/m$.

To solve $ax^2 + bx + c = 0$, with $a \neq 0$, take $x = \dfrac{-b \pm \sqrt{b^2 - 4ac}}{2a}$.

10.4 ROOT-FINDING APPROXIMATION METHODS

However, if $p(x)$ is a cubic polynomial (degree 3), we generally must hope that we can find its factors. Artificially nice textbook examples may condition you to think this is not so hard, but arbitrarily chosen cubic polynomials do not generally lend themselves to factorization "on sight."

There actually is a general method for solving cubic polynomial equations, as given by *Cardan's formulas*, named after Girolamo Cardan (1501-1576). We outline the step-by-step procedure below:

Method for solving cubic equations by Cardan's formulas

Step 1. Collect terms on one side of the equation and divide both sides by the appropriate constant to put the equation into the form

$$\frac{x^3}{3} + ax^2 + bx + c = 0.$$

Step 2. Compute

$$u = b - a^2 \quad \text{and} \quad v = \frac{2a^3 - 3ab + 3c}{2}.$$

Step 3. Compute

$$r = \sqrt[3]{-v + \sqrt{u^3 + v^2}} \quad \text{and} \quad s = \sqrt[3]{-v - \sqrt{u^3 + v^2}}.$$

Step 4. The solutions are:

$$x = r + s - a$$

$$x = \frac{r+s}{2} + \frac{\sqrt{3}(r-s)i}{2} - a$$

$$x = \frac{r+s}{2} - \frac{\sqrt{3}(r-s)i}{2} - a,$$

where $i = \sqrt{-1}$, the complex square root of -1.

As you can see, this general method for solving cubic equations involves some rather complicated algebraic expressions. The general method for quartic (degree 4) equations is even more complicated.

With today's computer symbolic algebra systems, one might think that the complexity is a trivial concern—we need only program the exact solution procedure for each degree polynomial equation and the machine will produce the exact solutions.

Alas, it has been shown that it is *mathematically impossible* to derive a general solution procedure along the lines of the quadratic or Cardan's formulas for polynomials of degree 5 and higher! (That does not mean it is impossible to solve *any* such equations—it just means that there is not a general formula which will produce the solution to *every* fifth degree polynomial equation.)

Given the difficulty of solving polynomial equations, much less those involving rational powers or transcendental functions such as trigonometric, exponential, and logarithmic functions. (Even when it is practical to express the solution to such equations in closed form, the evaluation of the result will almost certainly involve computing an approximation.) How are we to approach the general problem of solving an equation of the form

$$f(x) = 0,$$

where f is a real-valued function? There are some numerical methods which can help us find approximate solutions in many cases, and in this section we discuss two of the most popular root-finding procedures.

The bisection method

The **bisection method** is an iterative procedure for finding or approximating a zero or root of a continuous function f. The idea behind the method is very simple, and relies on the observation we recorded as *Bolzano's Theorem:*

If f is continuous on the interval $[a, b]$, and if $f(a)$ and $f(b)$ have different signs (one is positive and the other is negative), then there must be at least one solution to the equation

$$f(x) = 0$$

which lies between a and b.

Let's suppose we have the situation just described. How might we try to find the root we know is located somewhere between a and b? The strategy of the bisection method is to choose the point x_1 *halfway between* a and b. (We can compute the value of x_1 by simply averaging the endpoints: $x_1 = (a+b)/2$.)

Now, if $f(x_1) = 0$, then we've found a solution. But $f(x_1) \neq 0$, then $f(x_1)$ must be either positive or negative. This means that f will change sign over *one* of the two new intervals formed: $[a, x_1]$ and $[x_1, b]$.

We can repeat the bisection strategy on this interval. That is, we choose x_2 to be the midpoint of the "half" over which f changes sign. If $f(x_2) = 0$, we're done. If $f(x_2) \neq 0$, we repeat the procedure again for the appropriate "quarter" interval.

By continuing this process we get a sequence of inputs

$$x_1, x_2, x_3, \ldots$$

10.4 ROOT-FINDING APPROXIMATION METHODS

The procedure stops when we find an exact root x_n, that is, when we find

$$f(x_n) = 0.$$

However, it's quite conceivable that the process could go on indefinitely without ever "hitting" an exact root. But we do have a firm way of telling how close to an exact root we must be at each step. By the nature of the bisection process, we narrow the length of the interval of possibilities in half at each step. Since the original interval $[a, b]$ has a length $b - a$, this means that our *nth* choice x_n will be within $(b - a)/2^n$ of a true root r. In other words, we can say that

$$x_n - \frac{b-a}{2^n} \leq r \leq x_n + \frac{b-a}{2^n}.$$

This gives us an effective way of predetermining the number of steps required to get within a given precision of an actual root.

EXAMPLE 38 Show that $f : x \longmapsto x^3 + x - 3$ has a root in the interval $[0, 2]$. Use the bisection method to find the first six approximations $x_1, x_2, \ldots, x_6$, and find the accuracy of the sixth approximation.

Solution Since f is a polynomial function, we know it is continuous, and since

$$f(0) = 0^3 + 0 - 3 = -3 \quad \text{and} \quad f(2) = 2^3 + 2 - 3 = 7,$$

f changes sign on the interval $[0, 2]$ and must have at least one root therein.

We try $x_1 = 1$ (the midpoint of $[0, 2]$) as a possible root:

$$f(x_1) = f(1) = 1^3 + 1 - 3 = -1.$$

Of the two intervals $[0, 1]$ and $[1, 2]$, f changes sign on $[1, 2]$, so we next try $x_2 = 3/2$. (Note that we can compute the midpoint of an interval by averaging the endpoints: $3/2 = (1 + 2)/2$.)

$$f(x_2) = f(1.5) = (1.5)^3 + 1.5 - 3 = 1.875$$

Now f changes sign on $[1, 1.5]$ so we try

$$x_3 = \frac{1 + 1.5}{2} = 1.25$$

and

$$f(x_3) = f(1.25) \approx 0.2.$$

Since $f(1)$ is negative and $f(1.25)$ is positive, we next try

$$x_4 = \frac{1 + 1.25}{2} = 1.125$$

and

$$f(x_4) = f(1.125) \approx -0.46.$$

This indicates that the next try should be

$$x_5 = \frac{1.125 + 1.25}{2} = 1.1875$$

and

$$f(x_5) = f(1.1875) \approx -0.14.$$

Using $x_6 = \frac{1.1875 + 1.25}{2} = 1.21875$, we have

$$f(x_6) = f(1.21875) \approx .03.$$

Now a true root must be between 1.21875 and 1.1875, so x_6 is no further away than

$$1.21875 - 1.1875 = .03125.$$

Note that we can tell in advance the precision of x_6 before we start the bisection method. In this case $a = 0$, $b = 2$, and $n = 6$, so

$$\frac{b-a}{2^n} = \frac{2-0}{2^6} = \frac{1}{32} = .03125.$$

is the accuracy of x_6 as an approximation to a root of $x^3 + x - 3$. ∎

EXAMPLE 39 Find a root, with an error of at most 0.001, of the equation $\sin(x) - 0.3 = 0$ by the bisection method.

Solution Defining $f: x \mapsto \sin(x) - 0.3$, we note that $f(0) = -0.3 < 0$ and $f(1) \approx 0.54147 > 0$. Since f is continuous on $[0, 1]$ we have the conditions to begin bisection, and taking either endpoint as an approximation gives a root estimate with error of at most $|1 - 0| = 1$.

The midpoint is 0.5 and $f(0.5) \approx 0.17943 > 0$ so our new interval is $[0, 0.5]$ and the new error is at most 0.5. The next midpoint is 0.25 and $f(0.25) \approx -0.052596 < 0$ so that the next interval is $[0.25, 0.5]$ and the error is at most 0.25.

Continuing in this way we can collect our results in Table 10.2.

10.4 ROOT-FINDING APPROXIMATION METHODS

left endpoint	midpoint	right endpoint	maximum error
f(0.0)<0	f(0.25)<0	f(0.5)>0	0.5
f(0.25)<0	f(0.375)>0	f(0.5)>0	0.25
f(0.25)<0	f(0.3125)>0	f(0.375)>0	0.125
f(0.25)<0	f(0.28125)<0	f(0.3125)>0	0.0625
f(0.28125)<0	f(0.296875)<0	f(0.3125)>0	0.03125
f(0.296875)<0	f(0.3046875)<0	f(0.3125)>0	0.015625
f(0.3046875)<0	f(0.30859375)>0	f(0.3125)>0	0.0078125
f(0.3046875)<0	f(0.30859375)>0	f(0.30859375)>0	0.00390625
f(0.3046875)<0	f(0.306640625)>0	f(0.30859375)>0	0.001953125
f(0.3046875)<0	f(0.3056640625)>0	f(0.306640625)>0	0.0009765625

Table 10.2 Bisecting intervals in search of a root of $\sin(x) - 0.3 = 0$.

The length of the final interval shown is less than 0.001 so that either endpoint (or the midpoint, for that matter) provides an estimate of the root accurate within 0.001. ■

Graphical estimation of roots is essentially a visual version of the bisection method. First, we obtain a viewing window in which the graph crosses the x-axis. Then we repeatedly zoom in, all the while keeping the x-intercept in the viewing window. One measure of our accuracy at any step is simply the width of the window. (However, keep in mind that machine graphics are based on numerical computations of finite precision.)

Newton's method

An especially important iterative method for finding roots is called **Newton's method**, named after Isaac Newton. To solve an equation $f(x) = 0$ by Newton's method, we need

1) a "seed" x_0 (our first guess or rough estimate of the root), and
2) the iterating function $x \longmapsto x - \dfrac{1}{f'(x)} f(x)$.

The sequence of approximations is found by feeding the seed x_0 as an input into this function, and then producing an iterative sequence by continuing to feed the output result back into the same function:

$$x_1 = x_0 - \frac{f(x_0)}{f'(x_0)}$$

$$x_2 = x_1 - \frac{f(x_1)}{f'(x_1)}$$

$$x_3 = x_2 - \frac{f(x_2)}{f'(x_2)}$$

$$x_4 = x_3 - \frac{f(x_3)}{f'(x_3)}$$

$$\vdots$$

To see why this is effective, suppose first that our original function f is linear (but not constant). Then $f(x) = mx + b$ with slope $m \neq 0$. In this case $f'(x) = m$, and the iterating function is

$$x \longmapsto x - \frac{mx+b}{m} = -\frac{b}{m}.$$

Notice that $-b/m$ is the root of our original linear function f!

Now, in general, suppose we have a differentiable function f. Then the iterating function finds the root of the best linear approximation at any input x where the first derivative $f'(x) \neq 0$. For example, the best linear approximation to f at $x = x_0$ is

$$y = f'(x_0)(x - x_0) + f(x_0).$$

If we find the root x_1 of this best linear approximation (the value x_1 making $y = 0$), we obtain

$$x_1 = x_0 - \frac{f(x_0)}{f'(x_0)}.$$

We can then repeat the process by finding x_2, the root of the best linear approximation to f at $x = x_1$, and so on.

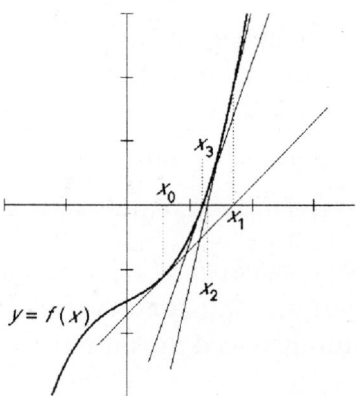

Figure 10.5 Newton's Method produces roots of best linear approximations to f.

Figure 10.5 illustrates Newton's method graphically. Starting with x_0, we find the tangent line to the graph of $y = f(x)$ at the point $(x_0, f(x_0))$. Our next input x_1 is where this tangent line crosses the x-axis. Then we find the tangent line at the point $(x_1, f(x_1))$ and the input x_2 where it crosses the x-axis. The input x_3 in Figure 10.5 almost coincides with the root, showing how quickly Newton's method can converge.

10.4 ROOT-FINDING APPROXIMATION METHODS

EXAMPLE 40 Find the third iterate x_2 to a solution of $x^2 = 3$ by Newton's method with an initial seed $x_0 = 5$.

Solution Rewriting this equation in the form $f(x) = 0$ where $f : x \longmapsto x^2 - 3$, we have $f'(x) = 2x$ and our iteration function is

$$x \longmapsto x - \frac{f(x)}{f'(x)} = x - \frac{x^2 - 3}{2x}$$

$$= \frac{2x^2}{2x} - \frac{x^2 - 3}{2x} = \frac{x^2 + 3}{2x}.$$

Using $x_0 = 5$ we obtain $x_1 = \dfrac{5^2 + 3}{2 \cdot 5} = 28/10 = 2.8$, and

$$x_2 = \frac{(2.8)^2 + 3)}{2(2.8)} = 1084/560 \approx 1.93571.$$

Note: the square root key on a calculator essentially uses Newton's method. ■

How do we judge how *good* an approximation is? One criterion may seem obvious:

How close is our approximation to the value we seek?

There is another criterion that may be more important for our practical purposes:

How well does our approximation satisfy our original requirements?

In the case of approximating roots of the equation $f(x) = 0$, the first criterion judges an approximation x_n by the distance $|x_n - r|$, where r is a true root—the smaller $|x_n - r|$ is, the closer x_n is to a root of the original equation. On the other hand, the second criterion judges an approximation x_n by measuring $|f(x_n)|$—the smaller $|f(x_n)|$ is, the closer x_n *behaves* like a root of the original equation. Which criterion is more important depends on what we need the approximation for.

EXAMPLE 41 From the previous example, $x_2 \approx 1.93571$, while $\sqrt{3} \approx 1.73205$. Hence, by the first criterion,

$$|x_2 - \sqrt{3}| \approx |1.93571 - 1.73205| \approx 0.20366.$$

By the second criterion,

$$|x_2^2 - 3| \approx |(1.93571)^2 - 3| \approx 0.74699.$$

■

When does Newton's method fail?

Graphically, we can see that Newton's method will fail if at any step the approximation term x_n produces a first derivative value $f'(x_n) = 0$: the tangent line in this case is *horizontal*, and hence does not cross the x-axis (unless x_n is already the root we seek). Note also that the output of the iteration function is undefined when $f'(x_n) = 0$.

Another situation where Newton's method may fail to converge is when the iterations oscillate or cycle periodically through the same set of values.

EXAMPLE 42 Investigate the behavior of Newton's method in solving $x^4 - 10x^2 = 8$ using the initial term $x_0 = 2$ and the initial term $x_0 = 3$.

Solution We need to find a root of the function $g : x \longmapsto x^4 - 10x^2 - 8$. Since $g'(x) = 4x^3 - 20x$, we have

$$x \longmapsto x - \frac{x^4 - 10x^2 - 8}{4x^3 - 20x}$$

as our iteration function.

First, with $x_0 = 2$, we have

$$x_1 = 2 - \frac{2^4 - 10(2)^2 - 8}{4(2)^3 - 20(2)} = -2,$$

and

$$x_2 = -2 - \frac{(-2)^4 - 10(-2)^2 - 8}{4(-2)^3 - 20(-2)} = 2.$$

We've bounced back to our original seed $x_0 = 2$. If we continue, our approximations will simply oscillate forever between 2 and -2 and never converge.

In contrast, if we start out with $x_0 = 3$, the iteration function produces a sequence that converges quickly to a value $x \approx 3.27789$ after only four steps. If we substitute this back into the original equation, we find

$$(3.27789)^4 - 10(3.27789)^2 \approx 8.00000.$$

In this case, changing the seed brought better fruit to our efforts. ∎

It is also possible for Newton's method to diverge away from a root.

10.4 ROOT-FINDING APPROXIMATION METHODS

EXAMPLE 43 The function $h : x \longmapsto x^{1/3}$ certainly has a root $x = 0$. However, if we form the iteration function, we obtain

$$x \longmapsto x - \frac{x^{1/3}}{(1/3)x^{-2/3}}.$$

For any $x \neq 0$, the iteration function simplifies to $x \longmapsto 3x$. That means for any seed $x_0 \neq 0$, the iteration function will produce

$$x_1 = 3x_0, \quad x_2 = 3x_1 = 9x_0, \quad x_3 = 3x_2 = 27x_0, \quad \ldots$$

Unless we hit the root with our very first guess, Newton's method will take us farther and farther away. ∎

Which technique is better—Newton's method or the bisection method? Newton's method has the advantage of very rapid convergence (when it does indeed converge). However, as some of the examples above show, it can sometimes fail miserably. The bisection method is slower, but it has the advantage of keeping the root "trapped" in smaller and smaller intervals. A strategy used by some machine root-finders is to use a combination of both methods.

Fixed points and the general iteration method

Newton's method of rootfinding is based on the general notion of a *contraction*. The idea goes like this: if the graph of the iteration function is nearly flat, then the outputs of the function will be closer together than the corresponding inputs. If we call this iteration function f, then starting with an initial seed x_0

$$x_0 \xrightarrow{f} x_1 \xrightarrow{f} x_2 \xrightarrow{f} \ldots$$

can give us a convergent sequence. Newton's method provides a function that often acts this way in converging to a root of a related function.

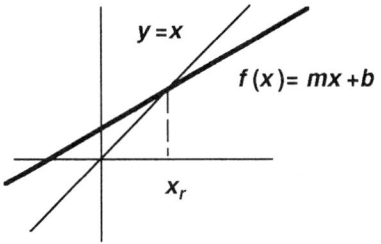

Figure 10.6 If $m \neq 1$, the graph of $x \longmapsto mx + b$ must intersect the diagonal.

To see how this works, it is helpful to first look at iterating any linear function $f : x \longmapsto mx + b$. If the slope $m \neq 1$, then the graph of f must intersect the diagonal (the graph of $y = x$) somewhere because the graphs are not parallel. Suppose that (x_r, x_r) is the point of intersection (see Figure 10.6). Then $f(x_r) = x_r$, and x_r is called a **fixed point** of the function f.

If we write the Taylor form of f at the point x_r, we obtain

$$f : x \longmapsto m(x - x_r) + f(x_r) = m(x - x_r) + x_r.$$

We can now find the convergence properties of the sequence

$$x_0 \xrightarrow{f} x_1 \xrightarrow{f} x_2 \xrightarrow{f} \ldots$$

First, we note that $f(x) - x_r = m(x - x_r)$ for any x. This makes it easy to compute the distance $d_n = |x_n - x_r|$ between x_n and the fixed point x_r:

$$d_0 = |x_0 - x_r|$$
$$d_1 = |x_1 - x_r| = |f(x_0) - x_r| = |m(x_0 - x_r)| = |m| d_0$$
$$d_2 = |x_2 - x_r| = |f(x_1) - x_r| = |m(x_1 - x_r)| = |m| d_1 = |m|^2 d_0$$
$$d_3 = |x_3 - x_r| = |f(x_2) - x_r| = |m(x_2 - x_r)| = |m| d_2 = |m|^3 d_0$$
$$d_n = |x_n - x_r| = |f(x_{n-1}) - x_r| = |m(x_{n-1} - x_r)| = |m| d_{n-1} = |m|^n d_0.$$

The distance is multiplied by a factor of $|m|$ at each iteration (a geometric sequence!). If $|m| < 1$, then the distance approaches 0, and the sequence $x_0, x_1, \ldots$ converges to x_r geometrically. On the other hand, if $|m| > 1$, the distance increases without bound and so the sequence diverges.

There is a simple but effective way of graphing the iteration process using the graph of the diagonal. Normally we think of an input x_0 to a function f as a point on the horizontal axis, and the output $f(x_0)$ as a point on the vertical axis. We graphically look up the output by connecting the input to the graph with a vertical line, and then a horizontal line to the vertical axis.

Now, if we want to use f as an iteration function, we then need to find the point on the horizontal axis corresponding to the output on the vertical axis (see illustration on the left in Figure 10.7) and repeat the process. On the other hand, if we think of input x_0 and output $f(x_0)$ both as points on the diagonal (x_0, x_0) and $(f(x_0), f(x_0))$, then graphing the iterates is easier. Starting with a point on the diagonal, draw a vertical line to the graph of f, and then a horizontal line back to the diagonal. The new point represents the output of f and is ready to be used as a new input (see illustration on the right in Figure 10.7).

10.4 ROOT-FINDING APPROXIMATION METHODS

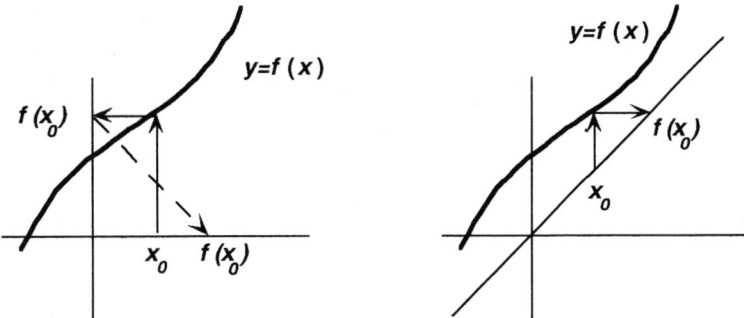

Figure 10.7 Graphing input to output relationship without and with use of the diagonal.

EXAMPLE 44 Graph three iterates of the linear function $f : x \mapsto .8(x-2)+2$ using initial term $x_0 = .5$.

Solution The graph is shown in Figure 10.8. Note that, because the slope 0.8 is smaller than 1 in absolute value, the iterates are marching toward the fixed point $x_r = 2 = f(2)$. ∎

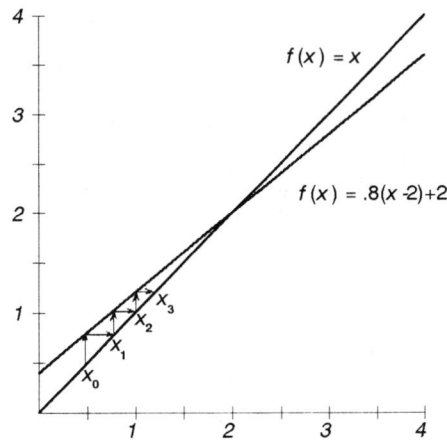

Figure 10.8 Graphing three iterates of $f : x \mapsto .8(x-2)+2$.

With this graphing technique in hand, we can look at four generic cases of iteration for linear functions:

1) the slope is between 0 and 1,
2) the slope is between -1 and 0,
3) the slope is greater than 1, and
4) the slope is less than -1.

These four cases are illustrated in Figure 10.9.

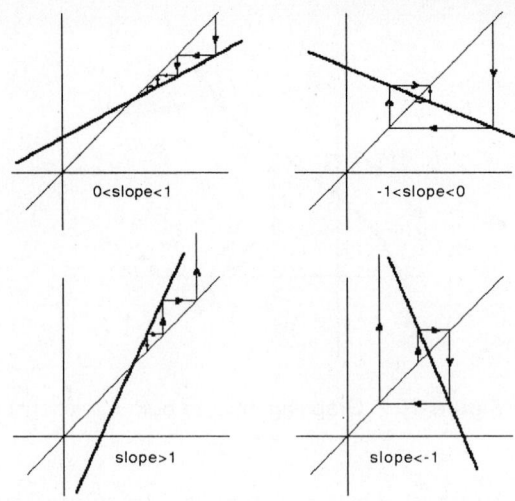

Figure 10.9 Four generic cases of iteration.

If the slope is positive, the iterates either "stairstep" toward or away from the fixed point, as the slope is less than or greater than 1, respectively. If the slope is negative, the iterates spiral toward or away from the fixed point, again as the slope is less than or greater than 1, respectively.

If we zoom in sufficiently on the graph of a differentiable function, we expect to see a straight line. This suggests we can use the information above even for nonlinear functions, as long as we restrict our attention to a small enough neighborhood of a fixed point. Unfortunately, if we knew enough about the function to be able to zoom in on the fixed point, we wouldn't need iteration as a method for finding the fixed point! Fortunately, you don't need to get very close to the fixed point for this analysis to work, as shown by the following theorem.

Theorem 10.3

Fixed Point Theorem.
Hypothesis 1: g is continuous and differentiable on $[a,b]$.
Hypothesis 2: $g(x) \in [a,b]$ for all $x \in [a,b]$.
Hypothesis 3: $|g'(x)| < 1$ for all $x \in [a,b]$.
Conclusion: There is exactly one point $a \leq x_r \leq b$ such that $g(x_r) = x_r$ and any iteration sequence starting with $x_0 \in [a,b]$ converges to x_r.

10.4 ROOT-FINDING APPROXIMATION METHODS

Reasoning Hypothesis 1 tells us that the function graph will be smooth and unbroken over the interval $[a, b]$. Geometrically, Hypothesis 2 tells us that the graph of g over $[a, b]$ lies entirely within the square box with corners (a, a), (b, b), (a, b), and (b, a) (see Figure 10.10).

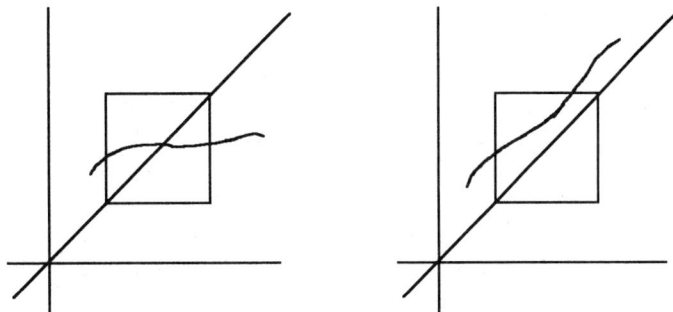

Figure 10.10 Geometric interpretation of hypothesis 2 and a violation of it.

Notice that the graph at the left endpoint must be on or above the diagonal, and the graph at the right endpoint must be on or below the diagonal. Since the function is continuous, the graph must cross the diagonal, and the function must have *at least one fixed point*.

Now, why couldn't g have *two* or more fixed points in this interval $[a, b]$? Here's where Hypothesis 3 is important. If $g(x_1) = x_1$ and $g(x_2) = x_2$, then the Mean Value Theorem tells us that there is some point $x_1 < c < x_2$, such that

$$g'(c) = \frac{g(x_2) - g(x_1)}{x_2 - x_1} = \frac{x_2 - x_1}{x_2 - x_1} = 1.$$

That would violate Hypothesis 3, so we can conclude that under these three hypotheses, the function g must have exactly one fixed point in the interval $[a, b]$.

If we start with an initial seed $x_0 \in [a, b]$, then Hypothesis 3 also means that the graph is flat enough to essentially force the resulting iteration outputs to become closer and closer to each other. This sequence must converge to the single fixed point x_r. □

The Fixed Point Theorem can be used to obtain criteria for the convergence of Newton's method. Suppose we are trying to solve

$$f(x) = 0$$

using Newton's method. In this case, we are searching for a fixed point of the iteration function

$$g(x) = x - \frac{f(x)}{f'(x)}.$$

For this function g to satisfy the Fixed Point Theorem over an interval $[a,b]$, we need

1) g differentiable on $[a,b]$,

2) $g(x) \in [a,b]$ for all $x \in [a,b]$,

3) $|g'(x)| < 1$ for all $x \in [a,b]$.

Let's translate each of these requirements into requirements for f:

1) For $g'(x) = 1 - \dfrac{(f'(x))^2 - f(x)f''(x)}{(f'(x))^2} = \dfrac{f(x)f''(x)}{(f'(x))^2}$ to be defined on $[a,b]$, we must require $f'(x) \neq 0$ and $f''(x)$ to be defined on $[a,b]$.

2) We must require $x - f(x)/f'(x)$ to be between a and b for all $a \leq x \leq b$.

3) Since $|g'(x)| < 1$ for all $x \in [a,b]$, we must require

$$\left| \dfrac{f(x)f''(x)}{(f'(x))^2} \right| < 1.$$

If we can guarantee these conditions for f on an interval $[a,b]$, then we know that Newton's method will succeed in finding a root in that interval.

EXERCISES

In each of exercises 1-10, find three additional successive approximations to the roots of the given equation by the bisection method, starting with the indicated endpoints.

1. $x^2 - 3.2 = 0$; $x = 1$, $x = 2$

2. $x^3 - 23 = 0$; $x = 2$, $x = 3$

3. $\sin(x) - 0.7 = 0$; $x = 0.7$, $x = 0.8$

4. $x^5 + 3.6x^4 - 2.51x^3 - 22.986x^2 - 28.24x - 10.4$; $x = 2$, $x = 3$

5. $x + \sqrt{x} + \sqrt{x^3} - 7 = 0$; $x = 1.8$, $x = 2.6$

6. $x\exp(x^2) - \sqrt{x^2 + 1} = 0$; $x = 0.4$, $x = 1$.

7. $\ln(x^2 + \ln(x))$; $x = 0.625$, $x = 1.35$

8. $x^3 - \sqrt{x^2 + 1} + x - 3 = 0$; $x = 1.1$, $x = 1.6$

9. $3\cos(\cos(x)) - 2x = 0$; $x = 1.1$, $x = 2.2$

10. $\sqrt[4]{x} + x^3 - x - 0.5 = 0$; $x = 0.2$, $x = 1.2$

In exercises 11-20, use Newton's method to find five successive approximations to a root of the given equation in the given interval.

10.4 ROOT-FINDING APPROXIMATION METHODS

11. $x^2 - 3.2 = 0$; $[1, 2]$

12. $x^3 - 23 = 0$; $[2, 3]$

13. $\sin(x) - 0.7 = 0$; $[0.7, 0.8]$

14. $x^5 + 3.6x^4 - 2.51x^3 - 22.986x^2 - 28.24x - 10.4 = 0$; $[2, 3]$

15. $x + \sqrt{x} + \sqrt{x^3} - 7 = 0$; $[1.8, 2.6]$

16. $x \exp(x^2) - \sqrt{x^2 + 1} = 0$; $[0.4, 1]$.

17. $\ln(x^2 + \ln(x)) = 0$; $[0.625, 1.35]$

18. $x^3 - \sqrt{x^2 + 1} + x - 3 = 0$ $[1.1, 1.6]$

19. $3\cos(\cos(x)) - 2x = 0$; $[1.1, 2.2]$

20. $\sqrt[4]{x} + x^3 - x - 0.5 = 0$; $[0.2, 1.2]$

Exercises 21-26 refer to iterating the function $g : x \longmapsto x + bx(7-x)$ for various values of the parameter b, and various choices of initial seed x_0.

21. Find two values for b for which the iteration sequence of g with $x_0 = 1$ converges to 7.

22. Find a value for b for which the iteration sequence of g with $x_0 = 1$ diverges to $+\infty$, or show that there aren't any.

23. Find a value for b for which the iteration sequence of g with $x_0 = 1$ diverges to $-\infty$, or show that there aren't any.

24. Find two values for b for which the iteration sequence of g with $x_0 = 1$ neither converges nor diverges, but rather oscillates regularly.

25. Find a value for $b \neq 0$ for which $1 = x_0 = x_2$. Does this give a convergent sequence?

26. Find two values for b for which the iteration sequence of g with $x_0 = 1$ seems to oscillate chaotically.

---

27. Use Cardan's formula to find the exact root of $x^3 + x - 3$ that lies in the interval $[0, 2]$. Then find the first six Newton's method approximations starting with $x_0 = 1$. Compare the speed of convergence with that of the bisection method.

28. Use Newton's method to find the three smallest positive roots of the equation $x + \tan(x) = 0$.

CHAPTER

11

Series and Function Approximations

In the last chapter, we examined sequences of real numbers and how they arise naturally in methods for finding approximations of roots. In this chapter, we continue with the themes of infinite processes and approximation by examining the limiting behavior of some very special sequences of real numbers, and by examining some important polynomial approximation techniques for *functions*.

With any sequence is associated a **series**, which can be thought of as the "sum" of all the terms in the sequence. Of course, it is physically impossible to add up infinitely many numbers in a finite amount of time, so mathematically, we consider a series as the limit of the running totals, called *partial sums* of the terms in the sequence. The questions we ask about sequences in general can be asked with regards to series:

Does the series (limit of partial sums) *converge* to a particular real number?

If so, what is that limit?

How many steps does it take to achieve a given level of accuracy?

A polynomial is a sum of terms that are multiples of powers of x. Just as a series can be thought of as the "sum" of infinitely many real numbers, a **power series** can be thought of as a "polynomial" with infinitely many terms. Power series can be used to define functions, and we'll be especially interested in determining the domain of valid inputs for such functions.

We've already seen **Taylor polynomials** arise as useful approximations for functions. The tangent line approximation to a function at a specific point is the first degree Taylor polynomial approximation, and it represents the *best linear approximation* to a function at that point. Similarly, the second degree Taylor polynomial represents the *best quadratic approximation*. We'll examine more closely what we mean by "best" and look at the behavior of higher degree Taylor polynomials. The limit of the infinite sequence of polynomial approximations we obtain in this way is naturally called a **Taylor series**.

11.1 SERIES

We can roughly think of a series as the sum of the infinitely many terms in a sequence. Before embarking on our study of series, let's first take some time to get better acquainted with summation notation, as we'll be utilizing it extensively in our discussion.

Summation notation

A convenient notation for a sum is the **sigma-notation**, and we have already made use of it in our discussion of Riemann sums.

An example of this notation is

$$\sum_{n=1}^{4} n^2.$$

The capital Greek letter "sigma" ($\sum$) corresponds to the letter S and stands for a sum. The letter n is called an **index variable**. It serves to label the individual terms in the sum. In this case, the notation tells us that the starting value of the index is $n = 1$, and the ending value is $n = 4$. Each integer value n from 1 through 4 corresponds to a separate term of the sum. In this case, each term is given by the expression n^2, and the notation is really shorthand for the sum of four squares:

$$\sum_{n=1}^{4} n^2 = 1^2 + 2^2 + 3^2 + 4^2 = 1 + 4 + 9 + 16 = 30.$$

EXAMPLE 1 $\sum_{n=5}^{14}(8 - n)$ represents the sum of 10 consecutive integers (in descending order) starting with 3:

$$3 + 2 + 1 + 0 + (-1) + (-2) + (-3) + (-4) + (-5) + (-6) = -15.$$

∎

Here, it was not difficult to write down all the terms of the sum explicitly. The convenience of the sigma-notation is most evident when we have many terms to sum or perhaps an unspecified number of terms to sum.

EXAMPLE 2 $\sum_{n=1}^{100}(2n - 1)$ represents the sum of the first 100 odd positive integers:

$$1 + 3 + 5 + 7 + \cdots + 195 + 197 + 199 = 10000.$$

∎

EXAMPLE 3 $\sum_{n=1}^{N} 2n$ represents the sum of the first N even positive integers:

$$2 + 4 + 6 + \cdots + 2(n-2) + 2(n-1) + 2n.$$

∎

The index n is sometimes referred to as a *dummy* index, because the letter plays no other role than to label the terms. Any other letter can serve this purpose just as well.

EXAMPLE 4 $\sum_{i=1}^{4} i^2 = 1 + 4 + 9 + 16 = 30.$ ∎

The index may not even be involved in the expression for the nth term.

EXAMPLE 5 $\sum_{n=1}^{5} 3 = 3 + 3 + 3 + 3 + 3 = 15.$ ∎

The Archimedean property of real numbers

Suppose we take a positive real number r and start adding it to itself over and over again. No matter how small r is ($r > 0$), this process will produce arbitrarily larger and larger sums. This fact is known as the **Archimedean property of real numbers** and we can state it more precisely as follows:

If r, M are any two positive real numbers, then there exists a positive integer N such that $Nr > M$.

The quantity Nr represents a sum:

$$Nr = \sum_{n=1}^{N} r = r + r + r + \ldots r \qquad (N \text{ addends of } r).$$

The ancient Greek mathematicians did not have a language or notation that included something like our symbol ∞ in their mathematics, but we can write the Archimedean property as follows:

$$\text{For any } r > 0, \quad \lim_{n \to \infty} nr = \infty,$$

meaning that the quantity nr grows without bound as $n \to \infty$.

Zeno's paradox

We might think of the Archimedean property as saying that a "sum" of *infinitely many* terms $r > 0$ is infinite. Is there any way that the "sum" of infinitely many positive numbers could not be infinite? Many of the ancient Greek mathematicians did not think so, because of the Archimedean property. However, this belief was shaken by what is known as **Zeno's Paradox**.

Zeno's paradox goes like this: Suppose a person runs along a racecourse of length 1 mile. Before the race is completed, the person must first cover half the distance. Now, of the half-mile remaining, the person must cover half of it (or $1/4$ of a mile) to reach the $3/4$ mile mark. Before finishing the final quarter-mile, the person must first cover half of it ($1/8$ mile) to reach the $7/8$ mile mark. This process occurs over and over: given any distance remaining, the person must first cover half of the distance before the race is completed (see Figure 11.1).

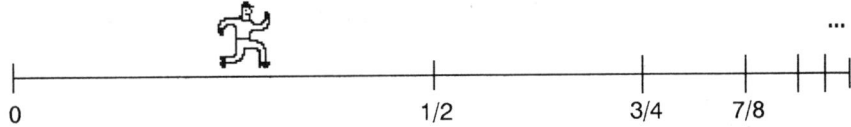

Figure 11.1 Zeno's racecourse.

Each of these distances is positive, and the person must cover infinitely many of them. The Archimedean property might suggest that the total distance is infinite, and the person cannot possibly complete the race. On the other hand, it is clear that people complete such races regularly. This paradox forces us to admit that it is reasonable to write something like

$$\frac{1}{2} + \frac{1}{4} + \frac{1}{8} + \cdots = 1.$$

The word "paradox" usually refers to a statement or situation contradicting itself or some firmly established truth. Zeno's paradox is not really a paradox, for it does *not* contradict the Archimedean property. Unlike the situation of adding the *same* positive number r to itself infinitely often, the racecourse example has us adding infinitely many *different* positive numbers, each one smaller (by the factor one-half) than the number preceding it. Evidently, in this different situation, it may be reasonable to assign a finite total to a sum of infinitely many positive terms.

What is a series?

The notion of a series is convenient for describing situations like Zeno's paradox or any other infinite process producing a sequence of running totals. In fact, using the terms in any sequence of real numbers $\{a_n\}$, we can build a new sequence of running totals. A **series** is the limit of this new sequence.

The notation for a series uses Σ-notation for sums, and we write

$$\sum_{n=1}^{\infty} a_n$$

as shorthand for

$$a_1 + a_2 + a_3 + a_4 + \cdots + a_n + \ldots$$

It is physically impossible to add infinitely many numbers together by actually summing them, so mathematically the series is the limit of the sequence of the **partial sums**:

$$S_1 = a_1, \quad S_2 = a_1 + a_2, \quad S_3 = a_1 + a_2 + a_3, \quad S_4 = a_1 + a_2 + a_3 + a_4, \quad \ldots$$

where, in general, the Nth partial sum is

$$S_N = a_1 + a_2 + \cdots + a_N.$$

Using summation notation,

$$\sum_{n=1}^{\infty} a_n = \lim_{N \to \infty} \sum_{n=1}^{N} a_n = \lim_{N \to \infty} S_N.$$

EXAMPLE 6 Find the first four partial sums of the series $\sum_{n=1}^{\infty} \frac{1}{2^n}$.

Solution

$$S_1 = \frac{1}{2^1} = \frac{1}{2}$$

$$S_2 = \frac{1}{2^1} + \frac{1}{2^2} = \frac{3}{4}$$

$$S_3 = \frac{1}{2^1} + \frac{1}{2^2} + \frac{1}{2^3} = \frac{7}{8}$$

$$S_4 = \frac{1}{2^1} + \frac{1}{2^2} + \frac{1}{2^3} + \frac{1}{2^4} = \frac{15}{16}.$$

Can you come up with a formula that gives the Nth partial sum S_N? ∎

11.1 SERIES

A series is said to **converge** if the sequence of partial sums converges; otherwise we say the series **diverges**. Finding the sum of a convergent infinite series means finding the limit of the sequence of its partial sums.

We can numerically investigate the convergence or divergence of a series in much the same way we investigate the limiting behavior of any sequence—by actually evaluating the partial sums S_N and examining their behavior as N grows larger and larger without bound.

While we may have a nice closed-form or recursive description of the Nth *term* a_N of the sequence $\{a_n\}$, it may be difficult or impossible to find a nice description of the Nth *partial sum* S_N of the series $\sum a_n$.

Examples of series

A **telescoping series** gets its name from the behavior of its partial sums. Usually, the terms of a telescoping series can be written as a sum of differences of quantities, all of which cancel except for the first and last, collapsing (like a telescope) to a single difference. Thus, we can find a closed-form for the Nth partial sum and find the limiting behavior of it.

EXAMPLE 7 Does $\sum_{n=1}^{\infty} (\dfrac{1}{n} - \dfrac{1}{n+1})$ converge, and if so, what is its sum?

Solution The first few partial of the series are

$$S_1 = a_1 = \frac{1}{1} - \frac{1}{2} = \frac{1}{2}$$

$$S_2 = a_1 + a_2 = \left(1 - \frac{1}{2}\right) + \left(\frac{1}{2} - \frac{1}{3}\right) = 1 - \frac{1}{3} = \frac{2}{3}$$

$$S_3 = a_1 + a_2 + a_3 = \left(1 - \frac{1}{2}\right) + \left(\frac{1}{2} - \frac{1}{3}\right) + \left(\frac{1}{3} - \frac{1}{4}\right) = 1 - \frac{1}{4} = \frac{3}{4}$$

$$S_4 = a_1 + a_2 + a_3 + a_4 = \left(1 - \frac{1}{2}\right) + \left(\frac{1}{2} - \frac{1}{3}\right) + \left(\frac{1}{3} - \frac{1}{4}\right) + \left(\frac{1}{4} - \frac{1}{5}\right) = \frac{4}{5}.$$

Notice how the "internal terms" cancel. We can see that

$$S_N = a_1 + a_2 + \cdots + a_N = 1 - \frac{1}{N+1}$$

and $\sum_{n=1}^{\infty} a_n = \lim_{N \to \infty} 1 - \dfrac{1}{N+1} = 1.$ ∎

Geometric series

The terms of a **geometric series** form a geometric sequence—each term is a constant factor times the preceding term.

EXAMPLE 8 The series of Zeno's paradox $\sum_{n=1}^{\infty} \frac{1}{2^n}$ is a geometric series. ∎

In general, a geometric series has the form

$$a + ar + ar^2 + ar^3 + ar^4 + \ldots ar^n + \ldots$$

where a is the initial term and r is the constant factor. If we use the power of r as the index, then we can start with $n = 0$ and write the series as

$$\sum_{n=0}^{\infty} ar^n.$$

Let's examine how the limit of a geometric series depends on the factor r. If $r = 1$, then

$$\sum_{n=0}^{\infty} ar^n = a + a + a + a + \ldots$$

will *diverge* for any nonzero a. If $r = -1$, then

$$\sum_{n=0}^{\infty} ar^n = a - a + a - a + \ldots$$

also diverges because the partial sums oscillate between a and 0:

$$S_0 = a, \quad S_1 = 0, \quad S_2 = a, \quad S_3 = 0, \quad \text{and so on.}$$

Now, let's look at the cases where $r \neq \pm 1$. The Nth partial sum is

$$S_N = a + ar + ar^2 + \cdots + ar^N.$$

Multiplying both sides by r, we obtain

$$rS_N = ar + ar^2 + ar^3 + \cdots + ar^{N+1},$$

and then subtracting the two sides from the original equation for S_N, we have

$$S_N - rS_N = a - ar^{N+1}.$$

We can solve for S_N, provided $r \neq 1$:

$$S_N = \frac{a(1 - r^{N+1})}{(1 - r)}.$$

11.1 SERIES

With this closed form for S_N we can determine the limit of the geometric series

$$\sum_{n=0}^{\infty} ar^n = \lim_{n \to \infty} \frac{a(1 - r^{N+1})}{1 - r} \qquad (r \neq 1).$$

If $r > 1$ or $r < -1$, then r^{N+1} will become larger and larger in magnitude and the limit will not exist.

If $-1 < r < 1$ then $r^{N+1} \to 0$ as $N \to \infty$ and

$$\sum_{n=0}^{\infty} ar^n = \frac{a(1 - 0)}{1 - r} = \frac{a}{1 - r}.$$

Summarizing all the cases, we have the following theorem.

Theorem 11.1

If $a \neq 0$, then the geometric series $\sum_{n=0}^{\infty} ar^n$ *diverges* if $r \geq 1$ or $r \leq -1$ and *converges* to $\frac{a}{1 - r}$ if $-1 < r < 1$.

□

EXAMPLE 9 Does the geometric series

$$7 + \frac{7}{5} + \frac{7}{25} + \frac{7}{125} + \cdots$$

converge or diverge? If the series converges, what is the sum?

Solution This series can also be written in summation notation as

$$\sum_{n=0}^{\infty} 7(\frac{1}{5})^n.$$

The first term is $a = 7$, and the factor is $r = 1/5$. Since $-1 < 1/5 < 1$, we know that the series must *converge*, and

$$\sum_{n=0}^{\infty} 7(\frac{1}{5})^n = \frac{7}{1 - 1/5} = \frac{7}{4/5} = \frac{35}{4}.$$

■

EXAMPLE 10 Does the series

$$\sum_{n=0}^{\infty}(-3)^{1-n}$$

converge or diverge? If the series converges, what is the sum?

Solution If we write out the first few terms of the series, we may more easily recognize that it is a geometric series:

$$-3+1-\frac{1}{3}+\frac{1}{9}-\frac{1}{27}+\frac{1}{81}\cdots,$$

where the first term is $a = -3$, and the factor is $r = -1/3$.

Since $-1 < -1/3 < 1$, we know that the series must *converge*, and

$$\sum_{n=0}^{\infty}(-3)^{1-n} = \frac{-3}{1-(-1/3)} = \frac{-3}{4/3} = \frac{-9}{4}.$$

∎

EXAMPLE 11 Does the series $.01 + .02 + .04 + .08 + \ldots$ converge or diverge? If the series converges, what is the sum?

Solution This is a geometric series with first term $a = .01$ and factor $r = 2$ (the terms double at each step). Since $r \geq 1$, the series *diverges*. ∎

EXAMPLE 12 Does the series $\sum_{n=2}^{\infty} \frac{5}{4^n}$ converge or diverge? If the series converges, what is the sum?

Solution Note that the starting index is $n = 2$. If we write out the first few terms of the series, we have

$$\frac{5}{4^2}+\frac{5}{4^3}+\frac{5}{4^4}+\cdots$$

so the first term is $a = 5/16$ and the factor is $r = 1/4$.

Since $-1 < 1/4 < 1$, the series *converges*, and

$$\sum_{n=2}^{\infty}\frac{5}{4^n} = \frac{5/16}{1-1/4} = \frac{5}{12}.$$

∎

11.1 SERIES

 Be careful to distinguish between the limit of a *sequence* $\{a_n\}_{n=1}^\infty$ **and the the associated** *series* $\sum_{n=1}^\infty a_n$. **It is quite possible that the sequence of individual** *terms* **converges to 1, while the sequence of** *partial sums* **diverges.**

A simple example of this is illustrated by the constant sequence

$$1, 1, 1, 1, \ldots$$

which certainly converges, while the associated series

$$1 + 1 + 1 + 1 + 1 + 1 + \ldots$$

diverges (by the Archimedean property of real numbers).

Harmonic series

The **harmonic series** is obtained by summing the terms of the harmonic sequence:

$$\sum_{n=1}^\infty \frac{1}{n} = 1 + \frac{1}{2} + \frac{1}{3} + \frac{1}{4} + \ldots$$

While the *harmonic sequence converges to* 0, the *harmonic series diverges*. One way to see this is to look at the partial sums $S_1, S_2, S_4, S_8, S_{16}, \cdots, S_{2^n}$. The first few of these partial sums are

$$S_1 = 1$$

$$S_2 = 1 + \frac{1}{2} = 1.5$$

$$S_4 = 1 + \frac{1}{2} + \frac{1}{3} + \frac{1}{4} > 2$$

$$S_8 = 1 + \frac{1}{2} + \frac{1}{3} + \frac{1}{4} + \frac{1}{5} + \frac{1}{6} + \frac{1}{7} + \frac{1}{8} > 2.5$$

$$S_{16} = 1 + \cdots + \frac{1}{16} > 3.$$

Each time we double the number of terms in the partial sum, we have added at least 1/2 to the running total. This means that

$$S_{2^n} = 1 + \frac{1}{2} + \frac{1}{3} + \cdots + \frac{1}{2^n} > 1 + \frac{n}{2}.$$

The partial sums will grow larger and larger without bound, since we can keep doubling the number of terms indefinitely. Hence, the harmonic series *diverges*.

The **alternating harmonic series** is obtained by summing the terms of the alternating harmonic sequence:

$$\sum_{n=1}^{\infty} \frac{(-1)^{n+1}}{n} = 1 - \frac{1}{2} + \frac{1}{3} - \frac{1}{4} + \frac{1}{5} - \cdots$$

It is unclear whether or not the alternating harmonic series converges or diverges from looking at the first few partial sums. Interestingly, it is known that the alternating harmonic series *converges* to the limit $L = \ln(2)$.

Properties of series

Because a series is the limit of a sequence of partial sums, some of the limit properties of sequences correspond to properties of series.

If $\sum a_n = L_1$ and $b_n = L_2$, then all of the following are true:

$$\sum (a_n + b_n) = L_1 + L_2$$

$$\sum (a_n - b_n) = L_1 - L_2$$

$$c \sum a_n = cL_1 \quad (c \text{ a constant}).$$

If $\sum a_n$ diverges and b_n converges, then

$$\sum (a_n + b_n) \quad \text{diverges}$$

$$\sum (a_n - b_n) \quad \text{diverges}$$

$$c \sum a_n \quad \text{diverges if } c \neq 0.$$

EXERCISES

Find the sums indicated in exercises 1-8.

1. $\sum_{n=1}^{10} 2^n$

2. $\sum_{n=0}^{5} 2n^2 - 3n + 1$

3. $\sum_{i=1}^{4} i^i$

4. $\sum_{k=0}^{5} 2k^2 - 3k + k^3$

5. $\sum_{n=0}^{100} (-1)^n$

6. $\sum_{i=0}^{5} 2i^2 - 3i + i^3$

7. $\sum_{n=1}^{17} 5$

8. $\sum_{n=1}^{1000} n$

(Hint: $1 + 1000 = 1001$, $2 + 999 = 1001$, $3 + 998 = 1001$, ...)

11.1 SERIES

In exercises 9–24, determine whether the given series converges or diverges. If the series converges, find the sum.

9. $\sum_{n=1}^{\infty} (-1/3)^n$

10. $\sum_{n=0}^{\infty} (1/3)^n$

11. $\sum_{n=1}^{\infty} e^{-n}$

12. $\sum_{n=0}^{\infty} \frac{3^n + 2^n}{4^n}$

13. $\sum_{n=1}^{\infty} (-1)^n$

14. $\sum_{n=0}^{\infty} 1/2$

15. $\frac{1}{11} - \frac{10}{11^2} + \frac{100}{11^3} - \frac{1000}{11^4} \cdots$

16. $\sum_{n=1}^{\infty} n$

17. $.001 - .003 + .009 - .027 \ldots$

18. $\sum_{n=1}^{\infty} \frac{2}{n}$

19. $\sum_{n=2}^{\infty} e^n / 3^{n-2} \ldots$

20. $\sum_{n=1}^{\infty} (-1)^n / 2n$

21. $\sum_{n=3}^{\infty} \frac{1}{n-1} - \frac{1}{n+1}$

22. $\sum_{n=1}^{\infty} \frac{3}{2n-1} - \frac{3}{2n+1}$

23. $\sum_{n=1}^{\infty} \frac{1}{3^{n+1}} - \frac{1}{3^n}$

24. $\sum_{n=0}^{\infty} \sin(\pi n)$

---

25. Find two different geometric series, both of which converge to 17/3.

26. If a geometric series converges to 1, then the sum of its first term a and the factor r is 1. Why? Does the converse hold?

27. Find the first ten partial sums of $\sum_{n=0}^{\infty} 1/n!$, where $n! = 1 \cdot 2 \cdot 3 \cdots n$ for $n \geq 1$ and $0! = 1$. Does the apparent limiting value look familiar?

28. Find S_{10}, S_{100}, and S_{1000} for the alternating harmonic series and compare these values to $\ln(2)$.

29. A ball is dropped from a height of six feet and begins bouncing up and down. The height of each bounce is 3/4 the height of the previous bounce. Find the total vertical distance travelled by the ball.

30. Let

$$c_n = 1 + \frac{1}{2} + \frac{1}{3} + \frac{1}{4} + \cdots + \frac{1}{n} - \ln(n).$$

Find c_{40}, c_{50}, and c_{60}. The sequence $\{c_n\}$ converges to a number known as **Euler's constant** $\gamma \approx 0.5772$.

---

31. Cantor's middle-third set, named for the mathematician Georg Cantor (1835-1918) is constructed as follows:

First, the open interval $(\frac{1}{3}, \frac{2}{3})$ is "erased" from the closed interval $[0, 1]$. This leaves two closed intervals $[0, \frac{1}{3}]$ and $[\frac{2}{3}, 1]$.

Next, the middle third of each of these two remaining closed intervals is erased. That is, the open intervals $(\frac{1}{9}, \frac{2}{9})$ and $(\frac{7}{9}, \frac{8}{9})$ are erased, leaving four closed intervals $[0, \frac{1}{9}]$, $[\frac{2}{9}, \frac{1}{3}]$, $[\frac{2}{3}, \frac{7}{9}]$, and $[\frac{8}{9}, 1]$.

We continue in the same way, erasing the open middle third of each remaining closed subinterval to obtain eight new closed subintervals. The illustration below shows the original interval $[0, 1]$ and what remains after each of the first three stages.

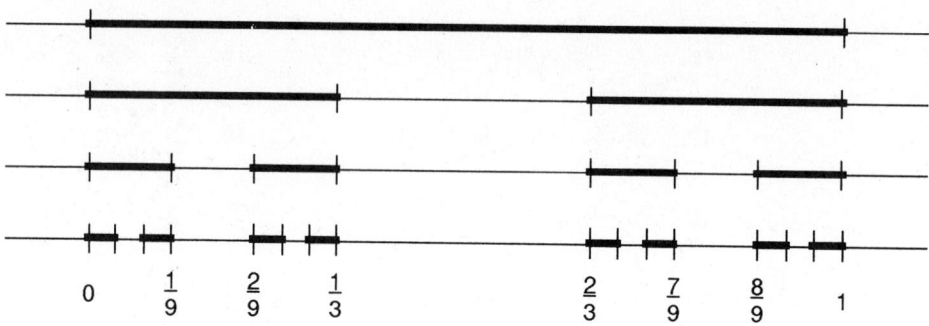

We continue this process indefinitely, at each stage erasing the open middle third of each remaining closed subinterval. Find the total length of all the subintervals erased (use a geometric series). Give an example of a point that is *never* erased. How many points are left in the Cantor set?

11.2 CONVERGENCE TESTS FOR SERIES

It may be difficult to find a closed form that describes the Nth partial sum $S_N = \sum_{n=1}^{N} a_n$ of a series $\sum_{n=1}^{\infty} a_n$. With the use of computers or calculators, we can certainly calculate many of these partial sums quickly and for large index values N to get a feel for the behavior of the series and to approximate its limit if it converges.

However, note that any partial sum will have a finite sum (perhaps large), so the convergence or divergence of a series may be difficult to judge just by the values of these partial sums. For example, the harmonic series

$$\sum_{n=1}^{\infty} \frac{1}{n}$$

diverges, but very slowly. Indeed, the harmonic series will appear to converge if we calculate the partial sums using a machine with fixed precision.

If we could determine in advance that a given series converges, then we may be able to approximate its limit closely with a partial sum. There

11.2 CONVERGENCE TESTS FOR SERIES

are a variety of these **convergence tests** for series. Sometimes a convergence test can also give us information regarding how many terms are necessary to include in a partial sum to achieve a given level of accuracy. In this section and the next, we'll discuss several of these convergence tests.

For each of the tests we describe, it is important to pay special attention to

1) the requirements that must be fulfilled for the test to be applied,

2) the criteria for making a decision regarding convergence or divergence,

3) any accuracy estimates possible for approximating the sum of a convergent series.

For each of the tests described in this section, we'll provide some discussion of how the test works, and illustrate it with some examples.

Nth term test

Requirements:

The N**th term test** can be applied to any series.

Criteria:

If $\lim_{n \to \infty} a_n \neq 0$, then $\sum a_n$ diverges.

If $\lim_{n \to \infty} a_n = 0$, then the Nth term test provides *no information*.

Discussion:

For the partial sums to converge to a single limiting value, the individual terms a_n must approach 0. (A convergent sequence of partial sums must be a Cauchy sequence, and the individual terms a_n give us the differences between consecutive partial sums.) The Nth term test simply relies on this observation.

EXAMPLE 13 $\sum_{n=1}^{\infty} n$ *diverges* since $\lim_{n \to \infty} n = \infty \neq 0$. ■

EXAMPLE 14 $\sum_{n=1}^{\infty} (-1)^n$ *diverges* since $\lim_{n \to \infty} (-1)^n$ does not exist.

Note that $(-1)^n$ oscillates between 1 and -1 depending on whether n is even or odd. ■

EXAMPLE 15 $\sum_{n=1}^{\infty} (1 - \frac{1}{n})$ *diverges* since $\lim_{n \to \infty} (1 - \frac{1}{n}) = 1 \neq 0$. ■

EXAMPLE 16 The Nth term test provides no information regarding the harmonic series

$$\sum_{n=1}^{\infty} \frac{1}{n}$$

since $\lim_{n\to\infty} \frac{1}{n} = 0$. However, we do know that this series diverges. ∎

EXAMPLE 17 The Nth term test provides no information regarding the geometric series

$$\sum_{n=1}^{\infty} \frac{1}{2^n}$$

since $\lim_{n\to\infty} \frac{1}{2^n} = 0$. However, we do know this series *converges* to 1. ∎

As the last two examples show, $a_n \to 0$ is a *necessary* but not *sufficient* condition for the series $\sum a_n$ to converge. The Nth term test can only help us determine that a series diverges—it can never tell us that a series converges.

Integral Test

Requirements:

The **integral test** can be applied to any series whose terms $a_n = f(n)$, where f can be considered a *continuous*, *positive*, *decreasing* function of a real variable x.

Criteria:

If $\int_1^{\infty} f(x)\,dx$ converges, then $\sum_{n=1}^{\infty} a_n$ converges.

If $\int_1^{\infty} f(x)\,dx$ diverges, then $\sum_{n=1}^{\infty} a_n$ diverges.

Discussion:

The integral test allows us to decide the convergence or divergence of a series by comparing it to an improper integral. If all the terms a_n in

$$\sum_{n=1}^{\infty} a_n$$

are positive, we can represent them as the *areas* of rectangles setting atop the x-axis (see Figure 11.2). The base of each rectangle is 1, so the height of the nth rectangle must be a_n.

11.2 CONVERGENCE TESTS FOR SERIES

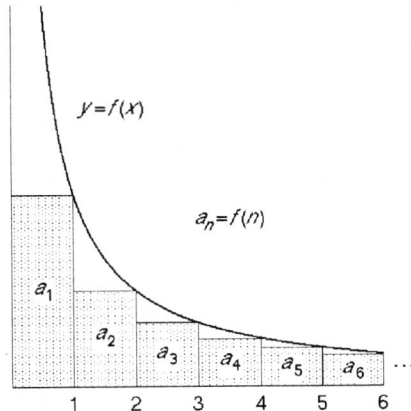

Figure 11.2 If $\int_1^\infty f(x)\,dx$ is finite, so is the total area $\sum \int_1^\infty a_n$.

If $a_n = f(n)$, then the graph $y = f(x)$ passes through the upper right corner of each rectangle. If the function f is decreasing, then all of the rectangles starting with the second one fit *under* the graph. Hence, if the improper integral

$$\int_1^\infty f(x)\,dx = L,$$

then

$$\sum_{n=1}^\infty a_n \leq L + a_1.$$

Since each term a_n makes a positive contribution to the total, the partial sums S_N are increasing as $N \to \infty$. Thus, the partial sums form a *monotonically increasing bounded sequence*, which must converge.

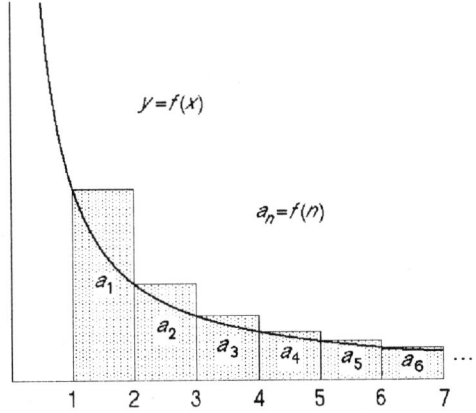

Figure 11.3 If $\int_1^\infty f(x)\,dx$ is infinite, so is the total area $\sum \int_1^\infty a_n$.

Now, suppose we shifted all our rectangles one unit to the right, as in Figure 11.3. Now the graph $y = f(x)$ passes through all the upper left corners of the rectangles and

$$\int_1^\infty f(x)\,dx \leq \sum_{n=1}^\infty a_n.$$

Hence, if $\int_1^\infty f(x)\,dx$ *diverges*, then so must $\sum_{n=1}^\infty a_n$.

EXAMPLE 18 Use the integral test to test the convergence of the harmonic series $\sum_{n=1}^\infty \frac{1}{n}$.

Solution The integral test can be applied because $f : x \longmapsto 1/x$ is positive, continuous and decreasing over the interval $[1, \infty)$.

$$\int_1^\infty \frac{1}{x}\,dx = \lim_{b \to \infty} \int_1^b \frac{1}{x}\,dx = \lim_{b \to \infty} \ln|b| = \infty$$

This improper integral diverges, so the harmonic series $\sum_{n=1}^\infty \frac{1}{n}$ diverges. ∎

EXAMPLE 19 Use the integral test to test the convergence of the series $\sum_{n=1}^\infty \frac{1}{n^2}$.

Solution The integral test can be applied because $f : x \longmapsto 1/x^2$ is positive, continuous and decreasing over the interval $[1, \infty)$.

$$\int_1^\infty \frac{1}{x^2}\,dx = \lim_{b \to \infty} \int_1^b \frac{1}{x^2}\,dx = \lim_{b \to \infty} \frac{-1}{b} + 1 = 1.$$

This improper integral converges, so the series $\sum_{n=1}^\infty \frac{1}{n^2}$ converges. (Note: The *series* does not converge to 1, however.) ∎

In the section on improper integrals, we noted that

$$\int_1^\infty \frac{1}{x^p}\,dx$$

converges for $p > 1$ and diverges for $p \leq 1$.

We can apply the integral test to "p-series" of the form

$$\sum_{n=1}^\infty \frac{1}{n^p}$$

11.2 CONVERGENCE TESTS FOR SERIES

and conclude that $\sum_{n=1}^{\infty} \frac{1}{n^p}$ converges for $p > 1$ and diverges for $p \leq 1$.

EXAMPLE 20 The three series

$$\sum_{n=1}^{\infty} \frac{1}{n^3}, \quad \sum_{n=1}^{\infty} \frac{1}{n^{3/2}}, \quad \sum_{n=1}^{\infty} \frac{1}{n^{1.01}}$$

all *converge* ($p = 3$, $p = 3/2$, $p = 1.01$ respectively).

On the other hand, the three series

$$\sum_{n=1}^{\infty} \frac{1}{\sqrt{n}}, \quad \sum_{n=1}^{\infty} \frac{1}{\sqrt[3]{n}}, \quad \sum_{n=1}^{\infty} \frac{1}{n^{0.99}}$$

all *diverge* ($p = \frac{1}{2}$, $p = \frac{1}{3}$, $p = 0.99$ respectively). ∎

EXAMPLE 21 The integral test cannot be applied to the alternating harmonic series

$$\sum_{n=1}^{\infty} \frac{(-1)^{n+1}}{n}$$

because the terms are not all positive (check the requirements for the integral test). ∎

Integral Test Estimate:

When the integral test indicates that a series $\sum_{n=1}^{\infty} a_n$ converges, we can get an estimate of how good an approximation the partial sum

$$S_N = \sum_{n=1}^{N} a_n = a_1 + a_2 + \cdots + a_N$$

is to the actual sum of the series. Let's use the notation R_N to indicate the remaining "tail" of the series:

$$R_N = \sum_{n=N+1}^{\infty} a_n = a_{N+1} + a_{N+2} + \cdots$$

so that

$$S_N + R_N = \sum_{n=1}^{\infty} a_n.$$

From Figure 11.2 we can see that

$$R_N \leq \int_N^{\infty} f(x)\, dx.$$

When we shift the rectangles one unit to the right (Figure 11.3), we can also see that

$$\int_{N+1}^{\infty} f(x)\,dx \leq R_N.$$

Combining these together means that we have "trapped" the remainder tail of the series:

$$\int_{N+1}^{\infty} f(x)\,dx \leq R_N \leq \int_{N}^{\infty} f(x)\,dx.$$

If we use the lower bound $\int_{N+1}^{\infty} f(x)$ as an approximation for R_N, then the worst error possible is the difference between the lower bound and upper bound:

$$\int_{N}^{\infty} f(x)\,dx - \int_{N+1}^{\infty} f(x)\,dx = \int_{N}^{N+1} f(x)\,dx.$$

This all means that

$$\sum_{n=1}^{\infty} a_n \approx \sum_{n=1}^{N} a_n + \int_{N+1}^{\infty} f(x)\,dx$$

where the estimate on the right is *under* the true value of the sum by

$$\text{error} \leq \int_{N}^{N+1} f(x)\,dx.$$

EXAMPLE 22 Estimate $\sum_{n=1}^{\infty} \dfrac{1}{n^2}$ to within .0001 of the true value of its sum.

Solution   The approximation we need is

$$\sum_{n=1}^{\infty} \frac{1}{n^2} \approx \sum_{n=1}^{N} \frac{1}{n^2} + \int_{N+1}^{\infty} \frac{1}{x^2}\,dx.$$

We need to choose N large enough so that the

$$\text{error} \leq .0001.$$

Now,

$$\int_{N}^{N+1} \frac{1}{x^2}\,dx = \left.\frac{-1}{x}\right]_{N}^{N+1} = \frac{-1}{N+1} + \frac{1}{N} = \frac{1}{N(N+1)},$$

and when $N = 100$, we have

$$\frac{1}{N(N+1)} = \frac{1}{(100)(101)} < .0001.$$

11.2 CONVERGENCE TESTS FOR SERIES

This indicates that we can use $N = 100$ to make our estimate:

$$\sum_{n=1}^{\infty} \frac{1}{n^2} \approx \sum_{n=1}^{100} \frac{1}{n^2} + \int_{101}^{\infty} \frac{1}{x^2}\, dx.$$

We use a calculator to compute

$$\sum_{n=1}^{100} \frac{1}{N^2} \approx 1.63498$$

to five decimal places, and

$$\int_{101}^{\infty} \frac{1}{x^2}\, dx = \lim_{b \to \infty} \frac{-1}{b} + \frac{1}{101} = \frac{1}{101} \approx .0099.$$

Thus,

$$\sum_{n=1}^{\infty} \frac{1}{n^2} \approx 1.64488,$$

accurate to within .0001. ∎

Sometimes we can judge the convergence or divergence of one series by comparing it to another series whose behavior is known. The next two tests are of this type.

Comparison Test

Requirements:

The two series $\sum a_n$ and $\sum b_n$ must each have nonnegative terms. That is,

$$a_n \geq 0, \quad b_n \geq 0 \quad \text{for all } n.$$

Criteria:

If $a_n \leq b_n$ for each n, and $\sum b_n$ converges, then $\sum a_n$ converges.

If $b_n \leq a_n$ for each n, and $\sum b_n$ diverges, then $\sum a_n$ diverges.

Discussion:

The requirement that the terms a_n and b_n all be nonnegative guarantees that the partial sums of both series are *monotonically increasing*, and hence either series will converge *if and only if* its partial sums are *bounded*.

The way this test is used in practice is as follows: If the behavior of a given series

$$\sum a_n$$

is unknown and each term $a_n \geq 0$, then we search for (or devise) a comparison series $\sum b_n$ with $b_n \geq 0$.

If we suspect that $\sum a_n$ converges, then we search for a "larger" series $\sum b_n$ (in other words, $b_n \geq a_n$ for each n) that we know converges. If we suspect that $\sum a_n$ diverges, then we search for a "smaller" series $\sum b_n$ (in other words, $b_n \leq a_n$ for each n) that we know diverges. In either case, success in finding the series $\sum b_n$ with the property we desire means that we can use the comparison test to deduce the convergence or divergence of the original series $\sum a_n$.

What are good series $\sum b_n$ to look for? In general, any series whose behavior you know is a candidate. Geometric series and p-series are nice because they provide us with a large storehouse of examples of both convergent and divergent series.

How do you choose a particular series $\sum b_n$? If you're looking for a larger series, try making some simplifying changes to the terms a_n that will result in larger terms (for example: decreasing denominators, increasing numerators, adding a positive amount, etc.). If you're looking for a smaller series, try making simplifying changes that will result in smaller terms (increasing denominators, decreasing numerators, subtracting a positive amount, etc.).

EXAMPLE 23 Does the series $\sum_{n=1}^{\infty} \frac{2^n}{3^n + n^4}$ converge or diverge?

Solution To get a feel for the series, we might write out the first few terms in the sum:

$$\frac{2}{5} + \frac{4}{25} + \frac{8}{108} + \frac{16}{337} + \cdots$$

These terms appear to be decreasing in size quite quickly, and the form of the series suggests we try comparing it to a geometric series. Since we suspect that the series converges, we look for a larger series that we know converges.

If we remove n^4 from the denominator then we certainly have

$$a_n = \frac{2^n}{3^n + n^4} \leq \frac{2^n}{3^n} = b_n$$

for each n.

11.2 CONVERGENCE TESTS FOR SERIES

The geometric series $\sum_{n=1}^{\infty}(\frac{2}{3})^n$ converges $(-1 < r = 2/3 < 1)$, so

$$\sum_{n=1}^{\infty} \frac{2^n}{3^n + n^4}$$

converges by the comparison test. ∎

EXAMPLE 24 Does $\sum_{n=1}^{\infty} \frac{1}{n + \sqrt{n}}$ converge or diverge?

Solution If we remove $\sqrt{n}$ from the denominator, we get a larger series

$$\sum_{n=1}^{\infty} \frac{1}{n}$$

that diverges. Notice, though, that this is of no help to us. (Similarly, finding a smaller series that converges will tell us no information about the original series). On the other hand, if we replace $\sqrt{n}$ by n we obtain a *smaller* series

$$\sum_{n=1}^{\infty} \frac{1}{n + n} = \sum_{n=1}^{\infty} \frac{1}{2n} = \frac{1}{2} \sum_{n=1}^{\infty} \frac{1}{n}$$

that we know diverges. We can conclude that $\sum_{n=1}^{\infty} \frac{1}{n + \sqrt{n}}$ *diverges* by the comparison test. ∎

EXAMPLE 25 The comparison test cannot be used on the alternating harmonic series

$$\sum_{n=1}^{\infty} \frac{(-1)^{n+1}}{n}$$

because some of its terms are negative. ∎

Since any partial sum S_N is necessarily finite (adding finitely many real numbers always yields a finite result), it is always the remainder tail R_N that decides whether or not a series converges.

This means that the requirements and criteria of the comparison test need only be satisfied by all the terms in a common remainder tail of the two series being compared.

EXAMPLE 26   Consider the two series

$$\sum_{n=1}^{\infty} \frac{1}{2^n + 1} = \frac{1}{3} + \frac{1}{5} + \frac{1}{9} + \frac{1}{17} + \cdots$$

and

$$\sum_{n=1}^{\infty} \frac{1}{n^2 - 2} = -1 + \frac{1}{2} + \frac{1}{7} + \frac{1}{14} + \cdots .$$

Note that the first term of the second series is negative and smaller than the first term of the first series. However, starting with the next term, the second series has consistently larger terms. This means that the comparison test can be used with these two series. That is, if the first diverges, so does the second. If the second converges, so does the first. (By the way, which is it?) ∎

Limit Comparison Test

Requirements:

Both series $\sum a_n$ and $\sum b_n$ must have positive terms

$$a_n, b_n > 0$$

for each n.

Criteria:

If $\lim_{n\to\infty} \frac{a_n}{b_n} = \infty$ and $\sum b_n$ diverges, then $\sum a_n$ diverges.

If $\lim_{n\to\infty} \frac{a_n}{b_n} = 0$ and $\sum b_n$ converges, then $\sum a_n$ converges.

If $\lim_{n\to\infty} \frac{a_n}{b_n} = L \neq 0$ (L a real number), then $\sum a_n$ and $\sum b_n$ behave the same. That is, either both series converge or both diverge.

Discussion:

Recall that a necessary (but not sufficient) condition for a series to converge is that the individual terms must approach 0 as $n \to \infty$ (the Nth term test). Even if the terms do approach 0, they might not approach 0 "fast enough" for the series to converge (the harmonic series is the best example of this phenomenon).

The limit comparison test looks at the ratio of the terms $\frac{a_n}{b_n}$ from two series. If both $a_n \to 0$ and $b_n \to 0$, then

$$\lim_{n\to\infty} \frac{a_n}{b_n}$$

gives us a measure of the relative speeds that a_n and b_n approach 0.

11.2 CONVERGENCE TESTS FOR SERIES

If $\lim_{n\to\infty} \dfrac{a_n}{b_n} = \infty$ then $a_n \to 0$ much "slower" than $b_n \to 0$.

If $\lim_{n\to\infty} \dfrac{a_n}{b_n} = 0$ then $a_n \to 0$ much "faster" than $b_n \to 0$.

If $\lim_{n\to\infty} \dfrac{a_n}{b_n} = L \neq 0$, then $a_n \to 0$ and $b_n \to 0$ at "comparable" speeds.

Like the comparison test, the limit comparison test is used to decide the divergence or convergence of a series $\sum a_n$ by choosing or devising a comparison sequence $\sum b_n$ whose behavior is known. If a_n has a closed-form formula, then a good choice for b_n is to use only the dominant (fastest growing) parts of the formula. Constant coefficients can usually be changed to $+1$ or -1 depending on the sign.

EXAMPLE 27 Does $\displaystyle\sum_{n=1}^{\infty} \dfrac{6}{\sqrt{17n^3+5}}$ converge or diverge?

Solution The fastest growing term in the denominator under the radical sign is $17n^3$. We'll choose $b_n = \dfrac{1}{\sqrt{n^3}}$, changing the coefficients 6 and 17 to 1.

Now,

$$\lim_{n\to\infty} \frac{a_n}{b_n} = \lim_{n\to\infty} \frac{\dfrac{6}{\sqrt{17n^3+5}}}{\dfrac{1}{\sqrt{n^3}}} = \lim_{n\to\infty} \frac{6\sqrt{n^3}}{\sqrt{17n^3+5}} = \lim_{n\to\infty} 6\sqrt{\frac{n^3}{17n^3+5}}.$$

If we use L'Hôpital's Rule we can calculate

$$\lim_{n\to\infty} \frac{n^3}{17n^3+5} = \frac{1}{17}.$$

Hence,

$$\lim_{n\to\infty} \frac{a_n}{b_n} = \frac{6}{\sqrt{17}} \neq 0.$$

The limit comparison test tells us that both $\sum a_n$ and $\sum b_n$ have the same behavior. Since

$$\sum_{n=1}^{\infty} b_n = \sum_{n=1}^{\infty} \frac{1}{\sqrt{n^3}} = \sum_{n=1}^{\infty} \frac{1}{n^{3/2}}$$

converges, we conclude that $\displaystyle\sum_{n=1}^{\infty} \dfrac{6}{\sqrt{17n^3+5}}$ also *converges*. ∎

EXAMPLE 28 Again, the limit comparison test cannot be used on the alternating harmonic series

$$\sum_{n=1}^{\infty} \frac{(-1)^{n+1}}{n}$$

because some of the terms are negative. ∎

Alternating Series Test

Requirements:

The terms in the series are alternately positive and negative. Such a series can be written

$$\sum_{n=1}^{\infty} (-1)^{n+1} c_n \qquad (c_n > 0)$$

if the first term is positive, or

$$\sum_{n=1}^{\infty} (-1)^{n} c_n \qquad (c_n > 0)$$

if the first term is negative.

Criteria:

If the magnitudes of the terms are strictly decreasing, so that

$$c_n > c_{n+1} > 0$$

for every n, and if

$$\lim_{n \to \infty} c_n = 0,$$

then the alternating series *converges*.

If $\lim_{n \to \infty} c_n \neq 0$, then the series *diverges* by the Nth term test.

11.2 CONVERGENCE TESTS FOR SERIES

EXAMPLE 29 Does the alternating harmonic series $\sum_{n=1}^{\infty} \frac{(-1)^{n+1}}{n}$ converge or diverge?

Solution Here $c_n = \frac{1}{n}$. The alternating series test applies since the terms alternate signs and

$$\frac{1}{n} > \frac{1}{n+1} > 0$$

for every n.

Since $\lim_{n \to \infty} \frac{1}{n} = 0$, the alternating series test tells us that the alternating harmonic series *converges*. ∎

Discussion:

Let's compute the first few partial sums of the alternating harmonic series:

$$S_1 = 1$$

$$S_1 = 1 - \frac{1}{2} = \frac{1}{2}$$

$$S_3 = 1 - \frac{1}{2} + \frac{1}{3} = \frac{5}{6}$$

$$S_4 = 1 - \frac{1}{2} + \frac{1}{3} - \frac{1}{4} = \frac{7}{12}$$

$$S_5 = 1 - \frac{1}{2} + \frac{1}{3} - \frac{1}{4} + \frac{1}{5} = \frac{47}{60}$$

$$S_6 = 1 - \frac{1}{2} + \frac{1}{3} - \frac{1}{4} + \frac{1}{5} - \frac{1}{6} = \frac{37}{60}.$$

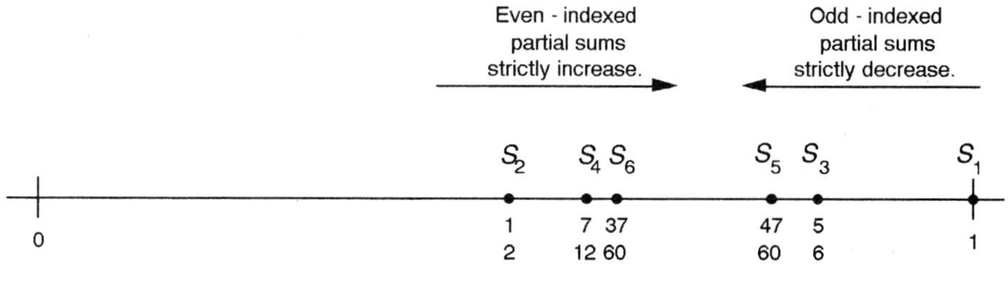

Figure 11.4 Partial sums of $\sum_{n=1}^{\infty}(-1)^n/n$.

If we plot these on a number line, we see that the even partial sums $S_2, S_4, S_6, \ldots$ strictly increase while the odd partial sums strictly decrease (see Figure 11.4).

A similar configuration of positions for partial sums will be the case for any alternating series satisfying the criteria of the alternating series test. From the starting point S_1 we take a sequence of smaller and smaller steps, reversing direction each time. At any point in our "trip," all future positions will be between our last two consecutive positions. Since the distance between our last two positions S_{n-1} and S_n is simply the size of our last step c_n, and $c_n \to 0$, we must converge on a single point.

Alternating Series Estimate:

This also means that we have a definite idea of how close we are to the limit L after any step, because the limit must be between our present position and our next position. In other words, L is between S_n and S_{n+1} for every n, and if we use S_n as an approximation for L, then the remainder has magnitude

$$|R_n| < c_{n+1}.$$

We can tell whether S_n is on the right or left of the limit by the sign of the $(n+1)$st term.

EXAMPLE 30 Find the limit of the alternating harmonic series within .001.

Solution We choose $N = 1000$ so that

$$c_{N+1} = \frac{1}{N+1} = \frac{1}{1001} < .001.$$

Now we compute

$$\sum_{n=1}^{1000} \frac{(-1)^{n+1}}{n} \approx 0.69265$$

(to five decimal places). We mentioned that it is known that

$$\sum_{n=1}^{\infty} \frac{(-1)^{n+1}}{n} = \ln(2).$$

We note that $\ln(2) \approx 0.69315$ to five decimal places. ∎

Absolute and conditional convergence

When a series has both positive and negative terms, as is the case with an alternating series, the notions of *absolute convergence* and *conditional convergence* become relevant.

11.2 CONVERGENCE TESTS FOR SERIES

Definition 1

Suppose a series $\sum a_n$ converges.
If the series $\sum |a_n|$ converges, we say $\sum a_n$ is **absolutely convergent**.
If the series $\sum |a_n|$ diverges, we say $\sum a_n$ is **conditionally convergent**.

EXAMPLE 31 The alternating harmonic series is conditionally convergent because

$$\sum_{n=1}^{\infty} \frac{(-1)^{n+1}}{n}$$

converges, but

$$\sum_{n=1}^{\infty} \left| \frac{(-1)^{n+1}}{n} \right| = \sum_{n=1}^{\infty} \frac{1}{n}$$

diverges. ■

EXAMPLE 32 The geometric series

$$\sum_{n=1}^{\infty} (-1/2)^n$$

is absolutely convergent because

$$\sum_{n=1}^{\infty} |(-1/2)^n| = \sum_{n=1}^{\infty} (1/2)^n$$

converges. ■

A useful result is the following theorem.

Theorem 11.2

Hypothesis: $\sum |a_n|$ converges. **Conclusion:** $\sum a_n$ converges.

In other words, if we replace each term in a series by its absolute value and the resulting series converges, then the original series must converge. In fewer words, *an absolutely convergent series converges.*

Reasoning Suppose we have a series $\sum a_n$ and we know that
$$\sum |a_n|$$
converges. If we let $b_n = a_n + |a_n|$, we'll either get $b_n = 0$ (when a_n is negative) or $b_n = 2|a_n|$ (when a_n is positive). Either way, we can say
$$0 \leq b_n = a_n + |a_n| \leq 2|a_n|.$$

Using the comparison test, we can see that $\sum b_n$ converges because $\sum 2|a_n|$ converges. Finally, that means $\sum a_n = \sum (b_n - |a_n|)$ must converge also. □

This theorem is useful because we have several tests requiring all our terms to be nonnegative (the integral test, the comparison test, and the limit comparison test). If we have a series $\sum a_n$ not satisfying this requirement, we can try replacing a_n by $|a_n|$. All the terms of $\sum |a_n|$ are nonnegative, so there are more tests that are available to use. If we find $\sum |a_n|$ converges, then this theorem tells us that $\sum a_n$ must also converge. However, if we find $\sum |a_n|$ diverges, we really don't have any more information about $\sum a_n$, because it could be conditionally convergent.

EXERCISES

1. Decide whether $\sum_{n=1}^{\infty} \dfrac{1}{3+5^n}$ converges or diverges by using the comparison test with the geometric series $\sum_{n=1}^{\infty} \dfrac{1}{5^n}$.

2. Decide whether $\sum_{n=1}^{\infty} \dfrac{e^{1/n}}{n}$ converges or diverges by using the limit comparison test with $\sum_{n=1}^{\infty} \dfrac{1}{n}$.

Apply the alternating series test to each of the series in exercises 3-6.

3. $\sum_{n=1}^{\infty} (-1)^n \ln(1 + \dfrac{1}{n})$
4. $\sum_{n=1}^{\infty} (-1)^n \dfrac{\ln n}{n}$
5. $\sum_{n=1}^{\infty} (-1)^n \dfrac{\tan^{-1} n}{n^3}$
6. $\sum_{n=1}^{\infty} \dfrac{(-1^n)n}{2n^2 - 1}$

Apply the integral test to each of the series in exercises 7-10.

7. $\sum_{n=1}^{\infty} 3e^{-n}$
8. $\sum_{n=1}^{\infty} ne^{-n}$

11.2 CONVERGENCE TESTS FOR SERIES

9. $\sum_{n=1}^{\infty} \dfrac{\ln n}{n}$

10. $\sum_{n=1}^{\infty} \dfrac{1}{4n^2 + 9}$

Why doesn't the integral test apply to the series in exercises 11 and 12?

11. $\sum_{n=1}^{\infty} e^{-n} \sin n$

12. $\sum_{n=1}^{\infty} \dfrac{(-1)^n}{n}$

---

13. Given the p-series $\sum_{n=1}^{\infty} \dfrac{1}{n^6}$, find the least positive integer N such that the remainder tail $R_N = \sum_{n=N+1}^{\infty} \dfrac{1}{n^6}$ is less than 2×10^{-11}. Find S_N, and give an upper and lower bound for the sum of the series.

14. Given the series $\sum_{n=0}^{\infty} \dfrac{(-1)^n}{n!}$, find the least positive integer N such that $|R_N| < .000005$. The value S_N will approximate the series sum accurate to 5 decimal places. (Note: It is known that $\sum_{n=0}^{\infty} \dfrac{(-1)^n}{n!} = \dfrac{1}{e}$. Use this to check your result.)

15. Show that the series $\sum_{n=1}^{\infty} \ln(1 + \dfrac{1}{n})$ diverges using the integral test. Use integration by parts, letting $u = \ln(1 + \dfrac{1}{x})$ and $dv = dx$. (Note: When evaluating the integral, you will need to use L'Hôpital's rule, since the indeterminate form $0 \cdot \infty$ appears.)

16. Determine whether $\sum_{n=1}^{\infty} (-1)^{n+1} 3e^{-n}$ converges absolutely, converges conditionally, or diverges.

17. Find the sum of the series $\sum_{n=1}^{\infty} \dfrac{2^n}{3^{n-1}}$.

In exercises 18-20, indicate whether the given statement is true or false. In each case, provide an example that supports your decision.

18. If $\lim_{n \to \infty} a_n = 0$, then the positive term series $\sum_{n=1}^{\infty} a_n$ converges.

19. If the sequence $\{a_n\}$ converges, then the series $\sum_{n=1}^{\infty} a_n$ may not converge.

20. If $\sum_{n=1}^{\infty} |a_n|$ diverges, then the alternating series $\sum_{n=1}^{\infty} a_n$ also diverges.

For each of the series in exercises 21-42, indicate whether it converges absolutely, converges conditionally, or diverges. Indicate what test or tests you use in each case.

21. $\sum_{n=1}^{\infty} \dfrac{n^2}{\sqrt{n^5}}$

22. $\sum_{n=1}^{\infty} \dfrac{1}{(-5)^n}$

23. $\sum_{n=1}^{\infty} \dfrac{1}{n + 1000000}$

24. $\sum_{n=1}^{\infty} (-e)^{-n}$

25. $\sum_{n=1}^{\infty} \dfrac{\sin^2 \frac{1}{n}}{2^n}$

26. $\sum_{n=1}^{\infty} \dfrac{1}{2^{1/n}}$

27. $\sum_{n=1}^{\infty} \dfrac{n^n}{n!}$

28. $\sum_{n=1}^{\infty} (2 - e)^n$

29. $\sum_{n=1}^{\infty} \dfrac{3^n}{n!}$

30. $\sum_{n=1}^{\infty} \dfrac{n \ln n}{n^2 + 1}$

31. $\sum_{n=1}^{\infty} \dfrac{1}{\sqrt{n}}$

32. $\sum_{n=1}^{\infty} \dfrac{(-1)^n}{\sqrt{n}}$

33. $\sum_{n=1}^{\infty} \dfrac{(-1)^n}{\sqrt[n]{10}}$

34. $\sum_{n=1}^{\infty} \dfrac{e^n}{2^n}$

35. $\sum_{n=1}^{\infty} \sin(n\pi/2)$

36. $\sum_{n=1}^{\infty} \dfrac{n^2}{n^3 + n}$

37. $\sum_{n=1}^{\infty} \dfrac{(-1)^n}{\ln n}$

38. $\sum_{n=1}^{\infty} \left(\dfrac{n-1}{n} \right)^n$

39. $\sum_{n=1}^{\infty} \dfrac{3^n}{2^n + 4^n}$

40. $\sum_{n=1}^{\infty} n \sin(1/n)$

41. $\sum_{n=1}^{\infty} \dfrac{(-5)^n}{n!}$

42. $\sum_{n=1}^{\infty} n e^{-n^2}$

11.3 RATIO AND ROOT TESTS—POWER SERIES

The last two convergence tests we'll discuss are used to test for absolute convergence.

Ratio test

Requirements:

$\sum a_n$ can be any series.

Criteria:

If $\lim_{n \to \infty} |\frac{a_{n+1}}{a_n}| = L$, then

$\sum a_n$ is absolutely convergent when $L < 1$,

$\sum a_n$ is divergent when $L > 1$,

and the ratio test *fails to provide information* when $L = 1$.

Discussion:

When $\lim_{n \to \infty} |\frac{a_{n+1}}{a_n}| = L < 1$, this tells us that the terms $|a_n|$ are decreasing fast enough for the series $\sum |a_n|$ to converge. (Essentially, for some remainder tail of the series $\sum |a_n|$, we can find a larger convergent geometric series by choosing a factor $L < r < 1$ and use the comparison test.) If $\lim_{n \to \infty} |\frac{a_{n+1}}{a_n}| = L > 1$, then the terms a_n are not approaching 0 and the series $\sum a_n$ must diverge.

The ratio test is often useful when the index n appears as an exponent in the formula for a_n (as it does for geometric series). It is also useful when the formula for a_n involves $n!$ (read n **factorial**). This is a shorthand notation for

$$n! = 1 \cdot 2 \cdot 3 \cdots n.$$

EXAMPLE 33 $1! = 1$, $2! = 1 \cdot 2 = 2$, $3! = 1 \cdot 2 \cdot 3 = 6$, $4! = 1 \cdot 2 \cdot 3 \cdot 4 = 24$, and $5! = 120$. By agreement, we say $0! = 1$. (This turns out to be a nice convention for use in a variety of formulas.) ■

EXAMPLE 34 Use the ratio test to test for the absolute convergence of the series

$$\sum_{n=0}^{\infty} \frac{(-1)^n}{n!}.$$

Solution The first few terms in the sum are

$$1 + (-1) + \frac{1}{2} - \frac{1}{6} + \frac{1}{24} - \frac{1}{120} + \cdots$$

This is an alternating series whose terms strictly decrease in magnitude. Since

$$\lim_{n \to \infty} \frac{1}{n!} = 0$$

the series converges. To check for absolute convergence, we first form the ratio

$$\left| \frac{a_{n+1}}{a_n} \right| = \left| \frac{\frac{(-1)^{n+1}}{(n+1)!}}{\frac{(-1)^n}{n!}} \right| = \frac{n!}{(n+1)!} = \frac{1}{n+1}$$

(If the last step in the above equality seems confusing to you, try evaluating $\frac{n!}{(n+1)!}$ for $n = 1, 2, 3, 4$ to see the pattern.)

Now, $\lim_{n \to \infty} \left| \frac{a_{n+1}}{a_n} \right| = \lim_{n \to \infty} \frac{1}{n+1} = 0 < 1$. So the ratio test tells us that the series $\sum_{n=0}^{\infty} \frac{(-1)^n}{n!}$ absolutely converges. ∎

EXAMPLE 35 Use the ratio test to test for the absolute convergence of the series $\sum_{n=1}^{\infty} \frac{2^n}{n^5}$.

Solution The first few terms in the sum are

$$2 + \frac{1}{8} + \frac{8}{243} + \frac{16}{1024} + \cdots$$

and at first glance, these terms appear to be approaching 0 rapidly. However, if we apply the ratio test, we find

$$\lim_{n \to \infty} \left| \frac{a_{n+1}}{a_n} \right| = \lim_{n \to \infty} \frac{\frac{2^{n+1}}{(n+1)^5}}{\frac{2^n}{n^5}} = \lim_{n \to \infty} 2 \cdot \left(\frac{n}{n+1} \right)^5 = 2 > 1.$$

The ratio test tells us that this series diverges. ∎

11.3 RATIO AND ROOT TESTS—POWER SERIES

EXAMPLE 36 Use the ratio test to test the absolute convergence of the alternating harmonic series $\sum_{n=1}^{\infty} \frac{(-1)^{n+1}}{n}$.

Solution

$$\lim_{n \to \infty} \left| \frac{a_{n+1}}{a_n} \right| = \lim_{n \to \infty} \left| \frac{\frac{(-1)^{n+2}}{(n+1)}}{\frac{(-1)^{n+1}}{n}} \right| = \lim_{n \to \infty} \frac{n}{n+1} = 1.$$

The ratio test fails to provide us any information in this case (though we know that the alternating harmonic series conditionally converges). ∎

Root test

Requirements:

$\sum a_n$ can be any series.

Criteria:

If $\lim_{n \to \infty} \sqrt[n]{|a_n|} = L$, then

$\sum a_n$ is absolutely convergent when $L < 1$,

$\sum a_n$ is divergent when $L > 1$,

and the root test *fails to provide any information when* $L = 1$.

Discussion:

The root test is more powerful than the ratio test in the following sense: whenever the ratio test provides information, so will the root test; however, there are instances where the ratio test fails but the root test does not. In practice, you will probably find the ratio test much easier to apply. (Essentially, if $L \neq 1$, then the root test follows from applying the limit comparison test to the series $\sum |a_n|$ and the geometric series $\sum r^n$, where r is chosen to be some number between 1 and L.)

EXAMPLE 37 Use the root test to test for the absolute convergence of the series

$$\sum_{n=1}^{\infty} \frac{1}{n^n}.$$

Solution $\lim_{n \to \infty} \sqrt[n]{|a_n|} = \lim_{n \to \infty} \sqrt[n]{1/n^n} = \lim_{n \to \infty} 1/n = 0 < 1$. The root test tells us that this series absolutely converges. ∎

Summary of Convergence Tests

Test	Form	Converges	Diverges
Nth Term	$\sum_{n=1}^{\infty} a_n$	no information	$\lim_{n \to \infty} a_n \neq 0$
Integral	$\sum_{n=1}^{\infty} a_n$ where $a_n = f(n)$ (f must be continuous, positive, and decreasing)	$\int_{1}^{\infty} f(x)\, dx$ converges	$\int_{1}^{\infty} f(x)\, dx$ diverges
Integral Test Estimate:		$\sum_{n=1}^{\infty} a_n = \sum_{n=1}^{N} a_n + \int_{N+1}^{\infty} f(x)\, dx + \text{error}$ where the error $\leq \int_{N}^{N+1} f(x)\, dx$	
Comparison	$\sum_{n=1}^{\infty} a_n$	$0 \leq a_n \leq b_n$, and $\sum_{n=1}^{\infty} b_n$ converges	$0 \leq b_n \leq a_n$, and $\sum_{n=1}^{\infty} b_n$ diverges
Limit Comparison	$\sum_{n=1}^{\infty} a_n$ $(a_n, b_n > 0)$	$\lim_{n \to \infty} \frac{a_n}{b_n} = L \geq 0$ and $\sum_{n=1}^{\infty} b_n$ converges	$\lim_{n \to \infty} \frac{a_n}{b_n} = L > 0$ or ∞ and $\sum_{n=1}^{\infty} b_n$ diverges
Alternating	$\sum_{n=1}^{\infty} (-1)^{n+1} c_n$ $0 < c_{n+1} < c_n$	$\lim_{n \to \infty} c_n = 0$	$\lim_{n \to \infty} c_n \neq 0$
Alternating Series Estimate:		$\sum_{n=1}^{\infty} (-1)^{n-1} c_n = \sum_{n=1}^{N} (-1)^{n-1} c_n \pm \text{error}$ where the error $\leq c_{N+1}$	
Ratio	$\sum_{n=1}^{\infty} a_n$	$\lim_{n \to \infty} \left| \frac{a_{n+1}}{a_n} \right| < 1$	$\lim_{n \to \infty} \left| \frac{a_{n+1}}{a_n} \right| > 1$
Root	$\sum_{n=1}^{\infty} a_n$	$\lim_{n \to \infty} \sqrt[n]{|a_n|} < 1$	$\lim_{n \to \infty} \sqrt[n]{|a_n|} > 1$

11.3 RATIO AND ROOT TESTS—POWER SERIES

Some Special Series

Series	Form	Converges	Diverges
Telescoping	$\sum_{n=1}^{\infty}(b_n - b_{n+1})$	$\lim_{n\to\infty} b_n = 0$	$\lim_{n\to\infty} b_n \neq 0$
Geometric	$\sum_{n=0}^{\infty} ar^n$	$\lvert r \rvert < 1$	$\lvert r \rvert \geq 1$
p-series	$\sum_{n=1}^{\infty} \dfrac{1}{n^p}$	$p > 1$	$p \leq 1$

Power series

A real **power series** has the form

$$\sum_{n=0}^{\infty} c_n x^n = c_0 + c_1 x + c_2 x^2 + c_3 x^3 + \cdots + c_n x^n + \cdots$$

where each c_n is a real number and x is a variable. Roughly speaking, a power series is a polynomial with infinitely many terms. More precisely, it is the limit of a sequence of polynomials:

$$p_0(x) = c_0$$
$$p_1(x) = c_0 + c_1 x$$
$$p_2(x) = c_0 + c_1 x + c_2 x^2$$
$$p_3(x) = c_0 + c_1 x + c_2 x^2 + c_3 x^3$$
$$\vdots$$

where $p_n(x) = c_0 + c_1 x + c_2 x^2 + c_3 x^3 + \cdots + c_n x^n$ is a polynomial of degree n. These polynomials play the role of partial sums, and

$$\sum_{n=0}^{\infty} c_n x^n = \lim_{n\to\infty} p_n(x).$$

We'll see that power series may be manipulated much like polynomials. Polynomial functions are particularly simple to work with in calculus, and it is remarkable that so many non-polynomial functions (like the trigonometric functions) can be represented as power series.

Sometimes it is convenient to have a particular real number a as a point of reference. In this case, we may wish to write a power series in terms of $(x - a)$ instead of x:

$$\sum_{n=0}^{\infty} c_n(x - a)^n = c_0 + c_1(x - a) + c_2(x - a)^2 + c_3(x - a)^3 + \cdots + c_n(x - a)^n + \cdots$$

A power series written this way is said to be represented *about* $x = a$. A power series written in terms of powers of x is said to be represented about $x = 0$.

EXAMPLE 38 $\sum_{n=0}^{\infty} x^n = 1 + x + x^2 + x^3 + \cdots$ is a power series represented about $x = 0$.

$\sum_{n=0}^{\infty} \dfrac{(x-1)^n}{n} = 1 + (x - 1) + \dfrac{(x-1)^2}{2} + \dfrac{(x-1)^3}{3} + \cdots$ is about $x = 1$. ∎

To evaluate a power series $\sum_{n=0}^{\infty} c_n x^n$ at a specific input value $x = x_0$, we substitute the value x_0 and obtain a "regular" series:

$$\sum_{n=0}^{\infty} c_n x_0^n$$

that may or may not converge to finite value.

This raises the question: for exactly which input values x does the power series $\sum_{n=0}^{\infty} c_n x^n$ converge?

Radius and interval of convergence

If $\sum_{n=0}^{\infty} c_n x^n$ is a power series represented about $x = 0$, then there is always one value x that guarantees convergence, namely $x = 0$. Note that substituting $x = 0$ makes all the terms 0 except perhaps the constant c_0.

Similarly, if $\sum_{n=0}^{\infty} c_n(x - a)^n$ is a power series represented about $x = a$, then substituting $x = a$ guarantees convergence. To find the entire set of input values for which a power series converges, the ratio or root test may be used.

11.3 RATIO AND ROOT TESTS—POWER SERIES

EXAMPLE 39 Find the set of input values x for which the power series $\sum_{n=0}^{\infty} \frac{x^n}{3^{2n}}$ converges.

Solution Using the ratio test, we find that

$$\lim_{n \to \infty} \left| \frac{x^{n+1}/3^{2(n+1)}}{x^n/3^{2n}} \right| = \lim_{n \to \infty} |x| \frac{3^{2n}}{3^{2n+2}} = \lim_{n \to \infty} \frac{|x|}{9} = \frac{|x|}{9}.$$

This limit value depends on x, and the ratio test tells us that the power series

$$\sum_{n=0}^{\infty} \frac{x^n}{3^{2n}} \text{ converges if } |x|/9 < 1,$$

$$\sum_{n=0}^{\infty} \frac{x^n}{3^{2n}} \text{ diverges if } |x|/9 > 1,$$

and the behavior is as yet undetermined for $|x|/9 = 1$, or equivalently, for $x = \pm 9$.

We can determine the behavior when $x = \pm 9$ by substituting these two values directly back into the power series. For $x = 9$, we have

$$\sum_{n=0}^{\infty} \frac{9^n}{3^{2n}} = \sum_{n=0}^{\infty} 1,$$

which *diverges*. For $x = -9$, we have

$$\sum_{n=0}^{\infty} \frac{(-9)^n}{3^{2n}} = \sum_{n=0}^{\infty} (-1)^n,$$

which also *diverges*. We can conclude that the power series converges precisely when $|x|/9 < 1$, or equivalently, on the set

$$\{x \ : \ -9 < x < 9\}.$$

We note that the same conclusion can be reached by using the root test. In this case, we examine

$$\lim_{n \to \infty} \sqrt[n]{|x^n/3^{2n}|} = \lim_{n \to \infty} |x|/3^2 = |x|/9.$$

The criteria for the root test are the same as for the ratio test, so the analysis is exactly the same. ∎

The power series $\sum_{n=0}^{\infty} \frac{x^n}{3^{2n}}$ of this example is represented about $x = 0$. Note that the set of values for which the power series converges is an *interval* $(-9, 9)$ *centered* at 0. This is no coincidence.

Theorem 11.3

> **Hypothesis:** $\sum_{n=0}^{\infty} c_n(x-a)^n$ is a power series about $x = a$.
>
> **Conclusion:** The set of values for which the power series converges is an *interval* centered at a.

This interval is called the **interval of convergence** for the power series. The radius R of this interval (in other words, the distance from the center a to either endpoint) is called the **radius of convergence**.

The interval of convergence may take any of the forms

$$[a-R, a+R], \quad [a-R, a+R), \quad (a-R, a+R], \quad (a-R, a+R)$$

depending on which endpoints, if any, are included.

Reasoning If we use the root test on any power series $\sum_{n=0}^{\infty} c_n(x-a)^n$, our criteria for convergence requires us to find the values x such that

$$\lim_{n \to \infty} \sqrt[n]{|c_n(x-a)^n|} = \lim_{n \to \infty} \sqrt[n]{|c_n|}\, |x-a| < 1.$$

If $\lim_{n \to \infty} \sqrt[n]{|c_n|} = \infty$, then the power series converges only for $x = a$. In this situation, $R = 0$ is the radius of convergence, and the interval of convergence is the single point $\{a\}$.

If $\lim_{n \to \infty} \sqrt[n]{|c_n|} = 0$, then the power series converges for *any* value $x \in (-\infty, \infty)$. In this situation, we say $R = \infty$ is the radius of convergence, and the interval of convergence is the whole real line $(-\infty, \infty)$.

If $\lim_{n \to \infty} \sqrt[n]{|c_n|} = L \neq 0$, then the power series converges for those values x satisfying

$$L\,|x-a| < 1 \quad \text{or} \quad |x-a| < R,$$

where we have written $R = 1/L$. In addition, the power series may or may not converge at the two values $x = a - R$ or $x = a + R$. In any case, the set of values in this situation is a bounded interval centered at $x = a$. □

In general, we can use the root or ratio test to find an open interval of convergence $(a - R, a + R)$, and then check the endpoints by direct substitution into the power series.

11.3 RATIO AND ROOT TESTS—POWER SERIES

EXAMPLE 40 Find the interval and radius of convergence for the power series

$$\sum_{n=0}^{\infty} \frac{3^n(x+2)^n}{n!}.$$

Solution Using the ratio test, we have

$$\lim_{n \to \infty} \frac{|3^{n+1}(x+2)^{n+1}/(n+1)!|}{|3^n(x+2)^n/n!|} = \lim_{n \to \infty} \frac{3|x+2|}{n+1} = 0$$

for any value x. Therefore, the interval of convergence is $(-\infty, \infty)$ and the radius of convergence is $R = \infty$. ∎

EXAMPLE 41 Find the interval and radius of convergence for the power series

$$\sum_{n=0}^{\infty} \frac{3^n(x+2)^n}{n+1}.$$

Solution Using the ratio test, we have

$$\lim_{n \to \infty} \frac{|3^{n+1}(x+2)^{n+1}/(n+2)|}{|3^n(x+2)^n/(n+1)|} = \lim_{n \to \infty} \frac{3(n+1)|x+2|}{n+2} = 3|x+2|$$

since $(n+1)/(n+2) \to 1$ as $n \to \infty$.

By the ratio test, the power series definitely converges for any value x satisfying

$$3|x+2| < 1 \quad \text{or} \quad -7/3 < x < -5/3.$$

We must check the behavior at the endpoints $x = -7/3$ and $x = -5/3$ by direct substitution into the power series $\sum_{n=0}^{\infty} \frac{3^n(x+2)^n}{n+1}$.

For $x = -7/3$, we have $\sum_{n=0}^{\infty} \frac{3^n(-1/3)^n}{n+1} = \sum_{n=0}^{\infty} \frac{(-1)^n}{n+1}$. This series satisfies the requirements of the alternating series test and *converges*. Therefore, we include $x = -7/3$ in the interval of convergence.

For $x = -5/3$, we have $\sum_{n=0}^{\infty} \frac{3^n(1/3)^n}{n+1} = \sum_{n=0}^{\infty} \frac{1}{n+1}$. This is just the harmonic series, which we know diverges. Therefore, we do *not* include $x = -5/3$ in the interval of convergence.

Hence, the interval of convergence is $[-7/3, -5/3)$, with center at $a = -2$ and radius of convergence $R = 1/3$. ∎

EXAMPLE 42 Find the interval and radius of convergence of $\sum_{n=0}^{\infty}(-1)^n n! x^n$.

Solution By the ratio test, we have

$$\lim_{n\to\infty} \frac{|(-1)^{n+1}(n+1)!x^{n+1}|}{(-1)^n n! x^n} = \lim_{n\to\infty} |x|(n+1)$$

which is infinite for any $x \neq 0$. Hence the interval of convergence $\{0\}$ contains only one point $x = 0$, and the radius of convergence is $R = 0$. ∎

Calculus and power series

If we use a power series to define a function f, so that

$$f(x) = \sum_{n=0}^{\infty} c_n(x-a)^n,$$

then the domain of f is the interval of convergence for the power series. A function defined by a power series in this way is continuous on the interval of convergence. In fact, it can be differentiated and antidifferentiated term-by-term just like a polynomial!

Theorem 11.4

Hypothesis: The function f is defined by a power series:

$$f(x) = \sum_{n=0}^{\infty} c_n(x-a)^n.$$

Conclusion: The derivative

$$f'(x) = \sum_{n=0}^{\infty} nc_n(x-a)^{n-1}$$

and the antiderivative

$$\int f(x)\,dx = C + \sum_{n=0}^{\infty} \frac{c_n(x-a)^{n+1}}{n+1}$$

have the same radius of convergence as f.

11.3 RATIO AND ROOT TESTS—POWER SERIES

EXAMPLE 43 Suppose $f(x) = 1 + x + x^2 + x^3 + \cdots = \sum_{n=0}^{\infty} x^n$.

Then $f'(x) = 0 + 1 + 2x + 3x^2 + \cdots = \sum_{n=0}^{\infty} nx^{n-1}$, and

$$\int f(x)\, dx = C + x + \frac{x^2}{2} + \frac{x^3}{3} + \frac{x^4}{4} + \cdots = C + \sum_{n=0}^{\infty} \frac{x^{n+1}}{n+1},$$

where C is an arbitrary constant.

The radius of convergence for $f(x) = \sum_{n=0}^{\infty} x^n$ is $R = 1$. By the theorem, the derivative f' and antiderivative $\int f(x)\, dx$ have the same radius of convergence. (However, the *interval* of convergence for f and f' is $(-1, 1)$, while the antiderivative also converges for $x = -1$.) ∎

 Be careful! Two power series can have different intervals of convergence, even if the center a and radius of convergence R are the same. This is because the behavior at the *endpoints* $x = a - R$ and $x = a + R$ may be different for the two power series.

Functions defined by power series

We have seen that we can define a function using a power series. The domain of the function is the interval of convergence for the power series. Many familiar functions may be represented by a suitable power series.

EXAMPLE 44 Find the domain and a closed form formula for the function f defined by

$$f(x) = \sum_{n=0}^{\infty} x^n.$$

Solution This power series can be thought of as a geometric series with factor $r = x$. As such, it converges to $1/(1-x)$ for $-1 < x < 1$, and we can write

$$f(x) = \frac{1}{1-x} \qquad \text{with domain } (-1, 1).$$

∎

EXAMPLE 45 If we differentiate f to obtain

$$f' : x \longmapsto \frac{1}{(1-x)^2},$$

we already have a power series representation for f' valid for $-1 < x < 1$, namely

$$\frac{1}{(1-x)^2} = 1 + 2x + 3x^2 + \cdots = \sum_{n=0}^{\infty} nx^{n-1}.$$

If we antidifferentiate f, we obtain the power series

$$\int f(x)\,dx = C + x + \frac{x^2}{2} + \frac{x^3}{3} + \frac{x^4}{4} + \cdots = C + \sum_{n=0}^{\infty} \frac{x^{n+1}}{n+1},$$

which must also converge for $-1 < x < 1$. Note that

$$F : x \longmapsto -\ln(1-x)$$

is an antiderivative for f satisfying the initial condition $F(0) = 0$, and whose formula is valid for $-1 < x < 1$. If we set $C = 0$, we have found a power series representation for F:

$$-\ln(1-x) = x + \frac{x^2}{2} + \frac{x^3}{3} + \frac{x^4}{4} + \cdots = \sum_{n=0}^{\infty} \frac{x^{n+1}}{n+1}.$$

∎

Since a function f defined by power series has a derivative f' that is also represented as a power series, we can continue differentiating for higher order derivatives f'', f''', $f^{(4)}$, and so on indefinitely. In other words, f is infinitely differentiable, and all its derivatives have the same radius of convergence. For this reason, it is particularly nice to have a power series representation of a function. In the next section, we'll see how we can build power series for most of the basic functions we work with in calculus.

EXERCISES

In exercises 1-5, determine whether the given series converges or diverges using the indicated test.

1. $\displaystyle\sum_{n=0}^{\infty} (-1)^n \frac{e^n}{e^{2n}+1}$ (Use ratio test.)

2. $\displaystyle\sum_{n=1}^{\infty} \frac{(1-\pi)^n}{2^{n+1}}$ (Use root test.)

11.3 RATIO AND ROOT TESTS—POWER SERIES

3. $\sum_{n=1}^{\infty} \dfrac{2^{3n}}{7^n}$ (Use root test.)

4. $\sum_{n=1}^{\infty} \dfrac{n! n^2}{(2n)!}$ (Use ratio test.)

5. $\sum_{n=1}^{\infty} \dfrac{(-1)^{n+1} 2^{3n+2}}{n^n}$ (Use root test.)

Consider the series

$$\sum_{n=1}^{\infty} \dfrac{n}{2n^2 - 1}.$$

In exercises 6-9, apply each of the tests indicated. Explain why the result in each case is convergence, divergence, or "no information."

6. comparison test with $\sum_{n=1}^{\infty} \dfrac{1}{2n}$

7. limit comparison test with $\sum_{n=1}^{\infty} \dfrac{1}{n}$

8. ratio test

9. root test

---

10. Use the ratio test to show the series $\sum_{n=0}^{\infty} \dfrac{(-1)^n}{(2n)!}$ converges absolutely. Find the smallest N such that S_N is accurate to 6 decimal places. Give the estimate for the sum of the series accurate to 6 decimal places.

11. Determine whether $\sum_{n=1}^{\infty} (-1)^{n+1} \dfrac{n^{2n}}{(3n^2 + 1)^n}$ converges absolutely, converges conditionally, or diverges.

12. Show that the ratio test is inconclusive when applied to the series

$$\sum_{n=1}^{\infty} (-1)^{n+1} \dfrac{\pi^{1/n}}{n}.$$

Then use any tests necessary to determine whether the series converges absolutely, converges conditionally, or diverges.

13. Suppose I have a "p-series" $\sum_{n=1}^{\infty} \dfrac{1}{n^p}$ where p is a specific positive number. Show that the ratio test *always* fails to give any information about the convergence or divergence of this series.

14. Suppose I have an "alternating p-series" $\sum_{n=1}^{\infty} \dfrac{(-1)^n}{n^p}$ where p is a specific positive number. For what values of p will this series converge?

15. Find the interval and radius of convergence for $\sum_{n=0}^{\infty} \dfrac{2^n x^n}{n!}$

16. Find the interval and radius of convergence for $\sum_{n=0}^{\infty} \dfrac{(-1)^n x^n}{3^n n}$

17. Find the interval and radius of convergence for $\sum_{n=0}^{\infty} \dfrac{(-1)^n (x-3)^n}{2^n}$.

Power series representations for new functions can be derived from known representations through substitution, differentiation, and antidifferentiation. In exercises 18-21 you will use the power series representation

$$f(x) = \frac{1}{1-x} = \sum_{n=0}^{\infty} x^n = 1 + x + x^2 + x^3 + \cdots \quad \text{for } -1 < x < 1$$

to find power series representations for other functions.

18. Find a power series representation for

$$g(x) = \frac{1}{1+x}$$

by substituting $-x$ for x. What is the interval of convergence for g?

19. Find a power series representation of $h(x) = \ln(1+x)$ by antidifferentiating the power series from exercise 18. (What's the initial condition?)

20. Use the ratio test to determine the interval of convergence for $h(x)$. Be sure to check the endpoints.

21. Let $x = 1$. Estimate $\ln(2)$ by adding the first 101 of the series from exercise 19. Use the alternating series remainder to determine the accuracy of the result. How well does this agree with the value of $\ln(2)$?

The function $f : x \longmapsto e^x$ is represented by the power series

$$e^x = \sum_{n=0}^{\infty} \frac{x^n}{n!} = 1 + x + \frac{x^2}{2} + \frac{x^3}{6} + \frac{x^4}{24} + \cdots$$

In exercises 22-27, you will use this power series representation.

22. Show that the interval of convergence for this power series is $(-\infty, \infty)$.

23. Find a power series representation for e^{-x}.

24. Find a power series representation for $1 - e^{-x}$.

25. Find a power series representation for $\dfrac{1 - e^{-x}}{x}$.

26. Find a power series representation for the antiderivative of $\dfrac{1-e^{-x}}{x}$.

27. Use your answer to the previous exercise to approximate the value of $\displaystyle\int_0^{1/2} \dfrac{1-e^{-x}}{x}\, dx$ to within 1×10^{-10}.

Here are power series representation of two trigonometric functions:

$$\sin(x) = x - \frac{x^3}{6} + \frac{x^5}{120} - \frac{x^7}{7!} + \cdots = \sum_{n=0}^{\infty} \frac{(-1)^{n+1} x^{2n+1}}{(2n+1)!}$$

$$\cos(x) = 1 - \frac{x^2}{2} + \frac{x^4}{24} - \frac{x^6}{6!} + \cdots = \sum_{n=0}^{\infty} \frac{(-1)^n x^{2n}}{(2n)!}.$$

In exercises 28-30, you will use these power series representation.

28. Show that the interval of convergence for both of these power series is $(-\infty, \infty)$.

29. Find a power series representation for the function

$$x \longmapsto \frac{x - \sin x}{x^3 \cos x}.$$

Graph the function.

30. Evaluate $\displaystyle\lim_{x \to 0} \dfrac{x - \sin x}{x^3 \cos x}$ using the result from the previous exercise. Does the graph support your answer? Notice that L'Hôpital's rule is appropriate here (but rather undesirable).

11.4 TAYLOR POLYNOMIALS AND SERIES

Consider the following functions:

$$x \longmapsto x^3 - 2x + 5 \quad x \longmapsto \sin(x) \quad x \longmapsto 2^x \quad x \longmapsto \log_3 x \quad x \longmapsto \arctan(x)$$

and suppose you were asked to evaluate each function's output at the input $x = 1.35$. For a polynomial function, such as $x \longmapsto x^3 - 2x + 5$, you need nothing more than addition, subtraction, multiplication, and (possibly) division: the operations provided by a four function calculator. All of the other calculations require either a more sophisticated machine or tables of values.

Machines that can evaluate transcendental functions (like trigonometric, exponential, logarithmic, and inverse trigonometric functions) are readily available, but polynomial functions are particularly nice for a variety of reasons—they are easy to compute, to compare, to differentiate, to integrate, and so on. Moreover, polynomials are, in a sense, the *only* functions you can calculate. Even scientific calculators' and computers' versions of the transcendental functions are, in fact, *approximations by polynomials*, or something very similar. The technique of approximating functions by polynomials is used in many settings. We have studied cubic splines as one example of a piecewise polynomial approximation.

We've already seen **Taylor polynomials** arise as useful approximations for functions. The tangent line approximation to a function at a specific point is the first degree Taylor polynomial approximation, and it represents the *best linear approximation* to a function at that point. Similarly, the second degree Taylor polynomial represents the *best quadratic approximation*. In this section we'll examine higher degree Taylor polynomials and their behavior. The limit of the infinite sequence of polynomial approximations we obtain in this way is called a **Taylor series**.

Goals of function approximation

When we approximate a number (as we do with root-finding techniques, for example) we generally have a choice of goals. We can either try to guarantee that our approximation is "close" to the desired number, or we can try to guarantee that it "behaves like" the desired number. For example, if we are approximating a root x_r of an equation $f(x) = 0$, then we can either look for a number x_0 that is close to x_r, or we can look for a number x_0 that behaves like x_r. To behave like the the root of the equation simply means that $f(x_0)$ is very close to 0, not necessarily that x_0 is very close to x_r. For many purposes, however, approximating behavior is perfectly adequate.

Similarly, when we want to approximate a target function f with another function like a polynomial p, we have a choice of goals. We can try to guarantee that all of the approximation function's outputs $p(x)$ are close to the target function's outputs $f(x)$, or we can try to guarantee that the approximation function "behaves" like the target function. Behaviors for a function include whether it is increasing or decreasing, concave up, concave down, or flat, and so on. These behaviors are all determined by the values of the function and *its derivatives* at a particular input.

The polynomial approximations we study in this section approximate the behavior of a function by matching as many of the function's derivatives as they can at a particular point $x = a$. These polynomials are called the **Taylor polynomials** of the function.

11.4 TAYLOR POLYNOMIALS AND SERIES

Determining the Taylor polynomial at a point $x = a$

To see how Taylor polynomials are determined, it's instructive to review the cases of the best linear and quadratic approximations.

Suppose we are given a function f that is differentiable at the point $x = a$. The best linear approximation p_1 can be written

$$p_1 : x \longmapsto c + m(x - a)$$

where the slope $m = f'(a)$ and $c = f(a)$.

This linear polynomial is designed to satisfy *two* criteria:

$$p_1(a) = c + m(a - a) = f(a) \quad \text{and} \quad p_1'(a) = m = f'(a).$$

In other words, p_1 has the same output and same derivative value at $x = a$ as the function f does. Indeed, this completely determines p_1.

If f is twice differentiable at $x = a$, then the best quadratic approximation p_2 at that point can be written

$$p_2 : x \longmapsto f(a) + f'(a)(x - a) + \frac{f''(a)}{2}(x - a)^2.$$

This quadratic polynomial is designed to satisfy *three* criteria:

$$p_2(a) = f(a) \qquad p_2'(a) = f'(a) \qquad p_2''(a) = f''(a).$$

We can extend this analysis to higher degree polynomials p_n. If f is n times differentiable at $x = a$, then an nth degree polynomial p_n can be designed to satisfy $n + 1$ criteria:

$$p_n(a) = f(a)$$
$$p_n'(a) = f'(a)$$
$$p_n''(a) = f''(a)$$
$$\vdots$$
$$p_n^{(n)}(a) = f^{(n)}(a).$$

The nth degree polynomial p_n has $n + 1$ coefficients that are determined by these requirements. If we write

$$p_n(x) = c_0 + c_1(x - a) + c_2(x - a)^2 + c_3(x - a)^3 + \cdots + c_n(x - a)^n$$

and compute the output of p_n and its first n derivatives at $x = a$, our requirements are

$$p_n(a) = c_0 = f(a)$$
$$p_n'(a) = c_1 = f'(a)$$
$$p_n''(a) = 2c_2 = f''(a)$$
$$p_n'''(a) = 3 \cdot 2 c_3 = f'''(a)$$
$$p_n^{(4)}(a) = 4 \cdot 3 \cdot 2 c_4 = f^{(4)}(a)$$
$$\vdots$$
$$p_n^{(n)}(a) = n \cdot (n-1) \cdots 3 \cdot 2 c_n = f^{(n)}(a).$$

Now we can solve for the coefficients and find that

$$p_n(x) = f(a) + f'(a)(x-a) + \frac{f''(a)}{2}(x-a)^2 + \frac{f'''(a)}{6}(x-a)^3 + \cdots + \frac{f^{(n)}(a)}{n!}(x-a)^n.$$

In fact, if we agree to write $f^{(0)}$ to mean the original function f, and since $0! = 1$, $1! = 1$ and $2! = 2$, we have a nice formula for any coefficient c_k, for $0 \leq k \leq n$:

$$c_k = \frac{f^{(k)}(a)}{k!}.$$

We summarize this analysis in the following theorem.

Theorem 11.5

Hypothesis: f is continuous and n-times differentiable at $x = a$.
Conclusion: The degree n Taylor polynomial approximation of f at $x = a$ is given by

$$p_n(x) = f(a) + f'(a)(x-a) + \frac{f''(a)}{2}(x-a)^2 + \cdots + \frac{f^{(n)}(a)}{n!}(x-a)^n$$
$$= \sum_{k=0}^{n} f^{(k)}(a) k! (x-a)^k.$$

$\square$

11.4 TAYLOR POLYNOMIALS AND SERIES

EXAMPLE 46 Find the fourth degree Taylor polynomial of $x \mapsto \sqrt{(x)}$ at $x = 9$ and graph both over the interval $[2, 16]$.

Solution The polynomial is

$$p_4(x) = c_0 + c_1(x - 9) + c_2(x - 9)^2 + c_3(x - 9)^3 + c_4(x - 9)^4$$

where

$$c_0 = \sqrt{9} = 3,$$

$$c_1 = \left[\frac{d}{dx}\sqrt{x}\right]\bigg|_{x=9} = \frac{1}{2} \cdot \frac{1}{\sqrt{9}} = \frac{1}{6}$$

$$c_2 = \frac{1}{2}\left[\frac{d^2}{dx^2}\sqrt{x}\right]\bigg|_{x=9} = -\frac{1}{2} \cdot \frac{1}{4} \cdot \frac{1}{9\sqrt{9}} = -\frac{1}{216}$$

$$c_3 = \frac{1}{3 \cdot 2}\left[\frac{d^3}{dx^3}\sqrt{x}\right]\bigg|_{x=9} = \frac{1}{3 \cdot 2} \cdot \frac{3}{8} \cdot \frac{1}{81\sqrt{9}} = \frac{1}{3888}$$

$$c_4 = \frac{1}{4 \cdot 3 \cdot 2}\left[\frac{d^4}{x^4}\sqrt{x}\right]\bigg|_{x=9} = \frac{1}{4 \cdot 3 \cdot 2} \cdot \frac{15}{16} \cdot \frac{1}{729\sqrt{9}} = -\frac{5}{279936}.$$

The graphs of $y = \sqrt{x}$ and $y = p_4(x)$ are both shown in Figure 11.5. Note that near the input $x = 9$, the two graphs are virtually indistinguishable.

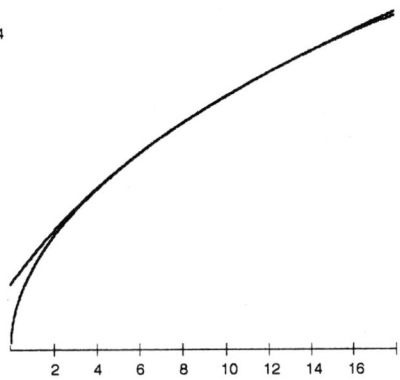

Figure 11.5 Graph of $y = \sqrt{(x)}$ and its fourth degree Taylor polynomial at $x = 9$.

■

We can now observe graphically how higher and higher order Taylor polynomials incorporate more and more of the behavior of the approximated function.

EXAMPLE 47 Find the first through sixth degree Taylor polynomials at $x = 1$ of the function $x \longmapsto \ln(x)$ Graph each one of the approximations along with $y = \ln(x)$.

Solution First, we compute the first six derivatives of $y = \ln(x)$, and evaluate each at $x = 1$:

$$y = \ln(x) \qquad y(1) = 0$$
$$y' = x^{-1} \qquad y'(1) = 1$$
$$y'' = -1 \cdot x^{-2} \qquad y''(1) = -1$$
$$y''' = 2x^{-3} \qquad y'''(1) = 2$$
$$y^{(4)} = -6x^{-4} \qquad y^{(4)}(1) = -6$$
$$y^{(5)} = 24x^{-5} \qquad y^{(5)}(1) = 24$$
$$y^{(6)} = -120x^{-6} \qquad y^{(6)}(1) = -120$$

Note that $y^{(k)}(1) = (-1)^{k-1}(k-1)!$, so in general, we have

$$c_k = \frac{y^{(k)}(1)}{k!} = \frac{(-1)^{k-1}}{k}$$

Now we have a general formula for p_n, the nth degree Taylor polynomial approximation:

$$\sum_{k=1}^{n} \frac{(-1)^{k-1} \cdot (k-1)!}{k!} (x-1)^k = \sum_{k=1}^{n} \frac{(-1)^{k-1}}{k} (x-1)^k.$$

We can use this formula to find p_1 through p_6:

$$p_1(x) = x - 1$$

$$p_2(x) = (x-1) - \frac{1}{2}(x-1)^2$$

$$p_3(x) = (x-1) - \frac{1}{2}(x-1)^2 + \frac{1}{3}(x-1)^3$$

$$p_4(x) = (x-1) - \frac{1}{2}(x-1)^2 + \frac{1}{3}(x-1)^3 - \frac{1}{4}(x-1)^4$$

$$p_5(x) = (x-1) - \frac{1}{2}(x-1)^2 + \frac{1}{3}(x-1)^3 - \frac{1}{4}(x-1)^4 + \frac{1}{5}(x-1)^5$$

$$p_6(x) = (x-1) - \frac{1}{2}(x-1)^2 + \frac{1}{3}(x-1)^3 - \frac{1}{4}(x-1)^4 + \frac{1}{5}(x-1)^5 - \frac{1}{6}(x-1)^6.$$

The graphs of p_1 through p_6 are shown in Figures 11.6, 11.7, and 11.8 (the vertical axis has been removed to avoid distraction).

11.4 TAYLOR POLYNOMIALS AND SERIES

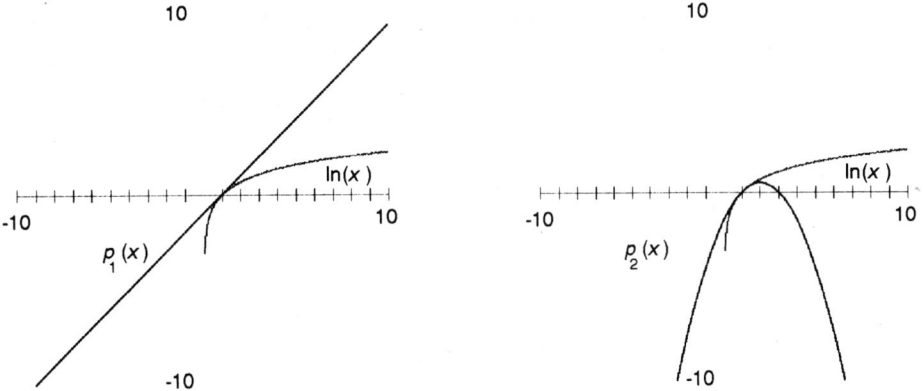

Figure 11.6 Degree 1 and 2 Taylor Polynomials at $x = 1$ of $\ln(x)$.

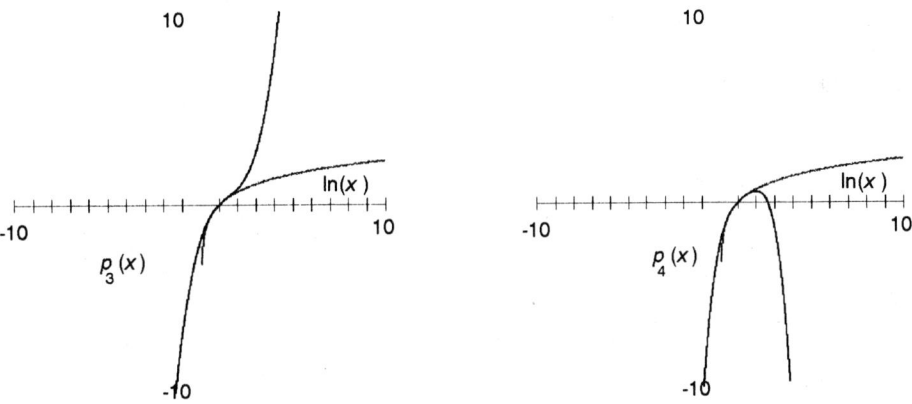

Figure 11.7 Degree 3 and 4 Taylor Polynomials at $x = 1$ of $\ln(x)$.

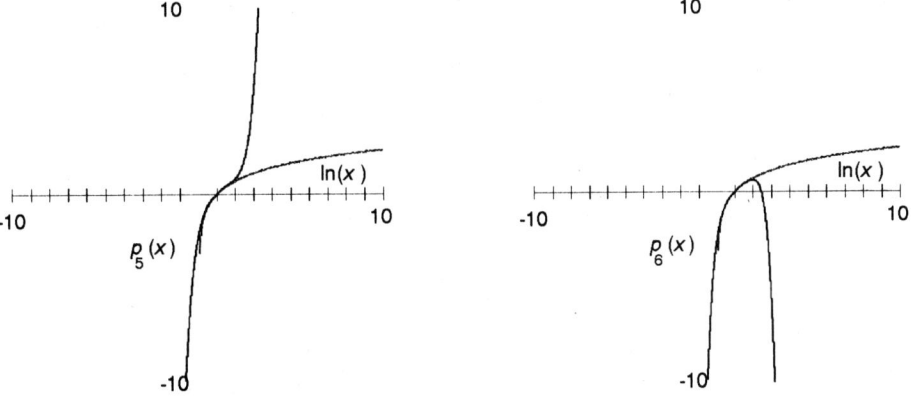

Figure 11.8 Degree 5 and 6 Taylor Polynomials at $x = 1$ of $\ln(x)$.

■

A Taylor polynomial approximation at $x = 0$ is sometimes known as a **Maclaurin polynomial**, named after the Scottish mathematician Colin Maclaurin (1698-1746).

EXAMPLE 48 Find the degree 1, 3, 5, 7, 9, and 11 Maclaurin polynomials (Taylor polynomials at $x = 0$) of the function $x \longmapsto \sin(x)$. Graph each one of the approximations along with $y = \sin(x)$.

Solution   The derivatives of $y = \sin(x)$ "cycle" as follows:

$$y' = \cos(x), \qquad y'' = -\sin(x), \qquad y''' = -\cos(x), \qquad y^{(4)} = \sin(x),$$

and we're back to our original function. If we evaluate the original function and its derivatives at $x = 0$, we have

$$y(0) = 0, \qquad y'(0) = 1 \qquad y''(0) = 0 \qquad y'''(0) = -1$$

and this pattern continues indefinitely.

This means that all the even coefficients $c_0, c_2, c_4, \ldots$ are 0, and the odd coefficients are $c_1 = 1$, $c_3 = -1/6$, $c_5 = 1/120$, $c_7 = -1/5040$, $c_9 = 1/9!$, and $c_{11} = -1/11!$. In general, the degree $2n + 1$ Maclaurin polynomial for $\sin(x)$ can be written

$$\sum_{k=0}^{n} \frac{(-1)^k}{(2k+1)!}(x-0)^{2k+1}.$$

Hence,

$$p_1(x) = x$$

$$p_3(x) = x - \frac{1}{6}x^3$$

$$p_5(x) = x - \frac{1}{6}x^3 + \frac{1}{120}x^5$$

$$p_7(x) = x - \frac{1}{6}x^3 + \frac{1}{120}x^5 - \frac{1}{5040}x^7$$

$$p_9(x) = x - \frac{1}{6}x^3 + \frac{1}{120}x^5 - \frac{1}{5040}x^7 + \frac{1}{9!}x^9$$

$$p_{11}(x) = x - \frac{1}{6}x^3 + \frac{1}{120}x^5 - \frac{1}{5040}x^7 + \frac{1}{9!}x^9 - \frac{1}{11!}x^{11}$$

The graphs are shown in Figures 11.9, 11.10, and 11.11.

11.4 TAYLOR POLYNOMIALS AND SERIES

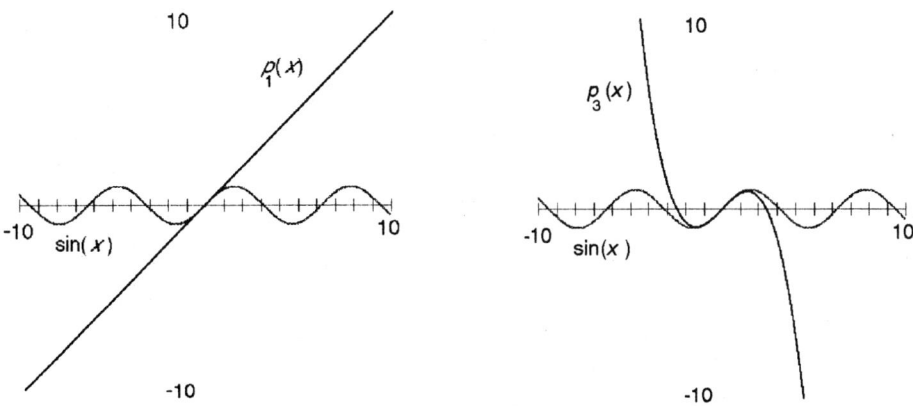

Figure 11.9 Degree 1 and 3 Taylor Polynomials at 0 of $\sin(x)$.

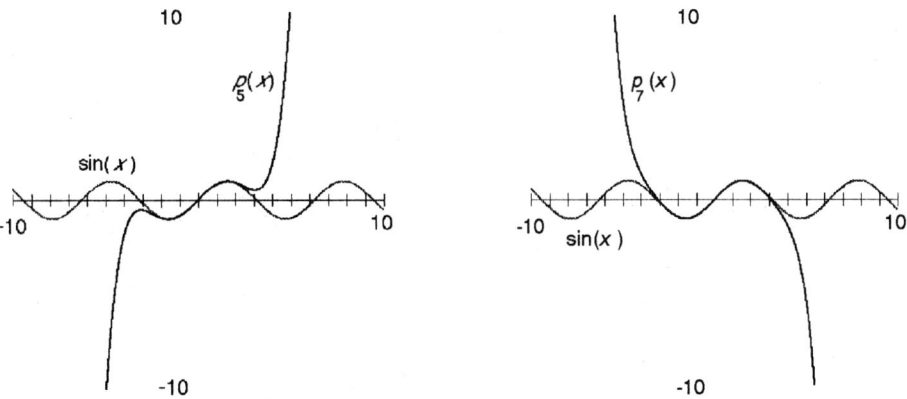

Figure 11.10 Degree 5 and 7 Taylor Polynomials at 0 of $\sin(x)$.

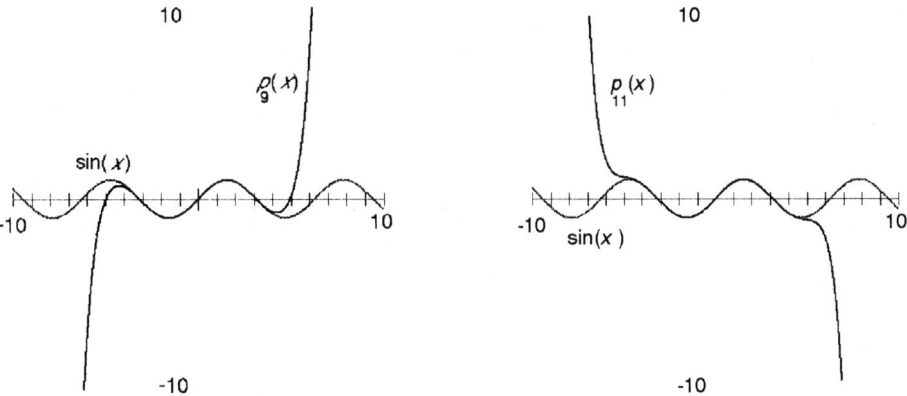

Figure 11.11 Degree 9 and 11 Taylor Polynomials at 0 of $\sin(x)$.

■

EXAMPLE 49 Find the first through sixth degree Maclaurin polynomials of the function $x \longmapsto \exp(x) + 1$. Graph each one of the approximations along with $y = \exp(x) + 1$.

Solution All of the derivatives are the same:
$$y^{(k)} = \exp(x) \qquad \text{for } k = 1, 2, 3 \ldots$$

Since $(\exp(0) + 1) = 2$, the constant term is 2 and we can write the degree n Maclaurin polynomial as
$$2 + \sum_{k=1}^{n} \frac{1}{k!} x^k.$$

Thus,
$$p_1(x) = 2 + x$$
$$p_2(x) = 2 + x + \frac{1}{2}x^2$$
$$p_3(x) = 2 + x + \frac{1}{2}x^2 + \frac{1}{6}x^3$$
$$p_4(x) = 2 + x + \frac{1}{2}x^2 + \frac{1}{6}x^3 + \frac{1}{24}x^4$$
$$p_5(x) = 2 + x + \frac{1}{2}x^2 + \frac{1}{6}x^3 + \frac{1}{24}x^4 + \frac{1}{120}x^5$$
$$p_6(x) = 2 + x + \frac{1}{2}x^2 + \frac{1}{6}x^3 + \frac{1}{24}x^4 + \frac{1}{120}x^5 + \frac{1}{720}x^6.$$

The graphs of these are shown in Figures 11.12, 11.13, and 11.14. Notice how the polynomial approximations diverge from the graph of $y = \exp(x) + 1$ as we move away from the input $x = 0$. For any given p_n of higher degree, we would still see this graphical behavior. The graph of our function $y = \exp(x) + 1$ has a horizontal asymptote $y = 1$, since
$$\lim_{x \to -\infty} \exp(x) + 1 = 1.$$

On the other hand, any polynomial p_n of degree $n \geq 1$ will exhibit unbounded behavior as $x \to -\infty$.

In some sense, the best approximation of $y = \exp(x) + 1$ for large magnitude negative values x is the *constant* Taylor polynomial
$$p_0(x) = 1.$$

11.4 TAYLOR POLYNOMIALS AND SERIES

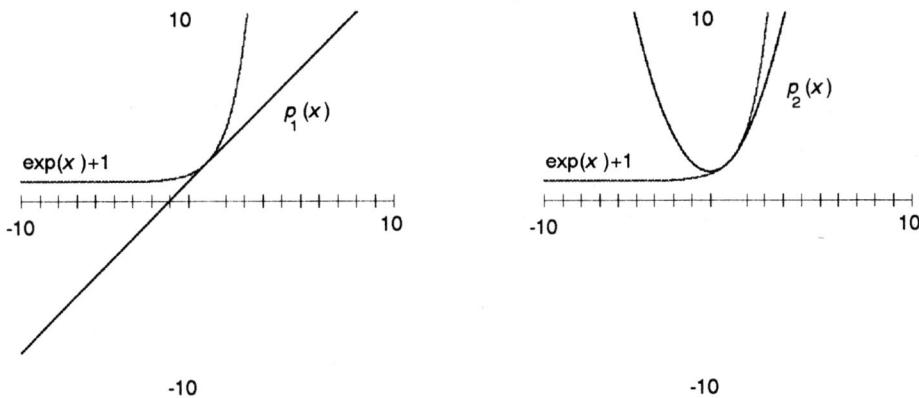

Figure 11.12 Degree 1 and 2 Maclaurin Polynomials of $\exp(x) + 1$.

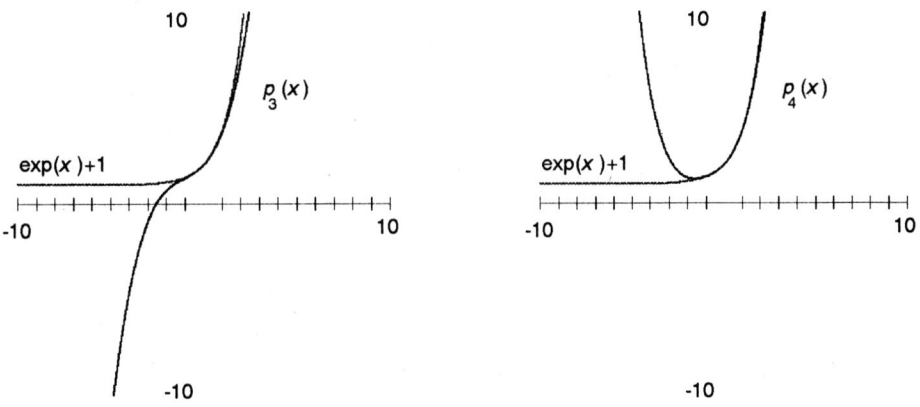

Figure 11.13 Degree 3 and 4 Maclaurin Polynomials of $\exp(x) + 1$.

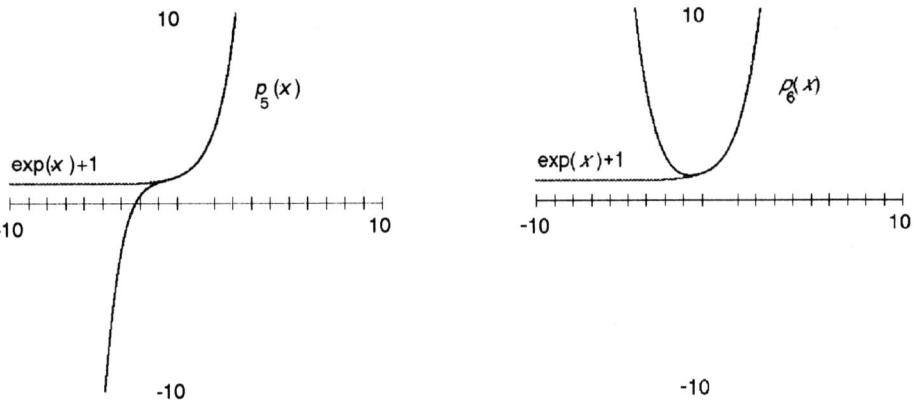

Figure 11.14 Degree 5 and 6 Maclaurin Polynomials of $\exp(x) + 1$.

■

EXAMPLE 50 Find the degree first through sixth degree Taylor polynomials at $x = 3$ of the function $x \longmapsto x^{-1}$. Graph each one of the approximations along with $y = x^{-1}$.

Solution Since

$$y' = -1 \cdot x^{-2}$$

$$y'' = \frac{d}{dx}(-1 \cdot x^{-2}) = (-2) \cdot (-1) \cdot (x^{-3})$$

$$y''' = \frac{d}{dx}(-2) \cdot (-1) \cdot (x^{-3}) = (-3) \cdot (-2) \cdot (-1) \cdot (x^{-4})$$

and so on, we have

$$y^{(k)} = (-1)^n \cdot (n!) \cdot (x^{-(n+1)}) \quad \text{and} \quad y^{(k)}(3) = (-1)^n \cdot (n!) \cdot (3^{-(n+1)}).$$

We can write the degree n Taylor polynomial at $x = 3$ as

$$\sum_{k=0}^{n} \frac{(-1)^k \cdot (k!) \cdot (3^{-(k+1)})}{k!} (x-3)^k = \sum_{k=0}^{n} (-1)^k \cdot (3^{-(k+1)}) \cdot (x-3)^k.$$

So

$$p_1(x) = \frac{1}{3} - \frac{1}{9}(x-3)$$

$$p_2(x) = \frac{1}{3} - \frac{1}{9}(x-3) + \frac{1}{27}(x-3)^2$$

$$p_3(x) = \frac{1}{3} - \frac{1}{9}(x-3) + \frac{1}{27}(x-3)^2 - \frac{1}{81}(x-3)^3$$

$$p_4(x) = \frac{1}{3} - \frac{1}{9}(x-3) + \frac{1}{27}(x-3)^2 - \frac{1}{81}(x-3)^3$$
$$+ \frac{1}{243}(x-3)^4$$

$$p_5(x) = \frac{1}{3} - \frac{1}{9}(x-3) + \frac{1}{27}(x-3)^2 - \frac{1}{81}(x-3)^3$$
$$+ \frac{1}{243}(x-3)^4 - \frac{1}{729}(x-3)^5$$

$$p_6(x) = \frac{1}{3} - \frac{1}{9}(x-3) + \frac{1}{27}(x-3)^2 - \frac{1}{81}(x-3)^3$$
$$+ \frac{1}{243}(x-3)^4 - \frac{1}{729}(x-3)^5 + \frac{1}{2178}(x-3)^6.$$

The graphs are shown in Figures 11.15, 11.16, and 11.17.

11.4 TAYLOR POLYNOMIALS AND SERIES

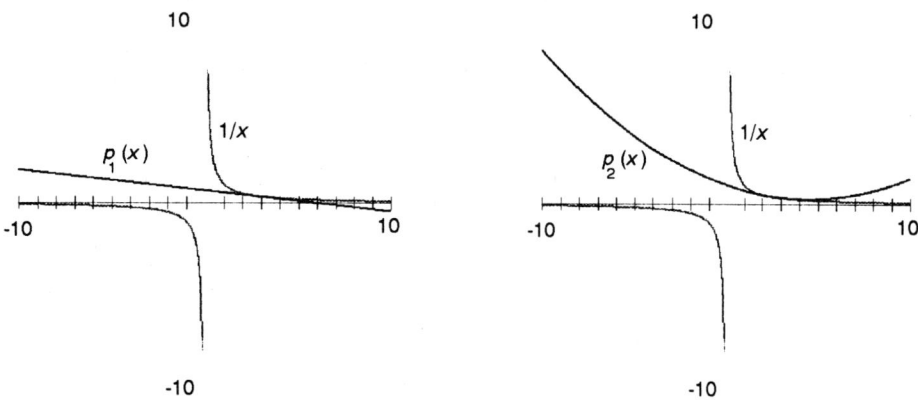

Figure 11.15 Degree 1 and 2 Taylor Polynomials at $x = 3$ of $1/x$.

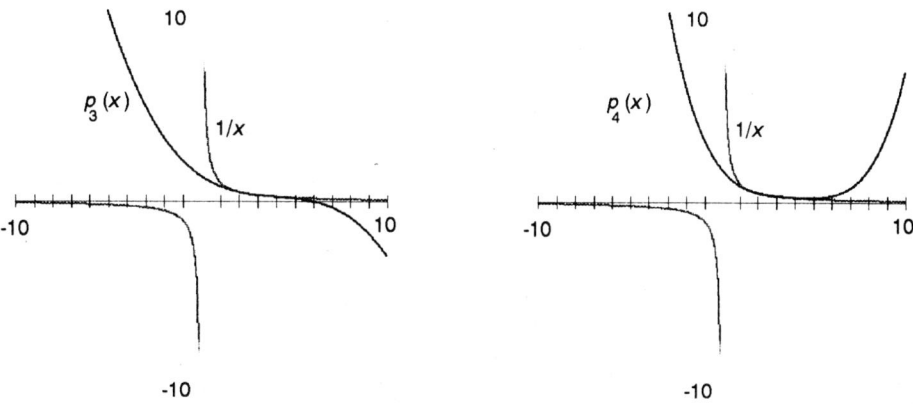

Figure 11.16 Degree 3 and 4 Taylor Polynomials at $x = 3$ of $1/x$.

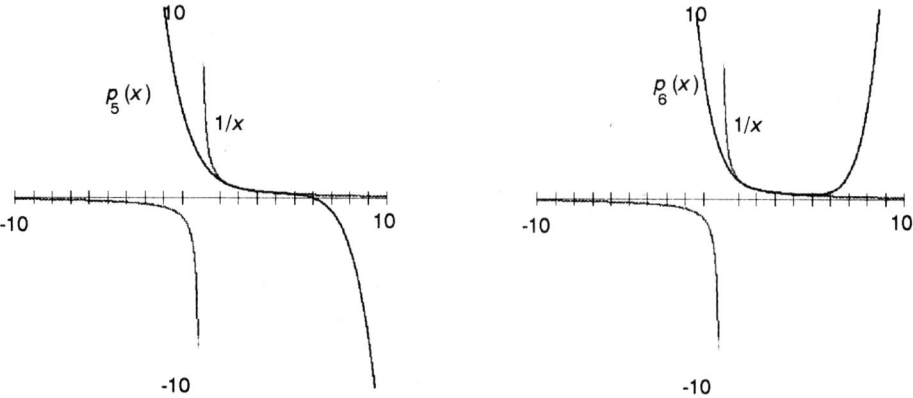

Figure 11.17 Degree 5 and 6 Taylor Polynomials at $x = 3$ of $1/x$.

■

Note in the graphs presented in these examples how more and more of the behavior, the "shape" of the function near the point of interest, is reflected in the higher and higher order Taylor and Maclaurin polynomials. Also note that, for some functions, the higher Taylor polynomials seem to get generally closer to the approximated function, while in other cases the higher order Taylor polynomials get closer and closer near the point of interest, but seem to move further away in other regions.

Suppose now that we are trying to use the Taylor polynomials at $x = a$ to estimate the function value of f at some other point, say $x = b$. Will it be true that higher order Taylor polynomials yield closer approximations to the function value? Absolutely not. The only sense in which the Taylor polynomials at $x = a$ are *guaranteed* to be on their best behavior is near $x = a$.

Taylor's Theorem

While it is true that a Taylor polynomial is not guaranteed to be a good approximation to a function other than very near the specified point $x = a$, we can get a handle on the error by means of a theorem which relates the error to the first "missing" derivative of the function.

Theorem 11.6

Taylor's Theorem.
Hypothesis 1: f is continuous and $(n+1)$-times differentiable on $[a, b]$.
Hypothesis 2: p_n is the nth degree Taylor polynomial approximation at $x = a$ to f.
Conclusion: There is at least one point c, with $a < c < b$, such that
$$\frac{f^{(n+1)}(c)}{(n+1)!} = \frac{f(b) - p_n(b)}{(b-a)^{n+1}}.$$
Similarly, the theorem holds when the order of a and b are reversed ($b < c < a$).

□

In the special case $n = 0$, we can consider the Taylor polynomial to be the constant function $p_0(x) = f(a)$. The conclusion becomes
$$f'(c) = \frac{f(b) - f(a)}{b - a},$$
which is simply a restatement of the Mean Value Theorem, and we can think of Taylor's Theorem as a generalization of the Mean Value Theorem.

11.4 TAYLOR POLYNOMIALS AND SERIES

The quantity $\dfrac{f^{(n+1)}(c)}{(n+1)!}(b-a)^{n+1}$ is often called the **Lagrange form** of the remainder (error term) $R_n(b)$ of the Taylor polynomial. Lagrange (1736-1813) was an Italian-French mathematician (who played a leading role in the introduction of the metric system in France). Using this notation, the conclusion of Taylor's theorem can be written

$$f(b) = p_n(b) + R_n(b) = \sum_{k=0}^{n} \frac{f^{(k)}(a)}{k!}(b-a)^k + R_n(b).$$

Taylor's theorem can be used to turn an estimate for the $(n+1)$st derivative of a function into an estimate of the error in the n degree Taylor polynomial approximation:

$$|f(b) - p_n(b)| = \left| \frac{f^{(n+1)}(c)}{(n+1)!}(b-a)^{n+1} \right|.$$

EXAMPLE 51 Suppose f is continuous and five times differentiable on $[1,3]$, and that

$$\left| f^{(5)}(x) \right| < 0.1$$

for $x \in [1,3]$. What is the biggest possible error if we use fourth Taylor polynomial approximation at $x = 1$ to approximate $f(3)$?

Solution

$$|f(3) - p_4(3)| = \left| \frac{f^{(5)}(c)}{(5)!}(3-1)^5 \right| \quad \text{for some } 1 < c < 3 \text{ so that}$$

$$|f(3) - p_4(3)| < \left| \frac{0.1}{120} 2^5 \right| \approx 0.026667.$$

∎

EXAMPLE 52 Use Taylor's Theorem to estimate the error at 0.5 for the fourth degree Maclaurin polynomial approximation to $x \longmapsto \sin(x)$.

Solution Since $|\sin(x)| \leq 1$ and $|\cos(x)| \leq 1$ for all x. Together these mean that we have a simple estimate for the nth derivative of f, namely $\left| f^{(n)}(x) \right| \leq 1$ for all x. Using this estimate in the remainder, we have

$$|\sin(0.5) - p_4(0.5)| \leq \left| \frac{1}{120}(0.5)^5 \right| \approx 0.00026.$$

∎

We can also express the **Integral form** of the remainder as

$$R_n(b) = \frac{1}{n!} \int_a^b (b-x)^n f^{(n+1)}(x)\, dx.$$

Taylor and Maclaurin Series

Suppose now that f is an infinitely differentiable function (as are almost all the functions we have dealt with), then we can extend the sequence of Taylor polynomials indefinitely to give the **Taylor Series** at $x = a$ for f

$$\sum_{k=0}^{\infty} \frac{f^{(k)}(a)}{k!}(x-a)^k$$

and, taking $a = 0$, the **Maclaurin Series** for f

$$\sum_{k=0}^{\infty} \frac{f^{(k)}(0)}{k!} x^k.$$

EXAMPLE 53 Compute the Maclaurin series for $f : x \longmapsto e^x$.

Solution Since $f(x) = f'(x) = f''(x) = \cdots = e^x$ and $f(0) = 1$, we have

$$\sum_{k=0}^{\infty} \frac{f^{(k)}(0)}{k!} x^k = \sum_{k=0}^{\infty} \frac{1}{k!} x^k.$$

∎

We can ask two questions regarding this power series:

For what values x does the Taylor (or Maclaurin) series converge?

In particular, for what values x does the series converge to $f(x)$?

The partial sums of a Taylor series are Taylor polynomials, so we are really asking about the convergence of the sequence of Taylor polynomials.

The convergence of the Taylor series to $f(x)$ is then equivalent to the convergence of the remainders $R_n(x)$ to 0. That is, the equality

$$f(x) = \sum_{k=0}^{\infty} \frac{f^{(k)}(a)}{k!}(x-a)^k$$

holds if and only if $\lim_{n \to \infty} R_n(x) = 0$.

11.4 TAYLOR POLYNOMIALS AND SERIES

EXAMPLE 54 For which values of x does the Maclaurin series for $f : x \longmapsto e^x$ converge to e^x?

Solution The remainder for the nth degree Maclaurin polynomial in this case is

$$R_n(x) = \frac{e^c}{(n+1)!} x^{(n+1)}$$

where c is between 0 and x (note that c depends on n). Since e^x is positive and increasing, we can estimate the remainder by noting that $e^c \leq e^{|x|}$ for any c between 0 and x. Using the ratio test, we find that the series converges for all x, so its terms must approach 0.

This means $\lim_{k \to \infty} \frac{x^k}{k!} = 0$, and since $|R_n(x)| \leq e^{|x|} \left| \frac{x^k}{k!} \right|$, we can conclude that $\lim_{k \to \infty} R_k(x) = 0$ *for any* x. ∎

A Taylor series need not converge everywhere in the domain of the function, as shown by the following example.

EXAMPLE 55 Graph the 5th and 10th order Taylor polynomials at 1 of $f : x \longmapsto \sqrt{x}$. Judge from this the region in which the corresponding Taylor series converges.

Solution

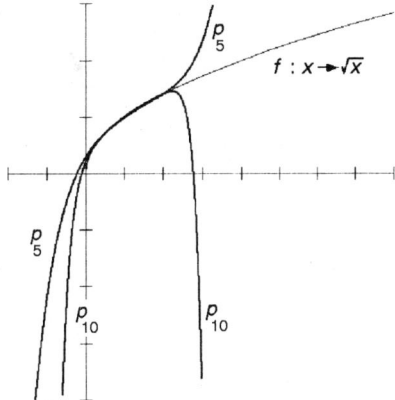

Figure 11.18 The 5th and 10th order Taylor polynomials at $x = 1$ of $\sqrt{x}$

It appears from the graph in Figure 11.18 that the series is diverging from the function outside the region $(0, 2)$ and converging for some region inside this one. You can check that the interval of convergence for this Taylor series is $(0, 2)$ by using the ratio test. ∎

A **point of validity** for a Taylor series is an input x for which the power series *converges to the function output* $f(x)$. For reference, here is a table of common Maclaurin series and their respective intervals of validity.

MACLAURIN SERIES	INTERVAL OF VALIDITY
$e^x = \sum_{k=0}^{\infty} \dfrac{x^k}{k!} = 1 + x + \dfrac{x^2}{2!} + \dfrac{x^3}{3!} + \dfrac{x^4}{4!} + \cdots$	$-\infty < x < +\infty$
$\sin x = \sum_{k=0}^{\infty} (-1)^k \dfrac{x^{2k+1}}{(2k+1)!} = x - \dfrac{x^3}{3!} + \dfrac{x^5}{5!} - \dfrac{x^7}{7!} + \cdots$	$-\infty < x < +\infty$
$\cos x = \sum_{k=0}^{\infty} (-1)^k \dfrac{x^{2k}}{(2k)!} = 1 - \dfrac{x^2}{2!} + \dfrac{x^4}{4!} - \dfrac{x^6}{6!} + \cdots$	$-\infty < x < +\infty$
$\ln(1+x) = \sum_{k=0}^{\infty} (-1)^k \dfrac{x^{k+1}}{k+1} = x - \dfrac{x^2}{2} + \dfrac{x^3}{3} - \dfrac{x^4}{4} + \cdots$	$-1 < x \leq 1$
$\tan^{-1} x = \sum_{k=0}^{\infty} (-1)^k \dfrac{x^{2k+1}}{2k+1} = x - \dfrac{x^3}{3} + \dfrac{x^5}{5} - \dfrac{x^7}{7} + \cdots$	$-1 \leq x \leq 1$
$\dfrac{1}{1-x} = \sum_{k=0}^{\infty} x^k = 1 + x + x^2 + x^3 + \cdots$	$-1 < x < 1$
$\sinh x = \sum_{k=0}^{\infty} \dfrac{x^{2k+1}}{(2k+1)!} = x + \dfrac{x^3}{3!} + \dfrac{x^5}{5!} + \dfrac{x^7}{7!} + \cdots$	$-\infty < x < +\infty$
$\cosh x = \sum_{k=0}^{\infty} \dfrac{x^{2k}}{(2k)!} = 1 + \dfrac{x^2}{2!} + \dfrac{x^4}{4!} + \dfrac{x^6}{6!} + \cdots$	$-\infty < x < +\infty$

11.4 TAYLOR POLYNOMIALS AND SERIES

The next example is a case where the Taylor series converges everywhere but only converges to the function at a single point. An alternate characterization of a Taylor polynomial will be useful here, and it highlights more precisely what we mean by saying that two functions have similar behavior. Instead of specifying that a degree n Taylor polynomial at $x = a$ must have its first n derivatives match those of the approximated function at $x = a$, we can say instead that the Taylor polynomial's outputs must approach the function's outputs *faster* than $(x - a)^n$ approaches 0. In other words, p_n must satisfy

$$\lim_{x \to a} \left| \frac{f(x) - p_n(x)}{(x-a)^n} \right| = 0.$$

Indeed, satisfying this single equality forces all the derivatives of p_n up to order n to match those of f (apply L'Hôpital's Rule repeatedly and solve for the coefficients c_n).

EXAMPLE 56 Find the Maclaurin series of the function

$$f : x \longmapsto \begin{cases} \exp(-\frac{1}{x^2}) & \text{for } x \neq 0 \\ 0 & \text{for } x = 0 \end{cases}$$

and determine for which inputs x the series converges to the function value.

Solution The simplest way to determine the coefficients of the Maclaurin series of this function is by means of the limit criteria we just discussed. The nth order Taylor polynomial p_n at $x = 0$ must satisfy

$$\lim_{x \to 0} \left| \frac{f(x) - p_n(x)}{x^n} \right| = 0.$$

The function outputs $f(x)$ approach zero so fast as $x \to 0$ that all the coefficients of p_n must be 0. Consider

$$\lim_{x \to 0} \left| \frac{f(x) - 0}{(x-0)^n} \right| = \lim_{x \to 0} \left| \frac{\exp(-\frac{1}{x^2})}{(x)^n} \right|$$

$$= \lim_{x \to 0+} \frac{\exp(-\frac{1}{x^2})}{(x)^n}$$

$$= \lim_{x \to 0+} \frac{\exp(-\frac{1}{x^2})}{\exp(n \ln(x))}$$

$$= \lim_{x \to 0+} \exp(-\frac{1}{x^2} - n \ln(x))$$

$$= \lim_{x \to 0+} \exp(\frac{-1 - x^2 n \ln(x)}{x^2}).$$

We know that $\lim_{x\to 0+} x^2 n \ln(x) = 0$ for any n (use L'Hôpital's rule on $\frac{n\ln(x)}{x^{-2}}$) so that

$$\lim_{x\to 0+} \exp\left(\frac{-1 - x^2 n \ln(x)}{x^2}\right) = \lim_{x\to 0+} \exp\left(\frac{-1}{x^2}\right) = 0.$$

The Maclaurin series is then

$$\sum_{n=0}^{\infty} 0 \cdot x^n = 0$$

and it converges for all x, but it converges to 0. On the other hand, $f(x) \neq 0$ for any x but $x = 0$. In conclusion, the Maclaurin series converges everywhere, but the only place it converges to the function value is at 0. Figure 11.19 shows a graph of the function f. ∎

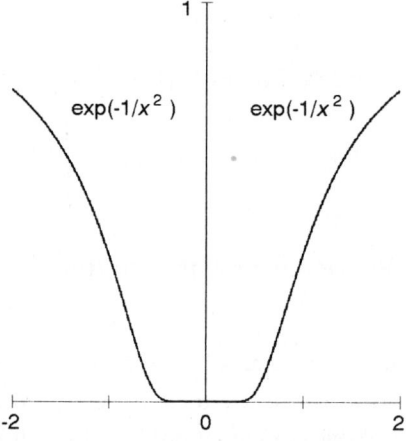

Figure 11.19 The Maclaurin series of f converges everywhere, but to f only at $x = 0$.

Using Taylor polynomials and series

While the Taylor polynomial approximation is sometimes useful for a function you know a lot about, it is more often used to approximate otherwise unknown functions. For example, we can use Taylor polynomials to approximate solutions to rather difficult equations. In this context, note that to find a Taylor polynomial approximation, we need only find the values of the derivatives of a function at a single point. Any method allowing us to produce these values provides an approximation to the function.

We will examine two such situations. In the first, we use implicit differentiation to inductively determine derivatives of a function satisfying an equation which can't be solved algebraically.

11.4 TAYLOR POLYNOMIALS AND SERIES

EXAMPLE 57 Find the first four terms of a Taylor series for a function $f : x \longmapsto f(x)$ satisfying the equation

(1) $$f(x) + \sin(f(x)) = x.$$

Solution The first step in the solution is to find an input $x = a$ and corresponding output value $f(a)$ such that

$$f(a) + \sin(f(a)) = a.$$

This will give us a point a around which to construct the Taylor series, the first term being $f(a)$. Nothing beyond guesswork and trial-and-error will help at this stage. However, trying the "obvious" guess, $a = 0$, and noting that $\sin(0) = 0$, we find that a possibility is $0 = a = f(a)$. With this as a starting point, we can systematically build as much of the Taylor series as we wish.

To get the next term in the Taylor series, $\dfrac{f'(0)}{1!}x$, we need to find $f'(0)$. To do this we implicitly differentiate equation (1) with respect to x:

(2) $$f'(x) + \cos(f(x)) \cdot f'(x) = 1.$$

We only need to know $f'(0)$ and we already know that $f(0) = 0$. Therefore, $f'(0)$ satisfies

$$f'(0) \cdot (1 + \cos(f(0))) = f'(0) \cdot (1 + \cos(0)) = 2f'(0) = 1$$

so $f'(0) = 0.5$, and we now know the first two terms of the Taylor series:

$$0 + 0.5x + \cdots.$$

To find the next term, we implicitly differentiate equation (2) to obtain

(3) $$f''(x) \cdot (1 + \cos(f(x))) - (f'(x))^2 \sin(f(x)) = 0.$$

Again, we only need to know $f''(0)$ and we already know $f(0) = 0$ and $f'(0) = 0.5$. so that $f''(0)$ satisfies

$$f''(0) \cdot (1 + \cos(0)) - (0.5)^2 \sin(0) = f''(0) \cdot (2) - 0 = 0$$

so $f''(0) = 0$, and we now know the first three terms of the Taylor series:

$$0 + 0.5x + \frac{0}{2!}x^2 + \cdots.$$

To find the fourth term, we implicitly differentiate equation (3):

$$f'''(x) \cdot (1 + \cos(f(x))) - 3f''(x)f'(x)\sin(f(x)) - (f'(x))^3 \cos(f(x)) = 0.$$

Again, we only need to know $f'''(0)$, and we already know $f(0) = 0$, $f'(0) = 0.5$, and $f''(0) = 0$. The value $f'''(0)$ must satisfy

$$f'''(0) \cdot (1 + \cos(f(0))) - 3f''(0)f'(0)\sin(f(0)) - (f'(0))^3 \cos(f(0)) = 0$$

or

$$f'''(0) \cdot (2) - (0.5)^3 = 0.$$

Hence, $f'''(0) = 0.625$, and we have the first four terms of the Taylor series:

$$0 + 0.5x + \frac{0}{2!}x^2 + \frac{0.625}{3!}x^3 + \cdots,$$

completing the problem.

Figure 11.20 shows the graph of the third degree Taylor polynomial represented by these known terms together with the graph of the "true" solution obtained by numerically solving, for each x plotted, $y + \sin(y) = x$ for y.

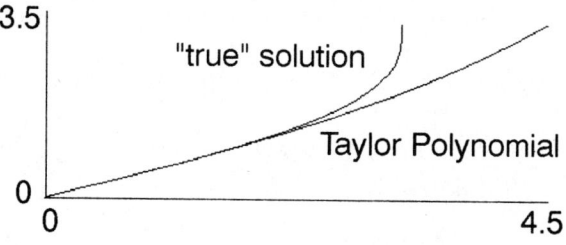

Figure 11.20 A Taylor approximation to the solution of $f(x) + \sin(f(x)) = x$.

∎

The method we have used to construct Taylor series was based on matching derivative values. This method can be used in general to find polynomial approximations to solutions of differential equations.

EXAMPLE 58   Find a third degree polynomial about $x = 0$ to approximate the solution of the differential equation

$$y' = x + y/2 \quad \text{with initial condition} \quad y(0) = 1.$$

Solution   A power series about $x = 0$ has the general form

$$y = c_0 + c_1 x + c_2 x^2 + c_3 x^3 + \cdots$$

whose derivative is $y' = c_1 + 2c_2 x + 3c_3 x^2 + \cdots$ Substituting into the given differential equation, we have

$$c_1 + 2c_2 x + 3c_3 x^2 + \cdots = x + \frac{c_0 + c_1 x + c_2 x^2 + c_3 x^3 + \cdots}{2}.$$

11.4 TAYLOR POLYNOMIALS AND SERIES

The initial condition $y(0) = 1$ tells us that $c_0 = 1$, and we have

$$c_1 + 2c_2 x + 3c_3 x^2 + \cdots = \frac{1}{2} + (1 + \frac{c_1}{2})x + \frac{c_2}{2}x^2 + \frac{c_3}{2}x^3 + \cdots.$$

We can equate coefficients of the two sides of the equation to see that

$$c_1 = \frac{1}{2} \qquad 2c_2 = 1 + \frac{c_1}{2} \qquad 3c_3 = \frac{c_2}{2}$$

or

$$c_1 = \frac{1}{2} \qquad c_2 = \frac{5}{8} \qquad c_3 = \frac{5}{48}.$$

Notice that we could extend this analysis to any degree n in a similar way. We have the coefficients we need for a third degree polynomial approximation

$$y(x) = 1 + x/2 + 5x^2/8 + 5x^3/48.$$

■

Figure 11.21 shows a graph of the actual solution to the differential equation and the graph of our third degree polynomial approximaton.

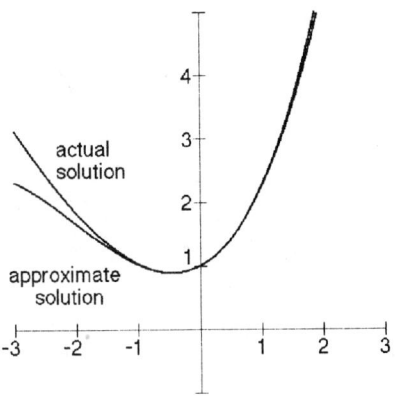

Figure 11.21 Actual and approximate solutions to $y' = x + y/2$ with $y'(0) = 1$.

EXERCISES

In exercises 1-10, find the first four terms of the Taylor series (that is, the third degree Taylor polynomial) of the given function about the given point.

1. $x \longmapsto \sin(2x + 1)$ $a = -1$
2. $x \longmapsto \cos(\frac{2\pi}{x^2+1})$ $a = 0$
3. $x \longmapsto \tan(x)$ $a = \pi/4$
4. $x \longmapsto \sec(x)$ $a = 0$
5. $x \longmapsto \csc(-x)$ $a = \pi/2$
6. $x \longmapsto -\cot(x)$ $a = 3\pi/4$
7. $x \longmapsto \exp(2x + 1)$ $a = -1$
8. $x \longmapsto 2/(x^2 + 1)$ $a = 0$
9. $x \longmapsto \sqrt{x^2 + 1}$ $a = 0$
10. $x \longmapsto 1/x$ $a = 4$

In exercises 11-20, graphically determine the point c between the given a and b that determines the remainder $R_3(b)$ of the third degree Taylor polynomial. In other words, find where the graph of $y = f^{(4)}(x)/24$ crosses the graph of the horizontal line

$$y = \frac{f(b) - p_3(b)}{(b-a)^4}$$

over the interval $[a, b]$.

11. $x \longmapsto \sin(2x + 1)$ $a = -1$, $b = -0.5$

12. $x \longmapsto \cos(\frac{2\pi}{x^2+1})$ $a = 0$, $b = 1$

13. $x \longmapsto \tan(x)$ $a = \pi/4$, $b = \pi/6$

14. $x \longmapsto \sec(x)$ $a = 0$, $b = 0.5$

15. $x \longmapsto \csc(-x)$ $a = \pi/2$, $b = 5\pi/6$

16. $x \longmapsto -\cot(x)$ $a = 3\pi/4$, $b = 7\pi/8$

17. $x \longmapsto \exp(2x + 1)$ $a = -1$, $b = 0$

18. $x \longmapsto \frac{2}{x^2+1}$ $a = 0$, $b = 1$

19. $x \longmapsto \sqrt{x^2 + 1}$ $a = 0$, $b = 1$

20. $x \longmapsto \frac{1}{x}$ $a = 4$, $b = 4.5$

In exercises 21-27, find the first four terms of the Taylor series (third order Taylor polynomial) about the given point, a, of a function f satisfying the given equation.

21. $x^2 - f(x) = \ln(f(x))$ $a = 1$
22. $f(x) \exp(x + f(x) - 1) = 1$ $a = 0$
23. $\sqrt{f(x) + x} + \sqrt{f(x) + x^2} - 2 = 0$ $a = 0$
24. $x \cos(f(x)) - f(x) = 0$ $a = 0$
25. $\exp(\ln(x) f(x)) = f(x)$ $a = 1$
26. $f(x)^5 - 3f(x)^3 + 3f(x) - x = 0$ $a = 1$

In exercises 27-30 find a second degree polynomial approximation about $x = 0$ to approximate the solution to the given differential equation with initial condition $y(0) = 0$.

27. $y' = x^2 - (0.1) \cdot (x + y)$
28. $y' = \cos(y)$
29. $y' = 1 - y^2$
30. $y' = x \cos(y/2)$

---

31. Suppose the second order Taylor polynomial at a of $f : x \longmapsto f(x)$ is

$$c_0 + c_1(x - a) + \frac{c_2}{2}(x - a)^2$$

and the second order Taylor polynomial at c_0 of $g : y \longmapsto g(y)$ is

$$q_0 + q_1(y - c_0) + \frac{q_2}{2}(y - c_0)^2.$$

What is the second order Taylor polynomial at a of $h : x \longmapsto g(f(x))$?

32. Calculate $\sin(0.5)$ with an error of at most 0.0001 using a Taylor polynomial at 0 of $x \longmapsto \sin(x)$.

33. Calculate $\ln(1.2)$ with an error of at most 0.001 using a Taylor polynomial at 1 of $x \longmapsto \ln(x)$.

34. Calculate $\sqrt{(1.2)}$ with an error of at most 0.001 using a Taylor polynomial at 1 of $x \longmapsto \sqrt{(x)}$.

35. Calculate $e = \exp(1)$ with an error of at most 0.001 using a Taylor polynomial at 0 of $x \longmapsto \exp(x)$.

ANSWERS TO SELECTED EXERCISES

Chapter 0

1. Those fractions whose denominators have prime factorization $2^n \cdot 5^m$ when written in reduced form.   3. 1   5. No.   7. 5   9. $\dfrac{\sqrt{n}}{2}$ 11. $\sin^{-1}\left(\dfrac{1}{\sqrt{101}}\right) \cong .099668$ 13. $n = 18$ 15. $-.00001667 < e < .00001667$ 17. $\dfrac{6\pi}{5280} \cong .00357$ miles

Chapter 1

1.1

1. $|x+5| \leq 3$ 3. $|x| > 3$ 5. $-3 \leq x \leq 3$ 7. $x < 1 - \dfrac{\sqrt{10}}{2}$ or $-\dfrac{1}{2} < x < 1 + \dfrac{\sqrt{10}}{2}$ 9. $x > \tfrac{7}{4}$ or $x < \tfrac{5}{4}$ 11. There are none.   13. Equality occurs when $a \leq b \leq c$.   15. $|a-b| + |b-c| \geq |a-c|$ is reduced to $0 + 0 \geq 0$.   17. $\left|x + \dfrac{5}{2}\right| \leq \dfrac{5}{2}$ 19. $\left|x - \dfrac{17}{2}\right| \leq \dfrac{21}{2}$ 21. $(-4, 10)$ 23. $(-1, 1)$ 25. both open and closed, unbounded   27. Not an interval   29. both open and closed, unbounded   31. both open and closed, unbounded   33. $[-3, 3]$ 35. $(-\infty, \infty)$ 37. $\left(-\infty, -\sqrt{11}\right] \cup \left[-\sqrt{7}, \sqrt{7}\right] \cup \left[\sqrt{11}, \infty\right)$ 39. $(-\infty, -1) \cup (0, 1)$ 45. $(-\infty, 0) \times (0, \infty)$ 47. $(0, \infty) \times (-\infty, 0)$ 49. $(-3, 3) \times (-\infty, \infty)$ 51. $[-2, 4] \times [-5, 3]$ 53. $[3.717, 3.723]$ or $|x - 3.72| \leq .003$

1.2

1. $P : [5/\sqrt{2}, \infty) \longrightarrow \mathbb{R}$, $P : \ell \longmapsto (2+\sqrt{2})\ell$ 3. $S : [4, 14] \longrightarrow \mathbb{R}$, $S : d \longmapsto \pi d^2$ 5. yes   7. Not a function.   9. Not a function.   11. $xy = 4$ does describe a function.   13. Each line has $(0, -1)$ in common.   15. $m = -\dfrac{1}{5}$ 17. $m = \dfrac{1}{2}$ 19. If $a \neq 0$, $y = ax^2 + bx + c$ describes a parabola. If $a = 0$, $y = ax^2 + bx + c = bx + c$ and describes a straight line with slope b and y-intercept c.   21. Changing b affects the location of the vertex.   23. If $d > 0$, f crosses the x-axis twice. If $d = 0$, f touches/crosses the x-axis in one place. If $d < 0$, f does not touch the x-axis at all.

1.3

1. The degree of the resulting polynomial is the sum of the degrees of the original polynomials.   3. Yes. For example: $(x^2+x+1)+(-x^2+2) = x+1$.   5. $(1,1)$ 7. $p = 1, q = 1$ or 3; $p = 2, q = 1, 2, 3$ or 4; $p = 3, q = 1$ or 3; $p = 4, q = 1, 2, 3$ or 4 9. $p = 1, q = 2$ or 4; $p = 3, q = 2$ or 4 11. One or both of p and q is even.   13. $p = 2, q = 3$ or 4 15. $p = 1, q = 1; p = 3, q = 3$ 17. p is even.   19. Odd, if $b = 0$ (b is y-intercept); Otherwise, neither.   21. $f : x \longmapsto x^{\frac{p}{q}}$; Odd, if $D = \mathbb{R}$ or $D = \{x : x \neq 0\}$ and p, q odd; Even if $D = \mathbb{R}$ or $D = \{x : x \neq 0\}$ and p even; Otherwise neither.   23. Sine and Cosine are bounded, the other four are unbounded.   25. Yes. $f : x \longmapsto 0$, since $f(x) = -f(x)$.   27. f odd $\Rightarrow f(-x) = -f(x) \Rightarrow f(-0) = f(0) = -f(0) \Rightarrow f(0) = 0$.   29. Constant functions.

1.4

1. $D = \{x : x \neq \pm\sqrt{3}\}$ 3. $D = \mathbb{R}$ 5. $D = \mathbb{R}$ 7. $g \circ f \circ h \circ j$ 9. $g \circ k \circ j \circ f$ 11. $h \circ g \circ g \circ f$ 13. 1 15. -6 17. undefined   19. 3 21. -2 23. $x = 0, 3$ 25. $x = 0$ 41. Yes. $f^{-1} : x \longmapsto x$ 43. Yes. $f^{-1} : x \longmapsto \frac{1}{x}$ 45. No.   47. If $m \neq 0$, then the inverse is $f^{-1} : x \longmapsto \frac{x-b}{m}$. If $m = 0$, no inverse exists.

Chapter 2

2.1

31. b) $\lim_{x \to 0+} f(x) = 0$, $\lim_{x \to 0-} f(x) = -1$ c) Does not exist.
33. b) $\lim_{x \to 0+} f(x) = 0$, $\lim_{x \to 0-} f(x) = 0$
35. b) $\lim_{x \to 0+} f(x) = 1$, $\lim_{x \to 0-} f(x) = -1$ c) Does not exist.
37. b) $\lim_{x \to 0+} f(x) = +\infty$, $\lim_{x \to 0-} f(x) = \infty$ c) Does not exist.
39. b) The one-sided limits do not exist. c) Limit does not exist.
41. b) $\lim_{x \to 0+} f(x) = 0$, $\lim_{x \to 0-} f(x) = 0$ c) $\lim_{x \to 0} f(x) = e \approx 2.718$
43. b) $\lim_{x \to 0+} f(x) = 0$, $\lim_{x \to 0-} f(x) = 0$ c) $\lim_{x \to 0} f(x) = 0$
45. b) $\lim_{x \to 0+} f(x) = e \approx 2.718$, $\lim_{x \to 0+} f(x) = e \approx 2.718$

2.2

17. $\lim_{x \to 0} f(x)$ does not exist.   19. $L = 0$, $[-.009, .009]$ 21. $\lim_{x \to 0} f(x)$ does not exist.   23. $\lim_{x \to 0} f(x)$ does not exist.   25. $\lim_{x \to 0} f(x)$ does not exist.   27. $L = 0$, $[-.009, .009]$ 29. $L = 0$, $[-.01, .01]$ 31. $L \approx 2.7$, $(L = e)$, $[-.005, .005]$

2.3

1. Continuous   3. Continuous   5. hole   7. jump   9. continuous at $x = -1$ 11. continuous at $x = -3$ 13. jump at $x = \pi$ 15. $A = 1$ 17. $A = \frac{\pi}{2}$ 19. $A = \frac{-4}{3}$ 21. $x = k\pi$, k any integer   23. $x = \frac{\pi}{2} + k\pi$, k any integer   25. $x = 1$, $x = 2$

2.4

1. continuous   3. continuous   5. removable discontinuity (hole)   7. essential discontinuity (jump)   9. continuous   11. continuous   13. jump discontinuity   15. $A = 1$ or $A = -2$ 17. $A = \pi/2$ 19. $A = -4/3$ 21. poles at $x = k\pi$, k any integer   23. poles at $x = k\pi + \pi/2$, k andy integer   25. poles at $x = 2$, $x = 1$ 27. f is not continuous on $[-1, 1]$.   29. f is not continuous on the closed interval $[0,1]$.   33. $\lim_{x \to \infty} f(x) = \frac{\pi}{2}$, $\lim_{x \to -\infty} f(x) = \frac{-\pi}{2}$ 35. $\lim_{x \to \infty} f(x) = 0$, $\lim_{x \to -\infty} = 0$ 37. $\lim_{x \to \infty} 2^x = \infty$, $\lim_{x \to -\infty} 2^x = 0$ 39. a) $x = 1$, $x = 5$ b) $x = 2$, $x = 3$ c) $\lim_{x \to 2^-} f(x) = -\infty$, $\lim_{x \to 2^+} f(x) = +\infty$, $\lim_{x \to 3^-} f(x) = +\infty$, $\lim_{x \to 3^+} f(x) = -\infty$ d) horizontal asymptote: $y = 1$ 41. a) No zeros.   b) $x = 1$, $x = -3$ c) $\lim_{x \to 1^-} f(x) = -\infty$, $\lim_{x \to 1^+} f(x) = +\infty$ $\lim_{x \to 3^-} f(x) = +\infty$, $\lim_{x \to -3^+} f(x) = -\infty$ d) Horizontal asymptote: $y = 1$ 43. a) $x = -1$ b) $x = \frac{1}{2}$, $x = -2$ c) hole: $(-2, \frac{1}{5})$, $\lim_{x \to \frac{1}{2}^-} f(x) = -\infty$, $\lim_{x \to \frac{1}{2}^+} f(x) = \infty$ d) Horizontal asymptote: $y = \frac{1}{2}$ 45. a) $x = 3$ b) $x = 3$ c) $\lim_{x \to 3^-} f(x) = -\infty$, $\lim_{x \to 3^+} f(x) = \infty$ d) Horizontal asymptote: $y = 1$

Chapter 3

3.1

1. $y = 3x + 13$ 3. $y = -\frac{1}{2}x - \frac{5}{2}$ 5. $y = -\frac{1}{2}(x - (-1.5)) + -\frac{7}{4}$ 7. $y = -\frac{2}{7}$ 9. $y = \frac{1}{3}(x - 6.5) + -6.5$ 11. $(f \circ g)(x) = -\frac{3}{2}x + \frac{11}{2}$ $(g \circ f)(x) = -\frac{3}{2}x - \frac{18}{2}$ Yes, No.   13. b. $y = -\frac{16x}{9} + \frac{16}{9}$ c. $y = -\frac{16}{9}(x - 1) + 0$ d. $-1.\overline{7}$ e. All.   15. b. $y = \frac{1}{3}x + \frac{1}{3}$ c. $y = \frac{1}{3}(x - 1) + \frac{2}{3}$ d. .3333333333, .3333333333 .333333334, .33333334 .33333334 3. x_1 and x_2 17. b. $y = \frac{1}{12345}x + 0$ c. $y = \frac{1}{12345}(x-1) + \frac{1}{12345}$ d. .00008100445525, .00008100445545, .00008100445525, .0000810044553, .000081004455.   e. x_1, x_2, and x_3.   19. $\tan \theta = m$ 21. $m = \tan \frac{3\pi}{4} = -1$, $y = -(x - (-3))$ 23. $((x-1)+1)^2 + ((x-1)+1) + 1 = (x-1)^2 + 3(x-1) + 3$ 25. $((x-1)+1)^4 = (x-1)^4 + 4(x-1)^3 + 6(x-1)^2 + 4(x-1) + 1$.   27. $y = .79(x - .05) + 5.715$ $[0, 1]$ 29. $y = 1.87(x - 2.5) + 3.135$ $[2, 3]$

3.2

1. $\dfrac{40}{1.0} = 40$ mph hours 3. slowing down 5. 1.1 hrs 7. 12.5 mph at $t = 0.5$ 100 mph at $t = 1$ hour 9. D 11. C 13. 23.3, 37.3, 31.25, 87.5 15. A or C 17. 10 min. 19. 17 min. 21. 0 min. 23. 4 min. 25. 14.5 min.

3.3

1. 4 3. $-\dfrac{1}{2}$ 5. 0 7. $\dfrac{1}{4}$ 9. a) 4, 4 b) 4 c) In both cases, we obtain the exact derivative. 11. a) $-.500500501$, $-.4995005$ b) $-.5000005$ c) The symmetric difference gives a more accurate estimate of the derivative. 13. a) 0, 0 b) 0 c) In both cases, we obtain the exact derivative. 15. .25, .25 b) .25 c) In both cases, we obtain the exact derivative. 17. $\dfrac{2}{3}$ 19. $\dfrac{4}{3}$ 21. $-\dfrac{1}{2}$ 23. $\dfrac{1}{2}$ 25. $2\ln 2$ 27. $\dfrac{\sqrt{2}}{2}, -\dfrac{1}{2}$ 29. $\dfrac{1}{2}, 4$ 31. $\dfrac{1}{\sqrt{2}}, -\dfrac{2}{3}$ 33. $1, \dfrac{\sqrt{3}}{2}$ 35. 0, 6 37. $\dfrac{2}{5}$ 39. 1 41. 15 meters 43. 14.3944 meters 45. 8.3644 meters 47. -10.04689 m/s 49. -16.7288 m/s 51. -4.0374 m/s 53. -20.0928 m/s 55. 0

3.4

1. $3x^2 + 2x + 1$ 3. $\dfrac{10}{3}x - \dfrac{35}{4}x^4$ 5. $-\dfrac{1}{3}x$ 7. $\dfrac{-\sin x - \cos x}{7}$ 9. $-216x^5 - 72x^3 + 231x^2 + 16$

3.5

1. 3.55 3. 6 5. 18 7. 6.5 9. 684 11. 7.5 13. 4.8 15. 24.4 17. $\dfrac{1}{3}$ 19. $-\dfrac{1}{2}$ 21. Undefined since $g'(3)$ is undefined. 23. $\dfrac{-x^4 + 6x^3 - 6x^2 + 16x - 24}{(x^3 + 8)^2}$ 25. $\dfrac{7x^2 + 14x + 7}{(x^2 + 3x + 2)^2}$ 27. $-\csc x \cot x$ 29. $3x^2 \sin(x^2 + 2x - 3) + (2x^4 + 2x^3 + 54x + 54)\cos(x^2 + 2x - 3)$ 31. $(4x - 5)\sec^2(2x^2 - 5x + 1)$ 33. $-(8 + 8x)\sin(6 - 4x - 2x^2)\cos(6 - 4x - 2x^2)$ 35. $\dfrac{1 - x}{(7 + 4x - 2x^2)^{\frac{3}{4}}}$ 37. $\dfrac{(11x^2 - 10x + 1)^{\frac{1}{2}}(33x^3 + 380x^2 + 644x - 335)}{3(3x^2 + 5x + 7)^{\frac{7}{3}}}$ 39. $8x(x^2 + 1)^3$ 41. $\dfrac{10x - 17}{3(5x^2 - 17x)^{\frac{2}{3}}}$ 43. $\cos(\sin(x))\cos x$ 45. 0 or 0 47. $\dfrac{1}{27}$ 49. $-2\sqrt{\pi}$ 51. $-\sqrt{3}/2$ 53. -5

Chapter 4

4.1

1. $\Delta V \approx 4\pi r^2 \Delta r$ 3. $\dfrac{dV}{dr} = 4\pi r^2$, $\left.\dfrac{dV}{dr}\right|_{r=3} = 36\pi \dfrac{\text{ft}^3}{\text{ft}}$

5. $V(r+\Delta r) - V(r) = \dfrac{4}{3}\pi(r+\Delta r)^3 - \dfrac{4}{3}\pi r^3 = \dfrac{4}{3}\pi[r^3 + 3r^2\Delta r + 3r\Delta r^2 + \Delta r^3] - \dfrac{4}{3}\pi r^3 = 4\pi r^2 \Delta 4\pi r \Delta r^2 + \dfrac{4}{3}\pi \Delta r^3$. 7. $P = 4x$, $\dfrac{dP}{dx} = 4$ $\left.\dfrac{dP}{dx}\right|_{x=3.6} = \dfrac{4\text{cm}}{\text{cm}}$ 9. $y = \sqrt{2}x$, $\dfrac{dy}{dx} = \sqrt{2}$, $\left.\dfrac{dy}{dx}\right|_{x=3.6} = \sqrt{2}\dfrac{\text{cm}}{\text{cm}}$ 11. $V = \dfrac{z^3}{3\sqrt{3}}$, $\dfrac{dV}{dz} = \dfrac{z^2}{\sqrt{3}}$ $\left.\dfrac{dV}{dz}\right|_{z=3.6} = 7.48\dfrac{\text{cm}^3}{\text{cm}}$ 13. $C = \pi d$, $\dfrac{dc}{dd} = \pi$, $\left.\dfrac{dc}{dd}\right|_{d=3.6} = \pi\dfrac{\text{cm}}{\text{cm}}$ 15. $A = \dfrac{C^2}{4\pi}$, $\dfrac{dA}{dC} = \dfrac{C}{2\pi}$, $\left.\dfrac{dA}{dC}\right|_{C=3.6} = \dfrac{1.8\,\text{cm}^2}{\pi\,\text{cm}}$ 17. a) $V(t) = -32t$, $V(0) = 0\dfrac{\text{ft}}{\text{s}}$ b) $V(t) = 0$ when $t = 0$ s c) $64000 = -16t^2 + 64000$, $t = 0$ s d) $s(0) = 64000$ ft 19. a) $V(t) = -\dfrac{t}{15}$, $V(0) = 0$ ft/s b) $V(t) = 0$ when $t = 0$ s c) $90 = 90 - \dfrac{t^2}{30}$, $\dfrac{t^2}{30} = 0$, $t = 0$ s d) $S(0) = 90$ ft 21. Degrees: $A = \dfrac{\pi}{10}\theta$ and $\dfrac{dA}{d\theta} = \dfrac{\pi}{10}$ in^2/deg Radians: $A = 18\theta$ and $\dfrac{dA}{d\theta} = 18$ in^2/rad 23. Degrees: $\dfrac{dA}{d\theta} = \dfrac{\pi}{10}$ in^2/deg Radians: $\dfrac{dA}{d\theta} = 18$ in^2/rad 25. $\dfrac{dH}{d\theta} == .006$ in^2/deg $= -1.083$ in^2/rad

4.2

1. $f(x) \approx 9 - 6(x+3)$

x	Approx	Actual	Error
-2	3	4	1
-4	15	16	1
-2.999	8.994	8.994001	.000001
-3.001	9.006	9.006001	.000001

3. $f(x) \approx 4 + \dfrac{1}{3}(x+8)$

-7	$4.\overline{3}$	3.65930571003	.6740276233
-9	$3.\overline{6}$	4.32674871093	.3264153776
-7.999	$4.000\overline{3}$	3.99966665972	.00066667361
-8.001	$3.999\overline{6}$	4.00033332639	.00066665972

5. $f(x) \approx \dfrac{\sqrt{3}}{2}\dfrac{1}{2}(x - \dfrac{\pi}{3})$

$\dfrac{\pi}{3}+1$	$\dfrac{\sqrt{3}+1}{2}$	.888651015008	.47737438872
$\dfrac{\pi}{3}-1$	$\dfrac{\sqrt{3}-1}{2}$	.04718003020	.31884537358
$\dfrac{\pi}{3}+.001$	.866525403785	.86652497069	.001499566905
$\dfrac{\pi}{3}-.001$	.865026503785	.865524970857	.001000432928

7. $f(x) \approx 1 + 2(x - \frac{\pi}{4})$

$\frac{\pi}{4} + 1$	3	−4.58803782493	7.58803782493
$\frac{\pi}{4} - 1$	−1	−.21795809846	.78204190154
$\frac{\pi}{4} + .001$	1.002	1.00200200267	.00000200267
$\frac{\pi}{4} - .001$	.998	.998001997338	.000001997338

9. $f(x) \approx x$

+1	1	1	0
−1	−1	−1	0
.001	.001	.001	0
−.001	−.001	.001	0

11. $5.16 + 4.2(x - 1.6)$ 13. $p(x) = 16$ 15. $p(x) = 1 + 5(x-1)$ 17.
$y = 4.2(x - 1.6) + 5.16$; $y = -\frac{1}{4.2}(x - 1.6) + 5.16$ 19. $y = 16$; $x = -2$

21. $y = 5(x-1) + 1$; $y = -\frac{1}{5}(x-1) + 1$ 23. $y = \frac{\sqrt{3}}{2}(x - \frac{\pi}{3}) + \frac{3}{4}$;
$y = -\frac{2}{\sqrt{3}}(x - \frac{\pi}{3}) + \frac{3}{4}$ 25. $y = 0$; $x = 0$

4.3

1. $f'(x) = 3x^2 - 1$ f increasing on $(-\infty, -\sqrt{3})$ and $\sqrt{3}, \infty)$ $f' = 0$ when $x = \pm\sqrt{3}$ decreasing on $(-\sqrt{3}, \sqrt{3})$.   Local min at $x = \sqrt{3}$ Local max at $x = -\sqrt{3}$.

3. $f'(x) = \frac{8}{(x+1)^3}$ f decreasing on $(-\infty, -1)$ No critical points   increasing on $(-1, \infty)$.

5. $f'(x) = \sec^2 x$ No critical points. f is increasing on $(\frac{k\pi}{2}, \frac{(k+2)\pi}{2})$, $k = \pm 1, \pm 3, \ldots$

7. $f'(x) = \sec x \tan x$ critical points $x = k\pi$, $k = 0, \pm 1, \pm 2$. f is increasing on $(2\pi k, (2k+1)\pi)$, $k = 0, \pm 1, \pm 2, \ldots$,   decreasing on $(\pi k, \pi(k+1))$, $k = 0, \pm 1, \pm 3, \pm 5, \ldots$.

9. $f'(x) = -2$ if $x < \frac{3}{2}$ f increasing on $(3/2, \infty)$, $f'(x) = 2$ if $x > \frac{3}{2}$, f decreasing on $(-\infty, 3/2)$ Critical point is $x = \frac{3}{2}$,   Local min at $x = \frac{3}{2}$.

11. Max= 4   Min = −.628   13. Max = 2   Min = −4   15. Max = .125   Min = −21   17. Max = 64,000 ft at $t = 0$s.   19. Max = 90 at $t = 0$ s.   21. Local min: $x = 3$ Local max: None   23. Local min: $x \approx 4.2$ Local max: $x = -2$ 25. Local min: $x \approx 2.8$ Local max: $x \approx -0.7$ 27. Local min: None   Local max: None

31. $C = 1$

4.4

1. a. $x = 0, 1$ b. $f(1) = 1$ is a local max c. $(0, \frac{2}{3})$ d. $(-\infty, 0)$ and $(\frac{2}{3}, \infty)$ e. $(0, 0)$ and $(\frac{2}{3}, \frac{16}{27})$ 3. a. $x = 0, 3$ b. $f(0) = 1$ local max c. $(1, 3)$ d. $(-\infty, 1)$ and $(3, \infty)$ e. $(1, -10)$ and $(3, -26)$ 5. a. $x = \frac{1}{3}, -1$ b. $f(\frac{1}{3}) = \frac{-37}{27}$ local min $f(-1) = 1$ local max c. $(-\frac{1}{3}, \infty)$ d. $(-\infty, -\frac{1}{3})$ e. $(-\frac{1}{3}, -\frac{5}{27})$ 7. a. $x = 0$ b. $f(0) = 0$ local max c. $(-\infty, -2)$ and $(2, \infty)$ d. $(-2, 2)$ e. None   9. $x = k\pi$ $k = 0, \pm 1, \pm 2, \ldots$ 11. $x = k\pi$ $k = 0, \pm 1, \pm 2, \ldots$ 13. None   15. Increasing on $(-1, 1)$ Concave up on $(0, 1)$ Concave down on $(-1, 0)$ Inflection point $(0, 0)$ 17. Increasing on $(-\infty, \infty)$ Concave up on $(-\infty, 0)$ Concave down on $(0, \infty)$ Inflection point $(0, 0)$ 19. Increasing on $(-\infty, \infty)$ Concave up on $(-\infty, \infty)$ 21. No inflection points Concave up on $(-6, 6)$ 23. Inflection points at $x = -4, \frac{1}{2}$ Concave up on $(-6, -4)$ and $(\frac{1}{2}, 6)$ Concave down on $(-4, \frac{1}{2})$ 25. Inflection point at $x = 2$ Concave down on $(-6, 2)$ Concave up on $(2, 5)$ 27. Inflection point at $x = 0$ Concave down on $(-6, 0)$ Concave up on $(0, 6)$ 39. All   41. None   43. 32   45. 35, 37   47. 31, 34, 35   49. $V(t) = -32t$ $a(t) = -32$ a. $a(0) = -32\frac{\text{ft}}{\text{s}^2}$ b. $a(t)$ never 0 c. $s(0) = 64000$ ft d. $a(t) = -32$ max for t in $[0, 60]$ 51. a. $a(0) = -\frac{1}{15}\frac{\text{ft}}{\text{s}^2}$ b. $a(t)$ never 0 c. $s(60) = 30$ ft. d. $a(t) = \frac{1}{15}$ max for t in $(30\sqrt{3}, 60]$
53. The cubic spline approximation is

$$f : x \longmapsto \begin{cases} (4\sqrt{2} - 6)(-x^3 + x^2) + 1 & \text{for } -\frac{\sqrt{2}}{2} \leq x \leq 0 \\ (4\sqrt{2} - 6)(x^3 + x^2) + 1 & \text{for } 0 \leq x \leq \frac{\sqrt{2}}{2} \end{cases}$$

55.

Spline: $x \longmapsto -.110739816362x^3 - .05738534102x^2 + x$;

Quadratic (linear in this case): $x \longmapsto x$;

Graph of spline and its error:

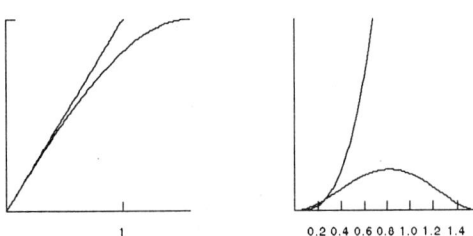

At, say, $4\pi/9$ the spline is better; at, say, $\pi/9$ the quadratic polynomial is better.

57.

Spline: $x \longmapsto 0.16x^3 - 0.52x^2 + 1$;

Quadratic: $x \longmapsto 1 - x^2$;

Graph of spline and its error:

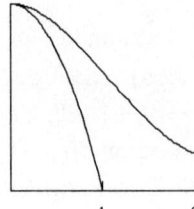 

At 1, the spline is better; at 0.2, the quadratic polynomial is better.

59.

Spline: $x \longmapsto 0$;

Quadratic: $x \longmapsto .5x^2$;

Graph of spline and its error:

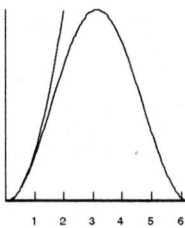 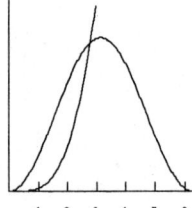

At $3\pi/4$, the spline is better; at $\pi/4$, the quadratic polynomial is better.

61.

Spline: $x \longmapsto 0.5x(2 - x)$;

Quadratic: $x \longmapsto x$;

Graph of spline and its error:

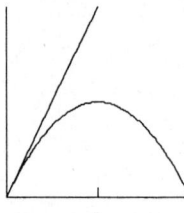 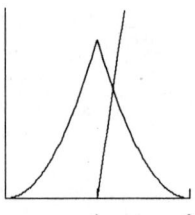

At 0.5 spline is better; at 1.8 quadratic polynomial is better.

Chapter 5

5.1

1. 276.5   3. 794.75   5. 859

7.

Interval	4 - 8	8 - 12	12 - 16	16 - 20	20 - 24	24 - 28	28 - 32
Avg. time increase	.0525	.04	.0325	.03	.025	.025	.0225

32 - 36	36 - 40	40 - 44	44 - 48	48 - 52	52 - 56
.0225	.02	.02	.0175	.0175	.0175

56 - 60	60 - 64
.0175	.015

9. $f'(h) = \dfrac{1}{8\sqrt{h}}$ 11. March   15. The rate of change is small.

17. Circumference $= \pi x = \pi(\sqrt{100 - h^2})$;   Surface area $= \pi h \sqrt{100 - h^2}$.

19. $V = 288h - \dfrac{h^3}{2}$ 21. $V = \dfrac{1}{3}\pi r^2 \left(\dfrac{12r^2}{r^2 - 36}\right) = \dfrac{4\pi r^4}{r^2 - 36}$.

23. $A = \tan\dfrac{\theta}{2} y^2 = \tan\dfrac{\theta}{2}\left(\dfrac{16}{\sin^2\frac{\theta}{2}} + \dfrac{32}{\sin\frac{\theta}{2}} + 16\right) = \theta \leq (0, \pi)$ 25. $\ell = 12$

5.2

1. 25×50 3. $\sqrt{\dfrac{50D}{3}} \times \dfrac{500\sqrt{3}}{\sqrt{50D}}$ 5. $\sqrt[3]{V} \times \sqrt[3]{V} \times \sqrt[3]{V}$ 7. $\left(\sqrt{\dfrac{24S}{T}} + 2S\right) \times \left(\dfrac{24\sqrt{T}}{\sqrt{24S}} + 2T\right)$ or $\left(\sqrt{\dfrac{24S}{T}} + 2S\right) \times \left(\sqrt{\dfrac{24T}{S}} + 2T\right)$ 9. $r = 1.128$ in   11. $\theta = \dfrac{\pi}{3}$ represents a maximum   13. 3.924 cm $\times 3.924$ cm   15. 4.514 cm $\times 4.514$ cm   17. Dimensions $r = \dfrac{10}{8 + \pi}$ ft; $h = \dfrac{20}{8 + \pi}$ ft   19. Lay the line 129 m under ground and then under water.   21. The wire should be cut every $\dfrac{1}{2}$ meter.   23. Height $= 20\tan(.615) = 20\left(\dfrac{1}{\sqrt{2}}\right) = 10\sqrt{2}$ m.   25. Order 490 gallons each time.   27. Maximum $= 288$ in^2; Minimum $= 0$.   29. Maximum $= 1152$ ft^2; Minimum $= 576$ ft^2 31. Maximum $= 12\sqrt{3}$ cm; Minimum $= 0$.   33. Maximum $= 12\sqrt{3}$; Minimum $= 0$.

5.3

1. $y - \sqrt{3} = \frac{2}{3\sqrt{3}}(x + \frac{3}{2})$ $y - \sqrt{3} = \frac{-3\sqrt{3}}{2}(x + \frac{3}{2})$. 3. $y - 8 = 3(x - 4)$ $y - 8 = -\frac{1}{3}(x - 4)$ 5. $y - 2 = -(x - 1)$ $y - 2 = x - 1$ 7. $y = 0, x = 0$ 9. $1 = \cos y \frac{dy}{dx}$ $\frac{dy}{dx} = \sec y$ 11. $1 = \sec^2 y \frac{dy}{dx}$ $\frac{dy}{dx} = \cos^2 y$ 13. $1 = -\csc y \cot y \frac{dy}{dx}$ $\frac{dy}{dx} = -\sin y \tan y$ 15. $1 - g'(y) = \frac{1}{\frac{dy}{dx}} = \frac{1}{f'(x)}$ 17. $\frac{df}{dt} = 4x^3 \frac{dx}{dt} + 6\cos\theta \frac{d\theta}{dt}$ 19. $\frac{df}{dt} = 15x^2 \frac{dx}{dt} - \frac{4}{\sqrt{1-y^2}} \frac{dy}{dt}$ 21. $\frac{df}{dt} = \pi \cos(x^2 \arcsin(y)) \left[\frac{x^2}{\sqrt{1-y^2}} \frac{dy}{dt} + 2x \arcsin y \frac{dx}{dt} \right]$ 26. 6π cm/s 28. $\frac{dv}{dt} = 4\pi(13.365cm)^2(3) \approx 6734$ cm^3/s $\frac{dE}{dt} = \frac{-1}{2(100)}(6734) = -33.67$ 30. 19 mph. 32. 3.84 m/s 34. $-.106$ ft/s. 36. $\frac{dK}{df}|_{f=5000} = -.00004$ $\frac{dK}{df}|_{f=10,000} = .00001$

5.4

1. $\frac{dy}{dt} = 3\cos t$ $\frac{dx}{dt} = -2\sin t$ $\frac{3\cos t}{-2\sin t}$ 3. $\frac{dy}{dt} = -10 + 2t$ $\frac{dx}{dt} = -1$ $\frac{dy}{dx} = 10 - 2t$ 5. $\frac{dy}{dt} = \sin t + t\cos t$ $\frac{dx}{dt} = \cos t - t\sin t$ $\frac{dy}{dx} = \frac{\sin t + t\cos t}{\cos t - t\sin t}$ 7. $\frac{dy}{dt} = 1$

$$\frac{dx}{dt} = \begin{cases} -1 & \text{if } 0 \le t < 5 \\ \text{undefined if } t = 5 \\ +1 \text{ if } 5 < t \le 10 \end{cases}$$

$$\frac{dy}{dx} = \begin{cases} -1 & \text{if } 0 \le t < 5 \\ \text{undefined if } t = 5 \\ 1 & \text{if } 5 < t \le 10 \end{cases}$$

9. negative, counterclockwise 11. $(\frac{3}{2}, \frac{3\sqrt{3}}{2})$ 13. $(\frac{3}{2}, \frac{-3\sqrt{3}}{2})$ 15. $(-\pi, 0)$ 17. $(0, 0)$ 19. $(5, .9273)$ $(5, 53.13°)$ 21. $(4.29, -.8622)$ $(4.29, -49.40°)$ 23. $(\sqrt{\pi}, \frac{\pi}{4})$ $(\sqrt{2\pi}, 45°)$ 25. $(1000, \frac{\pi}{2})$ $(1000, 90°)$ 27. $r = 2a\cos\theta$, $0 \le \theta \le \pi$ a is the radius of the circle and $(a, 0)$ is the center of the circle 29. $r = a(\cos\theta + 1)$ $0 \le \theta \le 2\pi$ $(2a, 0)$ is the furthest right-hand edge of the cardioid. $(0, \pm a)$ are the y-intercepts. 31. $r = a\sec\theta + 1$ $0 \le \theta \le 2\pi$ $x = a$ is the vertical asymptote. 33. $r = 1 + a\cos\theta$ $0 \le \theta \le 2\pi$ x-intercepts occur at $(a + 1, 0), (a - 1, 0)$ and if $a \ge 1$ at $(0, 0)$, y-intercepts are $(0, \pm 1)$ for all a. 35. $r = a\cos(2\theta\sec\theta)$ $0 \le \theta \le \pi$ x-intercepts occur at $(a, 0)$ and $(0, 0)$ $x = -a$ is an asymptote for the graph. 37. an ellipse for $e = \frac{1}{3}$ a hyperbola for $e = 3$ 39. an ellipse for $e = \frac{1}{3}$ a hyperbola for $e = 3$ 41. The outer tip of each leaf is a distance of a from the origin. 43. Using $\cos\theta$ puts the first leaf on the positive x-axis. Using $\sin\theta$ puts the first leaf in the direction $\theta = \frac{\pi}{2k}$. 45. If $r(\theta)$ is an even function, the graphs are identical. If $r(\theta)$ is an odd function,

the graphs are symmetric across the origin. 47. $\sin(\pi - \theta) = \sin\theta$ so for any equation with θ only in terms of $\sin\theta$ or $\csc\theta$, the new and old graphs are identical. $\cos(\pi - \theta) = -\cos\theta$ 49. If $r(\theta)$ is an odd function the graphs are identical. If $r(\theta)$ is an even function, the graphs are symmetric across the origin. 51. $\cos(\theta + \frac{\pi}{2}) = -\sin\theta, \sin(\theta + \frac{\pi}{2}) = \cos\theta$ 53. $3x^2y = x_2^2xy + y^2$ $x(t) = 3\sin t \cos^3 t$ $y(t) = 3\sin^2 t \cos^2 t$ 55. $y^2 = \frac{x^3}{5-x}$ $x(t) = 5\sin^2 t$ $y(t) = 5\sin^2 t \tan t$ 57. $\sqrt{x^2 + y^2} = \frac{-7x}{x-1}$ $x(t) = 1 - 7\cos t$ $y(t) = \tan t - 7\sin t$ 59. $x^2 + y^2 = x - 3\sqrt{x^2 + y^2}$ $x(t) = (-3 + \cos t)\cos t$ $y(t) = (-3 + \cos t)\sin t$

Chapter 6

6.1

1. 27   3. -4 5. $\frac{97}{3}$ 7. 3   9. -10 11. 12.5   13. -20 15. 0   17. 18.6875   19. 12.5   21. 33.6875   23. 32.5   25. 25   27. 62.5   29. .75   33. $\int_{-5}^{5} 2\,dx = 2(10) = 20$ $\int_{-5}^{5} d\,dx = 3(10) = 30$ 35. No.   37. 25.5   20.875

6.2

1. 2   3. .625   5. $L_5 \approx 1.11$ $U_5 \approx 1.192$ avg ≈ 1.151 worst error $\approx .041$ 7. $L_5 \approx 0$ $U_5 \approx .07896$ avg $\approx .0395$ worst error $\approx .0395$ 9. -2.5 11. -3.5 13. -5 15. 1   17. 0, 0, 0, 0   19. 0, 0, 0, 0   21. 0, -3, -2.5 23. $-8, 17, -19$ 25. $4, -1, -9$ 27. $3, 0, -5$ 29. 0   31. 0

6.3

1. 2.4834 miles   3. $t = 23$ minutes   5. $\frac{x^3}{3} + \frac{x^2}{2} + x + C$ 7. $\frac{x^5}{10} + C$ 9. $\frac{2}{3}x^{\frac{3}{2}} + C$ 11. $6\sqrt{x} + C$ 13. $\frac{x^4}{12} + \frac{x^{-2}}{6} + C$ 15. $\frac{3}{10}x^{\frac{10}{3}} + C$ 17. $x + C$ 19. $\frac{1}{4}\arctan(x) + C$

6.4

1. $F(x) = \int_1^x 2^{-t^2}\,dt$ 3. $F(x) = \int_{-\pi}^x \cos^2 t\,dt$ 5. $F(x) = \int_{\sqrt{17}}^x \frac{1}{t}\,dt$ 7. $\frac{\sin(2x)}{1+x^2}$ 9. $2x \int_2^x \frac{\sin(2t)}{1+t^2}\,dt + \frac{x^2 \sin(2x)}{12}$ 11. $\frac{2\sin(2x)}{1+x^2} \int_2^x \frac{\sin(2t)}{1+t^2}\,dt$ 13. $17\frac{1}{3}$ 15. $12\frac{2}{3}$ 17. 4π 19. .1963   29. This does not contradict the Second Fundamental Theorem because $\frac{1}{x^2}$ is not continuous on $[-1, 1]$.

Chapter 7

7.1 1. $y : x \longmapsto 7 + \int_2^x 2\sqrt{x}\, dx$ 3. $y : x \longmapsto 5 + \int_{-1}^x ((x+1)^2 + 1)^2\, dx$ 5. $y : x \longmapsto 4 + \int_1^x \frac{1}{x^2+1}\, dx$

7.

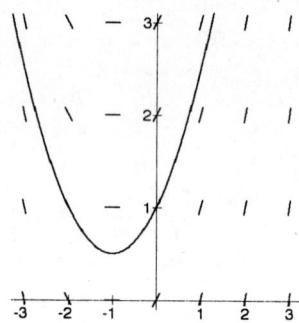

9.

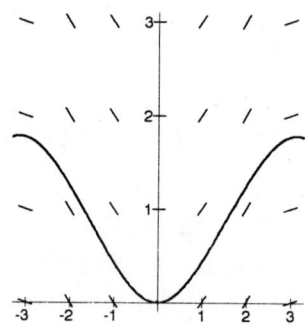

11.

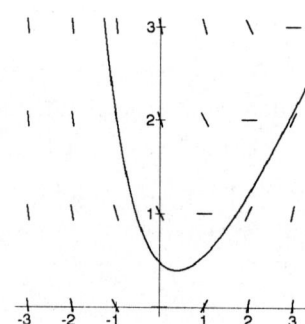

13. Let $u = -y^{-1}$. Then $u' = -y^{-2}y'$ and $y' = -x^2y^2$ is equivalent to $u' = x^2$ or $u = \frac{1}{3}x^3 + C$. $y(4) = 4$ implies $u(4) = -y(4)^{-1} = -\frac{1}{4}$ so $u = \frac{1}{3}x^3 - 1036/48$ and $y = -\frac{1}{\frac{1}{3}x^3 - 1036/48}$.

15. Let $u = y^{-2}$. Then $u' = -2y^{-3}y'$ and $y' = \frac{y^3}{x^2}$ is equivalent to $-\frac{1}{2}u' = \frac{1}{x^2}$ or $u = 2x^{-1} + C$. $y(3) = 2$ implies $u(3) = y(3)^{-2} = \frac{1}{4}$ so $u = 2x^{-1} - 5/12$ and $y = (2x^{-1} - 5/12)^{-\frac{1}{2}}$.

17. Let $u = \frac{1}{4}y^4 + \frac{1}{2}y^2$. Then $u' = (y^3 + y)y'$ and $y' = \frac{1}{y^3+y}$ is equivalent to $u' = 1$ or $u = x + C$. $y(1) = 1$ implies $u(1) = \frac{1}{4}y(1)^4 + \frac{1}{2}y(1)^2 = \frac{3}{4}$ so $u = x - \frac{1}{4}$ and $y = \sqrt{2\sqrt{x} - 1}$.

7.2

1. e^{-5a} 3. $e^{2\ln a}$ 5. $2.75^a = (0.9)^{a\log_{(0.9)} 2.75}$ 7. $2\ln a$ 9. $y = \frac{42}{e^4}e^{4x}$, $\mathbb{R}$ 11. $y = x$, $(0, \infty)$ 13. $y = \sqrt{2\ln|x| + 1 - 2\ln 2}$, $\left[\frac{2}{\sqrt{e}}, \infty\right)$ 15. $y = -\ln(x^2 + 1)$, $\mathbb{R}$ 17. $y' = e^{2x}$ $y(0) = 3$, $y(0) = -2$ 19. $y' = 5^{3x}$ $y(0) = 25$, $y(0) = 1$ 21. $y' = \frac{1}{x\ln 2}$ $y(1) = 0$, $y(1) = 5$

7.3

1. increasing exponentially ≈ 6928 3. linear ≈ 6361 5. decreasing exponentially ≈ 3737 7. $60.5°F$ 9. $61.5°F$, $101.0°$ 11. ≈ 5.76 years 13. 3 hours and 18 minutes. 15. 7.295, 12.8351535 17. 13 years and 1 quarter 19. 5.25% quarterly, 5.28445313%

7.4

1. $\frac{1}{2}\sin(x^2) + C$ 3. $\frac{1}{5}\ln|x^5 - 17| + C$ 5. $\frac{3}{60}(2x^3 + 7)^{\frac{5}{3}} - \frac{21}{2}(2x^3 + 7)^{\frac{2}{3}} + C$ 7. $-\frac{1}{2}e^{-x^2} + C$ 9. $\frac{1}{2}[\ln(x)]^2 + C$ 11. $x\arcsin(x) + \sqrt{1 - x^2} + C$ 13. $x^3e^x - 3x^2e^x + 6xe^x - 6e^x + C$ 15. $x\arccos(x) - \sqrt{1 - x^2} + C$ 17. $\frac{x^2\ln x}{2} - \frac{x^2}{4} + C$ 19. $-x^2\cos x + 2x\sin x + 2\cos x + C$ 25. $\cos h(x)$, $\cos h(x) + C$ 27. $\sec h^2 x$, $\ln(\cos h(x)) + C$ 31. $0, \pi^2, \frac{1}{2}\sin(\pi^2) \approx -.215$ 33. $-18, -16, \frac{1}{5}\ln(\frac{8}{9}) \approx -.0236$ 35. $5, 23, \approx 4.0516$ 37. $-1, -(\ln 2)^2, \approx -.125$ 39. $0, 1, \frac{1}{2}$

Chapter 8

8.1

1. 12.5 3. 1 5. 144 7. 1 9. ≈ 26.167 11. $2\sqrt{2} - \sqrt{3} - 1 \approx .096$ 13. $\approx .693$ 15. ≈ 1.66 17. $\sqrt[3]{16}$ 19. ≈ 110 21. 16 23. 23 25. 25 27. 9 29. 21.5625

8.2

1. $\dfrac{125}{3}$ 3. $\dfrac{\pi}{8}$ 5. 6   7. $\dfrac{4\pi}{3}, 4\pi$ 9. 24.56, 1.85   11. 23.43, 27.59
13. 7.24, 4.84   15. $\dfrac{\pi}{2}, 8\pi$ 17. $f(x) = \dfrac{2}{5}x, [0, 10]$ 21. 125π 23. $\dfrac{125}{3}\pi$

8.3

1. .30655   3. .7790   5. .3707   7. .1206   9. .1133   11. 26.44   13. 3.026   15. 3.09   17. 10.35   19. 24.40   21. 9.29   23. $4\sqrt{2}$ 25. 1.52   27. 6.66   29. 2.97   31. .23   33. .20   35. 2344.21   37. 18.19   39. 4.44

8.4

1. $A \approx 1.7533$ 3. $A \approx 1.6808$ 5. $A \approx 1.5901$ 7. $A \approx .3308$ 9. $A \approx 8.215 \times 10^{-2}$ 11. $A \approx 4.712$ $(A = \dfrac{3\pi}{2})$ 13. $A = \dfrac{3\pi}{2}$ 15. $A = \dfrac{3\pi}{2}$ 17. $A = \dfrac{3\pi}{2}$ 19. $A = \pi - \dfrac{2\sqrt{2}}{2} \approx .5435$ 21. $A = \pi - \dfrac{3\sqrt{2}}{2} \approx .5435$ 23. $A = \dfrac{\pi}{8} \approx .3927$ 25. $A = \dfrac{\pi}{12} \approx .2618$ 27. $A = \dfrac{\pi}{12} \approx .2618$ 29. $A = \dfrac{a^2\pi}{4k}$

Chapter 9

9.1

1. 3   3. ≈ 1.087 5. 21π 7. $19\dfrac{1}{3}$ 9. yes   11. .48   13. $\dfrac{1}{e-1} \approx .582$ 15. $\dfrac{\pi}{2}$ 17. $\dfrac{1}{2}$ 19. ≈ 9 21. 1 inch from the right angle   23. $\left(\dfrac{b}{3}, \dfrac{h}{3}\right)$ 25. $\sin x - \sin(x-1)$ 27. $x - 1$ 29. $-\dfrac{2}{3}\cos x^3 + \dfrac{2}{3}\cos(x - .5)^3$

9.2

1. $V(t) = \dfrac{1}{2}t^2 - \dfrac{1}{3}t^3 - 2$; $S(t) = \dfrac{1}{6}t^3 - \dfrac{1}{12}t^4 - 2t + 3$ 3. $V(t) = -\cos t$; $S(t) = -\sin t + 1$ 5. $V(t) = \dfrac{2}{3}t^{\frac{3}{2}} - 4$; $S(t) = \dfrac{4}{15}t^{\frac{5}{2}} - 4t + 2$ 7. 2.35, 2.35   9. 0, $2/\pi$ 11. 48 ft   13. 16   15. 115.2 inch.lbs   17. 24 inch.lbs   19. 5.625 ft.lbs   21. ≈ 8.57 ft.lbs, ≈ 4.86 ft, ≈ 5.14 ft.   23. $6\dfrac{1}{4}$ ft.lbs   25. 3187.5 ft.lbs   27. 2808.26 cm.lbs   29. 2106.19 cm.lbs   31. 1363.54 in.lbs   33. 1193.10 in.lbs   35. 21,875 ft.lbs   37. 410,156.25 ft.lbs   39. $kQ_1Q_2 \ln(10^7)$

9.3

1. $\frac{1}{2}$ 3. 1   5. 2   7. 100   9. diverges   11. converges for $p > 1$, diverges for $p \leq 1$ 13. diverges   15. $\frac{1}{2}$ 17. diverges   19. 2   21. $\int_0^\infty e^{px}\,dx$ converges when $p < 0$. $\int_{-\infty}^0 e^{px}\,dx$ converges when $p > 0$.   23. $\pi/2$ 25. diverges   27. diverges   29. diverges   31. diverges   33. diverges   35. diverges   37. $\int_0^1 \frac{1}{x^p}\,dx = \frac{1}{1-p}$ when $p < 1$ and diverges when $p \geq 1$.   39. $\frac{3}{2}\left(\sqrt[3]{9}-1\right) \approx 1.6201$ 41. $\sec(x)$ is undefined at $x = \frac{\pi}{2}$. $\int_0^\pi \sec^2(x)\,dx$ diverges.   43. $3\ln 3 - 2 \approx 1.2958$ 47. $1/(s^2+1)$

9.4

1. $Prob(X = 10) = \frac{32}{81}$; $Prob(X = 20) = \frac{24}{81}$; $Prob(X = 30) = \frac{16}{81}$; $Prob(X = 40) = \frac{8}{81}$; $Prob(X = 50) = \frac{1}{81}$.   3. $E(X^2) = 100\frac{32}{81} + 400\frac{24}{81} + 900\frac{16}{81} + 1600\frac{8}{81} + 2500\frac{1}{81} \approx 524.691$ $E(X^3) = 10^3\frac{32}{81} + 20^3\frac{24}{81} + 30^3\frac{16}{81} + 40^3\frac{8}{81} + 50^3\frac{1}{81} \approx 15962.963$ 5. $\frac{4}{9}; \frac{2}{9}; \frac{1}{9}; \frac{2}{9}$.   7. $\frac{\pi}{9}$.   9. ≈ 0.460.   11. ≈ 0.395.   13. $\approx 8.77 \times 10^{-3}$.   15. $\frac{1}{2}$ 17. valid prob. density; $\mu = e - 1$, $\sigma^2 = e^2/2 - 1/2 - e + 1$.   19. valid prob. density; $\mu = \frac{5}{4}$, $\sigma^2 = 0.95$.   21. 0.286   23. $a = 0.675$ 25. approximately 4.55%   27. approximately 16%   31. probability = 1/4

Chapter 10

10.1

1. $\{3 - 5n\}_{n=0}^\infty$; $a_n = a_{n-1} - 5$ 3. $\{\pi + 0n\}_0^\infty = \{\pi\}_{n=0}^\infty$; $a_n = a_{n-1}$ 5. $\{3(-5)^n\}_0^\infty$; $a_n = -5a_{n-1}$ 7. $\{\pi(0)^n\}_0^\infty$; $a_n = 0a_{n-1} = 0$ 11. 10th term.   13. $1, 1, 2, 3, 5, 8, 13, 21, 34, 55$ 15. $-\frac{1}{2}, \frac{2}{3}, -\frac{3}{4}, \frac{4}{5}, -\frac{5}{6}$ 17. $\sqrt{3}, 2.28, 2.615, 2.80, 2.899$ 19. $0, -1, -1.54, -1.57, -1.57, -1.57$ 21. $.5, .25, .0625, .0039, .0000152, \ldots$

10.2

1. True.   3. False, the alternating harmonic sequence is Cauchy, but not monotonic.   5. $2^{-1}, 2, 2^{-1}, 2, 2^{-1}$; this sequence is bounded, not monotonic, divergent, not Cauchy.   7. This sequence is unbounded, monotonic (strictly), divergent, not Cauchy.   9. Using $a_1 = 1$, the sequence is $1, .5, .6\overline{6}, .5\overline{9}, .625, .615, .619, \ldots$. This sequence is bounded, not monotonic, convergent and Cauchy.   11. 11th   13. 20th and 21st   19. $a_1 = 2, a_2 = 2.25, a_3 = 2.37, a_{1000} = 2.7169, a_{2000} = 2.71760, a_{3000} = 2.7178$. The sequence is bounded and monotonic with limit e. a_{3000} is within .0005 of the limit.   21. $\lim_{n \to \infty} a_n = 0$; 10th   23. $-\frac{1}{2}, \frac{2}{3}, -\frac{3}{4}, \frac{4}{5}, -\frac{5}{6}$ This sequence is bounded, not monotonic, and divergent.   25. 1.73, 2.28, 2.615, 2.80, 2.899   This sequence is bounded, monotonic and convergent.

10.3

1. $\frac{1}{2}$ 3. 1   5. 1   7. 1   9. 0, (slowly!)   11. $\frac{1}{3}$ 13. 1   15. 0

10.4

1. $[1.5, 2]$; $[1.75, 2]$; $[1.75, 1.875]$.   3. $[.75, .8]$; $[.775, .8]$; $[.775, .7875]$.   5. $[2.2, 2.6]$; $[2.2, 2.4]$; $[2.2, 2.3]$.   7. $[.9875, 1.35]$; $[.9875, 1.16875]$; $[.9875, 1.078125]$.   9. $[1.1, 1.65]$; $[1.925, 2.2]$; $[2.0625, 2.2]$.

11. (example) Iterator: $f : x \longmapsto x - 0.2(x^2 - 3.2)$;   Initial seed: 2.5;   next three iterates: 1.69, 1.75878, 1.78012.

13. (example) Iterator: $f : x \longmapsto x - 0.5(\sin(x) - 0.7)$; Initial seed: .75;   next three iterates: 0.75918, 0.76502, 0.76874.

15. (example) Iterator: $f : x \longmapsto x - 0.002(x + \sqrt{x} + \sqrt{x^3} - 7)$;   Initial seed: 2;   next three iterates: 2.07574, 2.12503, 2.15698.

17. (example) Iterator: $f : x \longmapsto x - 0.1(\ln(x^2 + \ln(x)))$;   Initial seed: 0.86; next three iterates: 0.91297, 0.94275, 0.96140.   19. (example) Iterator: $f : x \longmapsto x + 0.1(3\cos(\cos(x)) - 2x)$;   Initial seed: 1.65;   next three iterates: 1.61906, 1.59490, 1.57583.

21. Any b between 0 and ≈ 0.285 will work.   23. $b = 0.9$ will work.   25. $b \approx 1.128667$ or $b \approx -0.29533$. Neither will converge, they oscillate.

Chapter 11

11.1

1. 2046   3. 288   5. −1   7. 85   9. converges to −1/4   11. converges to $1/(e−1)$ 13. diverges   15. converges to $1/21$ 17. diverges   19. converges to $3e^2/(3−e)$ 21. converges to $5/6$ 23. converges to $−1/3$ 27. converges to e 29. 42 feet   31. total length of erased intervals is 1 unit; 0 is one point that is never erased.

11.2

1. converges   3. converges   5. converges   7. converges   9. diverges   11. not monotonically decreasing   17. 6   19. True. Example: the harmonic sequence converges, but the harmonic series diverges.   21. diverges   23. diverges   25. converges absolutely   27. diverges   29. converges absolutely   31. diverges   33. diverges   35. diverges   37. diverges   39. converges absolutely   41. converges absolutely

11.3

1. converges   3. diverges   5. converges   7. diverges by limit comparison test   9. no information from the root test   11. converges   15. interval of convergence: $(-\infty, \infty)$; radius of convergence: $R = \infty$ 17. interval of convergence: $(1, 5)$; radius of convergence: $R = 2$ 19. $\ln(1+x) = x - x^2/2 + x^3/3 - x^4/4 \cdots$ with interval of convergence $(-1, 1]$ 23. $e^{-x} = 1 - x + x^2/2 - x^3/6 + x^4/24 \cdots$ 25. $(1-e^{-x})/x = 1 - x/2 + x^2/6 - x^3/24 \cdots$ 27. 0.4438420791

11.4

1. $x \longmapsto \sin(-1) + 2\cos(-1)(x+1) - 2\sin(-1)(x+1)^2 - \frac{4}{3}\cos(-1)(x+1)^3$.   3. $x \longmapsto 1 + 2(x-\pi/4) + 2(x-\pi/4)^2 + \frac{8}{3}(x-\pi/4)^3$.   5. $x \longmapsto -1 - \frac{1}{2}(x-\pi/2)^2$.   7. $x \longmapsto \exp(-1) + 2\exp(-1)(x+1) + 2\exp(-1)(x+1)^2 + \frac{4}{3}\exp(-1)(x+1)^3$.   9. $x \longmapsto 1 + \frac{1}{2}x^2$.   11. $x \approx -0.89361$.   13. $x \approx 0.73887$.   15. $x \approx 1.89772$.   17. $x \approx -0.77015$.   19. $x \approx 0.21886$.   21. $1 + (x-1) + 0.75(x-1)^2 + 0.20833333(x-1)^3$.   23. $1 - 0.5x - .4375x^2 - 0.125x^3$.   25. $1 + (x-1) + (x-1)^2 + 1.5(x-1)^3$.   27. $y = \frac{1}{3}x^3 - (\frac{1}{2}x^2 + \frac{1}{12}x^4)(0.1) + \frac{1}{2}(\frac{1}{3}x^3 + \frac{1}{30}x^5)(0.1)^2$ 29. $y = x - \frac{x^3}{3} + \frac{2x^5}{15}$.

INDEX

absolute extrema, 198.
  maximum, 198.
  minimum 198.
absolute value, 9,15.
  notation for intervals, 15.
acceleration, 210.
  average, 210.
  average, 210.
alternating sequence, 517.
alternating harmonic series, 574.
angle of inclination, 123.
antiderivative, 298,329,333–336.
  estimation of, 333–336.
antidifferentiation, 298,329.
arc length, using definite integrals to measure, 451–456.
Archimedean property of real numbers, 566.
Archimedes (287-212 B.C.), 301.
area, using definite integrals to measure, 432–439.
argument, 29.
arithmetic sequence, 519.
asymptote, 93,100.
  vertical, 93.
  horizontal, 100.
asymptotic property (behavior), 100,102.
average rate of change, 127.
average value of a function, 463.

base of exponential function, 385.
best linear approximation, 186.
best quadratic approximation, 214.
binomial coefficients, 227.

bisection method, 550.
Bolzano, Bernard (1781-1848), 99.
Bolzano's Theorem, 99.
bounded function, 46.
bounded sequence, 530.
Buffon's Needle Problem, 512.

Cantor, Georg (1835-1918), 575.
Cantor's set, 575–576.
Cardan, Girolamo (1501-1576), 549.
Cardan's formulas, 549.
cardioid, 292.
Cartesian, 17–18.
  coordinate system, 18.
  plane, 17.
  product, 18.
Cauchy, Augustin-Louis (1789 - 1857), 531.
Cauchy sequence, 532.
Cavalieri, Bonaventura (1598 - 1647), 432.
Cavalieri's principle, 432.
ceiling function, 117.
center of mass, 466.
coefficient, 37.
  binomial, 227.
chain rule for derivatives, 168.
composition of functions, 48.
concavity, 211.
constant, 29.
constant function, 33.
constant sequence, 516.
continuous, 64,82–85,98.
  at a point, 82.
  from left or right, 83.

function, 64, 98.
continuously compounded interest, 406.
convergence of sequences, 528.
convergence of series, 569, 591.
    absolute convergence, 591.
    conditional convergence, 591.
convergence tests, 576–592.
    alternating series test, 588.
    alternating series test estimate, 590.
    comparison test, 583.
    integral test, 578.
    integral test estimate, 581–582.
    limit comparison test, 586.
    Nth term test, 577.
    ratio test, 595.
    root test, 597.
coordinates, 18, 285, 287.
    Cartesian (rectangular), 18.
    polar, 285.
    conversions between coordinate systems, 287.
critical point, 194.
Critical Point Theorem, 197.
cubic splines, 216.
cylindrical shells, 447.

decreasing function, 193.
dependent variable, 29.
degree measure, 41.
degree 35, 37.
definite integral, 304.
density, 467.
    probability, 505.
derivative, 107, 130, 134, 138, 143, 150–154, 157, 159, 162–164, 170, 198, 201, 209, 213.
    chain rule, 168.
    constant function, 150.
    cosine function, 154.
    definition of, 134.
    exponential function, 171, 395.
    first test for local extrema, 198.
    graphically estimating, 143.
    higher order, 209.
    left- and right-hand, 143.
    Leibniz notation, 154.
    linear function, 150.
    linearity of, 159.
    logarithmic function, 171, 395.
    Mean Value Theorem, 201.
    monomial function, 151.
    Newtonian notation, 157.
    numerically estimating, 138.
    operator notation, 157.
    polynomial function, 159.
    power rule, 169.
    product rule, 162–264.
    quotient rule, 166.
    rational power function, 170.
    second derivative test for local extrema, 213.
    sine function, 153.
    trigonometric functions, 167.
Descartes, René (1590-1650), 18.
difference quotient, 112–113, 140.
    symmetric, 140.
differentiation, 107, 263–266.
    implicit, 263–266.
    logarithmic, 397.
differential calculus, 107.
differential equations, 374–383,

628.

differentials, 155.

direction (slope) field, 336.

discontinuity, 90–92.
   essential (nonremovable), 92.
   removable, 91.

discrete mathematics, 2.

discriminant, 32.

distance travelled, net and total, 474.

divergence of sequences, 528–530.

divergence of series, 569.

domain of function, 23.

drug concentrations, 408.

eccentricity, 296.

element of a set, 11.

empty set, 12–13.

endpoints of an interval, 14.

essential (nonremovable) discontinuity, 92.

estimating derivatives numerically, 138.

estimating derivatives graphically, 141.

Euler's constant, 575.

even and odd functions, 36.

expected value, 503.

exponential functions, 44, 384–391.

exponential model, 403.

extrema, 196.
   absolute, 198.
   derivative test for local extrema, 198.
   local (relative), 196.
   second derivative test for local extrema, 213.

Extreme Value Theorem, 98.

factorial, 595.

Fibonacci (Leonardo de Pisa) (c. 1175 - c. 1250), 523.

Fibonacci sequence, 523.

fixed point, 558.

Fixed Point Theorem, 560.

function,
23, 25–29, 33–40, 43–44, 46, 48, 53, 55, 57, 63–64, 70, 98, 114, 115, 117,
193, 384–395.   argument, 29.
   average value of, 463.
   bounded, 46.
   ceiling, 117.
   composition, 48.
   constant, 33.
   continuous, 64, 98.
   decreasing, 193.
   defined by power series, 605.
   dependent variable, 29.
   even and odd, 36.
   exponential, 44, 384–391.
   floor, 117.
   graph of, 26.
   graphical representation of, 26.
   identity, 33.
   increasing, 193.
   independent variable, 29.
   inverse, 53, 60.
   inverse trigonometric, 57.
   limit of, 63–64.
   linear, 34, 108.
   locally linear, 115.
   logarithmic, 55, 384, 391–395.

monomial, 34.
monotonic, 193.
natural exponential function (exp), 390.
natural logarithmic function (ln), 391–392.
numerical representation of, 25.
periodic, 43.
piece-wise linear, 114.
piece-wise quadratic, 367.
polygonal, 114.
polynommial, 37.
rational, 38.
rational power, 39.
real-valued, 23.
signum, 70.
step, 117.
strictly decreasing, 193.
strictly increasing, 193.
strictly monotonic, 193.
symbolic representation of, 26.
transcendental, 374.
trigonometric, 40.
vector-valued, 23.
vertical line test for function graphs, 28.

fundamental period, 43.

Fundamental Theorems of Calculus, 347–356.
First Fundamental Theorem, 349–351, 377.
Second Fundamental Theorem, 354–356.

general form (of a linear function), 108.

geometric sequence, 520.

geometric series, 570.

graphical representation of a function, 26.

greatest integer function, 118.

half-life, 407.

harmonic sequence, 516.

harmonic series, 573.

heat transfer, 409.

higher order derivatives, 209.

hole, 91.

Hooke's law for springs, 478–479.

Horn of Gabriel, 498.

identity function, 33.

implicit differentiation, 263–266.

improper integrals, 487–494.
of horizontal type, 487–491.
of vertical type, 491–494.
convergence and divergence of, 488, 492.

increasing function, 193.

increment of a function, 113.

independent variable, 29.

indeterminate forms, 537–546.

index variable, 515, 565.

inductively defined sequence, 522.

infinite process, 4, 514.

infinity symbol (∞), 14.

inflection point, 212.

initial condition, 375.

instantaneous acceleration, 210.

instantaneous rate of change, 127.

instantaneous speed, 128, 133.

integral, 298, 304, 313, 315, 318, 322–325, 329,

349–351, 354–356, 360–369, 414, 415, 416, 420–424.
  computing and estimating, 313.
  definite, 304.
  First Fundamental Theorem of Calculus, 349–351.
  formal definition, 318.
  improper, 487–494.
  indefinite, 329.
  integration by parts, 420–421.
  limits of, 304.
  Mean Value Theorem, 464.
  method of substitution, 414.
  numerical techniques, 360–369.
  polar coordinates, 458–460.
  properties of definite integrals, 322–325.
  reduction formula, 424.
  Riemann, 315, 318.
  Second Fundamental Theorem of Calculus, 354–356.
  using tables of integrals, 423.
  variable of, 304.
integral calculus, 298.
integrand, 304.
Intermediate Value Theorem, 98.
Intermediate Zero Theorem, 99.
intersection of sets, 13.
interval, 13–15.
  bounded, 14.
  center of, 15.
  convergence of a power series, 602.
  closed, 14.
  endpoints of, 14.
  half-open or half-closed, 14.
  open, 14.
  radius of, 15.
  unbounded, 14.
inverse function, 53, 60.
inverse trigonometric function, 57.
irrational number, 3.
iteration, 524.
iterative process, 524.
iterative sequence, 522.

jump discontinuity, 92.

Lagrange (1736-1813), 623.
Lagrange's form of remainder, 623.
Laplace transform, 497.
left endpoint rule, 361.
left rectangle rule, 362.
left-hand derivative, 143.
Leibniz, Gottfried (1646-1716), 154.
Leibniz notation for derivative, 154.
L'Hôpital's Rule, 537–541, 543.
limit, 63–64, 70, 515.
  formal definition, 70
  left- and right-hand, 64.
  of a sequence, 515.
line of reflection, 36.
linear approximation, best, 186.
linear function, 34, 108.
linear model, 402.
linearity, detecting in data, 121.
local (relative) extrema, 196.
  derivative test for, 198.
  second derivative test for, 213.
local minimum and maximum, 196.

local slope, 115.
locally linear function, 115.
Location Principle, 99.
locus, 264.
logarithmic function, 55, 391–395.

Maclaurin, Colin (1698-1746), 616.
Maclaurin polynomials, 616.
Maclaurin series, 624.
mass, probability, 503.
mathematical modelling, 230, 236.
   linear model, 402.
   exponential model, 403.
   rates of change, 236.
maximum, 196, 198.
   absolute, 198.
   local (relative), 196.
mean of a random variable, 504.
Mean Value Theorem for Derivatives, 203.
Mean Value Theorem for Integrals, 464.
member of a set, 11.
method of exhaustion, 301.
methods of integration, 412–425.
   by parts, 420–421.
   by substitution, 414–416.
midpoint rule, 361.
minimum, 196, 198.
   absolute, 198.
   local (relative), 196.
moment, 467, 505.
monomial, 34–35.
   degree, 35.
   function, 34.
monotonic function, 193.

monotonic sequence, 531.
moving average, 470–471.

neighborhood of a point, 115.
Newton, Isaac (1642-1727), 154.
Newtonian notation for derivative, 157.
Newton's method, 553–557.
normal (Gaussian) density function, 508.
normally distributed probability, 509.
normal line, 189.
numerical representation, 25.
numerical techniques of integration, 361–369.
   left endpoint rule, 361.
   left rectangle rule, 362.
   midpoint rule, 361.
   right endpoint rule, 361.
   right rectangle rule, 362.
   Simpson's rule, 365.
   trapezoidal rule, 364.

optimization problems, 243–262.
ordered pair, 17.
origin, 17.
operator notation for derivative, 157.

p-series, 580.
Pascal's triangle, 227.
parameter, 29–30, 281.
parametric equation, 281.
parametrized curve, 281.
partial sums, 568.
partition, regular, 315.

period, 43.
periodic function, 43.
pi (π), 301.
piece-wise, 114, 116, 216, 233, 367.
   cubic (splines), 216.
   linear function, 114, 116.
   linear graph, 233.
   quadratic function, 367.
polar coordinates, 285.
polar graphs, 289.
pole, 94.
Polya, George (1887-1985), 228.
Polya's four steps in problem solving, 228.
polygon, 299.
polygonal function, 114.
polynomial, 37.
   degree of, 37.
   root or zero of, 37.
point of inflection, 211.
point of validity, 626.
population growth, 404.
power rule for derivatives, 169.
power series, 599–606.
   differentiating and integrating, 604.
   functions defined by, 605.
   interval of convergence, 602.
   radius of convergence, 602.
probability, 498–509.
   density, 505.
   expected value, 503.
   mass, 503.
   mean, 504.
   moment, 505.
   normal (Gaussian) density function, 508.
   normally distributed, 509.
   random variable, 503.
   standard deviation, 504.
   variance, 504.
product rule for derivatives, 162.

quadrants, 17.
quotient rule for derivatives, 166.

radian measure, 41.
radioactive decay, 407.
radius of convergence for a power series, 602.
random number generator, 502.
random variable, 503.
rational function, 38.
rational number, 3.
rational power function, 39.
real numbers, set of, 12.
real-valued function, 23.
rectangular coordinate system, 17.
recursively defined sequence, 522.
regular partition, 315.
related rates problems, 269–275.
relative (local) extrema, 196.
Riemann, Bernhard (1826-1866), 315.
Riemann integral, 315.
Riemann sum, 315, 318.
   lower and upper, 318.
right endpoint rule, 361.
right rectangle rule, 362.
right-hand deriative, 143.
Rolle's Theorem, after Rolle, M. (1652-1719), 202.

root of a function, 37.
root-finding techniques, 550–557.
   bisection method, 550.
   Newton's method, 553–557.
roster notation for sets, 12.
rule notation for sets, 12.

secant function, 40.
secant line, 113.
sequences, 514–534.
   terms of, 515.
   alternating, 517.
   arithmetic, 519.
   bounded, 530.
   Cauchy, 532.
   constant, 516.
   convergence of, 528.
   divergence of, 528–530.
   Fibonacci, 523.
   geometric, 520.
   harmonic, 516.
   index, 515.
   inductively defined, 522.
   iterative, 522.
   limit properties of, 533–534.
   monotonic, 531.
   monotonically decreasing, 531.
   monotonically increasing, 531.
   recursively defined, 522.
   strictly monotonically decreasing, 531.
   strictly monotonically increasing, 531.
   unbounded 530.
series, 565–606.
   absolute convergence of, 591.
   alternating harmonic, 574.
   conditional convergence of, 591.
   convergence of, 569, 591.
   divergence of, 569.
   geometric, 570.
   harmonic, 573.
   index variable, 565.
   p-series, 580.
   partial sums of, 568.
   power, 599–606.
   telescoping, 569.
series, convergence tests and estimates, 576–592.
   alternating series test, 588.
   alternating series test estimate, 590.
   comparison test, 583.
   integral test, 578.
   integral test estimate, 581–582.
   limit comparison test, 586.
   Nth term test, 577.
   ratio test, 595.
   root test, 597.
set notation, 11.
signum function, 70.
Simpson, Thomas (1710-1761), 365.
Simpson's rule, 365.
sine function, 40.
skip, 91.
slope, 34.
   local, 115.
slope (direction) field, 335, 376.
slope-intercept form, 108.
solid of revolution, 443.
squeezing principle, 77.

standard deviation, 504.

step function, 117.

strictly decreasing function, 193.

strictly increasing function, 193.

strictly monotonic function, 193.

strictly monotonically decreasing sequence, 531.

strictly monotonically increasing sequence, 531.

subset, 13.

summation notation, 565.

symbolic representation of a function, 26.

symmetric difference quotient, 140.

tangent function, 40.

tangent line, 188.

Taylor, Brook (1685-1731), 109.

Taylor forms, 109, 118–119.
   of linear function, 109.
   of quadratic polynomial, 111.

Taylor polynomials, 610–631.

Taylor series, 624.

Taylor's Theorem, 622.

Taylor's Remainder, 623–624.
   Lagrange's form, 623.
   Integral form, 624.

telescoping series, 569.

terms of a sequence, 515.

transcendental function, 374.

trapezoidal rule, 364.

triangle inequality, 20.

trigonometric functions, 40.

trigonometric identities, 43–44.

two-point formula for linear function, 108.

unbounded sequences, 530.

union of sets, 13.

variance, 504.

vector-valued function, 23.

velocity, 180.

vertical asymptote, 93–93.

vertical line test, 28.

viewing window, 19.

volume, using definite integrals to measure, 440–449.

work, 477–483.

Zeno's paradox, 567.

zero of a function, 37.

CALCULUS, VOLUME I

by Thomas Dick and Charles Patton
PRELIMINARY EDITION
Reviewer Survey

The Oregon State University Calculus Curriculum Project has developed this textbook with a great deal of input and feedback from manuscript reviewers and instructors at pilot test sites. We now seek student reaction. Your responses to the following questions will be of great help to both the authors and the publisher as a polished **First Edition** *is being assembled over the next few years.*

1. Please describe the ways in which you are expected to utilize technology in your calculus course. What materials and equipment are you required to purchase? What are supplied by your school or department?

_____
_____
_____

2. To what extent are technological tools incorporated in your class (i.e. daily, weekly, or monthly use; required for exercise sets or tests, etc.)?

_____
_____
_____

3. What are your impressions of the following elements in this textbook?

a. Writing style/explanations_____
_____

b. Level_____
_____

c. Exercises_____
_____

d. Illustrations_____
_____

4. What opinion do you have on the incorporation of technological methods in the text? Is it effective in teaching calculus concepts?

_____
_____
_____

5. Do you like the kinds of examples and applications in the text? Please comment briefly on their quantity, quality, and variety.
_____
_____
_____

6. What is your opinion of the *Technology in Calculus* supplement? Is it effective in teaching calculus concepts?
_____
_____
_____

7. Have you ever had previous training in calculus before? If so, when, and what are your impressions of this product juxtaposed to the one you used?
_____
_____

8. Are you glad that your instructor chose to use this text?
_____
_____
_____

9. How would you rate the overall quality of this product?

|————————|————————|————————|————————|————————|
Low Opinion   High Opinion

Any additional comments?
_____
_____
_____

Optional:

Name_____

Address_____

School_____

City_____ State_____ Zip_____

Please detach and return to:

   Steve Quigley, Senior Editor
   PWS-KENT Publishing Company
   20 Park Plaza
   Boston, Massachusetts 02116

Thank you for your time. We appreciate your assistance in the development of these materials.